Intelligent Multimedia Technologies for Networking Applications:

Techniques and Tools

Dimitris N. Kanellopoulos
University of Patras, Greece

Information Science
REFERENCE

Managing Director:	Lindsay Johnston
Editorial Director:	Joel Gamon
Book Production Manager:	Jennifer Yoder
Publishing Systems Analyst:	Adrienne Freeland
Development Editor:	Myla Merkel
Assistant Acquisitions Editor:	Kayla Wolfe
Typesetter:	Christy Fic
Cover Design:	Jason Mull

Published in the United States of America by
Information Science Reference (an imprint of IGI Global)
701 E. Chocolate Avenue
Hershey PA 17033
Tel: 717-533-8845
Fax: 717-533-8661
E-mail: cust@igi-global.com
Web site: http://www.igi-global.com

Library of Congress Cataloging-in-Publication Data

Intelligent multimedia techologies for networking applications: techniques and tools/Dimitris N. Kanellopoulos, editor.
    p. cm.
  Includes bibliographical references and index.
  Summary: "This book promotes the discussion of specific solutions for improving the quality of multimedia experience while investigating issues arising from the deployment of techniques for adaptive video streaming"--Provided by publisher.
  ISBN 978-1-4666-2833-5 (hardcover) -- ISBN 978-1-4666-2834-2 (ebook) -- ISBN 978-1-4666-2835-9 (print & perpetual access) 1. Multimedia communications. 2. Integrated services digital networks. 3. Image transmission. 4. Webcasting. I. Kanellopoulos, Dimitris N., 1967-
  TK5105.15I545 2013
  006.7--dc23
                        2012032520

British Cataloguing in Publication Data
A Cataloguing in Publication record for this book is available from the British Library.

Table of Contents

Section 1
Intelligent Multimedia Networking

Section 2
Adaptive Video Streaming and Video Coding Techniques

Chapter 18

Detailed Table of Contents

Section 1
Intelligent Multimedia Networking

This chapter provides an overview of the joint use of Network Coding (NC) and a set of current multimedia techniques that have shown promising results compared to traditional approaches, from distributed storage, content distribution with scalable coding, multiple description coding in P2P networks, to the applications in wireless networks and the recent trends in cross-layer optimization. NC is a new technique that can achieve the maximum information flow in the network by allowing nodes to combine received packets before retransmission.

In this chapter, the authors provide a broad introduction of the advanced IPv6 features and guide the readers from the basics of the new IP protocol family to its complex feature set and power to support multimedia communications in the mobility-centric Future Internet. They also present optimization techniques to further increase the adequacy of IPv6 for mobile multimedia.

Chapter 3

Alessandro Amirante, University of Napoli Federico II, Italy
Tobia Castaldi, University of Napoli Federico II, Italy
Lorenzo Miniero, University of Napoli Federico II, Italy
Simon Pietro Romano, University of Napoli Federico II, Italy

In this chapter, the authors focus on the complex interactions involving the various actors participating in a multimedia session over the Internet. More precisely, bearing in mind the current standard proposals coming from both the 3GPP and the IETF, they investigate some of the issues that have to be faced when separation of responsibilities comes to the fore. They highlight that protocol interactions become really complex under the depicted circumstances. Finally, they provide a survey of the current standardization efforts related to media control, together with a discussion of open issues and potential solutions.

Section 2
Adaptive Video Streaming and Video Coding Techniques

Chapter 4

Martin Fleury, University of Essex, UK
Laith Al-Jobouri, University of Essex, UK

This chapter considers several techniques for adaptive video streaming, including live HTTP streaming, bitrate transcoding, scalable video coding, and rate controllers. The chapter includes additional case studies of: (1) congestion control over the wired Internet using fuzzy logic; (2) statistical multiplexing to adapt constant bitrate streams to the bandwidth capacity; and (3) adaptive error correction for the mobile Internet. To guide the reader, the authors make a number of comparisons between the main techniques, for example explaining why currently pre-encoded video may be better streamed using adaptive simulcast than by transcoding or scalable video coding.

Chapter 5

Martin Fleury, University of Essex, UK
Ismail A. Ali, University of Essex, UK
Nadia Qadri, COMSATS Institute of Information & Technology, Pakistan
Mohammed Ghanbari, University of Essex, UK

This chapter presents intra-refresh techniques as these bring additional utility to the video stream. The authors compare periodic and gradual intra-refresh, each of which provides recovery points for the decoder and also allow stream switching or joining at these points. Thus, in intra-coding, the more normal temporal prediction is temporally replaced by spatial prediction, at a cost in coding efficiency but allowing a decoder in a mobile device to reset itself. After a review of research into this area, the authors provide a case study in non-periodic intra-refresh before considering possible future research directions.

This chapter covers the topic of the Region-of-Interest (ROI) processing and coding for multimedia applications. The variety of end-user devices with different capabilities has stimulated significant interest in effective technologies for video adaptation. Therefore, the authors make a special emphasis on the ROI processing and coding with regard to the relatively new H.264/SVC (Scalable Video Coding) standard, which has introduced various scalability domains, such as spatial, temporal, and fidelity (SNR/quality) domains. The authors' observations and conclusions are supported by a variety of experimental results, which are compared to the conventional Joint Scalable Video Model (JSVM).

Chapter 7

In this chapter, the authors perform a comprehensive review of the recent advances in computational complexity techniques for video coding applications. The chapter summarizes the recent advances in this field and provides explicit directions for the design of the future complexity-aware video coding applications. The computational complexity issue is critical for video applications implemented by relatively new video coding standards, such as the H.264/AVC (Advanced Video Coding), which has a large number of coding modes. One of the main reasons for the importance of providing an efficient complexity control in video coding applications is a strong need to decrease the encoding/decoding computational complexity, especially when the encoding and/or decoding devices are resource-limited, such as portable devices. In turn, the efficient complexity control enables the reduction of video coding processing time and the conservation of power resources during the encoding and/or decoding process.

Section 3
Multimedia Content Adaptation

This chapter proposes a content customization and distribution system for changing content consumption by adapting content according to target end-user profiles. The aim is to give Content Providers (CPs) ways to allow users and/or Service Providers (SPs) to configure contents according to different criteria, improving users' Quality of Experience and SP's revenue generation, and to possibly charge users and SPs (e.g. advertisers) for such functionalities. The author proposes to employ artificial intelligence techniques, such as mixture of Gaussians, to learn the functional constraints faced by people, objects, or even scenes on a movie stream, in order to support the content modification process. The solutions reported allow SPs to provide the end-user with automatic ways to adapt and configure the (on-line, live) content to their preferences—and even more—to manipulate the content of live (or off-line) video streams (in a way that photo editing did for images or video editing, to a certain extent, for off-line videos).

In this chapter, the authors propose a novel context modeling approach based on services where context information is linked according to explicit high-level constraints. In order to validate their proposal, they have used Semantic Web technologies by specifying RDF profiles and experiment their usage on several platforms.

Section 4
Intelligent Multimedia Applications and Services

This chapter shows the migration of mobility-enabled services to the cloud. It presents a SIP-based hybrid architecture for Web session mobility that offers content sharing and session handoff between Web browsers. The implemented system has recently evolved to a framework for developing different kinds of converged services over the Internet, which are similar to services offered by Google Wave and existing telephony APIs. The author shows efforts to migrate the SIP/HTTP application server to the cloud, which was necessitated by the need to include more functionalities (QoS and rich media support), as well as to provide large-scale deployment in a multi-domain scenario.

This chapter examines the PriorityQoE tool, which employs a methodology for hierarchizing packet video streaming based on objective QoE metrics for use in a set of intelligent mechanisms for packet video dropping in multimedia content transmission systems in the context of XaaS (X-as-a-service) paradigm. Hierarchization allows better decision-making in the packet dropping, and thus, there is less degradation of the QoE video streaming and there can be greater use of the link capacity for the transmission of multimedia content. In the Future Internet, an Integrated Cloud Service Management Platform is deployed within the XaaS (X-as-a-service) paradigm.

Chapter 12

This chapter presents an extensive study on existing Online Music Services (OMSs), which aims at identifying two principal characteristics: (1) the functionalities and interaction capabilities offered to their end-users; and (2) the tools of computational intelligence employed so as to enable these functionalities. The chapter also focuses on musical semantics, different methods for harvesting them, and approaches for exploiting them in existing OMSs.

Chapter 13

This chapter shows all steps to transmit the Internet Protocol Television (IPTV) service from the provider to the smart phone. The authors introduce the current hardware and operating systems for smart phones and describe the main parts of the IPTV architecture and the main protocols used for IPTV transmission in order to show how it works. Next, they show where intelligent systems can be deployed in the IPTV network in order to provide better QoE at the end-user side. Then, they discuss the limitations and requirements for IPTV reception on smart phones. Finally, they explain the IPTV implementation in smart phones and describe an IPTV player on Android.

In this chapter, the authors present an XML-based customizable multimedia solution for museums and exhibitions. The authors adopt a modular approach in order to provide a user interface abstraction and operation-business logic isolation from the data. The key advantage of the proposed solution is the separation of concerns for user interface, business logic, and data retrieval. Their solution allows the dynamic XML-based customization of museum multimedia applications to support additional data from new seasonal or one-time exhibitions at the same museum, re-arrangement of the exhibits in the museum halls, addition of new digitized halls with the respective multimedia data, and any additional documentation or multimedia extras for existing exhibits. The authors present a case study at the digital exhibition for the history of the ancient Olympic Games at the Older Olympia Museum.

Section 5
Multimedia Social Networks and Geo-Social Systems

Chapter 15

Andrew Laghos, Cyprus University of Technology, Cyprus

This chapter introduces Multimedia Social Networks and Online Communities, while it presents the key players of e-Learning in Multimedia Social Networks. Furthermore, it analyses social interaction research by concentrating on factors that influence social interaction, peer support, student-centered learning, collaboration, and the effect of interaction on learning. It also describes various methods and frameworks for analyzing multimedia social networks in e-Learning communities.

Chapter 16

Martina Deplano, University of Turin, Italy
Giancarlo Ruffo, University of Turin, Italy

In this chapter, the authors discuss the state-of-the-art of Geo-Social systems and Recommender Systems, which are becoming extremely popular for users accessing social media trough mobile devices. They introduce a general framework based on the interaction among those systems and the Game With A Purpose (GWAP) paradigm. Their proposed platform can help researchers to understand geo-social dynamics in order to design and test new services, such as recommenders of places of interest for tourists, real-time traffic information systems, personalized suggestions of social events, and so forth.

Section 6
Intelligent Image Processing and Image Retrieval

Chapter 17

Jakub Peksinski, West Pomerania University of Technology Szczecin, Poland
Michal Stefanowski, West Pomerania University of Technology Szczecin, Poland
Grzegorz Mikolajczak, West Pomerania University of Technology Szczecin, Poland

This chapter describes several methods for estimating the level of noise and presents a new method based on the properties of the smoothing filter. Actually, one of the significant problems in digital signal processing is the filtering and reduction of undesired interference. Due to the abundance of methods and algorithms for processing signals characterized by complexity and effectiveness of removing noise from a signal, depending on the character and level of noise, it is difficult to choose the most effective method. As long as there is specific knowledge or grounds for certain assumptions as to the nature and form of the noise, it is possible to select the appropriate filtering method so as to ensure optimum quality.

Chapter 18

Melih Soydemir, Bahçeşehir University, Turkey
Devrim Unay, Bahçeşehir University, Turkey

In this chapter, the authors present an image search system that permits search by a multitude of image features (content), and demographics, patient's medical history, clinical data, and ontologies (context). Moreover, they validate the system's added value in dementia diagnosis via evaluations on publicly available image databases. Comparison of multiple patients, their pathologies, and progresses by using image search systems may largely contribute to improved diagnosis and education of medical students and residents. Supporting image content information with contextual knowledge lead to increased reliability, robustness, and accuracy in search results.

Preface

As ubiquitous multimedia applications benefit from the rapid development of intelligent multimedia technologies, there is an inherent need to present frameworks, techniques, and tools that adopt these technologies to a range of networking applications. Intelligent multimedia technologies can solve various problems such as multimedia content adaptation. From another perspective, we meet innovative techniques that can be used in conjunction with current multimedia techniques in order to improve the quality of multimedia experience. For example, the *Network Coding* (*NC*) technique can achieve the maximum information flow in the network by allowing nodes to combine received packets before retransmission. NC can also be used in conjunction with current multimedia techniques in order to improve the quality of multimedia experience. In parallel, several techniques are available for adaptive video streaming including live HTTP streaming, bitrate transcoding, scalable video coding, and rate controllers. There are also various intra-refresh techniques for mobile video streaming that bring additional utility to the video stream. In addition, considerable processing and coding techniques concerning Region-of-Interest (ROI) with regard to the relatively new H.264/SVC (Scalable Video Coding) standard are being deployed. Besides, the computational complexity issue is becoming extremely critical for video applications implemented by relatively new video coding standards, such as the H.264/AVC (Advanced Video Coding) that has a large number of coding modes. Additionally, all these advances occur as mobility-enabled multimedia services are migrating to the Cloud.

BOOK OBJECTIVES AND INTENDED AUDIENCE

This book aims to provide relevant theoretical frameworks and the latest empirical research findings in the area of multimedia intelligence technologies. The main purpose of the book is to promote the discussion of and present specific solutions for adopting intelligent multimedia techniques and tools to a range of networking applications, such as e-Commerce, e-Learning, etc. Actually, the adoption of intelligent multimedia technologies to current networking applications can solve a variety of problems apart from the well-researched security-related problems, such as traffic surveillance and digital crime identification. Furthermore, the role and issues arising from the deployment of intelligent multimedia systems are investigated.

The ultimate goal of this publication is to be a scholarly edition suitable for practitioners and researchers in the area of multimedia intelligence with a focus on multimedia networking, processing, and media content adaptation.

The target audience of this book will be composed of professionals and researchers working in the field of multimedia technologies in various disciplines, e.g., multimedia communication, information and communication sciences, e-Commerce, e-Learning, administrative sciences and management, computer science, and information technology.

ORGANIZATION OF THE BOOK

After more than a decade of development, substantial advances have been achieved in the diverse areas of intelligent multimedia computing, and a number of promising research directions are springing up. Following an open call for chapters and a few rounds of extensive peer-review, 18 chapters of good quality have been finally accepted, ranging from technical review and literature survey on a particular topic, solutions to some technical issues, to the implementation of an intelligent multimedia system, as well as perspectives of promising applications. All the contributions have been reviewed, edited, processed, and placed in the appropriate order to maintain consistency with the intention that any reader irrespective of his/her level of knowledge in intelligent multimedia systems would get the most out of the book.

According to the scope of those chapters, this book is organized into six sections. The organization ensures the smooth flow of material as successive chapters build on prior ones. In particular, the topics of the book are the following:

- Network coding for multimedia communications.
- Mobility solutions for multimedia networking in IPv6.
- Protocol interactions among user agents, application servers, and media servers.
- Techniques and tools for adaptive video streaming.
- Intra-refresh techniques for mobile video streaming.
- Region-of-Interest: Processing and coding techniques.
- Computational complexity techniques for video coding applications.
- Adaptation and distribution of personalized multimedia content.
- Multimedia document adaptation.
- Converged applications and services via the Cloud.
- A tool for improving the quality of experience in video streaming.
- Online music services through intelligent computing.
- Intelligent IPTV distribution for smart phones.
- An XML-based customizable multimedia solution for museums and exhibitions.
- Multimedia social networks.
- Geo-social systems.
- Estimating the level of noise in digital images.
- Context-aware medical image retrieval.

Below, we briefly summarize the chapters in each section.

Section 1 (Chapter 1 to Chapter 3) presents network coding, mobility solutions in IPv6, and protocol interactions among user agents, application servers, and media servers.

- In Chapter 1 titled "Network Coding for Multimedia Communications," the authors provide an overview of the joint use of Network Coding (NC) and a set of current multimedia techniques that have shown promising results compared to traditional approaches, from distributed storage, content distribution with scalable coding, multiple description coding in P2P networks, to the applications in wireless networks and the recent trends in cross-layer optimization. NC is a new technique that can achieve the maximum information flow in the network by allowing nodes to combine received packets before retransmission.

- In Chapter 2 titled "Review of Advanced Mobility Solutions for Multimedia Networking in IPv6," the authors provide a broad introduction of the advanced IPv6 features and guide the readers from the basics of the new IP protocol family to its complex feature set and power to support multimedia communications in the mobility-centric Future Internet. They also present optimization techniques to further increase the adequacy of IPv6 for mobile multimedia.

- In Chapter 3 titled "Protocol Interactions among User Agents, Application Servers, and Media Servers: Standardization Efforts and Open Issues," the authors focus on the complex interactions involving the various actors participating in a multimedia session over the Internet. More precisely, bearing in mind the current standard proposals coming from both the 3GPP and the IETF, they investigate some of the issues that have to be faced when separation of responsibilities comes to the fore. They highlight that protocol interactions become really complex under the depicted circumstances. Finally, they provide a survey of the current standardization efforts related to media control, together with a discussion of open issues and potential solutions.

Section 2 (Chapter 4 to Chapter 7) addresses adaptive video streaming and video coding techniques.

- In Chapter 4 "Techniques and Tools for Adaptive Video Streaming," the authors consider several techniques for adaptive video streaming including live HTTP streaming, bitrate transcoding, scalable video coding, and rate controllers. The chapter includes additional case studies of: (1) congestion control over the wired Internet using fuzzy logic; (2) statistical multiplexing to adapt constant bitrate streams to the bandwidth capacity; and (3) adaptive error correction for the mobile Internet. To guide the reader, the authors make a number of comparisons between the main techniques, for example explaining why currently per-encoded video may be better streamed using adaptive simulcast than by transcoding or scalable video coding.

- Chapter 5 titled "Intra-Refresh Techniques for Mobile Video Streaming" presents intra-refresh techniques as these bring additional utility to the video stream. The authors compare periodic and gradual intra-refresh, each of which provides recovery points for the decoder and also allow stream switching or joining at these points. Thus, in intra-coding, the more normal temporal prediction is temporally replaced by spatial prediction, at a cost in coding efficiency but allowing a decoder in a mobile device to reset itself. After a review of research into this area, the authors provide a case study in non-periodic intra-refresh before considering possible future research directions.

- Chapter 6 titled "Region-of-Interest: Processing and Coding Techniques – Overview of Recent Trends and Directions" covers the topic of the Region-of-Interest (ROI) processing and coding for multimedia applications. The variety of end-user devices with different capabilities has stimulated significant interest in effective technologies for video adaptation. Therefore, the authors make a special emphasis on the ROI processing and coding with regard to the relatively new H.264/SVC (Scalable Video Coding) standard, which has introduced various scalability domains, such as spatial, temporal, and fidelity (SNR/quality) domains. The authors' observations and conclusions are supported by a variety of experimental results, which are compared to the conventional Joint Scalable Video Model (JSVM).

- In Chapter 7 titled "Recent Advances in Computational Complexity Techniques for Video Coding Applications," the authors perform a comprehensive review of the recent advances in computational complexity techniques for video coding applications. The chapter summarizes the recent advances in this field and provides explicit directions for the design of the future complexity-aware video coding applications. The computational complexity issue is critical for video applications implemented by relatively new video coding standards, such as the H.264/AVC (Advanced Video Coding), which has a large number of coding modes. One of the main reasons for the importance of providing an efficient complexity control in video coding applications is a strong need to decrease the encoding/decoding computational complexity, especially when the encoding and/or decoding devices are resource-limited, such as portable devices. In turn, the efficient complexity control enables the reduction of the video coding processing time and the conservation of power resources during the encoding and/or decoding process.

Section 3 (Chapter 8 to Chapter 9) focuses on multimedia content adaptation.

- In Chapter 8 titled "Intelligent Approaches for Adaptation and Distribution of Personalized Multimedia Content," the author proposes a content customization and distribution system for changing content consumption, by adapting content according to target end-user profiles. The aim is to give Content Providers (CPs) ways to allow users and/or Service Providers (SPs) to configure contents according to different criteria, improving users' Quality of Experience and SP's revenue generation, and to possibly charge users and SPs (e.g., advertisers) for such functionalities. The author proposes to employ artificial intelligence techniques, such as a mixture of Gaussians, to learn the functional constraints faced by people, objects, or even scenes on a movie stream in order to support the content modification process. The solutions reported allow SPs to provide the end-user with automatic ways to adapt and configure the (on-line, live) content to their preferences—and even more—to manipulate the content of live (or off-line) video streams (in a way that photo editing did for images or video editing, to a certain extent, for off-line videos).
- In Chapter 9 titled "A Semantic Generic Profile for Multimedia Documents Adaptation," the authors propose a novel context modeling approach based on services where context information is linked according to explicit high-level constraints. In order to validate their proposal, they have used Semantic Web technologies by specifying RDF profiles and experiment their usage on several platforms.

Section 4 (Chapter 10 to Chapter 14) focuses on intelligent multimedia applications and services.

- Chapter 10 titled "Provisioning Converged Applications and Services via the Cloud" shows the migration of mobility-enabled services to the Cloud. It presents a SIP-based hybrid architecture for Web session mobility that offers content sharing and session handoff between Web browsers. The implemented system has recently evolved to a framework for developing different kinds of converged services over the Internet, which are similar to services offered by Google Wave and existing telephony APIs. The author shows efforts to migrate the SIP/HTTP application server to the cloud, which was necessitated by the need to include more functionalities (QoS and rich media support) as well as to provide large-scale deployment in a multi-domain scenario.

- Chapter 11 titled "PriorityQoE: A Tool for Improving the QoE in Video Streaming" examines the PriorityQoE tool, which employs a methodology for hierarchizing packet video streaming based on objective QoE metrics for use in a set of intelligent mechanisms for packet video dropping in multimedia content transmission systems in the context of XaaS (X-as-a-Service) paradigm. Hierarchization allows better decision-making in the packet dropping, and thus, there is less degradation of the QoE video streaming and there can be greater use of the link capacity for the transmission of multimedia content. In the Future Internet, an Integrated Cloud Service Management Platform is deployed within the XaaS (X-as-a-Service) paradigm.

- Chapter 12 titled "The Realisation of Online Music Services through Intelligent Computing" presents an extensive study on existing Online Music Services (OMSs), which aims at identifying two principal characteristics: (1) the functionalities and interaction capabilities offered to their end-users and (2) the tools of computational intelligence employed so as to enable these functionalities. The chapter also focuses on musical semantics, different methods for harvesting them, and approaches for exploiting them in existing OMSs.

- Chapter 13 titled "Intelligent IPTV Distribution for Smart Phones" shows all steps to transmit the IPTV Internet Protocol Television (IPTV) service from the provider to the smart phone. The authors introduce the current hardware and operating systems for smart phones, and describe the main parts of the IPTV architecture, and the main protocols used for IPTV transmission in order to show how it works. Next, they show where intelligent systems can be deployed in the IPTV network in order to provide better QoE at the end-user side. Then, they discuss the limitations and requirements for IPTV reception on smart phones. Finally, they explain the IPTV implementation in smart phones and describe an IPTV player on Android.

- In Chapter 14 titled "An XML-Based Customizable Model for Multimedia Applications for Museums and Exhibitions," the authors present an XML-based customizable multimedia solution for museums and exhibitions. The authors adopt a modular approach in order to provide a User Interface abstraction and operation-business logic isolation from the data. The key advantage of the proposed solution is the separation of concerns for User Interface, business logic, and data retrieval. Their solution allows the dynamic XML-based customization of museum multimedia applications to support additional data from new seasonal or one-time exhibitions at the same museum, re-arrangement of the exhibits in the museum halls, addition of new digitized halls with the respective multimedia data, and any additional documentation or multimedia extras for existing exhibits. The authors present a case study at the digital exhibition for the history of the ancient Olympic Games at the Older Olympia Museum.

Section 5 (Chapter 15 to Chapter 16) deals with multimedia social networks and geo-social systems.

- Chapter 15 with the title "Multimedia Social Networks and E-Learning" introduces Multimedia Social Networks and Online Communities, while it presents the key players of e-Learning in Multimedia Social Networks. Furthermore, it analyses social interaction research by concentrating on factors that influence social interaction, peer support, student-centered learning, collaboration, and the effect of interaction on learning. It also describes various methods and frameworks for analyzing multimedia social networks in e-Learning communities.

- In Chapter 16 titled "GWAP as a Tool to Analyze, Design, and Test Geo-Social Systems," the authors discuss the state of the art of Geo-Social Systems and Recommender Systems, which are becoming extremely popular for users accessing social media trough mobile devices. They introduce a general framework based on the interaction among those systems and the *Game With A Purpose* (GWAP) *paradigm*. Their proposed platform can help researchers to understand geo-social dynamics in order to design and test new services, such as recommenders of places of interest for tourists, real-time traffic information systems, personalized suggestions of social events, and so forth.

Section 6 (Chapter 17 to Chapter 18) focuses on intelligent image processing and image retrieval.

- Chapter 17 titled "Estimating the Level of Noise in Digital Images" describes several methods for estimating the level of noise and presents a new method based on the properties of the smoothing filter. Actually, one of the significant problems in digital signal processing is the filtering and reduction of undesired interference. Due to the abundance of methods and algorithms for processing signals characterized by complexity and effectiveness of removing noise from a signal, depending on the character and level of noise, it is difficult to choose the most effective method. As long as there is specific knowledge or grounds for certain assumptions as to the nature and form of the noise, it is possible to select the appropriate filtering method so as to ensure optimum quality.
- In Chapter 18 titled "Context-Aware Medical Image Retrieval for Improved Dementia Diagnosis," the authors present an image search system that permits search by a multitude of image features (content), and demographics, patient's medical history, clinical data, and ontologies (context). Moreover, they validate the system's added value in dementia diagnosis via evaluations on publicly available image databases. Comparison of multiple patients, their pathologies, and progresses by using image search systems may largely contribute to improved diagnosis and education of medical students and residents. Supporting image content information with contextual knowledge leads to increased reliability, robustness, and accuracy in search results.

This book impacts the field of intelligent multimedia technologies by presenting the joint use of Network Coding (NC) and a set of current multimedia techniques, mobility solutions in IPv6, and protocol interactions among user agents, application servers, and media servers. It also addresses adaptive video streaming and video coding techniques. Further, it focuses on multimedia content adaptation and intelligent multimedia applications and services. In addition, this book deals with multimedia social networks and geo-social systems. Finally, it considers various issues concerning intelligent image processing and image retrieval.

Dimitris Kanellopoulos
University of Patras, Greece

Acknowledgment

This book came to light due to the direct and indirect involvement of many researchers, academicians, designers, developers, and industry practitioners. Therefore, I acknowledge and thank the contributing authors, research institutions, and companies whose papers, reports, notes, websites, and study materials have been referred to in this book. I hope that this book will serve as a valuable text for students and reference for researchers and practitioners working in the field of intelligent multimedia technologies and its emerging networking applications.

Dimitris Kanellopoulos
University of Patras, Greece

Section 1
Intelligent Multimedia Networking

Chapter 1
Network Coding for Multimedia Communications

Irina-Delia Nemoianu
TELECOM-ParisTech, France

Béatrice Pesquet-Popescu
TELECOM-ParisTech, France

ABSTRACT

Network Coding (NC) is an innovative technique that can achieve the maximum information flow in a network by allowing nodes to combine received packets before retransmission. This chapter provides an overview of the joint use of NC and a set of current multimedia techniques that have shown promising results compared to traditional approaches. The applications range from distributed storage, content distribution with scalable coding, multiple description coding in Peer-to-Peer (P2P) networks, to the applications in wireless networks and the recent trends in cross-layer optimization.

INTRODUCTION

In classical networks, multi-hop data transfer is performed at the intermediate nodes by relaying the received messages. The nodes forward the messages received on their input links to one or more of their output links, without modifying the content. If the messages belong to different data flows in the network, the node will assign them a priority (scheduling) and choose the output

DOI: 10.4018/978-1-4666-2833-5.ch001

link through which to send them (routing). This traditional view has changed with the arrival of *Network Coding* (NC) (Ahlswede, et al., 2000). With this new technique, each message sent on a node's output link is a function, or mixture, of the messages that arrive on the node's input links. Such a strategy of mixing packets or "coding" at the intermediate nodes, together with means of decoding at the receiver, has been shown to outperform traditional routing by improving the network throughput and minimizing the delay. Network coding can be applied to many forms of network communications. One case in which

NC has been proven to offer clear advantages is multicasting. The most celebrated result in network coding, the *Max-Flow-Min-Cut Theorem* for network information flow (Ahlswede, et al., 2000) states that coding within a network allows to multicast information at a rate approaching the smallest minimum cut between the source and any receiver, as long as a large enough field size is chosen for the coding symbols. Since this result cannot be achieved through traditional routing, it opens up a wide range of applications that could benefit from network coding, such as live broadcast, distributed information storage, content delivery through Peer-to-Peer (P2P) networks, and interactive communications such as multimedia conferencing. The topic of network coding has received a lot of attention from the research community, and has been approached from a multitude of disciplines, such as graph theory, information theory, channel coding theory, optimization theory, etc.

The purpose of this chapter is to present some of the most relevant results in network coding that can be used in conjunction with current multimedia techniques in order to improve the quality of multimedia experience. Because the field of network coding is still under development, we identify some of the challenges that may arise and discuss the scenarios in which NC is a viable solution and what performances are to be expected.

The remainder of the chapter is organized as follows. The next section introduces the theoretical aspects of network coding, and its fundamental theorems and algorithms. After that, we discuss results of using network coding for different multimedia applications along with some of the protocols proposed so far. In addition, we present some improvements achieved by using a cross-layer approach. Subsequently, we present methods for making network coding resilient to errors. Finally, the last section concludes the chapter.

NETWORK CODING THEORY

The demand for high quality multimedia content anytime anywhere is greater than ever with the development of the Internet, and the wide deployment of wireless technologies and more computational capable devices. The benefits of network coding make it very suitable for the diffusion of information in real networks. Protocols for future communications may benefit from incorporating these new techniques in order to meet the increasing demands of multimedia content as well as the constraints of the networks.

The Butterfly Network

The concept of "network coding" first appeared in the paper of Ahlswede et al. (2000) starting from the famous example of the butterfly network, presented in Figure 1. The problem of multicasting in a wireline network was considered, with two sources S_1 and S_2 wanting to deliver their respective messages b_1 and b_2 to two destination (sink) nodes D_1 and D_2. All links are assumed to have a capacity of one message per second. If intermediate nodes R_1 and R_2 only forward the received messages, at every sending opportunity they can either deliver b_1 to D_2 or b_2 to D_1. Thus, the link between them will become a bottleneck. If node R_1 can send a combination of b_1 and b_2 (like the bitwise XOR) as shown in the figure, both receivers can obtain both messages sooner. D_1 can decode b_2 by XOR-ing $b_1 \oplus b_2 \oplus$ with message b_1 previously received on the direct link from S_1, and D_2 can recover b_1 in a similar way. Therefore, network coding can obtain a multicast throughput of two messages/second, which is better than the routing approach that can achieve at best 1.5 messages/second.

Ahlswede et al. (2000) showed that in the case of a point-to-point communication network in which a source multicasts information to a certain set of destinations, if network coding is used, a

Figure 1. The butterfly network. Every arc represents a directed link with capacity of 1 message/ second. Source nodes S_1 and S_2 wish to transmit messages b_1 and b_2 to both sink nodes, D_1 and D_2

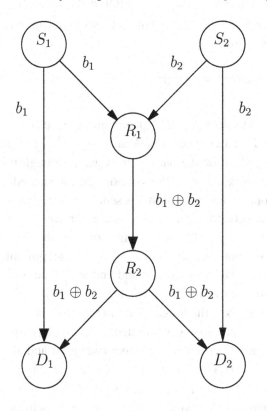

multicast capacity equal to the min-cut capacity of the network can be achieved. Such capacity is not possible, if traditional routing schemes are used.

Max-Flow Min-Cut Theorem for Network Coding

The most important result in network coding theory is known as the *Max-Flow Min-Cut Theorem for Network Information Flow*, which expands the result of the max-flow min-cut theorem in graph theory (Bondy & Murty, 2008). The network can be modeled as a direct acyclic graph $G = (V, E, c)$, where V is the vertex set corresponding to the nodes in the network, E is the edge set corresponding to the links in the network and c are the capacities

associated with the links. A simplified model considers that all links have unit capacity, which is not far from reality, as a link with integer capacity c can be replaced by c links with unit capacities.

An *s-t cut* in a graph is a partition of the vertex set V into sets A and B such that $s \in A$, $t \in B$, $A \cup B = V$ and $A \cap B = \Phi$.

The *capacity of the cut* refers to the sum of the capacities of the edges going from A to B. An *s-t cut* with minimum capacity is called a *minimum s-t cut* and also represents the minimum number of links that have to fail in order to interrupt the communication from s to t. The minimum *s-t cut* is an important characterization of a network, as it represents the bottleneck in the communication between the two nodes s and t.

The max-flow min-cut theorem states that in a flow network, the maximum amount of flow passing from the source s to the sink t is equal to the capacity of the minimum *s-t* cut. If this capacity is h, then the maximum flow in the network can be obtained using the Ford-Fulkerson algorithm (Ford & Fulkerson, 1956), which finds the h edge disjoint paths from s to t which will carry the flow.

Ahlswede et al. (2000) point out that in order for a source to multicast information at a rate approaching the smallest minimum cut between the source and any receiver, some sort of network coding has to be allowed at intermediate nodes. Codes usually used in source/channel coding can be easily employed for network codes. A data unit can be represented by an element of a base field F_q, with q the size of the field (in the butterfly network F is the Galois Field GF[2]). A message consisting of h data units can be represented as a vector $x \in F_q^h$. The propagation of message x in the network is represented by the transmission of a symbol $f_e(x) \in F_q^h$ over any channel/link e in the network. The encoding mechanism for every channel specifies the network code. With network coding, the max-flow in the network can be obtained as long as the symbols are coded in a finite field F_q whose size q is large enough.

This result was first presented as a conjecture, but subsequent work by Li et al. (2003) showed that, in a directed acyclic graph (DAG), the multicast capacity can be achieved, and it suffices for the encoding functions to be linear. Also, a finite field size can be chosen for the symbols. The proof is in fact a constructive algorithm for the encoding functions, named a generic *Linear Code Multicast* (LCM). If the links are ordered in topological order, then the encoding vector assigned to the current link should be linearly independent with respect to all previously assigned coding vectors.

The next step in investigating the new technique of network coding was to formulate and solve the multicast problem, *i.e.* finding necessary and sufficient conditions so that a given set of connections can be achievable over a given network. This was achieved by Koetter and Médard (2003) by introducing an algebraic framework for linear networks. Their result is a formulation of the feasibility of a multicast problem and the validity of a network coding solution in terms of transfer matrices. The result was also extended to arbitrary networks.

The algebraic framework was based on the observation that, if the network is linear over F_2^m, the relationship between an input vector **x** and an output vector **z** can be described by a *transfer matrix* M: $z = xM$, where M has coefficients from F_2^m. An outline of the results is given in the following.

When the network is represented as a directed graph G, a *topological ordering* is an ordering of nodes $v_1, ..., v_n$ such that if edge $(v_i, v_j) \in G$ then $i < j$, *i.e.* an edge can only point from a node with lower index to one with higher index. Any graph can be topologically ordered by renaming the vertices. Then, the adjacency matrix of the edges in the network, the $|E| \times |E|$ matrix F, will be upper triangular.

If there are μ sources in the network and ν sink nodes, then the transfer matrix M can be expressed in terms of matrices: A (of dimension $\mu \times |E|$, that specifies the transformation from the sources to the edges of the network), F (the adjacency matrix, of dimension $|E| \times |E|$), and matrix B (of dimension $v \times |E|$, that specifies the transformation from the edges of the network to the outputs) in the form:

$$M = A(I - F)^{-1} B^T$$

Matrices A and B specify how the data enters and leaves the network. Matrix $(I - F)^{-1}$ should specify how the data is propagated through the network, *i.e.* how the symbol sent on each edge contributes to the symbols sent on other edges in the network. This is true since, for any pair of edges (e_i, e_j), the contribution of e_i to the linear combination carried by e_j is given by all possible paths that start from e_i and end at e_j. The paths are captured in the sum: $I + F + F^2 + ... + F^N$, where N is the longest path in the network.

As previously mentioned, if the graph is topologically ordered, F is upper triangular and also $F^N = 0$. Thus, $(I - F)^{-1} = I + F + F^2 + F^N$, as $(I - F)(I + F + F^2 + ... + F^N) = I - F^{N+1} = I$. If matrices A and B have the same dimensions $h \times |E|$, with h the minimum cut in the network, the output vector **z** can be recovered if the transfer matrix is invertible, *i.e.* det(M) $\geq$ 0.

It can also be shown that
$$\det(M) = \pm \det \begin{bmatrix} A & 0 \\ I - F & B^T \end{bmatrix}.$$

The *multicast theorem of network* coding is restated as: Given a directed acyclic graph G with unit capacities, that has a single source node s (with h messages) and a set of terminal nodes T, the multicast property with rate h is said to be satisfied if *max-flow*$(s, T_i) \geq h$; for all $T_i \in T$. If G satisfies the multicast property, a network code that supports the multicast rate h is guaranteed to exist as long as the field size is larger than |T|.

Proof: For all the terminal nodes T_i to be able to recover the h source messages, the following has to hold:

$\prod_{i=1}^{|T|} \det(M_i) \neq 0$. If the coefficients of matrices A, B^T, and I – F are denoted by α, β, and ε, then the product of determinants is a polynomial in variables α's, β's, and ε's. Since each variable in $\begin{bmatrix} A & 0 \\ I - F & B^T \end{bmatrix}$ appears at most once, then in the determinant each variable also appears at most once. Therefore, in the product of determinants $\prod_{i=1}^{|T|} \det(M_i) \neq 0$ the maximum degree of any variable is at most |T|. Using the *sparse zero's lemma (Schwartz-Zippel lemma)*, one can show that an assignment of the variables such that the polynomial evaluates to a non-zero value exists as long as the size of the field is greater than |T|.

Subsequent work dealt with finding algorithms capable of solving the multicast problem and that could be implemented in practice. Jaggi et al. (2005) considered networks in terms of acyclic delay-free graphs and studied the single-source multicast problem. They provided centralized deterministic and randomized polynomial-time algorithms for finding network coding solutions by considering sub-graphs consisting of flows to each receiver. If the maximum rate allowed in the single-source multicast scenario is h, the algorithm first finds for each receiver t the s-t flows in the network. This can be done with a classical *Ford-Fulkerson algorithm* (Ford & Fulkerson, 1956) that finds the h edge-disjoint paths from the source to each receiver. Due to the fact that many s-t flows will overlap, the number of required operations is reduced compared to LCM by assigning coding vectors only to the links serving multiple flows. A detailed presentation of the Linear Information Flow (LIF) algorithm is presented in Jaggi et al. (2005).

Random Linear Network Coding

Coding strategies (like the one previously presented on the butterfly network) imply that there is certain knowledge of the network topology and that dedicated nodes are responsible for performing the same encoding operations. The scheme is thus centralized and fixed. However, in real networks, the structure, topology, and traffic demands may change quickly and drastically, while the information about those changes propagates with a certain delay. In wired networks, the edge capacities may vary due to changing traffic conditions and congestion. In peer-to-peer overlay networks, many nodes may join or leave the network in a short interval. A wireless network may vary in time due to fading channels, interference, and node mobility. In wireless ad-hoc networks, where the nodes are self-organizing, the participating nodes also have limited resources in terms of communication and computation, and transmission quality may vary due to node mobility. In the approaches presented so far, the network conditions are considered fixed over a fairly large period of time, but in practice each change in the network would imply computing new optimal combination operations. In networks with cycles and delays, network codes can be obtained in a distributed manner, by performing *random linear network coding* (Ho, et al., 2003), *i.e.* the coefficients of the linear encoding functions are chosen independently and randomly from the finite field.

In the standard framework, as previously explained, the network is modeled by an acyclic graph $G(V,E)$ having unit capacity edges E, a source $s \in V$ and a set of receivers T. The multicast capacity h is the capacity of the minimum cut between the source and any of the receivers. $x_1,...,x_h$ are the h source messages that are symbols over a field F_q where the order (size) of the field q is finite.

The probability that a set of coefficients chosen randomly from the finite field F_q does not ensure decodability at the receiver is given by:

$$p_e = 1 - (1 - \frac{|T|}{q})^{|E|}.$$

This result shows that the error probability in the case of random network codes does not depend on the maximum multicast rate h. Moreover, by working with a field size q large enough, the probability of error can be made negligible.

The problem then rests in transmitting the coding operations that take place with each packet. A practical approach was presented by Chou et al. (2003). Since then, it has become the de-facto method for Random Linear Network Coding (RLNC), and has been successfully applied to solve a series of problems in multimedia applications, described in Section "Network Coding for Multimedia Applications" briefly present it in the following.

Any intermediate node $v \in V$ will have a set of incoming edges $\Gamma_{in} = \{e': out(e') = v\}$ and a set of outgoing edges $\Gamma_{out} = \{e: in(e) = v\}$, as depicted in Figure 2. Each edge $e \in E$ going out of node v carries a symbol $y(e)$, which is computed as a linear combination of the symbols $y(e'_i)$ on the incoming edges e' of node v, i.e. $y(e) = \sum_{e' \in \Gamma_{in}} m_e(e')y(e')$, with $m_e(e') \in F_q$. The coefficient vector $m(e) = [m_e(e')]$ represents the local encoding vector along edge e.

By induction, the symbol $y(e)$ on any edge can be computed as a linear combination of the source symbols, i.e. $y(e) = \sum_{i=1}^{h} g_i(e)x_i$. The coefficients form a global encoding vector $g(e) = [g_1(e),..., g_h(e)]$, which is updated at each coding operation using the local encoding vectors $m(e)$. The global encoding vector $g(e)$ represents the code symbol $y(e)$ in terms of the source symbols $x_1,...,x_h$.

When a sink node $t \in T$ receives symbols $y(e_1),...,y(e_h)$, they can be expressed in terms of the source symbols as:

$$\begin{bmatrix} y(e_1) \\ \vdots \\ y(e_h) \end{bmatrix} = \begin{bmatrix} g_1(e_1) & \cdots & g_h(e_1) \\ \vdots & \ddots & \vdots \\ g_1(e_h) & \cdots & g_h(e_h) \end{bmatrix} \begin{bmatrix} x_1 \\ \vdots \\ x_h \end{bmatrix} = G_t \begin{bmatrix} x_1 \\ \vdots \\ x_h \end{bmatrix}$$

where G_t is the global encoding matrix and the ith row of G_t is the global encoding vector corresponding to $y(e_i)$. Sink t can recover the h source symbols as long as G_t is invertible, i.e. G_t has rank h.

This is true with high probability as long as the local encoding vectors are chosen randomly from a finite field that is sufficiently large. If the coefficients of matrix G_t are chosen uniformly at random from a field of size q, then the matrix would be full rank with high probability (Ho, et al., 2003). Simulation results (Chou, et al., 2003) show that even small field sizes (Galois Field GF[2^8] or GF[2^{16}]) work well in practice and the probability of error becomes negligible.

To generalize, each packet in the network can be considered as a vector of symbols. The above algebraic relationships remain valid for packets, if the source symbols are grouped into packets. In order to be able to invert the code at any receiver,

Figure 2. Intermediate node in the network. $y(e_i')$ represent the symbols carried by incoming edges e_i in the node, $y(e)$ is the symbol transmitted by the node after the encoding function

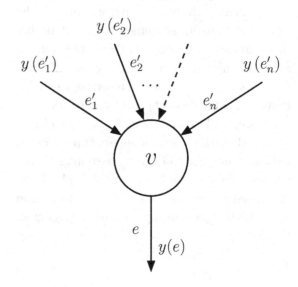

the global encoding vectors should be obtainable from the arriving packets themselves. This is done by prepending the ith source packet x_i with the ith unit vector.

$$\begin{bmatrix} y(e_1) \\ \vdots \\ y(e_h) \end{bmatrix} = G_t \begin{bmatrix} 1 & \cdots & 0 & x_1 \\ \vdots & \ddots & \vdots & \vdots \\ 0 & \cdots & 1 & x_h \end{bmatrix}$$

The benefit of these tags is that the global encoding vectors can be found within the packets themselves. The sinks can compute G_t without knowing the network topology or packet-forwarding paths. Thus, the network can be dynamic with nodes and edges being added or removed in an ad-hoc way.

In practice, the data stream is divided into generations with each generation consisting of h data packets. Intermediate nodes only mix packets coming from the same generation. The source generates h new data packets by prepending the h-dimensional i^{th} unit vector to the corresponding i^{th} data packet. Intermediate nodes store the received packets into buffers sorted by generation number and when a transmission opportunity occurs, the packet to be sent is generated by random network coding of all the packets in the current generation. The original packets are obtained at the receivers by performing *Gaussian elimination* as soon as h independent coding vectors have been received.

Since, in order to decode, a sink node has to wait for a number of independent packets equal to the generation size, this parameter clearly affects the delay in the network, so it is desirable to keep it as small as possible. In addition, the generation size gives the size of the overhead of transmitting the coefficients in the packet header, and it should also be kept small. As an example, an IP packet over the Internet has a maximum size of 1500 bytes. If every byte is treated as a symbol over $GF(2^8)$ and the generation size is h=50, then the overhead is approximately 50/1500=3.3%. This aspect might become important for wireless

communications, where the packets are smaller and the overhead may become important. On the other hand, the size of the generation affects how well packets are "mixed." The decoding matrix would be full rank with a probability at least $\left(1 - \frac{1}{q}\right)^h$, where h is the generation size, so it is desirable to have a fairly large generation size.

NETWORK CODING FOR MULTIMEDIA APPLICATIONS

In the previous section, the subject of network coding has been approached from a graph and information theory perspective. From the networking perspective, the OSI or the TCP/IP stack protocols offer several options regarding the level where NC should be implemented. In content distribution networks, NC can be implemented at the application layer with little effort. In self-organizing networks, it can be used for the wide dissemination of important data, as a substitute for routing at the network layer. In general packet networks, there is a high interest in integrating NC with the current TCP/IP protocol, while investigating the coding opportunities at the physical layer in wireless networks can be an interesting topic of research. For multimedia applications though, the recent trend is to build protocols that are implemented as a cross-layer between the application and network layer, so that they can be tailored to the specific media content to be delivered. The NC approaches can also be divided in two broad categories:

- **Intra-Session Network Coding:** If coding is allowed only among packets belonging to the same multicast session, this coding is called *intra-session network coding* and is based on the practical implementation of RLNC.
- **Inter-Session Network Coding:** It includes the butterfly network, and allows

for the combination of packets from different multicast or unicast sessions. For this reason, it is referred to as *inter-session network coding*.

Hereafter, we discuss some of the existing NC approaches for multimedia applications.

NC-Based Distributed Storage Systems

Distributed storage systems are a solution for the next generation multimedia applications that require increased storage space, ease of access and reliable recovery. As an example, more and more companies offer cloud services for distributed computing and storage. In order to allow information to be spread over multiple, unreliable nodes situated at different locations, such systems have to ensure the redundancy needed for reliable recovery in case of node failures. Hence, the bandwidth a replacing node requires in order to obtain the information needed for the reconstruction of lost data (also known as the *repair bandwidth*) is an important parameter of the system. Other parameters that influence the system's capability to recover, such as the delay in access and transmission, can be reduced by taking into account the geographical distribution of resources. In a distributed storage system, the storage elements are placed into nodes, which function independently of each other, and thus exhibit independent failure patterns. These nodes are often connected through a network with arbitrary topology, and the information objects are stored in specific nodes according to a mapping function. The use of NC for distributed storage systems has been studied (Dimakis, et al., 2010; Wu, 2009), and has been compared to the results obtained by using traditional erasure codes. In distributed storage systems that are over networks, the nodes may fail or leave the system quite often and the codes have to be maintained over time. Therefore,

the problem rests in finding a tradeoff between the storage capability and repair bandwidth.

If in the system an (N,M) erasure code is used, each information object is divided into M distinct blocks, which are encoded using appropriate erasure codes like the popular Reed-Solomon code. Encoding of the M original blocks generates $N \geq M$ encoded blocks and the entire object can be recovered from any $M\gamma$ out of the total N pieces, with $\gamma \geq 1$. The code has a rate of $R = \dfrac{M}{N} < 1$, and the relative redundancy introduced equals $\dfrac{N-M}{M}$.

If the encoded blocks are stored at different nodes in the system, when a node fails, recovery can be obtained if a new node downloads subsets of data stored at a number of surviving nodes and uses them to reconstruct the lost data. The data would not be able to be recovered only if the number of nodes that fail is greater than the introduced redundancy. The amount of storage needed to obtain a similar resilience by means of simple replication would be prohibitive. However, erasure coding and in general coding-based methods achieve best results in scenarios with medium/low node availability, as in mobile mesh or ad-hoc networks, and the complexity introduced by maintaining these codes is quite high. Although erasure codes offer a good tradeoff between redundancy and error tolerance, in practical storage systems a critical resource is also the *network bandwidth*. Minimizing the network bandwidth needed by repair operations is another important aspect to be considered, when designing the code.

In order to formulate the repair problem, Dimakis et al. (2010) introduce the *information flow graph* to represent the evolution of information flow as nodes join and leave the system. Each storage node is represented in the flow graph by a pair of nodes connected by an edge, whose capacity is the storage capacity of the node. A source *s* is supposed to have the entire file. The requests to reconstruct the original data from a subset of nodes are called *data collectors* and are represented as sink nodes whose input edges have infinite capacity. Through the information flow graph, the original storage problem can be formulated as a network communication problem, where the source *s* multicasts the file to the set of all possible data collectors. Performance bounds for codes can be derived by analyzing the connectivity in the information flow graph. Network coding can be employed over the resulting multicast graph to achieve the max-flow of the network, as described in the previous section. If the repair operations are posed as a max-flow/min-cut problem between source and any arriving node in the induced connectivity graph, repair bandwidth requirements can be derived so that the min-cut *k* of the graph would suffice to eventually recover the stored objects by just querying *k* storage nodes. However, if the minimum cut *k* between *s* and a data collector *t* is less than the size of the original file, then it is impossible for the data collector to reconstruct the original file by accessing only *k* storage nodes, regardless of the code used.

P2P-Based Content Distribution

Intra-session network coding, based on the practical implementation of RLNC, has been proven beneficial for large-scale content distribution in Peer-to-Peer (P2P) overlay networks (Gkantsidis & Rodriguez, 2005). In such a network, the server splits the file into small blocks or chunks, which are downloaded by end-users in parallel from different nodes. Once a user has downloaded a given block, that device can act as a server (seed) for anyone else interested in that file. The most popular of such cooperative architecture, BitTorrent, uses a *rarest-first* block download policy. It attempts a uniform distribution of parts among the nodes to prevent users, who have all but a few pieces from waiting too long to finish their download. However, some blocks often remain "rare," so when nodes are close to finishing their download,

they may attempt to obtain them from the server, causing unnecessary server overloading. Other inefficiencies in traditional P2P systems are more pronounced in large heterogeneous networks during flash crowds, in environments with high churn (many peers connecting or disconnecting in a short period of time) or when cooperative incentive mechanisms are in place.

The problem of block scarcity can be solved by allowing nodes to send a liner combination of all their available blocks. A first advantage of this network-coding scheme is that the system becomes more robust to situations when the server and/or the nodes leave the system. Due to the uniform random distribution of coefficients in the linear combinations, all nodes are able to finish the download, even in extreme situations. A second advantage can be observed, when scheduling the blocks for transmission. In a large-scale network that does not have a central scheduler, optimal scheduling of packets is difficult to achieve, and nodes usually have to make a decision based on local information only, which can be suboptimal. Using network coding, this decision is greatly simplified. A node can determine if it can provide an innovative packet to a neighbor by comparing their coefficient matrices. However, if this information is not available to the sender, it can generate a linear combination of all the coefficient vectors previously received and send the resulting vector. The receiver can afterwards check if the received vector is linearly independent with its own coefficient vectors, and thus it can determine if the sender can provide new blocks.

The first protocol that uses RLNC in P2P networks was *Avalanche* (Gkantsidis & Rodriguez, 2005). In order to function as a complete P2P system, Avalanche has three types of participants: *peers*, *registrars*, and *loggers*. A peer can be a source or sink for the content to be distributed; if it sends content into the system, then it is called *server*. If a peer remains in the system after it finished downloading, then it becomes a *seed*. Registrars and loggers are responsible for the overlay management. A registrar enables peer discovery and provides nodes with a set of active peers. Every peer reports to the registrar, when it needs more neighbors, and also reports detailed statistics to the logger. Each peer maintains 4-8 connections to other peers, out of which a neighbor is periodically dropped at random to prevent formation of isolated clusters. Avalanche has been compared with schemes where information is sent un-encoded and to schemes where only the server generates and transmits encoded packets. In clustered topologies, network coding shows a clear benefit. In such scenarios, since peers may belong to different clusters, without coding, some packets travel several times over the links that connect the clusters, thus wasting capacity. In heterogeneous networks, where nodes have different upload and download capacities, the performance achieved by the fast peers is degraded without coding. This is caused by slow nodes that may spend their bandwidth downloading from the server blocks that are not useful to fast peers.

Live P2P streaming is considered in Wang and Li (2007) where results are based on an experimental testbed consisting of a cluster of 44 dedicated dual-CPU servers. In their framework, *Lava,* a node has two functionalities:

1. A network part that deals with the connections and emulates the upload and download capacities.
2. The algorithm part that implements the algorithms and protocols for live P2P streaming.

Alive session consists of a multimedia stream with a specific multimedia rate, each stream being divided into segments further divided into blocks. In order to evaluate the benefits of using network coding when compared to traditional protocols, the network coding algorithm was added as a plugin component to a standard P2P protocol called *Vanilla*. Nodes keep a playback buffer in which segments for a streaming session are stored in order and removed after they have been played.

If the segment is not available in time for playing, it will be skipped. A player starts to produce and serve new coded blocks after it receives $a \times n$ coded blocks, where n is the number of blocks in a segment and $0<a<1$ is a parameter called *aggressiveness*, which can be tuned in order to evaluate the performance of the system. As a measure of system performance, the authors use the percentage of playback skips and of discarded blocks, as they are given by the number of linearly dependent packets and by the number of obsolete packets in the system. Their result show that network coding can support a wide range of streaming rates, but the decoding process represents a bottleneck in the streaming process. In the flash crowd scenario, network coding shows a higher percentage of initial playback skips, but is more resilient to peer departures. Subsequent work by Wang and Li (2007) shows that better throughput can be achieved with a push-based protocol that uses information exchange among peers before generating a new coded packet. Other architectures that use peer information exchange are discussed in other section.

NC for Multimedia Applications in Wireless Networks

Wireless technologies have been adopted quickly thanks to their advantages in terms of mobility, low cost equipment, and ease of deployment compared to wired networks. In addition to maximizing throughput, network coding can offer other advantages in wireless networks, as it can be used to reduce the energy consumption by reducing the number of transmissions. The advantages of NC are more evident in *wireless ad-hoc networks*, as in the simple scenario of two nodes wanting to exchange messages through one relay node presented in Figure 3. Each of the nodes A and B safely transmits its message to the relay R, which will afterwards broadcast the XOR of the two messages. This scheme allows each node to decode the desired message after three time slots as compared to the traditional four time slots. Therefore, this is an important result as the energy efficiency (the amount of battery energy consumed to transmit bits across a wireless link) is a critical design parameter for wireless ad-hoc networks.

A wireless ad-hoc network is a decentralized type of wireless network that, unlike traditional wireless networks, does not rely on access points for management. Each node participates in routing by forwarding data for other nodes, and the decision on which nodes to forward is made dynamically based on the network connectivity. In wireless mesh networks, a packet travels multiple wireless links before reaching a gateway node, which is connected to the wired Internet. Mesh networks can be preferable to single-hop access point networks as they can achieve the same coverage either with much lower transmission power or with lower deployment costs. The wireless routers in a mesh network form the backbone for clients. Clients can also act as routers and participate in forwarding packets, although typically they are end-user nodes with simpler wireless hardware and software. A network in which all nodes can transmit, receive as well as relay

data and where the topology may change over the time and space is called a *Mobile Ad-Hoc Network* (MANET).

Although wireless ad-hoc networks should be by definition cooperative, so far the communication has been implemented as point-to-point communication, similar to wireline or traditional access-point wireless networks. With current protocols, even though a sent packet is in fact broadcast and can reach all nodes in a neighborhood, the unintended recipients have to drop the packet. By allowing intermediate nodes to store unintended packets and then use them in network coding combinations, the throughput of the network can be increased, both in the case of multicast (Widmer, et al., 2005) and multiple unicasts (Katti, et al., 2008).

In the case of multicast, the participation in the forwarding process of a large percentage of nodes in the network and the use of random linear network coding can achieve high delivery of packets even when the packet loss rates are high (Widmer, et al., 2005). This makes NC much more energy efficient in wireless ad-hoc networks and could replace flooding, currently used for rapid dissemination of important messages in the network. The first NC protocol for wireless scenarios was the *COPE protocol*. It was proposed by Katti et al. (2008) and it uses simple binary XORs as its network coding operations in wireless mesh networks. COPE takes advantage of the broadcast nature of the wireless medium and allows nearby nodes to *overhear* a packet, which means a node can place the packet in its buffer even though the node is neither the destination for that packet nor a forwarding node in that particular transmission. Therefore, more nodes can retransmit the overheard packet or combine it with other packets, if this would be beneficial for other nodes in the network. COPE is designed to perform network coding for multiple wireless unicast flows, which raises other problems than the multicast cases studied so far. In unicast, packets from multiple unicast flows may get encoded at some intermediate node, but

Figure 3. Network coding for message exchange in one-hop wireless network. The number of total transmissions in reduced from 4 to 3, if the relay node R broadcasts a XOR of the two messages received from nodes A and B

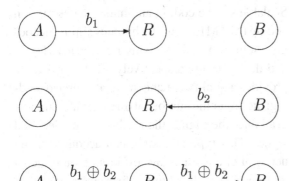

the paths will diverge later, at which point the flows need to be decodable. COPE also deals with many technical issues that may appear when implementing NC in today's wireless networks, shortly presented in the following.

Each node stores the overheard packets for a short time and also tells its neighbors which packets it has heard by annotating the packets it sends or by periodic exchange of status reports. When a node transmits a packet, it uses its knowledge of what its neighbors have heard to perform opportunistic coding; the node XORs multiple packets and transmits them as a single packet, if each intended next-hop neighbor has enough information to decode the encoded packet. This extends COPE beyond the case of two flows traversing the same node, as it allows to XOR more than two packets.

In the wireless environments, the main benefit of NC is that the encoded packets are broadcast and meant to be received by all neighboring nodes. However, the 802.11 broadcast has no collision detection or avoidance mechanism, as it requires no acknowledgement for the safely delivery, in contrast to the unicast case. The 802.11 unicast mode requires unicast packets to be immediately acknowledged by their intended next-hop. The lack of an ACK is interpreted as a collision by the MAC layer, which will react by backing off and allowing other nodes to share the medium before attempting to retransmit. In the broadcast case, the packet has many intended receivers and an ACK from all of them would add too much overhead to the communication. Rather than changing the MAC layer completely, the COPE protocol tries to work with the existing 802.11 standard, by using *pseudo-broadcast*. In pseudo-broadcast, the packets meant to be broadcast are actually sent as unicast packets with the link layer destination field set to the MAC address of one of the intended receivers. An extra header is added after the link layer header in which all the intended recipients are listed. Since nodes work in the promiscuous mode, which allows them to check packets not

addressed to them, they can find out if they are among the intended recipients, while only the target node sends the ACK. By using the 802.11 unicast, the MAC layer can detect collisions and can backoff accordingly.

The results of using a NC protocol in wireless environments clearly show an improvement in network throughput, particularly as the number of flows increases. However, even though with the increase of network flows the coding opportunities increase, the congestion also increases and many reception reports may get lost. The best results are obtained by COPE together with opportunistic routing. This implies that an intermediate node coding packets together checks not only the reception reports but also the path chosen for the packet by a routing protocol. The path is added to the header of the packet, and if the node discovers that it is a forwarding node for the packet, it takes responsibility for retransmitting it. While NC ensures earlier delivery of packets and reduces the number of transmissions in the network, improvements to current protocols can be made. A protocol that further reduces the number of transmissions by taking into account the priority of packets in multimedia communication will be presented in the Section "Error Resilient Network Coding."

Content Distribution with Scalable Coding and Multiple Description Coding

Scalable source coding and multiple description coding (MDM) are two techniques aiming at coding a source into sub-streams, referred to as layers and descriptions, respectively. Each sub-stream provides a quality enhancement at source decoding: the larger the number of different sub-streams available, the higher the quality of the decoded signal. This type of coding can account for the needs of clients that have different requirements with respect to the quality of service received or whose network conditions are not reliable. NC has already been shown to improve the throughput in the network. It would be interesting to see how a joint design can further improve the performance in the network.

The scalable variant poses a number of constraints in the order the layers are to be received, as higher order layers can be used only if all previous layers have been received. On the other hand, with MDC each received description can be independently used to increase the perceived quality. This type of source encoding is appropriate in applications where variable quality is admissible and perceptible (e.g. video or audio), and transmission is non-reliable, as in wireless scenarios or congested wired networks. It is also recommended in applications where they receive bandwidth may vary across receivers, as in broadcasting or multicasting. In this way the downlink bandwidth of the most favored destinations can be exploited while satisfying a lower but acceptable reception quality at the less favored destinations.

Multimedia streaming over the Internet is challenging due to packet loss, delay, and bandwidth fluctuation. Thus, many solutions have been proposed, ranging from source and channel coding to network protocols and architectures. For example, to combat the fluctuating and limited bandwidth, a scalable video bit stream can be used to allow the sender to dynamically adapt its video bit rate to the available bandwidth. To reduce packet loss and the associated delay due to the retransmissions of the lost packets, Forward Error Correction (FEC) techniques have been proposed to increase reliability at the expense of bandwidth increase.

Scalable coding can also be implemented in systems with specific architecture, like a multi-sender multi-receiver streaming system. In P2P networks, multiple copies of a video or chunks of a video are often present at peers either through a coordinated distribution of the video from an original server, or through an uncoordinated propagation of contents in the network. A designated algorithm must be able to dynamically and optimally assign chunks of different lengths to different servers based on their available bandwidths. In the case of streaming, in order for the newly received packets to be useful, the order in which the packets are received should be in agreement with the coding order.

The advantages of using NC in P2P systems have been presented earlier in the chapter. If a server wants to distribute a file to a number of peers in a P2P network, to increase the throughput, the server first divides a file into n different chunks and randomly distributes these chunks to the peers. The peers then can exchange their chunks among each other in a random manner, making the time a peer needs to recover all n chunks much shorter than that of downloading the file directly from the server. This design scales well with the arrival of new peers since no coordination among the peers is required. However, because of the random exchange of packets (for reducing the effort of tracking the packets), some of the packets received at a peer may be duplicate, resulting in wasteful bandwidth. If each peer is allowed to use NC, each received packet will be a combination of the original file packets, and independent from the previously received packets. Each packet will thus increase the rank of the decoding matrix. If a network coded packet is a combination of n packets, then an end-user will have to receive at least n coded packets in order for it to recover any one of the original packets. This potentially introduces unnecessary delay for video streaming applications. Hence, there is a need for a network code structure that enables a receiver to recover the important data gracefully in presence of limited bandwidth, which causes an increase in decoding delay.

With *Hierarchical Network Codes* (HNC) (Nguyen, et al., 2007), when a small number of coded packets are received, with high probability, the most important data can be recovered. To illustrate this approach, let suppose a file is divided into a number of packets, and each packet belongs to one of the 3 classes: A, B, and C, with A and C being the most and least important, respectively.

We further assume that there are six packets in the file with a_1, a_2 belonging to A, b_1, b_2 belonging to B, and c_1, c_2 belonging to C. The packets are randomly coded using one of the following structures:

$$N_1 = f_1^1 a_1 + f_2^1 a_2$$

$$N_2 = f_1^2 a_1 + f_2^2 a_2 + f_3^2 b_1 + f_4^2 b_2$$

$$N_3 = f_1^3 a_1 + f_2^3 a_2 + f_3^3 b_1 + f_4^3 b_2 + f_5^3 c_1 + f_6^3 c_2$$

where f_i^j are the random non-zero elements of a finite field F_q. Using this structure, it is easy to see that the probability of recovering a_i's is always larger than that of b_i's and the probability of recovering b_i's is always larger than that of c_i's. This is because if one assumes that all packets arrive at a peer in a random manner, then only two packets of N_1 type are needed to recover a's while 4 packets of either N_1 or N_2 types are needed to recover a's and b's. To fine tune the probability of receiving a certain type of packets, one can control the number of packets belonging to a certain type. One can increase the probability of receiving packets of type N_1 by generating more packets of N_1 type.

The benefits of this scheme are evident, when compared to a non-network coding scheme and a simple random linear network coding scheme (Nguyen, et al., 2007). In the non-network coding scheme, the origin server distributes the entire video file to all the streaming servers. Each streaming server randomly sends packets belonging to a chunk c_i. When the client signals an "end" signal, the server then moves on to the next chunk, and randomly sends packets of this new chunk. In the network-coding scheme, the origin server has previously generated network-coded packets (linear combination of a number of packets equal to the generation size for each chunk) and distributed these packets to the streaming servers. Thus, for each chunk c_i, each streaming server keeps a frac-

tion of random network code packets that it then sends at random to a client.

Results show that for the unencoded scheme, the time to decode any layer is the largest due to high probability of getting duplicate packets. For the RLNC scheme, the time to decode all 3 layers is the shortest. The times to decode layers 1 and 2 are longer than those of the HNC due to the fact that, with RLNC, in order to decode any packet in the generation, the destination node has to receive a number of independent encoded packets equal to the generation size. However, once enough packets are received, they are all decoded simultaneously. HNC allows a receiver to recover the important packets early, but pays extra overhead to recover all the packets. This is suitable for scalable video streaming since, if not enough bandwidth is available, the receiver can instruct the servers to start sending packets from the next chunk, while the receiver can playback the important layers it has received.

Although previously proposed for the case of multi-source streaming, this hierarchical method can be used in the case of single-source multicast where the receivers have different capabilities or demands. If the generation in an RLNC scheme is divided into layers of priority, this HNC scheme can be used to provide Unequal Error Protection to the layers. Vukobratović and Stanković (2010) generalize the design of RLNC with unequal error protection and provide exact decoding probability analysis for the different importance classes (layers) of the source data. If the source block is divided into L importance classes according to the L importance layers, RLNC is applied on the first i classes, with i varying from 1 to L. The most important source layer is contained in all classes, which makes it better protected, while the other layers are progressively less protected. Associated with the classes is the selection distribution, which can be tuned in order to cater for the loss probabilities of the network channels.

One limitation of scalable source coding is that layers have to be received in a predetermined order.

Multiple description coding does not suffer from this limitation. This makes it particularly interesting to be used jointly with NC, as the independence of the descriptions, as opposed to the hierarchical dependence of the layers, allows for the design of a system taking into account fewer constraints. This has permitted Ramasubramonian and Woods (2010) to develop a framework to maximize the throughput of a multicast video streaming in lossless networks. They proposed to encode the video sequence into N descriptions, with the total rate for the encoding, *i.e.* the combined rate of all descriptions, chosen to match the maximum max-flow of the nodes. The redundancy among the descriptions is chosen depending on the number of descriptions that each node receives, transmitted to the source through a feedback channel. The N descriptions are then used to generate N linearly independent random combinations, and sent to the channel. If the source still has available capacity, linear combinations of the descriptions can be sent to fill this extra capacity.

Thanks to the joint use of MDC and NC, users with high max-flow capacity will be able to satisfy their demands, whatever bottleneck may appear in the network at intermediate nodes. This result would not be possible with classical routing, which does not assure that the max flow is achieved. Neither would it be possible using scalable coding, since in MDC *any* combination of a given number of descriptions delivers almost the same video quality, whereas missing a layer renders it and any following layers useless. This technique requires that the nodes of the network are able to determine their max-flow capacity, and to transmit it to the source, which will dynamically determine the optimal rate allocation for the descriptions. Unfortunately, in many scenarios this is not possible. The source could have to transmit a pre-encoded sequence, and not have sufficient computational power to re-encode it based on the changing conditions of the distribution network. In addition, nodes may not be able to determine their max-flow capacity or it could

change too frequently to be communicated to the source with little overhead, such as in the case of mobile networks.

To deal with these challenges, we proposed a joint MDC and NC framework for mobile networks (Nemoianu, et al., 2012). This framework allows, *in a distributed fashion,* instant decoding of the received packets, and maximizes the average video quality. In this framework, each forwarding node decides its coding policies according to the state of its neighbors' buffers, choosing the code that will give the best result according to a certain metric. In our case, it is the average video quality of the neighbors. In addition to that, the delivery of at least one description to most of the nodes is prioritized compared to the delivery of multiple descriptions to a small number of nodes. This framework was implemented over a recently proposed overlay protocol for the diffusion of multiple descriptions (Greco & Cagnazzo, 2011), which is capable of maintaining the overlay in an efficient fashion. Given the results obtained, we believe that the next step is to jointly design the overlay management protocol and the optimal choice of network coding coefficients.

CROSS-LAYER OPTIMIZATION

So far, the problem of network coding has been translated into improving the network throughput, *i.e.* ensuring that the received packets are innovative for the destination nodes. Most works previously presented have been proposed at the network layer, concerned with finding paths and assigning network codes for those paths so that more of the transmitted packets can be decoded by most receivers. However, in real networks the quality of service and the amount of traffic are an important issue, which is why a network-coding algorithm should take into account the relative importance of media packets and prioritize their delivery. Other works have tried to deal with the problem at the source, by adding different protection strengths to different priority levels, but without knowledge of the network conditions. Recent works, presented in this section, have shown that a cross-layer optimization approach achieves better performance in lossy networks such as overlay networks or multi-hop wireless networks.

In the wireless case, the proposed algorithms (Seferoglu & Markopoulou, 2009) build on the COPE framework, were described in the Section "Network Coding for Multimedia Applications." In COPE, several packets from different unicast streams are packed into a single code for transmission that can be decoded at the next-hop node, thus ensuring the increase of network throughput. In addition to that, the network code should also take into account the importance of packets (in terms of contribution to the overall quality and playout deadline) within the same stream. The combined approach from Seferoglu and Markopoulou (2009) is presented in the following, as it has been shown to improve the quality of video delivery, while maintaining the same level of throughput as the COPE protocol. Although the algorithms were proposed for video streams, they can be extended to any type of media applications. The pseudo-broadcast is implemented similar to the COPE protocol, on top of the 802.11 unicast. Nodes can learn about the contents of the virtual buffers of their neighbors either explicitly through periodic status reports or implicitly from the annotations in the packet headers. NC is implemented as a thin layer between IP and MAC, as in the original COPE. The algorithm has to first solve the code construction problem, *i.e.* deciding which packets can be coded together, and then the code selection problem, *i.e.* choosing a code that improves the quality with respect to a certain metric, in this case the video quality. The intermediate node maintains a transmission queue with incoming video packets, and at a given time slot a packet is chosen for transmission, called the primary packet, whose destination is called the target node. The primary packet is XOR-ed together with other side packets

that can be useful to nodes other than the target node. The coded packet is broadcast to all nodes in the neighborhood. The target node will be able to decode the packet due to the code construction and will broadcast an Acknowledgement (ACK). The other neighboring nodes will overhear the packet and store it in their virtual buffer until an ACK from the target node is received or until the packet deadline expires. From the time it sends a network code the intermediate node will wait for a mean Round-Trip Time (RTT) to receive the ACK, during which the packets that were part of the code will be marked as inactive in the transmission queue, making them available as side packets but not as primary packets. When an ACK is received, the primary packet is removed from the transmission queue. When the RTT time expires without receiving an ACK, the packet is marked as active again and the node will try to retransmit it until the deadline expires.

The first algorithm, called NCV, constructs all candidate network codes that include the primary packet, among which it chooses the code that maximizes the total video quality improvement. In order for the network code to be optimal in a rate-distortion manner, the distortion value of every packet is determined at the source and communicated to the intermediate nodes in the packet header. In addition to the importance of packets in the flow (Δ), the different flows may be of different importance (γ), so the total importance of the packet is considered to be the product of the two terms, $\gamma \Delta$.

By construction, the network code corresponding to the primary packet p_i has to be decodable by the target node, which means the side packets must be among the packets already decoded at the target node, but still useful to other nodes. For each possible code c_k, the improvement of video quality at each neighboring node n is given by:

$$I_k(n) = \sum_{l=1}^{L_k} (1 - P(l))\Delta(l)\gamma(l)g_l^k(n)d_l^k(n),$$

where L_k is the number of original packets included in the code (packets that can be useful to any node, while the target node can benefit from only 1 such packet), $d_l^k(n)$ and $g_k^k(n)$ are indicator functions that take the value 1 if code c_k is useful for node n and 0 otherwise, $\Delta(l)$ is the improvement in video quality if packet l is received correctly and on time at client n, $\gamma(l)$ is the priority of the flow the packet belongs to, $P(l)$ is the loss probability of the packet, due to either channel errors or late arrival.

For every primary packet p_i NCV chooses the code c_k that maximizes the video quality improvement at all clients. NCV selects as primary packet the active packet at the head of the queue, which in turn determines the optimal selection of side packets. To optimize the selection of primary packet, the NCV algorithm is extended to looking into the queue in Depth (NCVD), *i.e.* considering all packets in the queue as candidates for the primary packet. Although this algorithm increases the options for candidate codes, it also requires more computation for code construction and selection.

Seferoglu and Markopoulou (2009) further integrate NC with the well-known rate-distortion optimized packet-scheduling framework (RaDiO). Without NC, in classical RaDiO packet scheduling, the node would choose a policy π for the next transmission opportunity. For every packet p_i in the queue, the policy would indicate whether this packet is transmitted $\pi(j) = 1$ or not $\pi(j) = 0$, so as to minimize a weighted function of distortion and rate $J(\pi) = D(\pi) + \lambda R(\pi)$.

With NC, the goal is to find the optimal code transmission policy on all nodes Π^{valid}, so as to minimize the total distortion $D(\Pi^{\text{valid}})$, subject to the rate constraint $R(\Pi^{\text{valid}}) \leq R_{av}$, where R_{av} is the available bit rate. Using Lagrangian relaxation, the problem turns to finding the code transmission policy Π^{valid} such that $J(\Pi^{\text{valid}}) = D(\Pi^{\text{valid}}) + \lambda R(\Pi^{\text{valid}})$ is minimized. Instead of finding the optimal code transmission policy, each code can

be mapped into the packets it contains and the problem can be turned into finding the optimal packet transmission policy.

The equivalent problem is to choose the packet policy π such that $\min_{\pi,\lambda} J(\pi) = \min_{\pi,\lambda} \{D(\pi) + \lambda R(\pi)\}$, where $D(\pi)$ and $R(\pi)$ are the total distortion and rate over all nodes under the policy π.

The optimal policy decides which node n should transmit and what code c_u should be transmitted by choosing the maximum Lagrange multiplier $\max_{n,c_u}\{\lambda_n(c_u)\}$. In practice this can be achieved in two rounds: first, each node n compares $\lambda_n(c_u)$ for all possible codes and finds $\lambda_n = \max_{c_u}\{\lambda_n(c_u)\}$; then all nodes exchange their λ_n values with their neighbors. Finally, the node with the max λ is the one who transmits and this is repeated at each transmission opportunity. Although this method obtains the global optimum at each transmission, it cannot be implemented in practice, as it requires either complete knowledge of the network, or the exchange of messages among all nodes. Simulations show however that the previously presented algorithms, NVC and NCVD, can perform well in practice with less message exchange and that they can be considered efficient heuristics to the general NC-RaDiO problem.

For overlay networks, Thomos et al. (2011) proposed a receiver-driven video streaming solution for video packets belonging to different priority classes. The problem of choosing the network coding strategy at every peer is formulated as an optimization problem of determining the rate allocation between the different packet classes such that the average distortion at the requesting peer is minimized. The packet classes can correspond to layers in scalable video streams or can be constructed based on the contribution of each packet to the overall quality of the media content. Class c is defined as the set of packets that are linear random combinations of packets from the c most important classes. The class number is identified in a small header in each packet. The protocol follows two stages: first, children nodes u_i compute the optimal coding strategy that their parents v_j should follow based on the available bandwidth, importance of packets in each class, and expected loss probability of the link. Then, they send a request message to their parents specifying the number of packets they want to receive from each class. Parent nodes send random linear combinations of packets in the requested classes. Based on the state of its buffer and the local network status, the child node re-computes the optimal coding strategy and makes another request. In this way, the algorithm is receiver-driven and can adapt to the needs of each node and to changing network conditions. A child node u sends the same request to all its parents, which takes the form of a rate distribution vector $w = [w_1, ..., w_C]$, where w_c represents the proportion of packets from class C in the requested packets, so $\sum_{c=1}^{C} w_c = 1$ and $w_c \geq 0$. The expected reduction in video distortion $D(u)$ is a function of the number of classes the node u can decode, and can be written as:

$$D(u) = \sum_{c=1}^{C} d_c p_d(c),$$

where $p_d(c)$ is the probability that node u is able to decode c video classes, *i.e.* the probability that it receives enough packets to decode packets up to class c, but not the subsequent classes.

Each client node thus solves a Rate Allocation Problem (RAP), which is formulated as finding the optimal w^* distribution over the classes that minimizes the expected reduction in distortion:

$$w^* = \arg \max_w D(u), \text{ such that } \sum_{c=1}^{C} w_c = 1 \text{ and}$$
$w_c \geq 0$, for $c = 1, ..., C$.

The authors further show that the optimization function can be put in a log-concave form that can be solved by means of iterative algorithms used in convex optimization problems. This aspect is important since every client has to solve the RAP problem independently and the search space would be huge if they were to do an exhaustive search. The authors also propose a greedy algorithm that is able to find a solution in a finite number of steps. Experimental results for such cross-layer optimization schemes show that they can not only improve the quality of the service provided, but they adapt well to different network characteristics like size of the network, link capacity or packet loss probability.

ERROR RESILIENT NETWORK CODING

Network coding is based mostly on performing linear coding operations at intermediate nodes. If each sink node is aware of both the coding functions and the network topology, perfect decoding is possible by solving a system of linear equations provided that no errors have occurred in the network. However, the assumption of error-free networks is problematic, since various kinds of errors are likely to take place in real networks. For instance, in a wireless scenario packets may experience random errors due to noisy links. Furthermore, malicious nodes may intentionally inject corrupted packets in order to alter information packets. Since even a single error has the potential to affect the decoded messages at all sink nodes, methods presented in the previous sections perform network coding at the application or network layer, after the erroneous packets have been dropped at the MAC layer. However, the transmission efficiency could potentially be improved by employing error-correcting codes, and thus avoiding retransmission.

The problem of error-control in random linear network coding was considered by Koetter and Kschischang (2008) starting from the observation that linear network coding is vector space preserving. The original data (packets) injected by the source is modeled as a basis for a vector space V and the network itself is considered as a black box, *i.e.* a linear operator, which transforms the input space on a possibly different output space. If no errors occur, vector spaces are preserved under linear transformations, and if errors do occur, the received vector space U is close to the transmitted vector space V under a distance metric appropriately defined on vector spaces. In other words, if the input spaces (codewords) have a certain minimum distance regarding the number of non-intersecting dimensions, error-correction can be achieved at the decoder provided that the linear network operator is not too rank-deficient and, furthermore, the received space does not contain too many "malicious" dimensions due to error packets. It has been observed (Koetter & Kschischang, 2008) that low complexity Maximum Rank Distance (MRD) codes, introduced by Gabidulin (1985), can be applied for network coding error detection and correction. The approach introduced in Plass et al. (2008) originally targeted for crisscross error patterns, can be successfully applied for RLNC. Some notions of rank codes will be shortly introduced in the following, together with a Berlekamp-Massey algorithm for decoding of rank metric codes, as presented in Plass et al. (2008).

Fundamentals of Rank Codes

If x is a codeword of length n with elements from $GF(q^N)$, where q is a power of a prime, we can consider a bijective mapping from the codeword $x=(x_1,..,x_n)$ into an $N x n$ array A.

Rank Metric over $GF(q)$: The rank of x over q is defined as $r(x|q)= r(A|q)$.

The rank function $r(A|q)$ is equal to the maximum number of linearly independent rows or columns of A over $GF(q)$. The rank function can be shown to define a norm ($r(x|q) \geq 0$, $r(x|q) = 0 \Leftrightarrow x = 0$, $r(x+y|q) \leq r(x|q) + r(y|q), r(ax|q) = |a|r(x|q)$ is also true if we set $|a| = 0$ for $a = 0$ and $|a| = 1$ for $a \neq 0$).

Rank Distance: If x and y are two codewords of length n with elements from $GF(q^N)$, the rank distance is defined as $dist_r(x, y) = r(x-y|q)$.

Similar to the minimum Hamming distance, the minimum rank distance of a code C can be determined.

Minimum Rank Distance: For a code C the minimum rank distance is given by:

$$d_r = \min\{dist_r(x, y)|x \in C, y \in C, x \neq y\}$$

or when the code is linear:

$$d_r = \min\{r(x|q)|x \in C, x \neq 0\}$$

Let $C(n,k,d_r)$ be a code of dimension k, length n, and minimum rank distance d_r.

It was shown in Gabidulin (1985) that there also exists a Singleton-style bound for the rank distance.

Singleton-style Bound: For every linear code $C(n,k,d_r) \in GF(q^N)^n$, d_r is upper bounded by:

$$d_r \leq d_h \leq n-k + 1,$$

where d_h is the minimum Hamming distance.

A linear code $C(n,k,d_r)$ is called *Maximum Rank Distance(MRD)* code, if the Singleton-style bound is fulfilled with equality.

In Plass et al. (2008), a constructive method for the parity check matrix and the generator matrix of an MRD code is given as follows:

Construction of MRD Codes

A parity check matrix H that defines an MRD code and the corresponding generator matrix G can be written as seen in Box 1.

The decoding of Rank-Codes with the modified Berlekamp-Massey algorithm is based on linearized polynomials.

A linearized polynomial over $GF(q^N)$, is a polynomial of the form $L(x) = \sum_{p=0}^{N(L)} L_p x^{q^p}$, where

Box 1.

$$H = \begin{bmatrix} h_0 & h_1 & \cdots & h_{n-1} \\ h_0^q & h_1^q & \cdots & h_{n-1}^q \\ \vdots & \vdots & \ddots & \vdots \\ h_0^{q^{d-2}} & h_1^{q^{d-2}} & \cdots & h_{n-1}^{q^{d-2}} \end{bmatrix}, G = \begin{bmatrix} g_0 & g_1 & \cdots & g_{n-1} \\ g_0^q & g_1^q & \cdots & g_{n-1}^q \\ \vdots & \vdots & \ddots & \vdots \\ g_0^{q^{d-2}} & g_1^{q^{d-2}} & \cdots & g_{n-1}^{q^{d-2}} \end{bmatrix}$$

where the elements $h_0, \ldots, h_{n-1} \in GF(q^N)$, and $g_0, \ldots, g_{n-1} \in GF(q^N)$, are linearly independent over $GF(q)$.

$Lp \in GF(q^N)$, and $N(L)$ is the norm of the linearized polynomial (the largest p where $L_p \neq 0$).

Let $\otimes$ be the *symbolic product* of linearized polynomials defined as:

$$F(x) \otimes G(x) = F(G(x)) = \sum_{p=0}^{j} \sum_{i+l=p} (f_i g_l^{q^i}) x^{q^p}, \text{ with}$$

$j = N(F) + N(G)$.

The symbolic product is associative, distributive with respect to ordinary polynomial addition, but non-commutative.

Berlekamp-Massey Algorithm for Decoding Rank-Codes

Let c, r and e be the codeword vector, the received vector and the error vector of length n with elements from $GF(q^N)$. The received vector is $r = c + e$. Let $v = r(e|q)$ be the rank of the error vector e.

If $2v < d_r$, the codeword can be correctly decoded:

Syndrome s is given by:

$$s = r \bullet H^T = (c + e) H^T = e \bullet H^T \tag{1}$$

A $(v \times n)$ matrix Y of rank v is defined, whose entries are from the base field $GF(q)$.

Then e can be written in the form

$$e = (E_0, E_1, \ldots, E_{v-1})Y, \tag{2}$$

where $E_0, E_1, \ldots, E_{v-1} \in GF(q^N)$ are linearly independent over $GF(q)$.

Matrix Z is defined as:

$$Z^T = YH^T = \begin{bmatrix} z_0 & z_0^q & \cdots & z_0^{q^{d-2}} \\ z_1 & z_1^q & \cdots & z_1^{q^{d-2}} \\ \vdots & \vdots & \ddots & \vdots \\ z_{v-1} & z_{v-1}^q & \cdots & z_{v-1}^{q^{d-2}} \end{bmatrix} \tag{3}$$

where the elements $z_0, z_1, \ldots, z_{v-1} \in GF(q^N)$ are linearly independent over $GF(q)$.

The syndrome equation can be written as: $(S_0, S_1, \ldots, S_{d-2}) = (E_0, E_1, \ldots, E_{v-1}) Z^T$ or element-wise:

$$S_p = \sum_{j=0}^{v-1} E_j z_j^{q^p}, \text{ for } p = 0, \ldots, d-2 \tag{4}$$

By raising each side of the equation to the power of q^p and after doing the operations in $GF(q^N)$ we obtain:

$$S_p^{q^{-p}} = \sum_{j=0}^{v-1} E_j^{q^{-p}} z_j, \text{ for } p = 0, \ldots, d-2 \tag{5}$$

This is a system of $d-1$ equations with $2 \cdot v$ unknowns that are linear in $z_0, z_1, \ldots, z_{v-1}$. The rank v of the error vector is also unknown. It is sufficient to find one solution of the system because every solution of $E_0, E_1, \ldots, E_{v-1}$ and $z_0, z_1, \ldots, z_{v-1}$ results in the same error vector e.

The *row error polynomial* $\wedge(x) = \sum_{i=0}^{v} \wedge_j x^{q^j}$ is a linearized polynomial which has $\lambda_0 = 1$, and all linear combinations of E_0, E_1,..., E_{v-1} over GF(q) are its roots.

The linearized syndrome polynomial can be written as: $S(x) = \sum_{j=0}^{d-2} S_j x^{q^j}$.

The *key equation* can be defined as:

$$\Lambda(x) \otimes S(x) = F(x) \bmod x^{q^{d-1}} \tag{6}$$

where $F(x)$ is an auxiliary linearized polynomial that has norm $N(F) < v$.

In order to get the row error polynomial $\Lambda(x)$, the following system has to be solved, with $2 \cdot v < d$:

$$\sum_{i=0}^{p} \Lambda_i S_{p-i}^{q^i} = 0, \text{ for } p = v,...,2v-1.$$

Subtracting $S_p \Lambda_0$ from both sides and taking into account that $\Lambda_0 = 1$ and $\Lambda_i = 0$ for $i > v$, we obtain

$$-S_p = \sum_{i=1}^{v} \Lambda_i S_{p-1}^{q^i} = 0, \text{ for } p = v,...,2v-1,$$

which can be written in matrix form as

$$S\begin{bmatrix} \Lambda_v \\ \Lambda_{v-1} \\ \vdots \\ \Lambda_1 \end{bmatrix} = \begin{bmatrix} -S_v \\ -S_{v+1} \\ \vdots \\ -S_{2v-1} \end{bmatrix}, \text{ with } S = \begin{bmatrix} S_0^{q^v} & \cdots & S_{v-1}^{q^1} \\ S_1^{q^v} & \cdots & S_v^{q^1} \\ \vdots & \ddots & \vdots \\ S_{v-1}^{q^v} & \cdots & S_{2v-2}^{q^1} \end{bmatrix} \tag{7}$$

Matrix S can be shown to be non-singular, so the system of equations has a unique solution. The solution can be found using the modified Berlekamp-Massey algorithm presented in Plass et al. (2008). The overall steps of the decoding procedure can be summarized as follows:

1. Calculate the syndrome with equation (1).
2. Solve the key equation (7) with the modified Berlekamp-Massey algorithm described in Plass et al. (2008) to obtain $\Lambda(x)$.

3. Calculate the linearly independent roots E_0, E_1,..., E_{v-1} of $\Lambda(x)$. This can be done with the Berlekamp-Massey algorithm.
4. Solve the linear system of equations (5) for the unknown variables z_0, z_1,..., z_{v-1}.
5. Calculate the matrix Y using equation (3).
6. Calculate the error vector e by equation (2) and then the decoded code word c = r − e.

The benefits of random linear network coding have made it appealing for several practical applications. However, the effect that corrupted packets can have on such a scheme cannot be neglected. On the other hand, the approach presented in this section shows that error-correcting codes for such a scheme exist, and that they can be easily hardware integrated, similar to the widely used Reed-Solomon codes.

CONCLUSION

In this chapter, we presented the concept of *network coding*; an innovative paradigm alternative to classical routing that allows maximizing the throughput of a network by enabling intermediate nodes to send combinations of the received packets instead of mere copies. We discussed how network coding can be beneficial both to non-live applications, such as distributed storage and content distribution, and to delay-constrained applications, such as video streaming and P2P video. In particular, we reviewed several techniques that combine the features of network coding in general with source coding frameworks, characteristic of multimedia applications over unreliable channels, such as scalable coding and multiple description coding. We showed how these paradigms could be integrated and pointed out the benefits of their interaction, in particular when a joint, cross-layer optimization is performed.

REFERENCES

Ahlswede, R., Cai, N., Li, S.-Y., & Yeung, R. W. (2000). Network information flow. *IEEE Transactions on Information Theory*, *46*(4), 1204–1216. doi:10.1109/18.850663

Bondy, J., & Murty, U. (2008). *Graph theory*. Berlin, Germany: Springer. doi:10.1007/978-1-84628-970-5

Chou, P. A., Wu, Y., & Jain, K. (2003). Practical network coding. In *Proceedings of 51st Allerton Conference on Communication, Control and Computing*. Allerton Conference on Communication, Control, and Computing.

Dimakis, A., Godfrey, P., Wu, Y., Wainwright, M., & Ramchandran, K. (2010). Network coding for distributed storage systems. *IEEE Transactions on Information Theory*, *56*(9), 4539–4551. doi:10.1109/TIT.2010.2054295

Ford, L. R., & Fulkerson, D. R. (1956). Maximal flow through a network. *Canadian Journal of Mathematics*. Retrieved from http://www.cs.yale.edu/homes/lans/readings/routing/ford-max_flow-1956.pdf

Gabidulin, E. M. (1985). Theory of codes with maximum rank distance. *Problemy Peredachi Informatsii*, *21*(1), 3–16.

Gkantsidis, C., & Rodriguez, P. (2005). Network coding for large scale content distribution. In *Proceedings of 24th Annual Joint Conference of the IEEE Computer and Communications Societies*, (Vol. 4, pp. 2235-2245). IEEE Press.

Greco, C., & Cagnazzo, M. (2011). A cross-layer protocol for cooperative content delivery over mobile ad-hoc networks. *International Journal of Communication Networks and Distributed Systems*, *7*(1-2), 49–63. doi:10.1504/IJCNDS.2011.040977

Ho, T., Koetter, R., Médard, M., Karger, D. R., & Effros, M. (2003). The benefits of coding over routing in a randomized setting. In *Proceedings of IEEE International Symposium on Information Theory*, (p. 442). IEEE Press.

Jaggi, S., Sanders, P., Chou, P., Effros, M., Egner, S., & Jain, K. (2005). Polynomial time algorithms for multicast network code construction. *IEEE Transactions on Information Theory*, *51*(6), 1973–1982. doi:10.1109/TIT.2005.847712

Katti, S., Rahul, H., Hu, W., Katabi, D., Medard, M., & Crowcroft, J. (2008). XORs in the air: Practical wireless network coding. *IEEE/ACM Transactions on Networking*, *16*(3), 497–510. doi:10.1109/TNET.2008.923722

Koetter, R., & Kschischang, F. R. (2008). Coding for errors and erasures in random network coding. *IEEE Transactions on Information Theory*, *54*(8), 2579–3591. doi:10.1109/TIT.2008.926449

Koetter, R., & Médard, M. (2003). An algebraic approach to network coding. *IEEE/ACM Transactions on Networking*, *11*(5), 782–795. doi:10.1109/TNET.2003.818197

Li, S.-Y. R., Yeung, R. W., & Cai, N. (2003). Linear network coding. *IEEE Transactions on Information Theory*, *49*(2), 371–381. doi:10.1109/TIT.2002.807285

Nemoianu, I., Greco, C., Cagnazzo, M., & Pesquet-Popescu, B. (2012). A framework for joint multiple description coding and network coding over wireless ad-hoc networks. In *Proceedings of IEEE International Conference on Acoustics, Speech and Signal Processing*. IEEE Press.

Nguyen, K., Nguyen, T., & Cheung, S.-C. (2007). Peer-to-peer streaming with hierarchical network coding. In *Proceedings of the IEEE International Conference on Multimedia and Expo*, (pp. 396-399). IEEE Press.

Plass, S., Richter, G., & Han Vinck, A. J. (2008). Coding schemes for crisscross error patterns. *Wireless Personal Communications*, *47*(1), 39–49. doi:10.1007/s11277-007-9389-6

Ramasubramonian, A., & Woods, J. (2010). Multiple description coding and practical network coding for video multicast. *IEEE Signal Processing Letters*, *17*(3), 265–268. doi:10.1109/LSP.2009.2038110

Seferoglu, H., & Markopoulou, A. (2009). Video-aware opportunistic network coding over wireless networks. *IEEE Journal on Selected Areas in Communications*, *27*(5), 713–728. doi:10.1109/JSAC.2009.090612

Thomos, N., Chakareski, J., & Frossard, P. (2011). Prioritized distributed video delivery with randomized network coding. *IEEE Transactions on Multimedia*, *13*(4), 776–787. doi:10.1109/TMM.2011.2111364

Vukobratović, D., & Stanković, V. (2010). Unequal error protection random linear coding for multimedia communications. In *Proceedings of the IEEE International Workshop on Multimedia Signal Processing,* (pp. 280-285). IEEE Press.

Wang, M., & Li, B. (2007). Lava: A reality check of network coding in peer-to-peer live streaming. In *Proceedings of the 26th IEEE International Conference on Computer Communications*, (pp. 1082-1090). IEEE Press.

Widmer, J., Fragouli, C., & Le Boudec, J.-Y. (2005). Low-complexity energy-efficient broadcasting in wireless ad-hoc networks using network coding. In *Proceedings of Workshop on Network Coding, Theory, and Applications, (NetCod 2005)*. Riva del Garda, Italy: NetCod.

Wu, Y. (2009). Existence and construction of capacity-achieving network codes for distributed storage. In *Proceedings of the IEEE International Symposium on Information Theory*, (pp. 1150-1154). IEEE Press.

ADDITIONAL READING

Fragouli, C., Le Boudecet, J.-Y., & Widmer, J. (2006). Network coding: An instant primer. *ACM SIGCOMM Computer Communication Review*, *36*(1), 63–68. doi:10.1145/1111322.1111337

Fragouli, C., & Soljanin, E. (2006). Information flow decomposition for network coding. *IEEE Transactions on Information Theory*, *52*(3), 829–848. doi:10.1109/TIT.2005.864435

Ho, T., Medard, M., & Koetter, R. (2005). An information theoretic view of network management. *IEEE Transactions on Information Theory*, *51*(4), 1295–1312. doi:10.1109/TIT.2005.844062

Wu, Y., Chou, P. A., & Kung, S.-Y. (2005). Minimum-energy multicast in mobile ad hoc networks using network coding. *IEEE Transactions on Communications*, *53*(11), 1906–1918. doi:10.1109/TCOMM.2005.857148

Yeung, R. W. (2008). *Information theory and network coding*. Retrieved from http://iest2.ie.cuhk.edu.hk/~whyeung/book2

Yeung, R. W., Li, S.-Y. R., Cai, N., & Zhang, Z. (2005). *Network coding theory*. Retrieved from http://iest2.ie.cuhk.edu.hk/~whyeung/netcode/monograph.html

Zhu, Y., Li, B., & Guo, J. (2004). Multicast with network coding in application-layer overlay networks. *IEEE Journal on Selected Areas in Communications*, 22(1), 107–120. doi:10.1109/JSAC.2003.818801

KEY TERMS AND DEFINITIONS

Inter-Session Network Coding: Network coding performed on messages belonging to different flows in the network.

Intra-Session Network Coding: Network coding performed on messages belonging to the same session in the network, usually implemented as random linear network coding.

Minimum Cut: In graph theory, the partition of the vertex set into two disjoint sets such that the sum of the capacities of all the links between the two set is minimum.

Network Coding: It is a technique in which intermediate nodes in a network forward a combination of their previously received messages.

Random Linear Network Coding: It is a network coding scheme in which the encoding functions are linear and the coding coefficients are chosen randomly from a field.

Rank-Metric: It is a metric for blockcodes over extension fields $GF(q^m)^n$ or equivalently the rank of array codes that consist of arrays over base fields $GF(q^m)$.

Chapter 2
Review of Advanced Mobility Solutions for Multimedia Networking in IPv6

József Kovács
Hungarian Academy of Sciences, Hungary

László Bokor
Budapest University of Technology & Economics, Hungary

Zoltán Kanizsai
Budapest University of Technology & Economics, Hungary

Sándor Imre
Budapest University of Technology & Economics, Hungary

ABSTRACT

IPv6 is the new version of the Internet Protocol (IP) that is expected to be introduced for a wide audience in the forthcoming years. IPv6 comes with a huge amount of improvements compared to the currently widespread IP version (IPv4), while it keeps the same conceptual basics. For instance, IPv6 has a comprehensive and built-in scheme for mobility management with a great set of additional functionality, while IPv4 has only an extension for this purpose (and it is usually not implemented). Considering the evolution of telecommunication architectures toward a heterogeneous all-IP fixed-mobile convergent multimedia-provisioning system, it is now obvious that only the appearance of IPv6 could extend the infrastructure to cope with the emerging scenarios and use-cases. This chapter provides a broad introduction of the advanced IPv6 features and guides the readers from the basics of the new IP protocol family to its complex feature set and power to support multimedia communications in the mobility-centric Future Internet. Optimization techniques to further increase the adequacy of IPv6 for mobile multimedia are also presented along with the description of several research directions.

DOI: 10.4018/978-1-4666-2833-5.ch002

INTRODUCTION

The vision of "anytime and anywhere" has become a powerful concept for voice telephony, where it has been widespread as a global phenomenon and an essential infrastructure. However, nowadays mobile telecommunications aim to emerge beyond individualized voice services and converge to a much more complex system by having mass media content (text, voice, sound, images, video, etc.) within integrated service platforms such creating the phenomenon of mobile multimedia. Newspapers, magazines, books, Internet radio and TV channels, websites, portable music (e.g., in MP3 format) or portable/on-line electronic games, text and rich (incorporating voice/picture/video material) messages, real-time and on-demand video materials (e.g., video phone) and photos are taking part from the emerging new medium of ubiquitous mobile networking. Such mobile networking continuously creates new types of content, initiates new technologies and allows people to interact in novel ways. In order to make all the above advanced mobile media applications available for the wide audience, network operators are taking the challenge of combining mobile communications and the Internet. The convergence is not only observable in networks but also in devices and services, and also amplifies the essential need of networked information provisioning for users anytime and anywhere. Current trends place mobile Internet architectures into the focus point of the whole technological progress. With the development of various wireless network technologies such as WiFi, WiMAX, UMTS, HSPA, LTE, LTE-A, more and more users want to enjoy the benefits of seamless connectivity and ubiquitous Internet access. Vendors prognosticate that mobile networks will suffer an immense multimedia traffic explosion in the packet switched domain up to year 2020 (UMTS Forum, 2010; Cisco VNI, 2011). In order to accommodate the Future Internet to the anticipated demands and requirements, technologies applied in the radio

access and core networks must become scalable and appropriate to advanced future use cases. Network operators not only have to take care of the growing traffic volumes and mass of users, the heterogeneous, overlapping wireless access, and secure communication, but they have also to enforce certain policies in order to provide the necessary *Quality of Service* (QoS) to consumers, all considering the fact that majority of mobile traffic consists of multimedia content (Bokor, Faigl, & Imre, 2011).

The increasing number of consumers, the complexity of mobility scenarios, the technological convergence in telecommunication and information technology present a great challenge for the architecture of the Internet we use today, as such things were not envisioned in the 70s, when the still used IP protocol was designed: IPv4 does not allow the mobility of hosts, works with relatively small address space and lacks support for QoS. To address all these problems and serve the evolving trends of mobile communication, IPv6, a new version of the protocol was developed (Hinden & Deering, 2006; Deering & Hinden, 1998). In terms of multimedia requirements, IPv6 has a number of features that not only optimize current networking techniques for multimedia content transmission, but tries to keep up with the growing demand for services, especially in mobile environments.

Future generations of mobile and wireless technologies will provide virtually unlimited possibilities to the community of multimedia users to all over the world. Network technology innovations and architecture evolution will create the convergent environment in which every media is available, and networked resources are accessible anytime and anywhere, via any kind of connected device in any number. IPv6—as the common language of the Future Internet both in the fixed and mobile domains—could be one of the most important tools for mobile content service delivery, in which enlarged address space, advanced security, multicast and QoS capabilities are naturally integrated with efficient and extendable mobility management in order to support mobile multimedia services for every possible application scenario.

In this chapter, we summarize the feature set of IPv6 for enabling seamless, transparent and secure transmission of multimedia content over mobile IPv6 networks. Then, we introduce a new handover technique, which intends to increase the networking performance of mobile multimedia services.

IPV6 ESSENTIALS: THE BACKGROUND OF MOBILE EVOLUTION

Content delivery is shifting towards peer-to-peer networks, while the majority devices are becoming mobile. This is intensified by Machine-to-Machine (M2M) communications which also accommodate end-to-end communicating devices without human intervention for remote controlling, monitoring and measuring, road safety (e.g., traffic avoidance, enforcement, and control systems), security/identity checking, video surveillance, electronic healthcare delivery, personal locator services, etc. Predictions state that there will be 225 million cellular M2M devices by 2014 with little traffic per node, but resulting significant growth in total, mostly in uplink direction (Dohler, Watteyne, & Alonso-Zárate, 2010). Therefore, we can say that one of the most obvious features of IPv6 for future mobile multimedia is the large address space. With a four times increase in address length compared to IPv4, any IPv6 enabled mobile device will be reachable via a globally routable unique address, eliminating the need for address space saving techniques, such as NAT (Network Address Translation). There are 2^{128} different IPv6 addresses, as opposed to the 2^{32} possible addresses in IPv4, which opens up new possibilities for multimedia content delivery. Based on the directionality and the number of participants in the communication, IPv6 addresses are grouped into unicast, multicast

and anycast address groups. Due to its one-to-many directionality, multicast addressing is an efficient way of transporting multimedia content, which will be described in greater detail later in the chapter.

The distribution and assignment of unicast IPv6 addresses is another key feature for mobile environments. When a node connects to an IPv6 network, it receives *Router Advertisement* (RA) messages from the router present on the network (Narten, et al., 2007). These RA messages contain the prefix used on the network and the validity of the addresses among other information. After processing the message, the node generates a unique 64-bit identifier from its physical interface identifier. In case of 48-bit MAC addresses, the uniqueness is guaranteed by a simple mapping from 48-bit to the 64-bit EUI address format. The generated address together with the network identifier received from the router is the unique global IPv6 address for the given host. This address configuration method is called *stateless address autoconfiguration* (Thomson, Narten, & Jinmei, 2007), and is only available in networks with 64-bit or less prefix size. When the stateless method is not applicable on a given network due to prefix size or other reasons, different IPv6 address provisioning mechanisms may be used. Stateful address autoconfiguration, such as DHCPv6 (Droms, 2003), is a technique where address provisioning and accounting is managed by a dedicated node on the network. The faster address configuration shows its advantage in mobile environments, allowing fast handovers between access networks, while granting media carrying transport and application protocols to continue to work seamlessly, anytime and anywhere. The simplified header format offers several options to increase the performance of IPv6, when carrying multimedia content. The *Traffic Class* field, which marks the priority of packet delivery, is used to ensure QoS (Quality of Service). As media content takes up significant slice of the overall Internet traffic, the networking protocol needs to be prepared to ensure the quality of service and experience remains positive for the user. IPv6 has a number of ways to improve support for QoS (Rajahalme, Conta, Carpenter, & Deering, 2004; Ping & Desheng, 2010; Zhenhua, Qiong, Xiaohong, & Yan, 2010). The *Flow Label* field allows labeling of packets belonging to the same data stream, such as TCP stream. *Payload Length* marks the size of the payload carried in the IPv6 packet, while *Hop Limit* defines the max number of hops a packet is allowed to travel. Fragmentation related fields are missing from the IPv6 header as IPv6 does not fragment the payload. Instead, communication parties perform *Path MTU Discovery* (McCann, Deering, & Mogul, 1996) to determine the maximum payload size between the source and the destination. The protocol uses the *Next Header* field to mark the type of the next protocol in the packet, allowing the presence of multiple IPv6 extensions while making it possible to prioritize transport protocols more easily.

Communication in the open, packet-based Internet must consider also security aspects. It is much easier to capture voice information transmitted by a VoIP solution through the Internet than by PSTN operating on basics of circuit switching. The same applies to all multimedia traffic using IP-based architecture as transport medium. The level of threat is even more serious, if the medium is shared, as in case of wireless and mobile environments. That is why another significant advantage of the new IP protocol is the standardized and deeply integrated IPsec security framework, implementing flexible end-to-end media security in the network layer (Kent & Seo, 2005). IPsec has two different communication modes:

- The *transport mode* is used to secure the IPv6 payload between communication endpoints. The IPv6 header is left intact and data is encrypted through the ESP (Encapsulating Security Payload) protocol, which provides confidentiality and authenticity of the payload. This mechanism is perfectly suitable to secure confidential multimedia content over IPv6. When en-

cryption can be omitted, but authentication is still required, the AH (Authentication Header) can be used. The AH header is inserted between the IPv6 header and the payload. As both ESP and AH modify the original structure of the IPv6 packet, the value of the *Next Header* field is modified to reflect the changes in the payload so that it can be reassembled at the receiving end.

- The *tunnel mode* is used to protect traffic between a router and another communication node which could be either a host or a router. Unlike transport mode, in tunnel mode the entire IPv6 packet is encapsulated by ESP/AH and a new IPv6 header with different endpoint addresses is created. The tunneling mechanism along with the security features presented above is a powerful tool to create Virtual Private Networks (VPN) or secure packet delivery on unsecure links such as WLAN backbones.

Due to the rapid and widespread introduction of world-wide multi-play services, mobile IPTV started to grow significantly, fastly creating mobile video and TV services as an essential part of consumers' lives. Current data network infrastructure both on the wired and the wireless segments mainly uses unicast (one-to-one) communication for content delivery, but it is not effective for providing such bandwidth-hungry multimedia services. Contrarily, the multicast data communication paradigm (one-to-many media transmission) provides resource efficient solution for wired IPTV provision and also could help to handle the estimated amount of future mobile video and mobile IPTV traffic. However, the small address space of IPv4 makes hard to grant the necessary support and acceptance for universal multicast communication. Widely deployed multicast services can only be built on the enhanced features of IPv6 multicasting: the large address space and the use of scoped multicast addresses with sophisticated control mechanisms can serve as essential basis for resource-saving multimedia applications with efficient traffic engineering capabilities (Pike, et al., 2007). This promising toolset of IPv6 multicasting has also been seriously considered for organic integration into 3G networks and beyond, as the *Multimedia Broadcast Multicast Service* (MBMS) concept was created by 3GPP to establish a framework for the point-to-multipoint downlink bearer service for IP multimedia in current and future mobile Internet architectures (3GPP TS 23.246, 2011).

The multicast traffic in IPv6 is managed by employing the *Multicast Listener Discovery* (MLD) protocol that aims to define which nodes are supposed to receive the multicast data in a network (Deering, Fenner, & Haberman, 1999). The MLD protocol controls the flow of traffic in a network using multicast queriers (network devices sending query messages to find out which nodes are members of a given multicast group) and hosts (receivers sending report messages to inform the querier of their multicast membership information). Querier and host devices both use MLD reports to join and leave different multicast groups and also to begin the reception of group media traffic. Multicast routing protocols manage the information exchange between routers in order to construct and maintain multimedia distribution trees and also to forward multicast packets from the source to destination nodes. Because multicast addresses identify transmission sessions rather than specific physical destinations, multicast routing is more complex than in the unicast case. Protocol Independent Multicast – Sparse Mode (PIM-SM) is a good example for multicast routing. PIM-SM is an IPv6-compatible solution that can either use the underlying unicast routing information base or a separate multicast-capable routing information base to build unidirectional shared trees rooted at a special entity called the Rendezvous Point (RP) per group, and optionally creating shortest-path trees per source (Fenner, et al., 2006).

There are several solutions to provide multicast services to mobile hosts such as the results of Sang-Jo and Seak-Jae (2006) and Zheng (2006). However, the most elaborated and standardized solution is the *Multimedia Broadcast Multicast Service* (MBMS). The MBMS service was created to overcome the shortcomings of the Cell Broadcast Service (CBS) of cellular networks and to introduce more sophisticated multicasting and broadcasting in the packet switched domain (3GPP TS 23.246, 2011). The core concept of MBMS is to save radio resources by sharing them between users belonging to the same multicast group. The main 3G (and beyond) packet switched elements and the radio access nodes and controllers should be all MBMS-enabled to offer MBMS services, while user terminals also should support MBMS, and also a new functional entity called the BM-SC (Broadcast/Multicast Service Center) should be available. BM-SC serves as an ingress point for multicast content providers, and manages and sets up the MBMS transport services operator's network. The IPv6-aware standard family of MBMS extends the 3G/4G mobile network to enable any multimedia traffic that uses multicast or broadcast addressing scheme to reach mobile subscribers in a resource efficient and well scalable way.

IPV6 MOBILITY MANAGEMENT FOR ON-THE-MOVE MEDIA APPLICATIONS

All the above mentioned features are more or less achievable in the presence of IPv4 as well. Basically, IPv6 is a conceptual copy of the IPv4 protocol, with almost all the functionalities existing in IPv4. The real difference lies in the extended address space, the integrity of the standard and the advanced mobility support. When designing IPv6, the authors were aware of the existing functionalities of IPv4: they tried to integrate all functionalities (including mobility management capabilities) of IPv4 extensions into the basic IPv6 standards. Thus, the IPv6 standards are more complete, and thus IPv6-based mobility management took the leading role in mobility-oriented research and development. Its importance is even more specific in future wireless systems: as access networks are becoming more heterogeneous, the issue of vertical handovers, where a mobile node has to change its point of connection to the Internet among different access media types, must be solved. Also offloading techniques are becoming increasingly popular in the cellular world (3GPP TR 23.829, 2010), allowing mobile operators to perform various policy enforcements without affecting user experience and creating even more complicated mobility scenarios to be handled. Therefore, the role of the mobile IPv6 technologies is crucial. Without efficient management of different mobility events in evolved mobile scenarios and use cases, it will not be possible to provide multimedia services to mobile users in Future Internet architectures with reasonable QoS and QoE (Quality of Experience).

MOBILE IPV6 (MIPV6)

The Mobile IPv6 protocol (Perkins, Johnson, & Arkko, 2011) together with its extensions provide solution for all the above problems allowing hosts to have a topology independent unique IPv6 address that is independent from its point of attachment to the Internet. Using a temporary address—called *Care-of Address* (CoA)—taken from the visited network the Mobile Node (MN) establishes a bidirectional tunnel to a known central entity, known as the Home Agent (HA), allowing uninterrupted IPv6 communication in diverse mobility scenarios. Figure 1 shows the general architecture and main protocol operation of Mobile IPv6 networks, where each MN has a globally unique static Home Address (HoA) independent from its actual point of attachment to the Internet. When a MN is visiting a foreign network, it registers a binding at the Home Agent. With the binding containing the actual CoA taken

from the remote access network and the HoA of the MN, the Home Agent always knows the location of the Mobile Node. The binding is registered and updated in the Binding Update (BU) control message sent by the MN, and acknowledged by the HA with the Binding Acknowledgement (BA) message. As long as the binding is kept up-to-date, the bidirectional IPv6-in-IPv6 tunnel is kept alive between the MN and the HA. The tunnel, similarly to a VPN, uses the actual CoA as source address and the address of the HA as destination. The inner IPv6 header, containing the payload is addressed by the Home Address the address of the Correspondent Node (CN). The job of the Home Agent is to encapsulate and decapsulate the packets belonging to the Mobile Node by impersonating presence of the Mobile Node on the Home Network. Because of the above operation, a usually sub-optimal route containing the HA inside the MN-CN path will be used for communication. This so called *triangular routing phenomenon* introduces additional delays and unwanted overhead, but it can be eliminated by directly registering the MN at the CN with a Binding Update/Acknowledgement message pair. Of course, this needs the CN to have MIPv6 capabilities, and also to employ some additional security mechanisms: in order to provide the CN with some reasonable assurance that the MN is in fact addressable at its stated CoA as well as at its HoA, the return routability procedure (HoTI-CoTI-HoT-CoT) must be executed before the BU/BA sequence (see Figure 1).

While Mobile IPv6 with its security extensions is a viable solution to provide always-on connectivity for nodes on the move, as large part of the Internet still uses IPv4, content delivery would not be efficient without an IPv4-IPv6 transition mechanism. Dual-Stack Mobile IPv6 (DSMIPv6) (Soliman, 2009) is one of the techniques that extend the functionality of MIPv6 to the presence of IPv4 access networks. However, due to its complexity DSMIPv6 is not widely used.

Figure 1. Basics and architecture of mobile IPv6

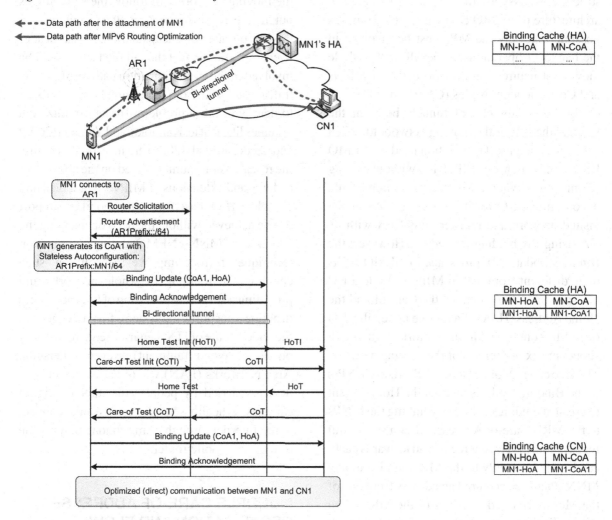

NETWORK MOBILITY BASIC SUPPORT (NEMO BS)

In order to support persistent connection of moving networks (e.g., trains with wireless hosts of passengers inside the carriages) to the Internet, the NEMO Basic Support protocol (Devarapalli, et al., 2005)—as an extension of MIPv6—was designed and approved as an RFC by the IETF. The main goal of this scheme is to preserve ongoing internal and external communication sessions of nodes attached to a moving network during the network's movement. Using the NEMOS BS protocol, the mobile node becomes a mobile router, providing

transparent, legacy network access to its Mobile Network Nodes, while performing mobility actions as a mobile node. Network Mobility is commonly used in Intelligent Transport Systems (ITS), where multiple mobile nodes move at the same time.

In NEMO BS terminology, a Moving Network (MNet) is defined as an entity handling several inside nodes and/or subnetworks as a whole whose Internet point of attachment changes in time. A moving network consists of one or more Mobile Routers (MR) and several Mobile Network Nodes (MNN). MR is the node that manages the tasks of internal routing within the moving network and connects the whole MNet to the external infrastructure. MNNs can either be fixed or mobile. The architecture óf NEMO BS (see Figure 2) makes possible that only the MR must be involved in the handover operations on behalf of the whole moving structure. Data traffic between MNNs and Correspondent Nodes (CNs) is managed by establishing bidirectional tunnels between the HA and the MR of the moving network to which the MNNs belong. The solution used by NEMO BS is similar to Mobile IPv6 but without routing optimization: when a MR leaves its home link, it configures a Care of Address (CoA) in the visited network and registers this CoA with its HA using the binding procedure. However, the Binding Update (BU) message in NEMO BS is quite different from that in MIPv6. While a BU message in MIPv6 contains the Care-of and the Home Address (HoA) of a mobile node, till a BU of an MR contains additional information: the IP subnet prefix or prefixes of the moving network. These so called Mobile Network Prefixes (MNPs) in the Binding Updates instruct the Home Agent to create a binding cache entry linking the MNPs to the MR's Care-of Address. After a successful registration, the HA intercepts and forwards packets destined not only to the MR, but also to any MNNs that have acquired an address from one of the Mobile Network prefixes of the MR. When the moving network changes its actual network point of attachment, only the MR configures new CoA and sends Binding Update (containing the MNPs) to the HA. Observing that the MNNs do not need to configure and bind new CoA as long as they are inside the moving network, signaling overhead can be reduced but it has its cost. A CN usually sends packets to a mobile node using the MN's HoA. Since the Home Addresses of the MNNs inside a moving network are associated with the MNPs registered in the HAs, the HA of the network's MR intercepts all the packets addressed to MNNs and forwards them towards the MR's CoA. The MR decapsulates the packets destined to MNNs and forwards them on its appropriate ingress interfaces. Packets originated from inside the moving network will follow the same routes but in the reverse direction. It is obvious that the big number of encapsulations cause header overhead, and the fact that all the HAs should be involved in the communication path results using traffic routes far from the optimal ones. In order to deal with these problems, route optimization schemes like Kafle, Kamioka, and Yamada (2006) and Calderón et al. (2006) are investigated within the research community. Based on the above procedures and extensions of MIPv6, a practical and complete IPv6-based network mobility support can be achieved without the need of changing the addresses of MNNs. NEMO routing optimization techniques further improve the solution, such enabling the roaming of whole networks and providing transparent provision of Internet access in public transportation systems for passengers, in the widest scale of ITS scenarios (e.g., road safety on the move entertainment) or even in Personal Area Networks (PAN) where various electronic devices carried by people (like tablets, digital cameras, e-health sensors, etc.) would connect to the Internet through a smartphone playing the role of the mobile router.

MULTIPLE CARE-OF ADDRESSES REGISTRATION AND FLOW BINDINGS

While the aforementioned protocols only allow the connection to one access network at a time, redundancy, handover delays and offloading techniques cannot be adopted in any of the mobility scenarios. To address this shortfall, a new extension called *Multiple Care-of Addresses Registration* (MCoA) (Wakikawa, 2009) was introduced to the Mobile IPv6 protocol family. By utilizing that mobile nodes or routers can connect to multiple access networks simultaneously, it is now possible to enhance handover latency, network redundancy and perform policy based routing.

Figure 2. Overview of NEMO BS

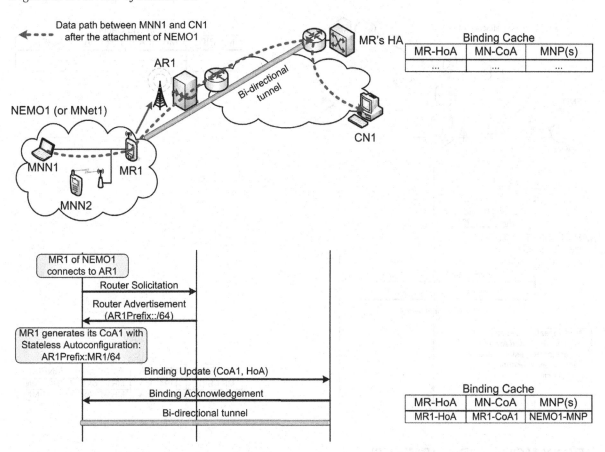

Figure 3 depicts a scenario, where the Mobile Router has two external interfaces. Each interface is connected to an access network with a CoA, and through each CoA a Mobile IPv6 tunnel is created to the Home Agent. While with NEMO BS, identifying a binding was enough using the CoA and the HoA, it is no longer the case with NEMO MCoA as each mobility tunnel endpoint uses the same Home Address on the MR. Using network layer information, the MR can no longer perform an exact routing decision to select an individual tunnel. To solve this issue, another identifier, known as Binding Identifier (BID), was introduced to identify the network interface over which the tunnel is established. As the BID is sent to the HA in the BU signaling message, the HA can differentiate between tunnels originating from the same MR. To identify and route packets toward the desired tunnel, policy routing must be used, which allows fine grained diversification among data packets and streams based on network layer and upper layer information. To avoid asymmetric routing where packets belonging to the same packet flow are routed on different tunnels, a flow binding mechanism has to be implemented. Using flow binding control messages, the MR registers flow descriptor and BID pairs at the Home Agent, so the HA would properly know which tunnel to use when it forwards packets of the data flow back to the mobile node (Tsirtsis, et al., 2011). Using the above introduced multihoming solution, routing of individual media streams can be easily solved, enhancing the experience for not only moving, but stationary mobile nodes as the presence of multiple egress interfaces makes content delivery more reliable and robust.

Figure 3. NEMO multihoming with MCoA

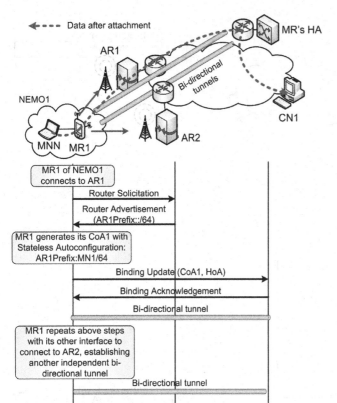

Binding Cache			
MR BIDs	MR-HoA	MN-CoA	MNP(s)
...	...	...	...
...	...	...	...

Binding Cache			
MR BIDs	MR-HoA	MN-CoA	MNP(s)
MR1-BID1	MR1-HoA	MR1-CoA1	NEMO1-MNP
MR1-BID2	MR1-HoA	MR1-CoA2	NEMO1-MNP

PROXY MOBILE IPV6 (PMIPV6)

Although Mobile IPv6 works logically and theoretically allows roaming to devices in wireless networks, in real mobile environments the performance of this protocol is not always satisfactory since the handover procedure can cause significant delay. As MIPv6 is a host-based solution, it requires implementation of the protocol's mechanisms in the kernel of the mobile (or even fixed) devices. This raises some serious problems, and therefore the deployment of MIPv6 in new devices could be very slow. The implementation of MIPv6 in end-user device kernels also provides an additional interface for security vulnerability. To avoid these problems, IETF created a working group called *Network-Based Localized Mobility Management* (NetLMM) to define network-based mobility protocols instead of host-based ones. A network-based protocol can manage MN handovers inside the mobile network core without involving or requiring anything from the MN itself. The main idea is to let the MN keep its IPv6 address during movements across multiple access routers and make this roaming transparent to the IP layer and above.

The proposed solution is Proxy Mobile IPv6 (PMIPv6) (Gundavelli, et al., 2008) and this name came from using proxy-like nodes to manage handovers on behalf of the mobile entities. The main advantage of PMIPv6 is that it needs no additional modifications on the MN (kernel and user space software), therefore it is transparent to the user devices. PMIPv6 is an access technology independent solution, so it can be used with WLAN, WiMAX, 3G UMTS, LTE, LTE-A or any other technology in the future. It provides fast handovers according to its localized nature, which means that the PMIPv6 has a well-defined domain area (Local Mobility Domain, LMD) where exchanging signaling messages is quite fast. PMIPv6 grants the same IP address to the MN during movement so it also provides session continuity within a single access technology domain, which means user space applications do not have to build up new sessions after a handover, because the IP address and the transport protocol ports remain the same. There are two new nodes defined in PMIPv6: (1) the Local Mobility Anchor (LMA) and (2) the Mobile Access Gateway (MAG). LMA acts as a Home Agent (HA) in MIPv6, it maintains a set of routes to every MN in the LMD and all the traffic from and to the MNs go through on this node. The LMA stores the Home Network Prefix (HNP) for every MN in its Binding Cache (BC) which is soft-state table and needs to be updated periodically. A MAG is the first hop router (access router) of the MNs attached to it and this node performs the mobility signaling on behalf of these MNs towards the LMA. The signaling messages are Proxy Binding Update and Proxy Binding Acknowledgement, which are the modifications of the original BU and BA messages from MIPv6. According to the MIPv6, in PMIPv6 we can also find a bi-directional tunnel, but not between the MN and the LMA (HA). This tunell is between the MAG and the LMA. For the same reason, there is a Proxy Care-of Address (Proxy CoA) for every MAG. This is the end point address of the tunnel towards the LMA. The architecture and main scenarios of PMIPv6 are depicted in Figure 4 which also emphasizes that the whole LMD seems to be a virtual link from the viewpoint of the MN, as roaming between the LMD's MAGs the MNs IPv6 address (and the opened sessions) remains the same. The first part of Figure 4(a) represents the signaling flow when a MN arrives in the LMD and attaches itself to the closest MAG. The second part (b) shows the signaling flow during a handover inside the PMIPv6 domain. This operator centric solution is a promising mobility manage-

ment candidate for future mobile systems: 3GPP adopted the scheme for beyond 3G architectures.

HIERARCHICAL MOBILE IPV6 (HMIPV6)

HMIPv6 (Soliman, et al., 2008) is an extension to MIPv6 with the straight purpose to decrease handover delay and make MN movements in the same domain transparent for the Correspondent Nodes (CN) and the HA by using micro-mobility. The main properties of this protocol are that some elemental MIPv6 signaling messages were modified (extended) to be able to be used in HMIPv6 architecture as well and this solution is independent from the underlying access layer technologies. HMIPv6 introduces a new network node called Mobility Anchor Point (MAP), which has the functionality of a HA, so it can store bindings between two IPv6 addresses. Two different types of addresses are used by the HMIPv6 protocol: the Regional Care-of Address (RCoA) and the On-link Care-of Address (LCoA). The second one, LCoA, has the same functionality as the CoA in MIPv6 and the name LCoA is only to distinguish it from RCoA. The RCoA is an address from the subnet of the MAP. After a HMIPv6-aware MN arrives to a domain, it generates an address for itself from the Router Advertisement of its default router and this will be the LCoA. The RA also contains information about MAPs in the domain. If it has one or more MAPs, the MN can decide whether to use HMIPv6 or just simply MIPv6 (with LCoA). When HMIPv6 is chosen the MN asks an RCoA from the MAP and then sends a local BU message with the address pair of LCoA and RCoA. The MAP processes the BU and stores the address pair in its Binding Cache and from this point it acts like a HA for the RCoA address: intercept packages sent to this address and sends it to the actual position of the MN. Then a Binding Acknowledgement is sent back to the MN and this initiates a build-up of a tunnel between the MAP

Figure 4. PMIPv6 architecture and operation

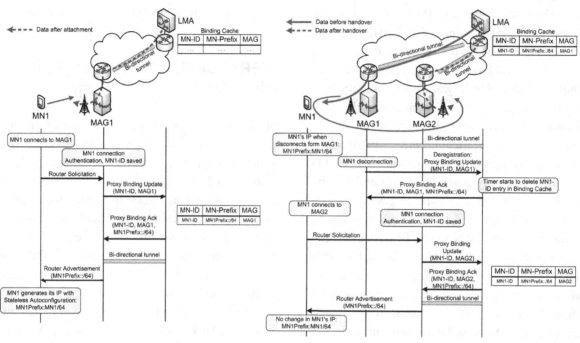

a) MN1 attaches to the PMIPv6 domain

b) MN1 handover between MAG1 and MAG2 inside PMIPv6 domain

and the MN. After this, the MN sends a BU to its real HA with the RCoA in the CoA field.

This means that within the domain managed by the chosen MAP the handovers are handled locally, with no need to send signaling messages to the maybe far away HA, and the movement of the MN is transparent for communication partners outside of the domain. Figure 5 shows the message flowchart of the scenario when a MN arrives in a HMIPv6 domain and establishes connection to its HA. Moving from one AR to another in the same MAP domain, the MN has to send a BU message only to the MAP containing the MN's new LCoA and its RCoA.

MOBILE IPV6 FAST HANDOVERS (FMIPV6)

FMIPv6 (Koodli, 2009) is also an extension of MIPv6 and independent from access layer pro-

tocols. The aim of FMIPv6 (see Figure 6) is to fasten up handovers and decrease the amount of lost packets, when the MN is moving from one AR to another. The first idea is to know the local environment in order to predict the next AR the MN will connect to during its movement and make it possible to get a new IPv6 address prior to connecting to the New AR (NAR). The second idea of this scheme is to use the Previous AR (PAR) to forward the packets addressed to the MN towards the NAR and by this way reduce the number of lost packets during the handover. FMIPv6 defines a new message called Router Solicitation for Proxy Advertisement (RtSolPr), which is sent by the MN to its AR (PAR) to get information about adjacent ARs. The PAR answers with a Proxy Router Advertisement (PrRtAdv) message. The MN chooses the appropriate NAR from the list and generates a New CoA (NCoA) according to the prefix used by in the subnet of the NAR.

Figure 5. HMIPv6 architecture and connection establishment

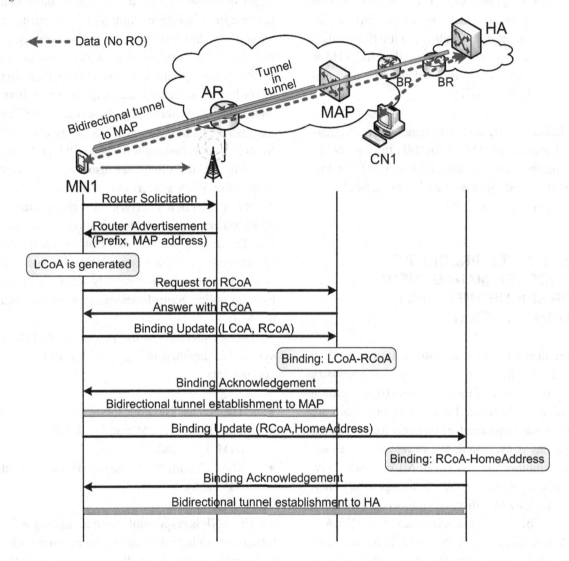

Based on the timing of the Fast BU (FBU) message, there two scenarios for the fast handover: (1) the predictive and (2) the reactive one.

- The *predictive method* requires from the MN to send the FBU message from its previous network to the PAR (the actual one) and wait for the Fast BAck (FBAck) there. The PAR sends a Handover Initiate (HI) message to the NAR, which acknowledges the handover by a Handover Ack (HAck) message. The PAR generates and sends

the FBAck message to the MN, when it receives the HAck from the NAR. In parallel, it starts forwarding the MN's packets to the new network. When the MN arrives to the new network, it sends an Unsolicited Neighbor Advertisement (UNA) to its NAR and from then a MN can immediately receive its packets from the CNs.

- The *reactive method* does not require sending the FBU from the previous network, right after receiving the PrRtAdv the MN can attach itself to the NAR by sending an

UNA message to it. Then, the FBU is sent from the new network to the PAR which initializes the handover with the method mentioned above except that the FBAck message is also forwarded to the new network of the MN.

In both cases, during the handover, packets are forwarded from PAR to the MN through NAR. For performance reasons, HMIPv6 and FMIPv6 are often used together (Lee & Ahn, 2006; Pérez-Costa, et al., 2002, 2003).

GNSS-AIDED PREDICTIVE HANDOVER MANAGEMENT FOR MULTIHOMED NEMO CONFIGURATIONS

The colorful palette of mobility solutions for IPv6 proves that transparent mobility in the network layer is a powerful tool for sensitive application protocols. However, lower layer protocols are usually not considered, when performance of such solutions is evaluated. We developed a method that combines the benefits of MCoA with a new prediction-based cross-layer management entity, which allows mobility solutions to operate using only the best available access networks (Kovács, Bokor, & Jeney, 2011). Predictive handover management is based on the following simple idea: as the node/network moves along a path, it records all access network related data in a database together with the geographical location information. The next time the node/network moves along the same path, based on the geographical information and speed vector, the stored information can be used to predict and prepare handovers before the actual availability of the networks based on calculated weighted performance parameters. When multiple interfaces are used, the above introduced MCoA and Flow Bindings solutions can be of use. The handover preparation consists of the following components. Flow Bindings are applied to

direct the whole traffic of the MR through one active egress interface. Although the benefits of redundancy are lost, we gain the possibility to use inactive interfaces for handover preparation: selecting appropriate access network, performing lower layer connections and acquiring new IPv6 addresses. The scheme requires several interfaces for operation. Some of the interfaces are used for normal communication (they will be referred as "active"), the others are used for handover preparation (they are termed as "inactive"). The activation of a new interface must be accurately synchronized with the deactivation of the old one. The activation/deactivation procedure means simultaneous reallocation of NEMO tunnels. It can be implemented by properly scheduled flow binding policy control messages on the HA and the MR.

The architecture of the proposed prediction system (as depicted in Figure 7) has three main components:

- The Access Network Predictor (ANP),
- The Handover Manager Mobile Router (HM-MR), and
- The Handover Manager Home Agent (HM-HA).

The ANP is responsible for maintaining a database containing information of access networks, and sending periodic prediction messages to the HM-MR module based on the current velocity vector and the contents of the database associated with the predicted geographical location. The database is kept up-to-date by the Measurement Unit residing in the Handover Manager, which passively monitors the available access networks via one of its passive interfaces, periodically sending network availability and performance indicators such as SNR and IPv6 prefix to the Access Network Predictor. Based on the predictions received from the ANP, the Connection Manager may decide that the currently active access network will no longer be the best available network in the predicted

Figure 6. FMIPv6 architecture and handover modes

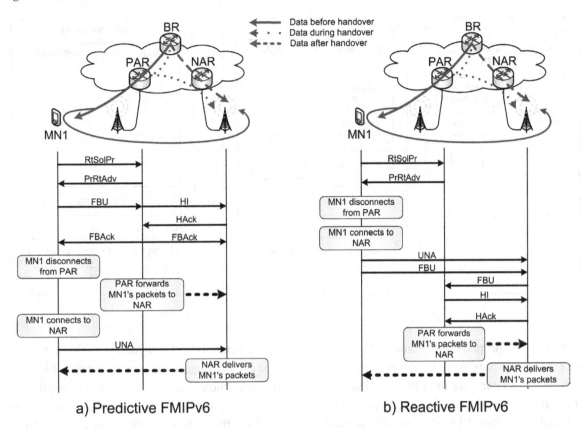

| a) Predictive FMIPv6 | b) Reactive FMIPv6 |

timeframe. When the HM decides to perform a handover, in order to use the benefits of MCoA, the following steps are executed. Using one of the inactive interfaces, the HM connects to the new access network and establishes a new Mobile IPv6 binding. At this stage, the current and new access networks are both connected and Mobility Tunnels are established between the MR and the HA. Handing over to the new access network is entirely based on flow-binding, which in this case means that all flows are moved from one interface to another. To avoid asymmetric routing, the MA and HA has to modify their bindings simultaneously. The schedule is communicated by the Flow Binding modules as an extension of the Flow Binding protocol. When the changes of flow bindings are executed, the new interface is marked as active, while the rest of the communication interfaces are set to inactive mode. Different Handover Policies

may have different effects on handover strategies. In our case, the implemented solution supports 3G and WLAN access networks, and WLAN is always preferred over 3G due to its advantageous bandwidth and latency properties. When multiple WLAN networks are available, the network with the best Signal-to-Noise Ratio (SNR) is selected.

A performance evaluation of the above introduced handover system is already published in Kovács, Bokor, and Jeney (2011). However, no application layer protocols were evaluated in the test setup. The results introduced in this chapter build on the same principles, extending the test environment with a media server as Correspondent Node, illustrated in Figure 8. To simplify the testing methodology, the database used by the ANP is predefined and actual movement is simulated by a prerecorded path using the *gpsfake* utility. The quality of WLAN access networks is adjusted

Figure 7. Prediction system architecture for GNSS-aided predictive mobility management for IPv6

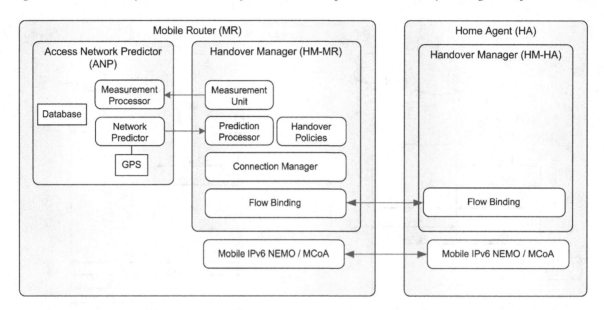

by the *txpower* property of the radios. The resulting handover points serve as heterogeneous set of use-cases to compare mobility solutions. The *tshark* utility was responsible for packet capture and analysis, while VLC was applied as media server. A sample 512 kbit/s CBR video stream was streamed over HTTP from the media server and playback experience was subjectively observed via buffering time periods and buffer underrun events, as the stream was played with VLC client on the Mobile Node.

Using MCoA handovers, the transport protocol performed within acceptable limits. This proved our assumptions, that when an inactive interface is used for connecting to the new network during a handover, the time duration of the actual handover is almost instantaneous. Figure 9 explains that although the mobile node spent time on the 3G medium as well, the average throughput had not degraded significantly. Allowing the node to use networks with poor performance properties, such as overloaded WLAN networks could be the bottleneck of this solution, as with low buffer sizes, the continuity of media playback could not be guaranteed. Comparing this solution

to Predictive NEMO MCoA, the selection of the best available access network is not possible when multiple choices are available. Using prediction, low quality networks were avoided, boosting the average throughput of the transport protocol. While the difference in average throughput may not be significant, when the overall path is evaluated, small disruptions in media streaming may occur due to sudden drops of available network bandwidth. Predicting the available access networks will allow the mobile node to choose the best available network and thereby maximize the user experience.

FUTURE RESEARCH DIRECTIONS

The currently standardized mobility management solutions introduced above rely on hierarchical and centralized architectures, which employ anchor nodes for mobility signaling and user traffic forwarding. In 3G UMTS and beyond, centralized and hierarchical mobility anchors are implemented by the entities in the architecture that handle traffic forwarding tasks using the apparatus of GPRS

Figure 8. Predictive MCoA handover

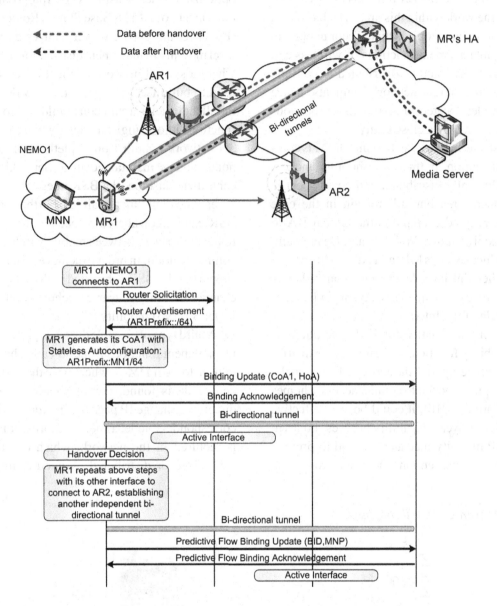

Tunneling Protocol (GTP). The similar centralization is noticeable when Mobile IPv6 is applied: the Home Agent administers mobile terminals' location information, and tunnels user traffic towards the mobile's current locations and vice versa. Up to this day, almost all the standardized enhancements and extensions of MIPv6 preserve the centralized and anchoring nature of the original scheme. This results in unscalable data and control plane with non-optimal routes, overhead and high end-to-end packet delay even in case of motionless users, centralized context maintenance and single point of failures. Anchor-based traffic forwarding and mobility management solutions also cause deployment issues for caching contents near the user. To solve all these problems and questions, novel - distributed and Dynamic Mobility Management (DMM) approaches must be envisaged, applicable to intra- and inter-technology mobility cases as well. The IETF DMM Working Group (formally known as the Mobility EXTensions for IPv6 WG) controls the work within this area. The basic idea of this hot research topic is that anchor nodes and mobility management functions of wireless and mobile systems could be distributed to multiple locations in different network segments, hence mobile nodes located in any of these locations could be served by a close entity.

A first alternative for achieving DMM is core-level distribution. In this case, mobility anchors are topologically distributed and cover specific geographical area but still remain in the core network. A good example is the Global HA to HA protocol (Thubert, Wakikawa, & Devarapalli, 2006), which extends MIP and NEMO in order to remove their link layer dependencies on the Home Link and distribute the Home Agents in Layer 3, at the scale of the Internet.

A second alternative for DMM solutions is when mobility functions and anchors are distributed in the access part of the network. For example, in case of pico- and femto cellular access schemes (FemtoForum, 2010), it could be very effective to introduce Layer 3 capability in access nodes to handle IP mobility management and to provide higher level intervention and even cross-layer

optimization mechanisms. A good proposal here is the concept of UMTS Base Station Router (BSR) (Bauer, et al., 2007), which realizes an access-level mobility management distribution technique where a special network element called BSR is used to build flat cellular systems. BSR merges the all the crucial architecture building blocks and functions into a single element: while a common 3G network is built from a plethora of network nodes and is maintained in a hierarchical and centralized fashion, the BSR integrates all radio access and core functionalities. Furthermore, the BSR can be considered a special wireless edge router that bridges between mobile/wireless and IP communication. In order to achieve this, mobility support in the BSR is handled at three layers: RF channel mobility, Layer 2 anchor mobility, and Layer 3 IP(v6) mobility.

A third type of possible distribution of mobility management functions is the so-called host-level or peer-to-peer DMM, where once the correspondent node is found, communicating peers can directly exchange IP packets. In order to find the correspondent node, a special information server is required in the network, which can also be centralized or distributed. A good example for

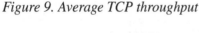

Figure 9. Average TCP throughput

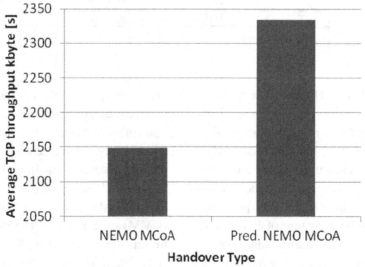

host-level schemes in the IP layer is MIPv6 which is able to bypass the user plane anchor (i.e., Home Agent) due to its route optimization mechanism, therefore providing a host-to-host communication method (Arkko, et al., 2007).

The three above DMM approaches can be applied together in an integrated manner for more flexibility and enhanced performance. PMIPv6 extension proposals, like Bernardos et al. (2012), are going on this path.

Another emerging area of IPv6 mobile multimedia delivery researches is the *flow mobility*. There are cases in multihoming Mobile IPv6 environments when flow mobility (or flow binding) is initiated by a central entity, such as the always available Home Agent. Operations like network-controlled flow binding revoking, moving, or provisioning are equally possible with this mechanism; making it possible to revoke an existing flow binding in case of an error, or move a flow from one interface to another on the MN side, or simply provide default flow settings for newly connected Mobile Nodes. The approach is not mutually exclusive with the MN initiated flow binding described in RFC 6089 (Tsirtsis, et al., 2011), it merely extends the mobility features it provides, meaning that flow bindings are not always initiated by the HA. There are drafts like Yokota et al. (2011) in which authors introduce a new Mobility Header and signaling messages based on the flow binding protocol implemented in RFC 6089. In addition, the PMIPv6 protocol extensions exist for this purpose (Bernardos, 2012). Possible application use cases of HA Initiated Flow Bindings may be default flow binding provisioning, traffic offloading and flow binding revocation. For example, default flow binding provisioning is used in an environment where a central entity wants to force Service Level Agreements (SLA) to a customer, e.g., forcing multimedia traffic through WLAN while allowing 3G access for HTTP traffic. The traffic offloading technique makes it possible to move certain data flows from one interface to another, e.g., in case of increasing

traffic load in 3G segment move video streams to the WLAN segment. Policies can be much complex based on the fact that the core network entities know about their actual traffic conditions. Flow binding revocation is useful, when due to an administrative decision; a certain flow binding is no longer valid for the MN.

The last group of research directions to be introduced here is about media handover optimization by applying cross-layer techniques. Several different mobility management schemes exist in the literature, but their optimization for heterogeneous access architectures just have been started. 802.21 Media Independent Handovers (MIH) (IEEE, 2009), and Access Network Discovery and Selection Function (ANDSF) (3GPP TS 23.402, 2011) are emerging methods for proactive handover control in heterogeneous architectures, but their ways of application in mobile environments and synthesis with different mobility execution mechanisms or with higher layer functions has not yet defined precisely. Integration of 802.21 MIH, and ANDSF and similar standards with existing mobility management schemes (e.g., Dual-Stack Mobile IP, Proxy Mobile IPv6) in order to reduce or even totally eliminate deteriorations during mobility events are still hot topics. Moreover, a hot research activity is the evaluation of mobility management schemes strongly relying on multiple existing host interfaces (i.e., multihoming) and integrate them with handover preparation/prediction mechanisms and cross-layer information provision. Such an integration is required in order to optimize access to heterogeneous access architectures, to benefit from overlapping coverages and also to build up a strong interworking between applications and handover procedures (e.g., to prepare a real-time mobile media flow for a handover event by proactively setting media codec parameters at the sender side).

CONCLUSION

In this chapter, we provided a comprehensive overview of the complex relation system between IPv6 and the multimedia driven future mobile Internet. We highlighted how the IPv6 standard family emerges with its suitability and applicability for mobile multimedia applications and services, and we introduced how IPv6 can serve as the main cornerstone for mobile architectures. In addition, we introduced a new method to improve the feasibility of Mobile IPv6 for multimedia content delivery. The discussion of the above areas together with the review of the most current research efforts hopefully guides the readers from the basics of IPv6 towards the most complex features of the protocol and power to build a novel Internet architecture for future multimedia-centric mobile communications.

ACKNOWLEDGMENT

The research leading to these results has received funding from the European Union's Seventh Framework Programme (FP7/2007-2013) under grant agreement n° 288502 (CONCERTO), and partly by the FP7 project ITSSv6. We would like to thank all participants and contributors, who were involved in the work.

REFERENCES

3GPP TR 23.829. (2010). *Local IP access and selected IP traffic offload, release 10, V1.3.0.* 3GPP Technical Report.

3GPP TS 23.246. (2011a). Multimedia broadcast/multicast service (MBMS) architecture and functional description, release 10, V10.1.0. 3GPP Technical Specification.

3GPP TS 23.402. (2011b). Architecture enhancements for non-3GPP accesses, release 10, V10.4.0. 3GPP Technical Specification.

Arkko, J., Vogt, C., & Haddad, W. (2007, May). *Enhanced route optimization for mobile IPv6.* IETF RFC 4866. Retrieved from http://www.ietf.org/rfc/rfc4866.txt

Bauer, M., Bosch, P., Khrais, N., Samuel, L. G., & Schefczik, P. (2007). The UMTS base station router. *Bell Labs Technical Journal, I. Wireless Network Technology, 11*(4), 93–111.

Bernardos, C. (2012). *Proxy mobile IPv6 extensions to support flow mobility.* Retrieved from http://draft-ietf-netext-pmipv6-flowmob-03

Bernardos, C., Oliva, A. D., Giust, F., Melia, T., & Costa, R. (2012). *A PMIPv6-based solution for distributed mobility management.* Retrieved from http://draft-bernardos-dmm-pmip-01

Bokor, L., Faigl, Z., & Imre, S. (2011). Flat architectures: Towards scalable future internet mobility. *Lecture Notes in Computer Science, 6656,* 35–50. doi:10.1007/978-3-642-20898-0_3

Calderón, M., Bernardos, C.J., Bagnulo, M., Soto, I., & Oliva, A. D. (2006). Design and experimental evaluation of a route optimization solution for NEMO. *IEEE Journal on Selected Areas in Communications, 24*(9), 1702–1716. doi:10.1109/JSAC.2006.875109

Cisco, V. N. I. (2011). *Global mobile data traffic forecast update, 2010-2015.* New York, NY: Cisco.

Deering, S., Fenner, W., & Haberman, B. (1999). *Multicast listener discovery (MLD) for IPv6.* IETF RFC 2710. Retrieved from http://www.ietf.org/rfc/rfc2710.txt

Deering, S., & Hinden, R. (1998). *Internet protocol, version 6 (IPv6) specification.* IETF RFC 2460. Retrieved from http://www.ietf.org/rfc/rfc2460.txt

Devarapalli, V., Wakikawa, R., Petrescu, A., & Thubert, P. (2005). *Network mobility (NEMO) basic support protocol.* IETF RFC 3963. Retrieved from http://tools.ietf.org/html/rfc3963

Dohler, M., Watteyne, T., & Alonso-Zárate, J. (2010). Machine-to-machine: An emerging communication paradigm. In *Proceedings of GlobeCom2010.* IEEE.

Droms, R. (2003). *Dynamic host configuration protocol for IPv6 (DHCPv6).* IETF RFC 3315. Retrieved from http://www.ietf.org/rfc/rfc3315.txt

FemtoForum. (2010). *Femtocells – Natural solution for offload – a Femto forum brief.* FemtoForum.

Fenner, B., Handley, M., Holbrook, H., & Kouvelas, I. (2006). *Protocol independent multicast - sparse mode (PIM-SM): Protocol specification (revised).* IETF RFC 4601. Retrieved from http://tools.ietf.org/html/rfc4601

Forum, U. M. T. S. (2010). *Recognising the promise of mobile broadband.* White Paper. Washington, DC: UMTS.

Gundavelli, S., Leung, K., Devarapalli, V., Chowdhury, K., & Patil, B. (2008). *Proxy mobile IPv6.* IETF RFC 5213. Retrieved from http://tools.ietf.org/html/rfc5213

Hinden, R., & Deering, S. (2006). *IP version 6 addressing architecture.* IETF RFC 4291. Retrieved from http://tools.ietf.org/html/rfc4291

IEEE. (2009). *IEEE standard for local and metropolitan area networks- Part 21: Media independent handover.* IEEE Std 802.21-2008. Retrieved from http://ieeexplore.ieee.org/xpl/articleDetails.jsp?tp=&arnumber=4769367&contentType=Standards&sortType%3Dasc_p_Sequence%26filter%3DAND%28p_Publication_Number%3A4769363%29

Kafle, V. P., Kamioka, E., & Yamada, S. (2006). MoRaRo: Mobile router-assisted route optimization for network mobility (NEMO) support. *IEICE Transactions in Information & Systems, E89-D*(1).

Kent, S., & Seo, K. (2005). *Security architecture for the internet protocol.* IETF RFC 4301. Retrieved from http://tools.ietf.org/html/rfc4301

Koodli, R. (2009). *Mobile IPv6 fast handovers.* IETF RFC 5568. Retrieved from http://tools.ietf.org/html/rfc5568

Kovács, J., Bokor, L., & Jeney, G. (2011). Performance evaluation of GNSS aided predictive multihomed NEMO configurations. In *Proceedings of the 2011 11th International Conference on ITS Telecommunications*, (pp. 293-298). St. Petersburg, Russia: ITST.

Lee, J., & Ahn, S. (2006). I-FHMIPv6: *A novel FMIPv6 and HMIPv6 integration mechanism.* Retrieved from http://draft-jaehwoon-mipshop-ifhmipv6-01.txt

McCann, J., Deering, S., & Mogul, J. (1996). *Path MTU discovery for IP version 6.* IETF RFC 1981. Retrieved from http://www.ietf.org/rfc/rfc1981.txt

Narten, T., Nordmark, E., Simpson, W., & Soliman, H. (2007). *Neighbor discovery for IP version 6 (IPv6).* IETF RFC 4861. Retrieved from http://tools.ietf.org/html/rfc4861

NetLMM. I. (2012). *Network-based localized mobility management (NetLMM) WG homepage.* Retrieved from http://datatracker.ietf.org/wg/netlmm/charter/

Pérez-Costa, X., Schmitz, R., Hartenstein, H., & Liebsch, M. (2002). A MIPv6, FMIPv6 and HMIPv6 handover latency study: Analytical approach. In *Proceedings of the IST Mobile & Wireless Telecommunications Summit (IST Summit).* Thessaloniki, Greece: IST Summit.

Pérez-Costa, X., Torrent-Moreno, M., & Hartenstein, H. (2003). A performance comparison of mobile IPv6, hierarchical mobile IPv6, fast handovers for mobile IPv6 and their combination. *ACM SIGMOBILE Mobile Computing and Communications Review, 7*(4), 5–19. doi:10.1145/965732.965736

Perkins, C., Johnson, D., & Arkko, J. (2011). *Mobility support in IPv6*. IETF RFC 6275. Retrieved from http://tools.ietf.org/html/rfc6275

Pike, T., Russell, C., Krumm-Heller, A., & Sivaraman, V. (2007). IPv6 and multicast filtering for high-performance multimedia application. In *Proceedings of the Australasian Telecommunication Networks and Applications Conference (ATNAC 2007)*, (pp. 146-150). ATNAC.

Ping, G., & Desheng, F. (2010). The discussions on implementing QoS for IPv6. In *Proceedings of the International Conference on Multimedia Technology (ICMT)*, (pp. 1-4). ICMT.

Rajahalme, J., Conta, A., Carpenter, B., & Deering, S. (2004). *IPv6 flow label specification*. IETF RFC 3697. Retrieved from http://www.ietf.org/rfc/rfc3697.txt

Sang-Jo, Y., & Seak-Jae, S. (2006). Fast handover mechanism for seamless multicasting services in mobile IPv6 wireless networks. *Wireless Personal Communications, 42*(4), 509–526.

Soliman, H. (2009). *Mobile IPv6 support for dual stack hosts and routers*. IETF RFC 5555. Retrieved from http://tools.ietf.org/html/rfc5555

Soliman, H., Castelluccia, C., Elmalki, K. E., & Bellier, L. (2008). *Hierarchical mobile IPv6 mobility management (HMIPv6)*. IETF RFC 5380. Retrieved from http://tools.ietf.org/html/rfc5380

Thomson, S., Narten, T., & Jinmei, T. (2007). *IPv6 stateless address autoconfiguration*. IETF RFC 4862. Retrieved from http://www.ietf.org/rfc/rfc4862.txt

Thubert, P., Wakikawa, R., & Devarapalli, V. (2006). *Global HA to HA protocol*. Retrieved from http://draft-thubert-nemo-global-haha-02

Tsirtsis, G., Soliman, H., Montavont, N., Giaretta, G., & Kuladinithi, K. (2011). *Flow bindings in mobile IPv6 and network mobility (NEMO) basic support*. IETF RFC 6089. Retrieved from http://tools.ietf.org/html/rfc6089

Wakikawa, R. E. (2009). *Multiple care-of addresses registration*. IETF RFC 5648. Retrieved from http://tools.ietf.org/html/rfc5648

Yokota, H., Kim, D., Sarikaya, B., & Xia, F. (2011). *Home agent initiated flow binding for mobile IPv6*. Retrieved from http://draft-yokota-mext-ha-init-flow-binding-01

Zheng, W. (2006). An efficient dynamic multicast protocol for mobile IPv6 networks. In *Proceedings of the 31st IEEE Conference on Local Computer Networks*, (pp. 913 - 920). IEEE Press.

Zhenhua, W., Qiong, S., Xiaohong, H., & Yan, M. (2010). IPv6 end-to-end QoS provision for heterogeneous networks using flow label. In *Proceedings of the 3rd IEEE International Conference on Broadband Network and Multimedia Technology (IC-BNMT)*, (pp. 130-137). IEEE Press.

ADDITIONAL READING

Davies, J. (2003). *Understanding IPv6*. New York, NY: Microsoft Press.

Groebel, J., Noam, E., & Feldmann, V. (2006). *Mobile media – Content and services for wireless communications*. Mahwah, NJ: Lawrence Erlbaum Associates, Inc.

Karmakar, G., & Dooley, L. S. (2008). *Mobile multimedia communications: Concepts, applications, and challenges*. Hershey, PA: IGI Global.

Li, Q., Jinmei, T., & Shima, K. (2007). *IPv6 advanced protocols implementation*. San Francisco, CA: Morgan Kaufmann Publishers.

Li, Q., Jinmei, T., & Shima, K. (2009). *Mobile IPv6: Protocols and implementation*. London, UK: Elsevier Inc.

KEY TERMS AND DEFINITIONS

Flow Mobility: If a mobile user runs several applications (e.g. file downloading, voice communication, video streaming, e-mail) on a device with multiple interfaces and simultaneously available access networks, and the actual connection (i.e., out- and inbound interface) of each flow is handled independently according to QoS requirements and environmental parameters, than we are talking about per-application mobility.

IPv6: Internet Protocol version 6 (IPv6) is the next-generation Internet Protocol version designed to overcome the imperfections of IPv4. The main motivation for the re-design of IPv4 was the presumptive IPv4 address exhaustion. IPv6 was firstly introduced in December 1998 in RFC 2460. IPv6 is a conceptual copy of the IPv4 protocol with several modifications and extensions to the basic standard.

Mobility Management in the Networking Layer: A mobility solution where the networking layer is responsible for handling various mobility scenarios, such as handover between different access network types, connecting to multiple access networks simultaneously, allowing the mobile node global reachability regardless of its current attachment to the Internet or the type of access medium in use.

Mobile Multimedia: Mobile multimedia denotes different types of media content that are either accessed or created by employing portable devices like Smartphones with sound and video playback capabilities, microphone and camera for mobile content creation, and wireless Internet access for on-the-move content reception and transmission.

Micromobility: If wireless networking domains are aggregated and a special protocol is responsible for the local mobility management of this group of domains in order to offer fast and seamless handover control over a limited geographical area, than we speak about micromobility, the aggregated group of domains is called micromobility domain, and the special control protocol is called micromobility protocol.

Multicasting: Delivery of information to a group of destinations simultaneously using the most efficient strategy to deliver the messages over each link of the network only once, creating copies only when the links to the destinations split.

Multimedia: Multimedia is a noun or adjective, introducing a medium which describes the usage of different types of content forms usually in the same time. A content form (media) can be: written text, still images, animation, audio, video, and interactivity. Some examples for multimedia: social networking (who-is-who websites); online journals and news sites; a blue-ray disc with video, audio, subtitles, and interactive menu points; or online gaming with other people.

Network Controlled Media Delivery: A special routing system, which utilizes the multihoming feature of Mobile IPv6 and the overall status of the network, allowing network operators to force network preferences to the host based on predefined routing policies and actual network status parameters.

Network Mobility: A special mobility scenario, which arises when a router—connecting a network to the Internet dynamically—changes its point of attachment to the fixed infrastructure, thereby causing the accessibility of the entire network to be changed in relation to the fixed Internet topology.

Chapter 3
Protocol Interactions among User Agents, Application Servers, and Media Servers:
Standardization Efforts and Open Issues

Alessandro Amirante
University of Napoli Federico II, Italy

Tobia Castaldi
University of Napoli Federico II, Italy

Lorenzo Miniero
University of Napoli Federico II, Italy

Simon Pietro Romano
University of Napoli Federico II, Italy

ABSTRACT

In this chapter, the authors focus on the complex interactions involving the various actors participating in a multimedia session over the Internet. More precisely, bearing in mind the current standard proposals coming from both the 3GPP and the IETF, they investigate some of the issues that have to be faced when separation of responsibilities comes to the fore. The scenario the authors analyze is one in which one or more user agents are put into communication with a media server through the mediation of an application server. In such scenario, the application server does play the role of a middlebox for all that concerns signaling, since it is responsible for the transparent negotiation of a session among the entities (the user agents on one side and the media server on the other) that will be exchanging media during the communication phase. In this chapter, the authors highlight that protocol interactions become really complex under the depicted circumstances. They provide a survey of the current standardization efforts related to media control, together with a discussion of open issues and potential solutions.

DOI: 10.4018/978-1-4666-2833-5.ch003

INTRODUCTION

Recently, advanced services have massively entered the Internet arena pushed by the revolutionary "global" approach envisaging the coexistence of a variegated portfolio of applications on top of an integrated IP-based network. Consequently, the Internet has become a place where an ever-increasing number of "dependent" or "correlated" transactions take place every day. This unexpected growth of complexity unavoidably unveils a number of less or more subtle issues that have to be faced when looking at the interactions among the various entities involved in the service delivery chain. Standardization bodies like the IETF (Internet Engineering Task Force) and the 3GPP (3rd Generation Partnership Project) are actively contributing both to the definition of an integrated framework for advanced service creation and deployment and to the solution of the above men-

tioned issues. As to the 3GPP, the consortium is currently standardizing the *IP Multimedia Subsystem* (IMS) architecture (see Figure 1), whose aim is to provide a common service delivery mechanism capable to significantly reduce the development cycle associated with service creation across both wireline and wireless networks.

The main objective of IMS resides in trying to reduce both capital and operational expenditures (i.e., CAPEX and OPEX) for service providers, at the same time providing operational flexibility and simplicity. Since the beginning, the IMS has chosen SIP (Session Initiation Protocol) (Rosenberg, et al., 2002) as the main signaling protocol among most of its components (3GPP, 2007). The envisaged portfolio of IMS services includes advanced IP-based applications like Voice over IP (VoIP), online gaming, videoconferencing, and content sharing. All such services are to be provided on a single, integrated infrastructure, ca-

Figure 1. The architecture of the 3GPP IP multimedia subsystem (IMS)

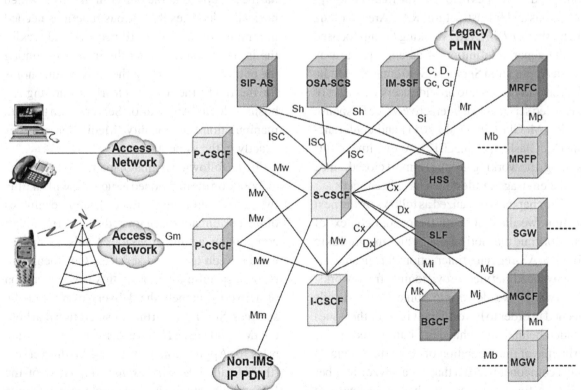

pable to offer seamless switching functionality between different services. It is worth noting that IMS is conceived as an access agnostic platform. This requirement clearly imposes a careful study of the core IMS components (such as Call/Session Control Function—CSCF, Home Subscriber Server—HSS, Media Resource Function—MRF, and Application Server—AS), which must be scalable and able to provide advanced features, like *five nine* reliability. A more-in-depth analysis of the IMS architecture is reported in Appendix A.

Similarly, the IETF is devoting a great effort to the definition of advanced frameworks for multimedia service delivery, starting from the effective utilization of the base functionality made available by the SIP protocol. SIP provides users with the capability to initiate, manage, and terminate communication sessions in an IP network. For a brief description of the SIP protocol and architecture, see Appendix B. The main working groups within the IETF involved in the standardization of advanced multimedia services belong to the Real-Time Applications and Infrastructure (RAI) Area. Among them, the *MEDIACTRL* Working Group focused on the general definition of an appropriate way for an Application Server (AS) to control a Media Server (MS) in order to provide users with a set of advanced services like Interactive Voice Response (IVR) (McGlashan, et al., 2011) and conferencing (McGlashan, et al., 2012). At the time of this writing, the working group is almost closed, since all the envisaged milestones have been met.

The chapter is organized as follows. In the next section, we illustrate both the overall context of our work and the motivations, which inspired us to focus on AS-regulated interactions for our analysis of advanced service provisioning frameworks. Afterwards, we delve into some of the details needed in order to have a clear vision of the issues hidden behind an architecture built based on the principle of the separation of concerns among its inner components. After that, we analyze a number of interesting scenarios, which help the reader to

understand how the issues identified can be dealt with as effectively as possible. Conclusions are provided in the last section. Finally, three appendixes expand a little bit on the IMS architecture and on the SIP and BFCP protocols.

CONTEXT AND MOTIVATION

From the brief discussion above, it comes out that the SIP protocol is actually paving ground for both 3GPP and IETF standardization efforts. Indeed, though work inside these two bodies is proceeding along independent tracks, there is currently a trend towards convergence at least at the level of the approach adopted to cope with both complexity and heterogeneity. This includes the choice of a fully distributed paradigm envisaging the definition of independent yet tightly interacting components, each responsible for a specific function inside the overall service architecture. As an example, the idea of separating responsibilities has been applied at the outset in the IMS, which basically identifies the various functions needed in a multimedia-capable IP network and tackles the issue of standardizing the interfaces among the entities implementing them. The same holds for the IETF: the idea of clearly separating the business logic (Application Server) from the data manipulation functionality (Media Server) goes exactly in the same direction.

In the following of this chapter, we will further elaborate on the mentioned issues. We will take the two referenced frameworks as leading examples of the current approach towards the definition of components acting, at various levels, as mediators between users and services. Our focus will be on a specific function of this new generation of networks, namely the delivery of multimedia services. Starting from the consideration that both the IMS and the IETF have identified two distinct roles for Application Servers and Media Servers (the so-called *Media Resource Function* of the IMS, which actually is a macro-component made of the two sub-elements called, respectively, *Media Resource Function Controller*—MRFC—and *Media Resource Function Processor*—MRFP), we will identify the Application Server as a critical entity acting as a middlebox which can highly influence the communication between users on one side and media servers on the other. We will show that such a role naturally imposes a trad-eoff between service composition flexibility and service management complexity, thus calling up network engineers in order to try and strike the balance between these two counteracting facets of any advanced service delivery framework.

CONTROLLING A REMOTE MEDIA SERVER: ARCHITECTURE AND PROTOCOLS

The *Media Resource Function* (MRF) is logically decomposed in the MRFC and MRFP, whereas the former acts as a controlling interface towards the MRF, while the latter takes care of actually manipulating the media resources according to the directives provided by the controller. The interface between the above-mentioned components has been only partially defined so far in IMS. In fact, while it has been specified that the AS and the MRFC need to involve SIP in their interaction, the protocol details have not been standardized yet.

To fill the gap due to such lack of standardization, the IETF Working Group called *MEDIACTRL* (Media Server Control) specified a complete architecture, called *Media Control Channel Framework* (MCCF) (Boulton, et al., 2011), for the interaction between Application Servers and Media Servers through the use of SIP. To map the work of the IETF onto the IMS architecture, the current specification maps the MCCF onto the MRF as a whole, meaning that it acts both as a controller and as a processor. Nevertheless, this distinction is not of great interest for the purpose of this chapter, since the MCCF actually acts as the SIP interface towards the MRFC as seen by IMS-compliant multimedia-aware Application Servers.

The architecture conceived in the MEDIAC-TRL specification is depicted in Figure 2. Please, note that the MCCF will be referred to as MS from now on.

The AS and the MS interact by means of two protocols: a dedicated protocol, called CFW (Control Framework), and SIP, as envisaged by the IMS specification. SIP is mainly used by the AS to attach UACs (User Agent Clients) needing multimedia resources to the MS, in order to let them negotiate media sessions: this is accomplished by making use of the 3rd Party Call Control mechanism (Rosenberg, et al., 2004). This call control mechanism envisages the AS as ter-

Figure 2. Protocols interaction between AS, MS, and UACs

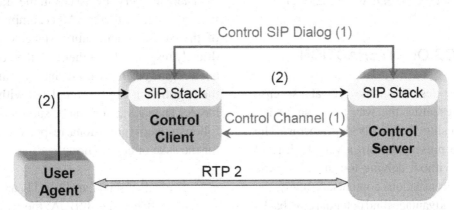

minating endpoint for what concerns signaling, both for the UACs ad for the MS (while media flow directly via RTP between the UACs and the MS). Instead, CFW is the protocol used between AS and MS to explicitly manipulate the negotiated resources, e.g., to add users to a conference mix or to present them with IVR menus. This protocol is also negotiated by means of SIP, within the context of a so-called COMEDIA negotiation.

The use of 3PCC to give UACs access to media resources, and of CFW to drive how these media resources can be accessed or need to be presented, makes it quite clear that all the application logic resides in the AS, while media processing is achieved in MS. To make it even clearer, the AS can be seen as the brain, whilst the MS is the arm. Of course, considering that the directives the AS sends to the MS through the control channel assume they both are aware of the available resources (including UACs and their negotiated media streams), the 3PCC mechanism and the CFW protocol transactions need to be properly synchronized in some scenarios. Some of these scenarios and the issues they can arise will be dealt with in the following section.

For the sake of completeness, we inform the reader that all this study has been accomplished by also referring to our implementation of the MEDIACTRL framework. Such implementation (at the time of this writing) is the only available implementation of the framework, and is completely open source (MEDIACTRL, 2012).

SCENARIOS OF INTERACTION

Hereafter, we describe some typical use case scenarios, presenting the way they can be accomplished using the involved protocols. The scenarios are presented from a very high-level perspective, without delving too much into the details of the transactions contents. Actually, the focus is on the advantages and potential drawbacks of each presented approach, with special emphasis on the middlebox role of the Application Server. Sequence diagrams are added for a better understanding of the interactions among the involved parties. The scenarios are presented in order of increasing complexity:

- First, the interaction between two User Agent Clients (UAC) and a "self-sustained" Back-to-Back User Agent (B2BUA) is presented.
- Then, the focus will move to the interaction between two UACs through an Application Server (AS), which relays media manipulation to a separate Media Server.
- Finally, signaling is complicated by the addition of a floor control protocol to the list of resources to be negotiated.

Self-Sustained B2BUA

The first scenario is by far the simplest one. In this scenario, a UAC (the caller) wishes to place a call to another UAC (the callee), and a B2BUA is in their way. The B2BUA presented in the scenario is a self-sustained one, in the sense that it acts as an integrated AS and MS: a typical example of such a middlebox is the popular open source PBX Asterisk2 (www.asterisk.org). The interaction between the caller and the callee is quite straightforward in this case, and is depicted as a sequence diagram in Figure 3.

There are very few issues in the signaling. In fact, considering that the AS is completely aware of the media functionality, since it actually is directly responsible for them, all the needed SDP answers and offers are immediately available, and the negotiation can be completed with no much hassle. The interaction can be summarized in the following steps (provisional responses are skipped for the sake of conciseness):

- The caller sends an INVITE addressed to the callee to the B2BUA; the INVITE con-

Figure 3. Phone call: back-to-back user agent

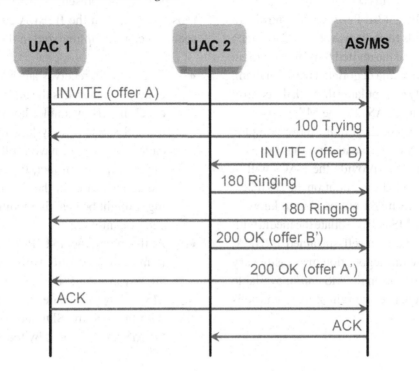

tains the SDP offer (A) with the media the caller wants to negotiate;

- The B2BUA matches the offer with its own capabilities (since it also acts as MS), and sends a new INVITE to the callee; this new INVITE carries as payload a new SDP offer (B), obtained by the previous match;
- In case the callee accepts the call (a 200 message), its SDP answer (B') is matched by the B2BUA as before; this answer is modified accordingly, if needed (A'), and then forwarded to the caller to complete the negotiation;
- As soon as the ACK from the caller is forwarded to the callee, the two UACs can interact; the B2BUA might have to act as a media transcoder between the two, if the negotiation presented such a need.

The call flow presents very few issues from a signaling point of view. The B2BUA has full control over both the signaling and the media

capabilities, and can take care of all the decisions according to its policies, application logic, and resources availability. However, such approach presents more than one drawback. In fact, it implies that the media functionality has to be replicated at every AS needing it, since each such AS has to act as a completely self-sustained B2BUA. Besides, even focusing on a single AS, such an approach completely lacks in terms of scalability and locates in the B2BUA a single point of failure.

AS and MS Separated

To somehow fill the gaps of the previously described B2BUA approach, a separation between the application logic and the media processing can be introduced. This can be done, as already explained in the introductory sections, with an approach a la MEDIACTRL. This is exactly the scenario that is presented hereafter. The provided example is again a phone call between two UACs, where the media flowing between them might need to be transcoded. In this scenario, the focus is on the so-called 3PCC (3rd Party Call Control). In fact, the AS needs to make use of 3PCC to attach both the calling and the invited UAC to the MS, in order to have the media negotiated between them, and subsequently manipulated through the control channel between the AS and the MS.

The 3PCC approach is quite more complex than the B2BUA one, considering that the SDP answers and offers to provide the UACs will in such case depend on the negotiation the AS relays to the MS: the AS might not be able to know in advance how the MS will negotiate the media with the UACs as it did as a self-sustained B2BUA, since in this case the negotiation may easily vary depending on policies in the MS and/or available resources. A sequence diagram of the scenario is depicted in Figure 4.

The call flow presents some more issues than those presented in the B2BUA case. Again, the scenario can be summarized in these steps:

- Once the AS receives an INVITE from the caller, it first invites the addressed callee to check if it is available; however, this time the AS sends a body-less INVITE to the callee, in order to have it offer all the media it supports; in fact, the AS cannot use the caller's offer in the INVITE, considering it might be heavily modified by the MS subsequently;

- At this point, the first 3PCC can take place; in fact, since an offer from the caller (A) is already available, the AS forwards it to the MS to have the caller negotiate its media with the MS; the SDP answer provided by the MS (A') is stored by the AS in order to

Figure 4. Phone call: allocating caller and callee

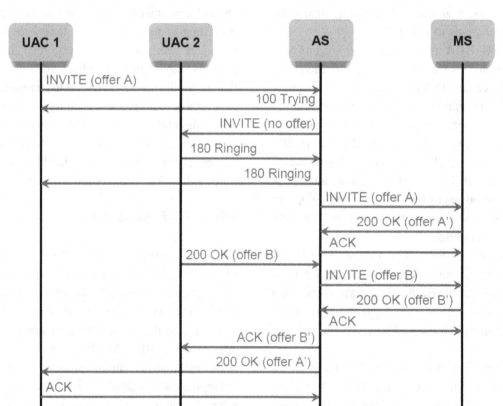

be able of relaying it later to the caller, that is as soon as the callee accepts the call;

- Once the callee accepts the call, providing in the 200 its offer (B), the second 3PCC can take place as well; the AS attaches the callee to the MS, and stores the negotiated answer (B');
- At this point, the AS has an SDP answer for both the caller and the callee. This means that the AS can complete the negotiation with both of them, sending the answer to the caller (A') in a 200 and the answer to the callee (B') in the final ACK;
- Now that the negotiation is complete, the AS can make use of the control channel to properly instruct the MS to attach the media connections of the two UACs with each other, thus allowing them to interact.

The additional complexity of the signaling in this scenario is immediately perceptible. It is worth noting that in this case the issue is not represented by the control channel interaction between the AS and the MS. In fact, even if such interaction always needs to be synchronized with the 3PCC in order to achieve the desired results, in the phone call case attaching the UACs can simply be done after both SIP negotiations have succeeded.

While the presented approach allows overcoming the limitations introduced by the B2BUA one, there are some drawbacks. One of them is related to the potentially premature allocation of resources. The caller, in the presented diagram, is allocated before knowing if the callee is available. This means that, if the callee rejects the call instead of accepting it as envisaged in the scenario, resources have been wasted, and must be de-allocated consequently. Moving the allocation of the caller after a 200 from the callee, solves this issue, but introduces a new one: in fact, the MS may be lacking the resources to allocate the UACs at that time, thus resulting in a media-less session between the UACs. Obviously, this is an undesirable behavior in an environment subject to billing policies.

Involving Moderation

The signaling can be further complicated by involving other protocols in the negotiation among the involved parties. In fact, all protocols in a session are often negotiated within the context of the same offer/answer. Besides, such protocols might need to be handled by the MS, meaning they would be part of the already mentioned 3PCC negotiation. An example of such protocols is the *Binary Floor Control Protocol* (BFCP) (Camarillo, et al., 2006). BFCP is a recently standardized floor control protocol, typically be involved in conferencing systems to allow moderated access to the available resources. More information about the BFCP protocol is reported in the Appendix C.

As specified in IMS, a MS may be invested with the additional role of floor control server. This means that the MS might not only have to deal with media in the SDP, but also with COMEDIA-based negotiations for BFCP as specified in Camarillo (2006). This negotiation is needed to provide the UAC with a list of attributes, including the transport address of the floor control server, as well as several BFCP-related identifiers. The problem introduced by this scenario is that, unlike media negotiation, the BFCP negotiation of a UAC with the floor control server can only be achieved after the UAC has been added to a conference through the control channel. However, the AS will not be able to correctly address the UAC in the control channel request before the UAC has been attached to the MS. This immediately suggests that the synchronization between 3PCC and the control channel transactions becomes of paramount importance. This section presents two possible approaches to deal with the described signaling issue (depicted in Figure 5 and Figure 6, respectively). Both use-cases present a single UAC interacting with the AS and the MS, instead of the caller and callee of the previous sections. In fact, in this case BFCP is assumed to be involved in a conferencing scenario, where the UAC places the INVITE to the AS to join a conference it is aware of. The approach (presented in Figure 5) relies on a re-INVITE generated by the MS to the UAC, in order to update the media session at a later moment.

Examining the signaling step by step:

- The UAC sends its INVITE to join the conference to the AS; the SDP offer includes BFCP-related lines as specified in Camarillo (2006); this INVITE is relayed to the MS as part of the 3PCC mechanism;
- As explained before, the BFCP negotiation is expected to fail, since no BFCP user identifier associated with the UAC is available yet; in fact, the UAC has not been added to the conference at the time of the INVITE; this results in the BFCP negotiation being refused;

- The SDP answer provided by the MS, which only includes the negotiated media is forwarded by the AS to the UAC;
- Now that the 3PCC negotiation is completed, the AS can add the UAC to the conference mix by means of the control channel; the control channel interaction between the AS and MS also includes directives related to BFCP, which results in a BFCP user identifier being associated with the UAC;
- At this point, the MS can trigger a re-INVITE addressed to the UAC, this time including all the BFCP-related identifiers as specified in Camarillo (2006); such re-

Figure 5. BFCP negotiation: reINVITE to the UAC

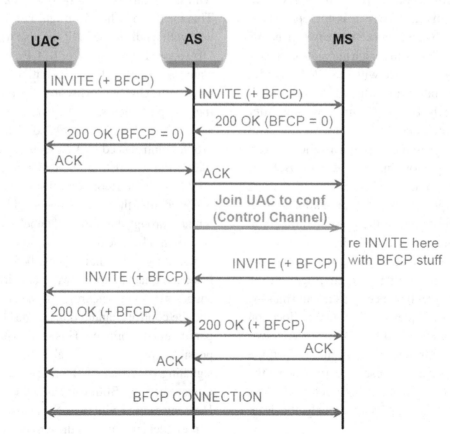

Figure 6. BFCP negotiation: transparent to the user

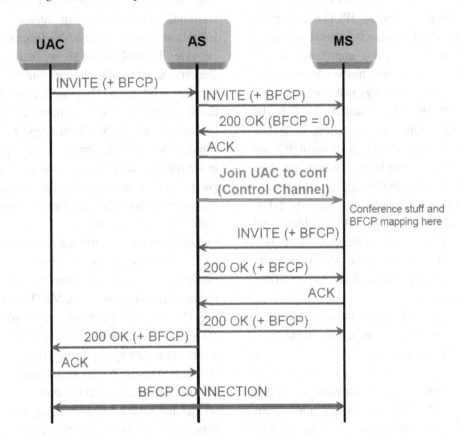

INVITE is then relayed by the AS to the UAC as part of a new 3PCC;

- Once this further negotiation is over, the UAC is able to open a BFCP connection with the floor control server in order to access its moderated resources.

This approach actually works (existing implementation efforts, e.g., our Confiance prototype [Buono, et al., 2007] seem to confirm this statement), and is how things would work anyway in any MS-generated re-INVITE involving BFCP. However, it is not transparent to the UAC, and so assumes that given functionality to be supported by the UAC, which can be seen, probably as a drawback, or at least as a limitation. Besides, the MS refusing the BFCP part of the first offer

implies that the UAC first gets an answer with the BFCP media line set to 0, which is ambiguous: the UAC cannot know whether the media line has been set to 0 because the BFCP identifier is not available yet, or simply because BFCP is not supported by the AS/MS at all. A way to make the negotiation less ambiguous and transparent to the UAC is illustrated in Figure 6.

In this approach, the AS only provides the UAC with a complete answer when all of the needed identifiers are made available. Examining the signaling step by step as before:

- The UAC sends its INVITE to join the conference to the AS; the SDP offer again includes BFCP-related lines as specified in

Camarillo (2006); this INVITE is relayed to the MS as part of the 3PCC mechanism;

- The BFCP negotiation is refused as before;
- Unlike in the previous approach, the AS does not forward this answer to the UAC yet; since the UAC has just been attached to the MS, the AS can add the UAC to the conference mix and configure its BFCP settings through the control channel; from the UAC's perspective; however, the join attempt is still proceeding;
- At this point, the MS can trigger a re-INVITE addressed to the UAC, including all the BFCP-related identifiers as before; this re-INVITE is used by the AS to complete the negotiation with the UAC as part of the original 3PCC;
- At this point, the UAC is attached to the conference and is immediately able to open the BFCP connection.

This approach also works in theory, and has the advantage of being completely transparent to the UAC. The ambiguous BFCP media line set to 0 is still involved, but in this case it can be considered less of a problem than before: in fact, it can quite safely be assumed that the AS is aware of the functionality the MS can provide, including floor control. One of the drawbacks is instead the slowed down signaling from the UAC perspective, considering that the negotiation is completed only after a series of intermediate steps. Besides, further input from the UAC (e.g., a SIP CANCEL during the inner AS-MS interaction) might raise additional issues, which would need to be taken care of accordingly in the application logic.

CONCLUSION AND FUTURE WORK

Many services currently available over the Internet involve complex multimedia interactions among the interested parties. In the IMS, such services are implemented by the Application Servers, which terminate the signaling originated by User Agent Clients willing to access the functionality they provide. Typical examples of such Application Servers include conferencing systems, voice-mail services, and so on. This implicitly suggests that these Application Servers have somehow to be multimedia-aware. While there exist Application Servers, which offer multimedia support as an inner functionality, this is not usually the case in Next Generation Networks for several reasons, including a potential functional redundancy as well as lack of scalability.

In this chapter, we introduced the use of remote Media Servers (as fostered by both 3GPP and the IETF), with special focus on the framework architecture proposed by the MEDIACTRL Working Group. Starting from our implementation efforts, we provided the reader with a set of real world scenarios typically involving the complex interaction envisaged by such approach, focusing on the critical role of the Application Server. In all the presented scenarios, Application Server acts as a middlebox for all the signaling. For each scenario, a different 3PCC-based approach has been presented, with its strengths and drawbacks. We have explained how the right choice for the signaling is never simple, and is often dependent upon the specifics or policies of the scenario itself, as well as the capabilities of the involved parties. This is even truer when involving non-multimedia resources in the signaling, as the presented case of BFCP and its COMEDIA-based negotiation. Strong standardization efforts, especially in researching best common practices regarding the protocols of interaction, become of paramount importance in such scenarios, considering how the choice of a specific pattern of signaling instead of another can lead to weaker results. Future work will definitely include further study of the presented scenarios and patterns of interactions, as well as the introduction of additional scenarios to deal with.

REFERENCES

3GPP. (2007). *Internet protocol (IP) multimedia call control protocol based on session initiation protocol (SIP) and session description protocol (SDP): Stage 3*. Technical Report. Retrieved from http://www.3gpp.org

Boulton, C., Melanchuk, T., & McGlashan, S. (2011). *Media control channel framework*. RFC 6230. Retrieved from http://www.rfc-editor.org/rfc/rfc6230.txt

Buono, A., Castaldi, T., Miniero, L., & Romano, S. P. (2007). Design and implementation of an open source IMS enabled conferencing architecture. In *Proceedings of the 7th International Conference on Next Generation Teletraffic and Wired/Wireless Advanced Networking (NEW2AN 2007)*. St. Petersburg, Russia: NEW2AN.

Camarillo, G. (2006). *Session description protocol (SDP) format for binary floor control protocol (BFCP) streams*. RFC 4583. Retrieved from http://tools.ietf.org/html/rfc4583

Camarillo, G., Ott, J., & Drage, K. (2006). *The binary floor control protocol (BFCP)*. RFC 4582. Retrieved from http://tools.ietf.org/html/rfc4582

Handley, M., Jacobson, V., & Perkins, C. (2006). *SDP: Session description protocol*. RFC 4566. Retrieved from http://tools.ietf.org/html/rfc4566

McGlashan, S., Melanchuk, T., & Boulton, C. (2011). *An interactive voice response (IVR) control package for the media control channel framework*. RFC 6231. Retrieved from http://tools.ietf.org/html/rfc6231

McGlashan, S., Melanchuk, T., & Boulton, C. (2012). *A mixer control package for the media control channel framework*. RFC 6505. Retrieved from http://www.rfc-editor.org/rfc/rfc6505.txt

MEDIACTRL. (2012). *IETF media server control prototype*. Retrieved from http://mediactrl.sourceforge.net

Rosenberg, J., Peterson, J., Schulzrinne, H., & Camarillo, G. (2004). *Best current practices for third party call control (3PCC) in the session initiation protocol (SIP)*. RFC 3725. Retrieved from http://tools.ietf.org/html/rfc3725

Rosenberg, J., Schulzrinne, H., Camarillo, G., et al. (2002). *SIP: Session initiation protocol*. RFC 3261. Retrieved from http://www.ietf.org/rfc/rfc3261.txt

ADDITIONAL READING

Ghandeharizadeh, S., & Muntz, R. (1998). Design and implementation of scalable continuous media servers. *Parallel Computing*, 24(1), 91–122. doi:10.1016/S0167-8191(97)00118-X

Hasswa, A., & Hassanein H. (2012). Utilizing the IP multimedia subsystem to create an extensible service-oriented architecture. *Journal of Computational Science*.

Liao, J., Qi, Q., Xun, Z., Li, T., Cao, Y., & Wang, J. (2012). A linear chained approach for service invocation in IP multimedia subsystem. *Computers & Electrical Engineering*, 38(4), 840–852. doi:10.1016/j.compeleceng.2012.03.010

Liao, J., Wang, J., Li, T., Wang, J., Wang, J., & Zhu, X. (2012). A distributed end-to-end overload control mechanism for networks of SIP servers. *Computer Networks*, 56(12), 2847–2868. doi:10.1016/j.comnet.2012.04.024

Luo, A., Lin, C., Wang, K., Lei, L., & Liu, C. (2009). Quality of protection analysis and performance modeling in IP multimedia subsystem. *Computer Communications*, 32(11), 1336–1345. doi:10.1016/j.comcom.2009.03.003

Pesch, D., Pous, M. I., & Foster, G. (2005). Performance evaluation of SIP-based multimedia services in UMTS. *Computer Networks*, 49(3), 385–403. doi:10.1016/j.comnet.2005.05.013

KEY TERMS AND DEFINITIONS

Application Server (AS): It is a component in charge of appropriately controlling a Media Server (MS) in order to provide advanced services to end-users. As such, AS is where all the application logic related to the services resides. To make a very simple example, AS can be seen as the brain in the architecture, viz. as the entity making decisions and controlling all the actions accordingly.

Binary Floor Control Protocol (BFCP): It is an IETF designed protocol used to handle moderation of resources. The protocol envisages the so-called *floor* as a token that can be associated with one or more resources. Queues and policies associated with such floors are handled by a Floor Control Server (FCS), which acts as a centralized node for all requests coming from Floor Control Participants (FCP).

IP Multimedia Subsystem (IMS): It is a standardized Next Generation Networking (NGN) architecture for telecom operators that want to provide mobile and fixed multimedia services. IMS uses a Voice-over-IP (VoIP) implementation based on a 3GPP standardized implementation of SIP, and runs over the standard Internet Protocol (IP).

Media Server (MS): It is a component conceived to take care of every facet of the media processing and delivery. Its operations are realized according to the directives coming from the controlling Application Server. MS can be seen as the arm in the architecture.

Media Server Control (MEDIACTRL): It is a Working Group of the IETF which aims at specifying an architectural framework to properly cope with the separation of concerns between Application Servers and Media Servers in a standardized way.

Session Description Protocol (SDP): It is an IETF designed protocol intended for describing multimedia communication sessions for the purposes of session announcement, session invitation, and parameter negotiation. SDP does not deliver media itself, but is used for negotiation between end points of media type, format, and all associated properties. The set of SDP properties and parameters constitute a *session profile*. SDP is designed to be extensible to support new media types and formats.

Session Initiation Protocol: (SIP): It is an IETF defined signaling protocol widely used for controlling communication sessions such as voice and video calls over Internet Protocol (IP). The SIP protocol can be used for creating, modifying, and terminating two-party (unicast) or multiparty (multicast) sessions. Sessions may consist of one or several media streams.

APPENDIX A: THE IP MULTIMEDIA SUBSYSTEM

Figure 1 shows the architecture of the 3GPP IP Multimedia Subsystem. Both IMS entities and IMS interfaces are showed in the picture. In the following of this appendix, we briefly expand on the components, which came into play for the work described in this chapter. The *User Equipment* (UE) implements the role of a participant and might be located either in the Visited or in the Home Network (HN). In any case, it can find the P-CSCF via the CSCF discovery procedure. Once done with the discovery phase, the UE sends SIP requests to the *Proxy-Call Session Control Function* (P-CSCF). The P-CSCF in turn forwards such messages to the Serving-CSCF (S-CSCF). In order to properly handle any UE request, the S-CSCF needs both registration and session control procedures (so to use both subscriber and service data stored in the *Home Subscriber Server* – HSS). It also uses SIP to communicate with the Application Servers (AS). An AS is a SIP entity hosting and executing services. The *IP Multimedia Service Control* (ISC) interface sends and receives SIP messages between the S-CSCF and the AS. The two main procedures of the ISC are: (1) routing the initial SIP request to the AS; (2) initiating a SIP request from the AS on behalf of a user. For the initiating request the SIP AS and the OSA SCS (*Open Service Access – Service Capability Server*) need either to access user's data or to know a S-CSCF to rely upon for such task. As we already mentioned, such information is stored in the HSS, so the AS and the OSA SCS can communicate with it via the *Sh* interface. In the scenario described in this work, the MRFC (*Media Resource Function Control*) shall regard the MRFP (*Media Resource Function Processing*) as a mixer. When the MRFC needs to control media streams (creating a conference, handling or manipulating a floor, etc.) it uses the *Mp* interface. This interface is fully compliant with the H.248 protocol standard. S-CSCF communicates with MRFC via *Mr*, a SIP based interface.

APPENDIX B: SIP – SESSION INITIATION PROTOCOL

The *Session Initiation Protocol* (Rosenberg, et al., 2002) is an end-to-end, client-server protocol. The design base was HTTP (HyperText Transfer Protocol) and SMTP (Simple Mail Transfer Protocol), two lightweight text-based protocols. SIP was originally used to establish, modify, and terminate multimedia sessions over the Internet. It has evolved to be able to set up a broad range of sessions, like multimedia (e.g., voice, video, etc.), gaming, Instant Messaging, and presence. SIP messages are either *requests* or *responses*, and may carry zero or more *bodies*. The most common body carried by SIP messages is an SDP payload. It is noteworthy that the *Session Description Protocol* (SDP) (Handley, et al., 2006) is used to describe the set of media streams, codecs, and other media-related parameters supported by either party in a multimedia session. All SIP implementations must support SDP. SIP runs on any transport protocol (UDP, TCP, TLS, SCTP); the specifications mandates UDP and TCP, while other transport protocols are optional. SIP provides the following functionality:

- User location.
- User availability.
- User capabilities.
- Session set up.
- Session management.

SIP does not provide services, but it enables the system to provide services in an easy way. The specifications envisage the following logical entities:

- **User Agent (UA):** An endpoint which can act by both a User Agent Client (UAS), when it sends requests and receives responses, and a User Agent Server (UAS), when it receives requests and sends responses;
- **Proxy Server:** A network host that proxies requests and responses. Hence, it acts as a UAC and a UAS;
- **Redirect Server:** A UAS that redirects requests to other servers;
- **Back-to-Back User Agent (B2BUA):** A UAS linked to a UAC. It acts as a UAS and as a UAC linked by some application logic;
- **Registrar:** A special UAS that accepts only registrations.

There are several types of SIP proxies, depending on the state they keep. A *stateless proxy* does not keep any state when forwarding requests and responses, while a *Transaction stateful proxy* stores state during the duration of the transaction. Finally, a *Call stateful proxy* stores all the state pertaining to a session (e.g., from the beginning to the end). SIP interactions usually happen by following the so-called *SIP trapezoid* depicted in Figure 7. We will not analyze the various SIP messages. The interested reader may refer to Rosenberg et al. (2002) for a detailed description.

Figure 7. The SIP trapezoid

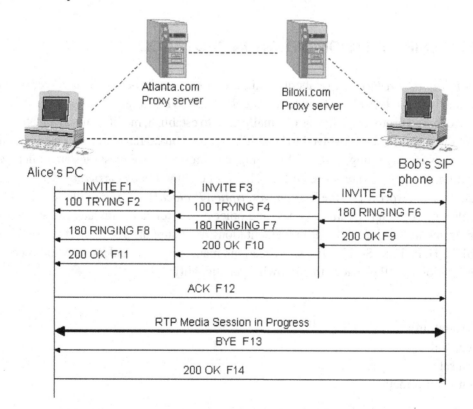

APPENDIX C: BFCP – BINARY FLOOR CONTROL PROTOCOL

As the name already suggests, *floor control* is a way to handle moderation of resources. In fact, a floor can be seen, from a logical point of view, as the right to access and/or manipulate a specific set of resources that might be available to end-users. Introducing means to have participants request such a right is what is called floor control. A typical example is a lecture mode conference, in which interested participants might need to ask the lecturer for the right to talk in order to ask a question. The IETF standardized, within the context of the XCON WG, a dedicated protocol to deal with floor control, the *Binary Floor Control Protocol* (Camarillo, et al., 2006). This protocol envisages the above mentioned floor as a token that can be associated with one or more resources. Queues and policies associated with such floors are handled by a Floor Control Server (FCS), which acts as a centralized node for all requests coming from Floor Control Participants (FCP). Decisions upon incoming requests (e.g., accepting or denying requests for a floor) can be either taken on the basis of automated policies by the FCS itself, or relayed to a Floor Control Chair (FCC), in case one has been assigned to the related floor. These decisions affect the state of the queues associated with the related floors, and consequently the state of the resources themselves. To go back to the lecture mode scenario example presented before, a participant who has been granted the floor (i.e., the right to ask a question to the lecturer) would be added to the conference mix, whereas participants without the floor (or with pending requests) would be excluded from the same mix, thus being muted in the conference. An example of BFCP interaction is depicted in Figure 8.

Figure 8. BFCP in action

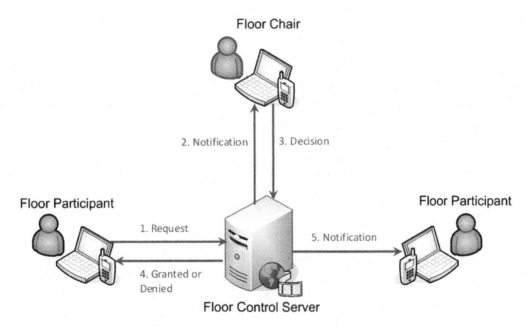

Section 2
Adaptive Video Streaming and Video Coding Techniques

Chapter 4
Techniques and Tools for Adaptive Video Streaming

Martin Fleury
University of Essex, UK

Laith Al-Jobouri
University of Essex, UK

ABSTRACT

Adaptive video streaming is becoming increasingly necessary as quality expectations rise, while congestion persists and the extension of the Internet to mobile access creates new sources of packet loss. This chapter considers several techniques for adaptive video streaming including live HTTP streaming, bitrate transcoding, scalable video coding, and rate controllers. It also includes additional case studies of congestion control over the wired Internet using fuzzy logic, statistical multiplexing to adapt constant bitrate streams to the bandwidth capacity, and adaptive error correction for the mobile Internet. To guide the reader, the chapter makes a number of comparisons between the main techniques, for example explaining why currently per-encoded video may be better-streamed using adaptive simulcast than by transcoding or scalable video coding.

INTRODUCTION

The scope of video streaming is currently expanding in two directions: towards delivery to mobile devices (Kumar, 2007; Schaar & Chou, 2007; Zhang, et al., 2008; Rupp, 2009), and towards streaming of *High-Definition (HD)* video (Park, et al., 2006; Zhu, et al., 2007; Bing, 2010). Both these developments imply adaptive streaming (Ortega & Wang, 2007) to cope with error-prone wireless channels and fluctuating bandwidths due to congestion. In fact, the two conditions may be combined, as laptops now exist that have HD displays, which compensate for the reduced viewing distance when reading a typical mobile

DOI: 10.4018/978-1-4666-2833-5.ch004

display. In the case of streaming over wireless channels, packet corruption can seriously damage video quality due to the temporal dependency of video compression. In the case of HD video, changes in bandwidth may make it impossible to stream at full resolution (either spatial, temporal resolution or *Signal-to-Noise Ratio [SNR]*—the video quality) while congestion persists. A further possibility (apart from spatial, temporal, and SNR adaptation) is to adapt the level of error protection (Al-Jobouri, et al., 2012) according to channel conditions. Thus, adaptive video streaming needs to be addressed in today's environment. An issue is which form of adaptive streaming to adopt.

In respect to pre-encoded video, the commercial world has largely opted for simulcast (Conklin, et al., 2001) as the means of adjusting the streaming rate delivered across access networks to the user's display. According to available bandwidth, the server can then switch between the streams at anchor frames. However, because the user chooses just one rate at streaming start-up, this method can result in service interruptions. To cope with changing conditions, variants of HTTP live streaming (Stockhammer, 2011) have been introduced and are on the point of standardization, for example as *Dynamic Adaptive Streaming over HTPP (DASH)* (Stockhammer, et al., 2011) (notice that 'Live Streaming' is not suitable for live video despite its marketing name). In that form of streaming, the streaming rate can be dynamically changed, though the granularity of rate changes is coarse (Cicco, et al., 2010).

Rather than switching at anchor frames (I-frames), streaming with switching frames, as introduced in the H.264/*Advanced Video Coding (AVC)* standard codec (Karczewicz & Kurceren, 2003), allows a more flexible form of rate switching. However, switching frames are not compatible with widely deployed multimedia players. Depending on the latency demands of the live streaming application, it may also be possible to change the streaming rate at the codec by altering the quantization parameter. In comparison, bit-rate transcoders (Assunção & Ghanbari, 1997; Ahmad, et al., 2005; Sun, et al., 2005) are able to change the rate of both live and pre-encoded video. However, most bitrate transcoders were developed for the MPEG-2 codec, while non-linearities within an H.264/AVC codec are a deterrent to bit-rate transcoding. It is an interesting coincidence that the in-loop de-blocking filter of H.264/AVC was made mandatory in the standard, and soon afterwards, the scalable extension of H.264/AVC was developed with the aim of removing the need for transcoders. Nevertheless, transcoding continues to be used. H.264/AVC spatial or temporal resolution transcoding remains entirely possible. The cost of a transcoder bank can also be traded against the need to store differing versions of the stream in the simulcast method. Trancoders can translate between different codec formats and are not limited by the multimedia player format at the receiver device.

On the other hand, to avoid the need for transcoder hardware, scalable video coding has been developed most recently in the Scalable Video Coding (SVC) extension to the H.264/AVC standard codec (Schwarz, et al., 2007), and it can be applied to interactive forms of video streaming, as latency is reduced. However, the current absence of hardware scalable codecs limits their extension to some mobile devices. Scalable video relative to simulcast reduces storage costs, as there is a single representation of the available bitrates, rather than multiple unicast versions, as in simulcast. On the other hand, as memory storage costs are currently very low (Kalva, et al., 2012), storage costs are *not* a big factor affecting commercial operations, while there is a reduction in bandwidth from using a somewhat more efficient non-scalable encoding.

A similar remark applies to the advantage of simulcast versus transcoding in terms of the low cost of storage compared to possible disadvantages of transcoding. Transcoding may also introduce a drop in video quality at each application (Chang & Eleftheriadis, 1994). Though one source of repetitive error (in the transform operation) has been addressed in H.264/AVC, errors may still accumulate in repetitive motion compensation or in rescaling when employing resolution transcoding. Therefore, the cost of transcoder banks and possible quality drops are a disadvantage of transcoding compared to simulcast methods.

With the possible exception of simulcast, all of these techniques may use intelligent forms of streaming control to govern the adaptation. In effect, the controllers (Widmer, et al., 2001) are the tools that make effective use of the adaptive streaming techniques possible. However, there are also many congestion controllers that do not alter the video rate itself but change the data rate. One way to achieve the latter is by widening the inter-packet gap (as in the case of *TCP-friendly Rate Control [TFRC]* [Handley, et al., 2003]), which has been widely adopted for the fixed Internet and has even been extended to wireless networks in various ways that will be described later, ways that are beyond its designers' intentions. To compensate for the latency introduced, significant buffering at the end device is necessary. Buffering in its trunk leads to pre-roll latency. It is also possible to change the packet size, and in the hybrid case study described in the Chapter using transcoded video, a fuzzy logic congestion controller is employed. The method can be applied both to Constant Bit-Rate (CBR) video, or constant quality video (Variable Bit-Rate—VBR) (or even truncated VBR in which there is a maximum limit to the VBR bitrate). However, in the second case study, the CBR rates of multiple, constituent video streams are altered dynamically, according to content coding complexity and in ratio to the available bandwidth. In the third case study for broadband wireless delivery, the *Forward Error Correction (FEC)* rate is adapted in response to feedback from the receiver device. As mentioned already, this method can be combined with prioritized data protection, and is suitable for VBR and CBR alike.

The rest of this chapter is organized as follows. The following Section considers the different forms of video streaming that exist, before specializing to emerging forms of commercial video streaming. Streaming of 3D video in its various forms (stereoscopic, multi-view, and free-viewpoint) is also touched upon. The section after that examines the main types of adaptive video streaming, i.e. adaptive simulcast, transcoding, scalable coding, and through rate controllers. As mentioned above, three case studies act as additional illustrations in the next Section. After considering possible future developments, the chapter concludes.

VIDEO STREAMING

Video Streaming Basics

In this section, video streaming basics are outlined. Notice that the alternative term 'multimedia streaming' may be used. The latter term may mean streaming of lower-quality video, as opposed to higher quality or broadcast-quality video. Multimedia streaming may also mean the inclusion of any associated audio streams. In which case, a multimedia container such as MPEG2 *Transport Stream (TS)* or the MPEG-4 container (Bing, 2010) can be employed or indeed the Real-time Transport Protocol (RTP) (Perkins, 2003) can carry associated audio streams. RTP can also transport MPEG2-TS packets as sub-packets within an RTP packet. Synchronization between video and audio can be provided within MPEG2-TS or RTP. However, because audio normally consumes much less bandwidth than video but has different characteristics, it is usually treated separately, and this chapter will be no different. For a general treatment that includes audio and video refer to Hwang (2009). The term 'media streaming' is also sometimes employed, instead of 'multimedia streaming' and video streaming. However, as 'media streaming' may also refer to software development for Microsoft products, it is also avoided in this chapter.

In true video streaming, a video aims to be displayed in real-time on the end user's device. In this context 'real-time' (or video rate) means at the same rate as it was originally captured, though there will generally be a short (and hopefully imperceptible) start-up delay. This chapter is mostly about true streaming. True streaming may be employed for on-demand streaming in which a user requests pre-recorded content. Pre-recorded content is normally pre-encoded in some form of secondary storage. True video streaming is also suitable for live streaming in which a copy of the same video is sent to every viewer at the same time. The Chapter concentrates on single stream or unicast streaming but it is obviously possible to multicast a video stream across a network, with potential saving in bandwidth. Multicasting is a large topic that deserves a separate treatment as in Wittmann and Zitterbart (2001). Compared to unicast, multicast improves bandwidth efficiency by sharing video packets delivered through a network. However, multicasting over a wireless network is subject to a physical channel that is error-prone and time varying. Consequently, users signed up to a multicast service in such a network can often suffer (Liu, et al., 2007) from diverse channel conditions. With multicast streaming there is a risk of a feedback implosion at the server, an overload of network resources due to the attempts of many receivers trying to send repair requests for a single packet. A number of approaches exist to avoid this implosion effect such as randomized timers, local recovery whereby receivers can also send repair packets, and hierarchical recovery (Tan & Zakhor, 2001). However, while such approaches are effective, providing reliability without implosion, they can result in significant and unpredictable delays, making them unsuitable for real-time applications. Other studies such as that of Rubenstein et al. (1998) have been devoted to hybrid schemes, combining *Forward Error Correction (FEC)* and *Automatic Repeat reQuest (ARQ)* to reliably deliver data with an emphasis on reducing delay and meeting real-time delivery constraints. However, there is a risk of unnecessary repair data being sent to those mobile receivers not affected by localized channel errors, such as through the effect of multipath.

In contrast to true video streaming, according to the 'download and play' approach, the whole compressed video file is transported across the network and stored on a user's device. In other words, the complete video file is downloaded before it is decoded and displayed. This allows a reliable transport protocol to be employed (such as *Transmission Control Protocol [TCP]*), which can improve coding efficiency in error-prone networks, as no error resilience measures are required. As TCP is a connection-oriented protocol, it is easy to manage at a firewall. However, all of the content is stored on the end device, exposing the content owners to multiple copies of their digital media. Another obvious disadvantage of download and play is that large files (such as for two-hour movies) not only result in a long start-up time but despite compression may overwhelm the storage at an end user's device. However, for short files of user-generated material, these restrictions do not arise. As an example the *Multimedia Messaging Service (MMS)* of the *Open Mobile Alliance (OMA)* is a parallel service to the text *Short Message Service (SMS)*. In MMS (Open Mobile Alliance, 2011), still images and video sequences are encoded prior to forwarding to an MMS store-and-forward server. Notice that the content may be automatically transcoded at the MMS server (called content adaptation) in order to match the target device's display resolution or the bandwidth capacity on the path to the device.

Progressive download and display or pseudo-streaming is a compromise between true video streaming and download and play. In basic terms, the content is split into separate physical or logical files, each of which contains a chunk or segment of compressed video. The chunks are then transported by a reliable transport protocol (normally TCP). As reliability implies the possibility of an unbounded delivery delay in best-effort IP networks or error-prone mobile networks, pseudo-streaming is unsuitable for interactive video streaming. As a small example, YouTube files are streamed in this way (Gill, et al., 2007; Abhari & Soraya, 2010), possibly after local caching in a *Content Distribution Network (CDN)*. As in MMS, a form of content adaptation through transcoding occurs before distribution, for example, limiting the rate to 1.25 Mbps (1.25 Mbps is the download rate of many domestic Asymmetric Digital Subscriber Line [ADSL] links).

In fact, it has been observed (Plissonneau, et al., 2008) that about 30% of YouTube pseudo downloads are at a rate of less than 340 kbps. This was attributed in Plissonneau et al. (2008) to a long delay at the local ADSL platform. However, ADSL is prone to error bursts from *Repetitive High-Level Impulse Noise (REIN)* (Broadband Forum, 2007). A long round-trip time and/or packet loss will cause a TCP source to staunch its rate. From (Plissonneau, et al., 2008) frequent abortions of the stream (around 40%) occur, as a user may be faced with poor throughput and only have a partial interest in a video clip. Similar behavior was reported in Alcock and Nelson (2011) but the diagnosis of the problem was shifted from the ADSL link to the packet scheduling policy of YouTube servers. An initial rapid transmission of packets from initial chunk(s) was observed, presumably to reduce start-up delay. After that, the chunk rate was more measured. (Chunks are called blocks in Alcock and Nelson [2011]) and it is unclear whether a self-contained, source-coded chunk coincides with a scheduling block.) However, when a chunk was sent, packets were rapidly scheduled, resulting in self-congestion at the access network. Thus, measurement studies of pseudo-streaming point to limitations to this form of streaming, whatever the current penetration in the marketplace. However, this form of streaming is associated with simulcasting, which is returned to as a form of adaptive streaming in the following section.

Direct streaming employing TCP transport without chunking can also be applied to pre-encoded or stored video. However, to do so successfully (Wang, et al., 2004) requires available bandwidth to be about twice the peak video rate and several seconds buffering is required. Successful streaming implies (Wang, et al., 2004) no more than 10^{-4} of packets are delayed beyond their play-out deadline, for a start-up penalty of 10 s. However, even a few seconds of buffering are a potential problem for click-and-view services such as YouTube, while the bandwidth requirements are severe for high-quality video. Given that the access network is the main bottleneck in current systems, direct transport of video by TCP requires the streaming rate to be half that of the access network capacity, which implies for most deployments lower-quality video streaming. On the other hand, as codec efficiencies improve TCP becomes more attractive. Improved coding efficiency from the High Efficiency Video Coding (HEVC) codec project is discussed in the Section on future developments.

Video Streaming Model

During the design of codecs, a theoretical model of the decoder and its input over a network is a convenient device. In MPEG-2, this model was called the *Video Buffer Verifier (VBV)* design (ISO/IEC, 2000), while the H.263 standard codec equivalent is called the *Hypothetical Reference Decoder (HRD)* (ITU-T, 1997). For live streaming, an encoder is connected through a video server to a decoder buffer servicing a decoder at the receiver. Figure 1 is an illustration of such a model. The model assumes a CBR channel between the server and the receiver. The encoder must then output at a rate that avoids underflow (causing display delay) or overflow (resulting in packet loss) at the decoder's buffer (notice that it is also often assumed that at a receiver there is an input buffer designed to smooth-out jitter and a further decoder buffer allowing the decoder to

Figure 1. Schematic model example used in specifying rate control for video encoders

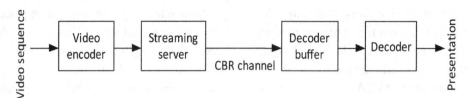

exploit predictive references). To avoid buffer violations (underflow or overflow), the models assume a leaky bucket model at the output of the encoder. In the leaky bucket model, a virtual hole in the bottom of the bucket creates a constant flow of water (bits) onto the channel. Any water (bits) dripping into the bucket while it is full will overflow. Consequently, the encoder must maintain a rate that avoids overflow. At the decoder, bits are removed from the buffer at a variable rate, according to the original coding complexity of the video frame. Start-up delay is: either fixed and sent with the initial parameters (as in the MPEG-2 CBR model); or determined by the time taken to fill the decoder buffer (as in the MPEG-2 VBR model); or specified by the size of the first compressed video frame (as in the H.263 model). By the relationship $delay = B/R$, where B is the decoder buffer size and R is the input CBR rate at the decoder buffer, fixing two of the variables determines the third. To conform to the VBV, HRD models, rate-control algorithms at the encoder for CBR and VBR video can then be proposed (Pao & Sun, 2001; Ma, et al., 2003; Zhao & Kuo, 2003).

In a network with access control, as in *Asynchronous Transfer Mode (ATM)* networks (Prycker, 1993) or where a virtual circuit can be established in an IP network, as for example by means of the *Resource Reservation Protocol (RSVP)* (Braden, et al., 1997), these models have a validity, apart from their convenience in specifying an encoding model. In best-effort, wired IP networks, the CBR channel assumption no longer holds due to factors such as traffic congestion and dynamic packet routing. As the source coding complexity of pictures varies over time, as does the type of encoding (spatial or temporal prediction), even with CBR encoding it is not possible to maintain a CBR. Instead, the output rate fluctuates around the target rate. Typically, a single-pass, closed loop CBR rate controller takes an estimate of the coding complexity of the input frame to set a *Quantization Parameter (QP)*. The resulting bitrate acts as a feedback control on the rate controller. The variation between QPs must also be limited in order to reduce disconcerting quality fluctuations over the space of a few frames. However, in HD video the fluctuations may still be large (Bing, 2010). Instead, open loop VBR encoding results in a much better rate-distortion performance (Lakshman, et al., 1998; Varsa & Karzewicz, 2001). The main advantage that CBR has it that it allows reservation of storage and bandwidth. However, against this must be set disadvantages such as less efficient storage, inefficient utilization of bandwidth, and more complex encoders.

The implication of the inability to meet the theoretical or hypothetical coding models is that video streaming should adapt to channel conditions. However, it is important to realize that unbridled bitrate adjustments according to available bandwidth may also be unwelcome to the viewer (Zink, et al., 2003). Thus, switching intervals may need to be restricted to around 3s and above (Ni, et al., 2010).

Future Streaming: 3D Video

Following on from the recent success of HDTV in Europe, the broadcasting industry has taken a renewed interest in 3D TV. Stereoscopic 3D has also become popular in the film world, after the outstanding commercial success of the film *Avatar* and its use in IMAX cinemas, leading to the launch of a number of pilot 3D TV stations. Stereoscopic (two view) 3D video is only one variety of three-dimensional video (Gürler, et al., 2011), as there is multi-view, providing a set of fixed viewpoints, and free-viewpoint, which allows the viewer to select their viewpoint. One version of free-viewpoint TV employs ray-tracing (Tanimoto, et al., 2011) to create a potentially infinite set of views from a limited number of camera images of a scene. There is a 3D extension to the H.264/AVC codec (Vetro, et al., 2011) and accompanying methods for transporting 3D video (Schierl & Narasimhan, 2011). The Blue-Ray Disc Association has adopted the Stereo High profile of the H.264/AVC 3D and multi-view extension. Though the film and broadcast industry has initially invested in 3D video, it could be Gürler et al. (2011) that networked streaming is most suited to the added flexibility that 3D video provides, especially in its multiple view varieties. This is because the end-user is able to select a view over the network rather than be restricted to a single, albeit 3D, viewpoint. The possibility has already been investigated (Savas, et al., 2011) of using peer-to-peer streaming to cope with the additional demands on the network in supplying multiple views.

In fact, the main impediment to streaming stereoscopic 3D is the need to adapt streaming of two views to existing network infrastructure. One way to do this is to store two reduced resolution views within one 2D frame, the frame-compatible approach. The side-by-side format subsamples the left and right view by 50%, allowing the two views to be placed as left and right images within a frame. The top-and-bottom format is similar except in the vertical direction. The latter two formats have been employed for High-Definition Multimedia Interface (HDMI) version 1.4a, as has the cable TV industry (however, notice that HDMI 1.4a also supports non-frame compatible 3D formats such as frame packing of alternating left and right views). Other frame compatible formats include row-interleaved and checkerboard but these can suffer from cross-talk artifacts and color bleeding across views (Vetro, et al., 2011). The interleaved format can increase the high frequency content and, hence, the compressed bitrate. Another way to reduce bandwidth requirements (Merkle, et al., 2007) is to code for one view together with a depth disparity map. However, two problems need to be resolved, sufficiently accurate depth maps camera output, and avoiding depth holes caused by occluded regions. Thus, extra depth layers (Zitnick, et al., 2004) are required to perform 'dis-occlusion' from the depth map or multiple depths can be coded for Shade et al. (1998). Another method available to reduce the bandwidth of stereoscopic 3D is to reduce the quality of one of the views, as provided the overall quality is above a threshold, the reduced quality of one of the views is not perceived by a viewer (Meesters, et al., 2004). In effect, in PSNR asymmetric coding, the low resolution view acts in a similar way to an enhancement layer in scalable video coding.

Multi-view coding is also necessary in order to reduce the bandwidth requirements that would arise were views to be sent independently of each other, with each sequence of frames predicted from other frames of the same view. It is not the computation involved in independent processing that is the problem, as that occurs anyway with simulcast coding but the need to send multiple independent streams. Alternatively, one view can be generated conventionally (in H.264/AVC format) with anchor frames to allow random access. Other views are predicted from the frames in the conventionally encoded view. However, in full prediction intra-view frames are also predicted internally from frames of the same view.

Alternatively, intra-view prediction is retained but prediction across views only occurs at anchor frames (Merkle, et al., 2007a). In fact, in the multi-view coding extension to H.264/AVC, additional frames can be selected as inter-view references but a list of these must obviously be supplied to the decoder through signaling. An additional frame type was specified in the standard (Vetro, et al., 2011), that is an inter-view anchor frame. As such, frames are anchors prediction from frames prior to these in temporal order is not permitted. It is also necessary to perform illumination compensation across views (Gürler, et al., 2011). The coding gains from inter-view prediction over simultaneously sending all views can be between 20% and 50% (Vetro, et al., 2011), depending on content and number of views. Asymmetric coding of multi-view video can also be applied in a similar way as for stereoscopic 3D video. Free-viewpoint video can operate similarly to multi-view by sending a set of views, provided the receiver is computationally equipped (Tanimoto, et al., 2011) to estimate depth and interpolate between the estimates in order to generate rays.

Streaming of 3D video itself has not apparently been investigated by a wide range of research groups, possibly due to the resources needed to conduct such tests. Because multi-view coding is unlikely to be performed in real-time due to its computational complexity, most streaming of multi-view video will be of pre-encoded video. Therefore, adaptive HTTP streaming may be more appropriate than true streaming. Streaming of multi-view video can also take place in a similar way to conventional single view streaming (Kurutepe, et al., 2007), provided inter-view dependencies, especially for asymmetric coding, are first signaled in order that the streaming client selects appropriate views for streaming. View selection prior to streaming remains a problem due to latencies in stream switching, were head tracking to be performed to automatically select the current views to stream. A possible solution is discussed in Kurutepe et al. (2007a) in which low-quality versions of views are transmitted while switching takes place.

TYPES OF ADAPTIVE STREAMING

The Introduction outlined the principle forms of adaptive streaming, which are:

1. Simulcast and adaptive simulcast (HTTP live streaming);
2. Transcoding;
3. Scalable video coding, including H.264/SVC (Schwarz, et al., 2007); and
4. Rate controllers with wired and wireless approaches.

Another form of adaptability is provided by *Multiple Description Coding (MDC)* (Goyal, 2001), which is now briefly considered. MDC involves encoding the same stream as multiple lower-quality versions of the original stream (called descriptions). However, unlike scalable video there is no scalable hierarchy of layers. Thus, if a single description is lost or temporarily corrupted in some way another description can act as a substitute. In contrast, in scalable video coding if the base layer is damaged in some way then the enhancements layers are no longer useful. Just like scalable video, though, if more than one description is received then the quality is enhanced. This opens up the possibility of exploiting path diversity to aggregate bandwidth and to protect against link outages. MDC is particularly appropriate for ad hoc networks (Chow & Ishii, 2007) as the possibility of multi-hop routing allows path diversity to be built into the distributed routing algorithm. Unfortunately, from the point of view of standardization, there is no common form of MDC, as it can take place in the spatial, temporal, or frequency domain. MDC codecs can also be complex (Wang, et al., 2005) and no public-domain codecs appear to be available. However, it is the specialist nature of MDC, the need for path diversity, which means MDC is not a mainstream form of adaptability. This chapter now considers the above options in some more detail.

Simulcast and Adaptive Simulcast

As mentioned in the Introduction, simulcast is largely a commercial solution to the problem of video streaming in which simplicity of implementation is traded off against the storage costs arising from switching between multiple bitrate versions of the same stream. As storage costs are currently negligible, amounting to $0.125 per GB in 2012 (Kalva, et al., 2012), simulcast becomes attractive for the streaming of pre-encoded video. Simulcast is not appropriate for live video streaming and is even less appropriate for conversational or interactive video applications such as video conferencing. This is because of the delay introduced by the need to create multiple versions at different qualities or bitrates. In simulcasting, different versions of the same pre-encoded stream are stored on a server, for example, about five versions (Cicco, et al., 2010) are stored on the Akamai *Content Distribution Network (CDN)*.

Of course, this involves considerable storage, and organization and management of that storage. This may be why two-pass CBR coding may well be used, as this allows the storage requirements to be fixed in advance and to reduce the impact on the network of increases sudden increases in bitrate, causing congestion and, hence, delivery delays. Of course, CBR encoding will result in quality fluctuations and coding latency, as well as inefficient storage (Bing, 2010). The pixel/frame resolution, rather than the video quality, is often varied, resulting in different bitrates to suit the bandwidth bottleneck that commonly exists over *ADSL* (called broadband in marketing literature) The BBC iPlayer's catch-up TV service is an example of simulcasting at different spatial resolutions. Streams can also be switched between different temporal resolutions. However, rather than simulcasting, the latter can also be created

by dropping reference-free B-pictures. Care must be taken to avoid a jerky display as a result of the missing frames. Decreasing the temporal resolution also reduces coding efficiency, as there is potentially more changes between one picture and the next and, hence, less exploitable redundancy.

Real Networks were pioneers (Conklin, et al., 2001) of simulcasting. In fact, streaming seems to have been introduced commercially by Real Networks with a pioneering audio broadcast of a baseball event in 1995. The company went on to introduce video streaming in 1997 so that by 2000, about 85% of streaming content in the Internet was apparently in the Real format. However, the Real Player, which now streams using a proprietary codec rather than their original H.263 standard codec, appears to have suffered a market decline, at least in respect to video streaming. One cause of this may have been the obtrusiveness of Real Player upgrades and auxiliary programs, or it could be the cost of the servers, or indeed competition with Adobe Flash Player (except on Apple devices). It is worth noting that though Real did prevent the writing of their streams to store, there now exists software to do this (de-streaming), thus allowing viewing times to be shifted. There are also filters to allow different players, other than the Real Player, to be used.

Commercial streaming also tends to use pseudo-streaming, which is also known as progressive download (see the earlier outline of pseudo-streaming). Pre-encoded bitstreams are broken into chunks, each of which is downloaded, while a previous chunk is played. Because browser plug-in compatibility is seen as important, connection-oriented TCP acts as the transport protocol for HTTP, or in Adobe's case as the basis of *Real Time Messaging Protocol (RTMP),* which in turn may be encapsulated in HTTP (RTMP is a proprietary protocol, which may have a vulnerable authentication mechanism). The widespread adoption of the Adobe Flash Player (present on about 95% of browsers in July 2011) acts as a non-technical constraint on the form of streaming. Microsoft

Silverlight also has about 53% penetration of browsers at the time of writing. Both of these players employ proprietary software.

As an aside, it is sometimes said that TCP is necessary to penetrate firewalls and *Network Address Translation (NAT)*. However, it is the negotiation of a connection that is needed for this purpose and a protocol such as *Datagram Congestion Control Protocol (DCCP)* can set up a connection before reverting to *User Datagram Protocol (UDP)* transport. The *Real-Time Transport Streaming Protocol (RTSP)* can also set up connections for RTP streams, and it is included in the *Internet Streaming Media Alliance (ISMA)'s* profiles. As is often observed, TCP, because of its built-in reliability, is unsuitable for a real-time service, as congestion can lead to frequent resends of the data and a slowing down of the transmission rate. The problem is worse in a wireless network as lost packets may be due to channel conditions. In which case, staunching the flow will not make the adverse channel conditions go away. Buffers can smooth out the delays. However, large buffers militate against 'click-and-view' type streaming, as they lead to long start-up delays. This is only partially mitigated by placing the content in caches nearer the user.

In recognition of these problems with pseudo-streaming, commercial operators have introduced adaptive pseudo-streaming in which the target download rate is not fixed at start-up time but can be adapted according to measured or estimated conditions on the network. Chunk sizes tend to be between 2s and 15s in duration. Unfortunately, this is unlikely to be suitable for wireless networks, as the channel can change over a time interval much smaller than a chunk's download time. On the other hand, if a chunk's size is reduced then compression-coding gain is reduced as well. This is because there may be just one I-picture per chunk, but with shorter chunks, there will be more I-pictures, which are encoded with less efficient spatial coding. Larger chunks suffer from bandwidth wastage, if the client switches to different content (Stockhammer, 2011) while the stream is being sent. Larger chunks also imply a longer start-up delay, especially as the initial chunk must normally contain streaming parameters. For these reasons, it appears that some systems first download advertisements to distract the viewer while this start-up delay occurs. In this way, it is hoped that once the stream of chunks starts then any network jitter will be absorbed while one chunk plays and another arrives.

The complexity of managing selection of chunks in the context of stateless HTTP is also a considerable burden. This is because for each change of rate request, a client must make an entirely new TCP connection and tear it down afterwards. On the plus side, the client is left in control of the connection. From the commercial point of view, the format allows the easy insertion of targeted advertising (Kim & Kang, 2011) and other personalized content (provided privacy legislation does not severely restrict this practice). More positively, it allows personalized video recording functionalities to be enabled and 'trick modes' such as seeking, fast rewind and forwarding, to be deployed.

A strong non-technical factor in favor of adaptive HTTP streaming is that it allows the widespread Web caching infrastructure to be utilized for streaming. For that reason, standardization is underway as the *Dynamic Adaptive Streaming over HTTP (DASH)* standard (Stockhammer, et al., 2011). DASH will also include provision for multiple languages and audio, as well as transport security and connection protection. The DASH architecture is shown in Figure 2. Notice that chunks become segments in the DASH nomenclature and that a set of simulcast video streams at different rates are called a media presentation. Commercial implementations of HTTP streaming, pre-date DASH and mainly include: Microsoft's Smooth Streaming (Zambelli, 2009), Apple's HTTP Live Streaming (Pantos & May, 2011) and Adobe's dynamic streaming (Adobe, 2010).

Figure 2. DASH architecture showing media presentations divided into multiple streams at different rates, with each stream subdivided into segments

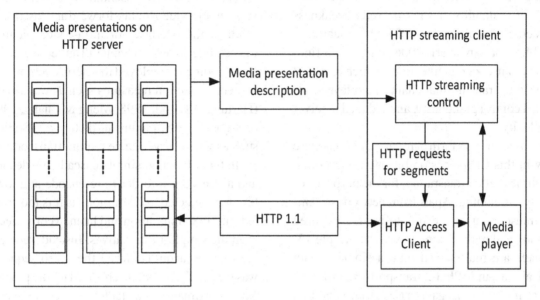

As has already been mentioned, adaptive streaming in this manner requires a huge logistic effort, apart from any need to locally cache the simulcast streams. This is apparent from the number of chunks (segments) that might need to be stored: A one-hour video encoded at five different quality levels and subdivided in chunks of six seconds each, requires 3000 files to be stored in one server for that video alone, if explicit storage of chunks is employed. The alternative, favored by Microsoft's Smooth Streaming, is to logically separate chunks in a single file. The client must first download a media presentation description containing the name of the desired streaming file and rules to create *Uniform Resource Identifiers (URIs)* to request chunks within the streaming file. In the case of separate files for each chunk, the media presentation description file will contain an ordered set of URIs for each file.

Cicco et al. (2010) report on Akamai high-definition video streaming. Akamai's CDN is also employed by YouTube to distribute its video clips. Though the streaming control algorithm appears to be activated every 2 s, adaptation time can take

as long as 150 s. Clearly, part of this time must be accounted for by a streaming adjustment algorithm, as too rapid changes, possibly caused by rate thrashing are disturbing to a viewer. However, part may also be caused by the time required to locate an appropriate chunk. An area of interest for researchers (assuming switching can take place quickly enough) is the adaptive algorithms required, as these may be driven by computational intelligence (Jarnikov & Özçelebi, 2011).

Transcoding

Video transcoders open up the possibility of sending a pre-encoded video bitstream at the maximum possible rate without overly exceeding the available network bandwidth. A transcoder (Assunção & Ghanbari, 1997), for a minimum acceptable picture quality, also sets a lower bound to the sending rate as a percentage of the pre-encoded input rate. In fact, it is quite possible to arrive at fewer packet losses or even avoid loss altogether by recompressing an already compressed bitstream by means of a transcoder. Hence, subsequent router buffering is able to cope with the output packet stream. A perceived weakness of transcoding, a lack of scalability, has not proved a weakness in practice (Kasai, et al., 2002). The research of Shanbleh and Ghanbari (2006) introduced fine-grained transcoder architectures, which removed the need to transmit redundant video layers, increased control granularity, and reduced receiver complexity.

Transcoding is a form of re-encoding. The case study in this Chapter is on bit-rate transcoding, sometimes called transrating, for the purposes of adaptive streaming. Apart from network streaming, transcoding of the bitrate is also required for statistical multiplexing in which multiple TV channels are multiplexed onto a single stream (typically as an MPEG-2 transport stream*)*. Statistical multiplexing gain arises from encoding in a dynamic manner a less complex stream at a lower rate than one that is more active and, hence, is more complex to code. Such an application may be required in satellite streaming of multiplexed TV channels.

However, it is worth noting that there are many other uses of transcoding other than transrating (Sun, et al., 2005). Transcoding may also be employed for conversion to a different codec format, most commonly when converting user-generated content to a common format on sites such as You-Tube. There is also a need to re-encode material encoded with a legacy codec, typically MPEG-2, to a more recent codec, such as MPEG-4 part 2 (Xin, et al., 2002) or H.264/AVC (Du, et al., 2008), which now has wide industrial support. New material may also be converted to an older codec, when older players are still in use. For example, it may be preferable to stream over a network after encoding with the more efficient H.264/AVC but convert to MPEG-2 at the set-top-box. Conversion to a new codec and a reduced spatio-temporal resolution may also be required (Shanableh & Ghanbari, 2000). This could involve conversion of *Standard Definition TV (SDTV)* material to HDTV format, which will also require the inclusion of

embedded information such as closed-captioning for sub-titles. For representations of faster speed versions of a video for fast forward and fast rewind (without the overhead of intra-coded pictures) transcoding is also of value (Tan, et al., 2002). Re-encoding is employed to insert covert or overt content protection measures such as watermarks (Hartung & Girod, 1998; Meng & Chang, 1998) or logos. Finally, picture-in-picture applications such as video surveillance require transcoding.

In the following, simply cascading a decoder and an encoder in order to re-encode at a lower bitrate is rejected as a solution, as it is too costly in terms of computation and latency compared to even the simplest alternatives. Instead, re-use of information acquired when the video sequence was originally encoded can speed up the process. Motion estimation is by far the most computationally complex encoding process (usually around 70% of encoding time) and, hence, this operation should be avoided at the transcoder. Open-loop transcoding is the simplest to implement, as it may be necessary to modify only the frequency transform coefficients (*Discrete Cosine Transform [DCT]* coefficients [Watson, 1994] in codecs prior to H.264/AVC, which itself uses an integer-valued approximation to the DCT to reduce round-off errors in computer arithmetic). The coefficients with amplitudes below a threshold value can be discarded or coefficients can be discarded in order to meet a desired bitrate (Eleftheriadis & Anastassiou, 1995). Alternatively, the transform coefficients can be inverse-quantized (Nakajima, et al., 1995) before re-quantization at a coarser quantization with more visually pleasing results. Notice that it is also necessary to re-quantize the chrominance coefficients as well as the luminance coefficients. However, in these operations it is normally necessary (Vetro, et al., 2003) to first perform entropy decoding (sometimes known as *Variable Length Decoding [VLD]*) to obtain the transform coefficients, before re-applying entropy coding (that is *Variable Length Coding [VLC]*). However, compared to motion estimation, entropy coding has a minor impact on the computational time. In Sun et al. (1996) even the VLD step was avoided by identifying high-frequency coefficient codewords in the bitstream.

Unfortunately, open-loop transcoding has a flaw, as it does not feedback the decoder's view of a de-compressed picture upon which to base re-quantization upon. Thus, any predictively-coded P-picture received at the transcoder that have been encoded by referencing a previous P-picture introduces error in an open-loop design. The decoder receives only the re-quantized versions of P-pictures to base its reconstructions upon, rather than the original P-pictures output by the encoder. The result of this is drift error that is visible as visual blurring until an I-picture arrives, as an I-picture is not temporally predicted. A solution to the problem of drift between the encoder and the decoder (Assunção & Ghanbari, 1997) as a result of the intervention of the transcoder is a closed-loop transcoder design. The transcoder receives a decoded version of its output with a frame's delay. In the original design of Assunção and Ghanbari (1997), the decoder's view of a previous picture was acquired in the spatial or pixel domain. This procedure requires first applying an inverse frequency transform prior to motion compensation of the macroblocks before transforming back into the frequency domain (for simplicity, subtraction to obtain residuals has been omitted from this description; see the original paper [Assunção & Ghanbari, 1997] or Vetro et al. [2003] for further details). It turns out that the required operations can be performed in the frequency domain (Chang & Messerschmidt, 1995), missing out the inverse and forward transform operations. These operations were found to provide (Sun, et al., 1997) equivalent video quality to that from spatial domain transcoding.

However, simplified structures for transcoding assume that operations omitted from processing behave linearly, implying that inverses exist. While this is the case for orthogonal transform operations, it is not for motion compensation and quantiza-tion. In addition, the H.264/AVC codec standard introduces an additional mandatory non-linear operation in the loop filter to counter blocking artifacts. In Lefol et al. (2006) it was found that motion compensation had a minimal effect, which is perhaps not surprising as its impact, though noticed, was not a source of concern for closed-loop MPEG-2 transcoders. However, the H.264/AVC mandatory in-loop de-blocking filter introduces severe effects by virtue of two thresholding operations. This is also unfortunate, as, compared to a cascaded encoder-decoder, the closed loop design applied to the relatively simple MPEG-2 codec, only dropped about 0.5 dB (Vetro, et al., 2003) (for the *Foreman* clip with CIF resolution at 30 fps). The more complex cascaded-pixel, closed-loop design appears to be the best prospect for an H.264/AVC bit-rate transcoder (Lefol, et al., 2006).

Scalable Video Coding

A scalable bit-stream is one in which parts of the stream can be removed but the remainder is still decodable. The stream becomes adaptive if a server is able to select in some way from a scalable bit-stream, so that the bitrate matches the bandwidth capacity and/or the display capabilities of the target device are met. This represents a considerable saving in storage over an adaptive simulcast solution. As an example (Bertone, et al., 2012), consider a single video providing six different quality levels doubling in size, ranging from 0.15 to 3 Mbps. Single-layer coding results in six different files for a total of 6.25 Mbps (0.15 + 0.3 + 0.5 + 0.8 + 1.5 + 3 Mbps). Scalable coding uses instead just a single file with a maximum size 20% bigger than the higher bitrate, requiring approximately 3.6 Mbps (-42% of storage space). However, against this must be balanced the very much reduced cost of storage space. However, in (Kalva, et al., 2012), as mentioned in the Introduction, it is shown that the overall monthly costs of streaming pre-encoded video depend not only on storage but on network streaming costs, which are greater for scalable video as a result of an overhead of between 10 and 20%. Thus, scalable video is only preferable in cost terms if the number of monthly sessions for streaming a film is low (around 16 sessions in Kalva, et al., 2012).

Provided a real-time codec is available (see Wien, et al., 2007, for a description of one), scalable video is appropriate for video conferencing (Eleftheriadis, et al., 2006) with a star-topology. Scalable encoding takes place at the end-points with base-layer transport through a privileged channel. Enhancement layers travel over a best-effort IP network, e.g. the Internet. Instead of a Multipoint Conferencing Unit, no decoding and re-encoding to form a composite picture takes place at the stream switching router, thus reducing delay and possible error accumulation. In that case, scalable video also allows adaptation to different device characteristics, as demonstrated by Vidyo Inc.'s video conferencing service in which incoming packets are combined to form a composite image at a participant's device. Notice that the low delay form of temporal scalability described below is required for video conferencing, as hierarchical B-pictures (described below) introduce unacceptable latency. A strong case has also been made (Wiegand, et al., 2009) for employing scalable coding for Internet Protocol TV (IPTV) that is streaming of TV content over a managed network with IP framing in the core and access network. However, that case is made against the use of transcoding at the user's device for resolution reduction, as this might have to operate at a speed to match the transfer capacity of the final link, which could be USB, Bluetooth (IEEE 802.15.1), Wifi (IEEE 802.11), and so on.

The *Scalable Video Coding (SVC)* extension (Schwarz, et al., 2007) to the H.264/AVC codec standard introduced a form of progressive encoding to H.264/AVC, similar to that of JPEG2000 (Taubman, 2000) (though with a wavelet transform in JPEG2000) for still images. In Wein et al. (2007), it was shown that, with optimized multi-layer encoder control (Schwarz & Wiegand,

2007), SVC's rate-distortion performance is about 0.5 dB below or a 10% rate increase over single-layer H.264/AVC (for a *Common Intermediate Format [CIF]* image with high temporal coding complexity). Notice though that there is a cost in coding complexity and single-pass coding is no longer possible. For a more detailed analysis of SVC complexity, refer to Wiegand et al. (2009).

An H.264/SVC bit-stream consists of a base layer and one or more nested enhancement layers. In fact, the base-layer of H.264/SVC is compatible with H.264/AVC, allowing the portion of the bitstream corresponding to the base layer to be extracted and sent to devices that do not support SVC. Scalability across the SNR (quality) layers is achieved by inter-layer reconstruction. Though this form of reconstruction improves coding efficiency considerably, it introduces dependencies between and within layers, which is a potential weakness. However, by allowing the exploitation of inter-layer dependencies the overhead of scalable coding is reduced. This overhead was identified (Wiegand, et al., 2009) as a principle reason why scalable coding options in prior standard codecs were not widely deployed.

Figure 3 shows an illustrative size-four SVC *Group of Pictures (GOP)* structure with hierarchical B-pictures (Schwartz, et al., 2006) between the key pictures. A key picture is a picture for which all previously coded pictures go before the key picture in display order. Unlike an *Instantaneous Decoder Refresh (IDR)* or I-picture, a key picture need not be intra-coded but may be predicted using a previous key picture as reference. Thus, an SVC GOP is different from an H.264/AVC GOP, as the GOP boundary may not be a point of random access. Key pictures form the coarsest temporal scalability layer (see below) and serve to delimit the extent of drift within motion-compensation decoding.

In H.264/SVC *Coarse-Grain Scalable (CGS)* quality layers, residual textural information (transform coefficients) from the prior layer is re-quantized at a finer resolution but there is a loss

Figure 3. Illustrative coding structure for GOP size of four with base and a quality enhancement layer, showing the prediction structure. Pictures appearing at 0, 4, and 8 are coded as key pictures.

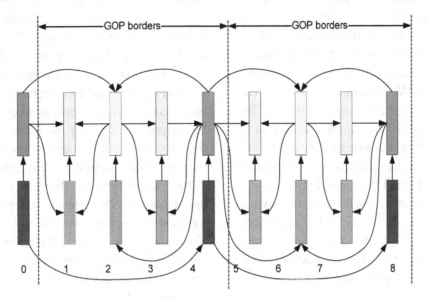

in rate switching flexibility, as layer switching may only occur at key pictures or IDR pictures (equivalent to intra-coded I-pictures in earlier standardized codecs). However, *Medium-Grained Scalability (MGS)* is additionally supported at the *Network Abstraction Layer Unit (NALU)* level. A NALU is an H.264/AVC virtual packet formed from data (Wenger, 2003) from the *Video Coding Layer (VCL)*. Thus, a NALU forms the payload of a network packet such as an RTP/UDP/IP packet. In MGS, switching between scalable layers is still restricted, as quality switching can only occur at the boundaries of NALUs. Notice also that MGS refinement takes place in the frequency domain, rather than in the spatial domain as for CGS (Wiegand, et al., 2009).

MPEG-4 part 2 (as opposed to MPEG-4 part 10 or H.264/AVC) also included a scalable mechanism called *Fine-Grained Scalability (FGS)* (Rahda, et al., 2001). FGS consisted of a base layer with one enhancement layer, which employed embedded scalability. A sender was able to select for each frame the base layer and additional bits of the FGS layer according to a desired bitrate.

However, though this arrangement results in a structure that is resilient to packet loss, as there are no inter picture dependencies at the enhancement layer, it introduces significant overhead. This is because motion-compensated prediction is only performed in the base layer. In fact, H.264/SVC also once included a fine-grained extension to MGS, in which the MGS bitstream within a NALU could be arbitrarily truncated. However, in Wien et al. (2007) it is shown that there is minimal gain in rate-distortion performance from including this added flexibility. Consequently, FGS was dropped from the SVC standard. Though MGS compared to MPEG-4 part 2 does improve on coding efficiency, it does introduce data dependencies between the layers, due to inter-layer prediction of motion vectors. In general, the consequence of more efficient coding (Auwera, et al., 2008) is that there is greater variation in the traffic rate of an H.264/SVC stream than an MPEG-4 part 2 stream. For example, in temporal scaling, this variability arises because the compressed size of I- and any P-pictures is much greater than intervening compressed hierarchical B-pictures (see below). Applying simple traffic smoothing is able to reduce the variability but not as much as for an equivalent MPEG-4 part 2. This may result in consequences for network dimensioning such as the need for larger buffers to absorb rate variability.

Temporal scalability in H.264/SVC was already present in H.264/AVC's hierarchical B-pictures. B-pictures are bi-predictive pictures that can be formed by forward or backward reference or a combination of both. Because of their coding efficiency in prior codecs, they acted as a kind of filler picture that could be dropped with no side effects on future pictures, while also improving the coding efficiency of the video sequence. However, in the H.264/AVC codec all pictures could be referenced. This allows a dyadic or other motion-compensated prediction structure to be built up within a GOP, terminated as previously mentioned by a key picture. Figure 3 illustrates a predictive structure using hierarchical B-pictures (though in practice larger GOP sizes of between 8 and 32 pictures are more efficient in coding terms in order to decrease the key frame rate). As previously mentioned, the key pictures form the coarsest temporal layer, with the pictures at positions 2 and 6 forming a temporal enhancement layer. Pictures at positions 1, 3, 5, and 7 form a higher temporal enhancement layer. Temporal prediction always occurs from the same or a coarser temporal layer. Notice that the B-pictures at positions 2 and 6 of the temporal enhancement layer are predicted from the key pictures of the previous GOP and the current GOP, instead of from adjacent pictures. In Figure 3, the encoding order would be pictures 0, 4, 2, 1, 3, and then in the second GOP 8, 6, 5, 7. The quantization parameter is also generally increased (resulting in lower quality) for each temporal enhancement layer. The changed encoding order adds to the decoding latency (which would be greater with a larger GOP size), making this form of coding structure unsuitable for the video conferencing application mentioned earlier. However, it is also possible to prevent forward references in order to

reduce latency (Wien, et al., 2007), thus employing hierarchical P-pictures rather than B-pictures.

It is also possible to 'decimate' the spatial resolution in passing from an enhancement layer to a lower layer. Higher spatial resolution layers are then intra predicted from lower spatial layers. Thus in Figure 3, the enhancement layer may be predicted from a lower-resolution base layer. At the base layer, not only may the frame rate be lower than that of other layers, but the spatial resolution may be coarser, as well as the quantization. This allows for combinations (Bertone, et al., 2012) of one or more of SNR, temporal, and spatial scalability between the layers, as is illustrated in Figure 4. The highest spatial layer, 720p representing 1280 × 720 pixels/frame progressively scanned HDTV, is at the same temporal rate as layer 3 (50/60 fps is more normal for HDTV). Layers 2, 3, and 4 have the same quantization level as layers 2, 3, and 4 with an illustrative-only spatial resolution. From this example, it is apparent that an enhancement might have the same scalability properties in two of the scalable dimensions but enhance the third scalable dimension.

However, there are restrictions to the extent of spatial scalability in one of H.264/SVC's profile configurations (Schwarz, et al., 2007). There are three profiles: Scalable Baseline, Scalable High, and Scalable High Intra. The Scalable High Intra profile allows only intra-code pictures, with no inter-picture predictive coding. It might be employed for transfer of video during studio editing. The Scalable High profile also has no spatial resolution restrictions and includes the possibility of cropping (Segall & Sullivan, 2007). In cropping, a spatial enhancement layer may contain additional parts of an image beyond that coded at the base layer. It is also possible to crop an enhancement-layer representation so that the base-layer spatial extent is greater than that of the enhancement layer. Of course, the spatial enhancement layer might also be coded with greater fidelity, resulting in region-of-interest coding suitable for (say) surveillance applications (Wien, et al.,

Figure 4. Illustration of mixed temporal, spatial, and SNR layering

Layer

4	Spatial: 720p @ 30 fps
3	Temporal: 360p @ 30 fps
2	SNR: 360p @ 15 fps
1	Spatial + Temporal: 360p @ 15 fps
0	Base layer: 240p @ 7.5 fps

2007). It is in the Scalable Baseline profile that there are restrictions on the spatial resolutions supported and the cropping support. In terms of H.264/AVC compatibility, Scalable Baseline's base layer corresponds to H.264/AVC's Baseline profile and similarly Scalable High corresponds to the High profile. However, notice that in the Scalable Baseline profile, apart for the above described scalable features, *Context Adaptive Binary Arithmetic Coding (CABAC)* is also assumed for resolutions above CIF, though this type of entropy coding is not present on most mobile devices.

It is probably in the reduction of overhead that H.264/SVC differs most from earlier standard scalable codecs. In effect, spatial scalability is implemented through a multi-resolution pyramid (Wien, et al., 2007), which delivers a scaled version of the input picture to each enhancement layer. An H.264/AVC encoder at each layer then codes a scaled version of the input picture, taking account of the temporal rate for that layer. However, coding is aided at higher layers by inter-layer communication. Thus, motion vectors for those pictures represented at a higher layer are passed to that higher layer. Similarly, the decoded output frames at a lower layer are passed up to a higher enhancement layer to aid in processing.

Rate Controllers

The dominant IP transport protocol, TCP, provides a reliable service, which unfortunately may result in large, possibly unbounded delays. Though normally regarded as unsuitable for video streaming, unless large buffers are available at the receiver to absorb delays, it is of interest as many experiments in congestion control have taken place under the cover of TCP. In TCP, packet loss traditionally signals the onset of network congestion, though employing loss to indicate congestion can be a performance bottleneck (Wei, et al., 2006). For transport of encoded video, packet loss is significant (Frossard & Vercheure, 2001), resulting in spatial error propagation within reference frames and spatial and temporal error propagation within other frame types. Re-synchronization points help the decoder. However, error concealment is implementation dependent and has itself latency implications for interactive applications. Nonetheless, TCP emulators (Widmer, et al., 2001) for video transport over UDP (to avoid unbounded delivery delay) do include feedback of a packet loss factor in their models. Over the years, streaming mechanisms such as Rate-Adaptive Protocol (RAP) (Rejaie, et al., 1999), TCP-Friendly Rate Control (TFRC) (Handley, et al., 2003), Loss-Delay Adaptation algorithm (LDA+) (Sisalem &

Wolisz, 2000), and TCP Emulation at Receivers (TEAR) (Rhee, et al., 2000) have in some measure based themselves on TCP's bandwidth probing mechanism, which results in packet loss before the rate is reduced.

Probe-based rate control (Wu, et al., 2000) tests the available bandwidth through acknowledgment-reported packet loss. One characteristic shared by rate-based methods is the aim to reproduce the average behaviour of TCP's *Additive Increase Multiplicative Decrease (AIMD)* congestion avoidance algorithm, to impart fairness amongst coexisting flows (Cai, et al., 2005). RAP and Binomial congestion control (Bansal & Balakrishnan, 2001) are examples of this approach. RAP uses acknowledgment packets to detect packet loss and infer *Round-Trip Times (RTTs)*. Every smoothed RTT, RAP implements an AIMD-like algorithm with the same thresholds and increments as TCP. Because this would otherwise result in TCP's 'sawtooth'-like rate curve, with obvious disruption to video streams, RAP introduces fine-grained smoothing, which takes into account short- and long-term RTT trends (by forming the ratio of the two). As RAP does not take account of time-outs for single loss detection, RAP's output is greater than equivalent TCP traffic during heavy congestion. RAPs form of rate control implies fixed-size packets, which are unsuitable for VBR video.

In TFRC, the sending rate is made a function of the measured packet loss rate during a single RTT duration measured at the receiver. The sender then calculates the sending rate according to the TCP throughput equation (Padhye, et al., 1998), using receiver measurements. TFRC was designed to produce smooth multimedia flows but because it assumes constant-sized large (*Maximum Transport Unit [MTU]* sized) packets it introduces a bias against variable-sized packets (and constant-sized small packets). TFRC's throughput model is sensitive to the loss probability and RTT, which are difficult to predict or measure. Another potential weakness is the response to short-term TCP flows, typically HTTP traffic, which never

develop long-term TCP flow behavior. TEAR also emulates TCP's sliding congestion window protocol, not directly but through smoothing the output of a window, to avoid TCP's sawtooth-like sending rate behaviour.

In LDA+, the AIMD increase and decrease parameters are adjusted by probing the available bandwidth through measurement of the inter-packet gap of two packets sent out back-to-back. Many measurement techniques (Dovrolis, et al., 2004) calculate either available bandwidth or capacity from packet inter-arrival times. These schemes rely on the network imposing some structure on the distribution of packet inter-arrival times, from which the network bandwidth characteristics can be inferred. The disadvantage of active probing schemes is the additional traffic, which not only further loads the network but can also distort the interpretation.

CASE STUDIES

This Section describes a number of brief case studies of adaptive video streaming in the experience of the authors.

Fuzzy Logic Adaptive Bitrate Control

Computational intelligence has been experimented with Jammeh et al. (2008) as an alternative to traditional video streaming rate controllers. Figure 5 is a block diagram of a Fuzzy-Logic Controller (FLC), with two inputs, a delay factor, *df*, and delay samples to form a trend. These inputs are converted to fuzzy form, whereby their membership of a fuzzy subset is determined by predetermined membership functions. This conversion takes place in the fuzzifier and trend analysis units of Figure 5. The fuzzy outputs are then combined in the inference engine through fuzzy logic. Fuzzy logic is expressed as a set of rules, which take the form of linguistic expressions. These rules express experience of tuning the controller and,

Figure 5. FLC delay-based congestion controller

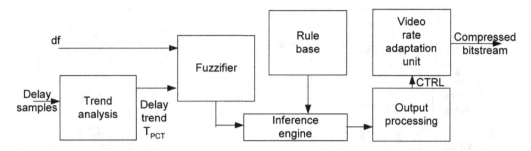

in the methodology, are captured in a rule base. The inference engine block is the intelligence of the controller, with the capability of emulating the human decision making process, based on fuzzy-logic, by means of the rule base, consisting of embedded rules for making those decisions. The output processing block converts inferred fuzzy control decisions from the inference engine to a crisp or precise value, which is converted to a control signal. Lastly, the rate of video is altered in the video rate adaptation unit. The compressed video bitstream is subsequently packetized prior to sending over the Internet.

Figure 6 shows a streaming architecture in which fuzzy logic controls the sending bit rate. The Congestion Level Determination (CLD) unit finds the congestion state of the network from measured delay samples and delay trend made by the timer module. The congestion state data are relayed to the sender and compared with the output inter-packet gaps. A FLC employs this delay information to compute a new sending rate that is a reflection of the current sending rate and the level of network congestion. The video rate adaptation unit (either by a bit-rate transcoder adapting pre-encoded video or by an encoder adapting the rate of live video through its quantization parameter) changes the sending rate to that computed by the fuzzy controller. The current implementation of the bitrate transcoder changes the quantization level of a frequency-domain transcoder (Assunção & Ghanbari, 1998) for VBR video.

The controller was evaluated through simulation with the ns-2 network simulator (version 2.28 used) (Issariyakul & Hossain, 2009). The simulated network, with a typical dumbbell topology, Figure 7, had a tight link between two routers and all side link bandwidths were provisioned such that congestion would only occur at the tight link. The one-way delay of the tight link was initially set to 5 ms and the side links' delays were set to 1 ms. The tight link router buffer's queuing policy was defaulted to be FIFO (droptail) and the queue size was set to twice the bandwidth-delay product, as is normal in such experiments to avoid packet losses from too small a buffer (The bandwidth-delay product measures the maximum amount of data that can occupy a link in one direction. A link is able to transmit in both directions; hence twice this product. A properly dimensioned buffer should be able to absorb any data while a link is cleared but not be so large that it causes delay or so small that arriving data is rejected because the link is fully occupied.). Sources in Figure 7 provide other traffic that competes with the video stream of interest.

Feedback was returned to the fuzzy controller after at least 40 ms had elapsed at the receiver, which corresponds to every frame at rate of 25 fps. As becomes evident in the reported tests, for larger network path delays (at a trans-continental scale), more frequent feedback messages would be needed to reduce the overall latency and improve the response. Some consideration to the stability of the control system might also be necessary in

Figure 6. Fuzzy logic congestion controlled video server

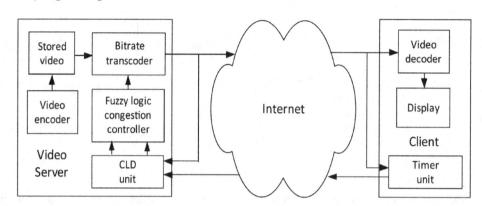

Figure 7. Classic dumbbell simulation topology used

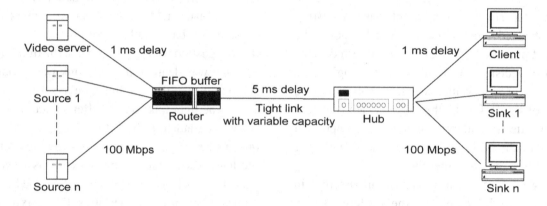

those circumstances. However, for expected streaming services within medium-sized networks, these measures are unlikely to be necessary.

The performance of the scheme was tested by changing the available bandwidth during a streaming session by injecting CBR background traffic at various rates. In Figure 8, it is clear that the FLC responds to the changes by varying the sender rate in a timely manner, after a brief period of adjustment. The ability to track a changing available bandwidth is illustrated by Figure 9. This is the same test as previously in Figure 8 but with a VBR *News* clip rather than a CBR stream under congestion control and with the objective video quality plotted at the video sink rather than the sending rate. *News* consists of a presenter and changing background, with

moderate motion. The videos were encoded using the MPEG-2 codec at an original bit-rate (before transcoding) of 1 Mbit/s. European SIF-format was used (progressive, 25frames/s, 352 × 288 pixel/ frame). The pictures were divided into eighteen per row of macroblock slices, with one slice in each packet for error resilience purposes. Also shown in Figure 9 is the CTRL output from the FLC module (see Figure 5), plotted against the left-hand vertical axis. Again, apart from small over-rides at the time of an abrupt upward available bandwidth change and some flattening out of the rate after a downward change, tracking accuracy is maintained. In practice, available bandwidth would not change as abruptly as happens in this test. Optimization of the fuzzy models will also further adjust for these small inconsistencies.

Ten experiments were conducted for a bottleneck capacity of 1 Mbps, with a 5 ms delay across the link. In the first experiment, only one TCP long-lived source was present as background traffic. In the second experiment, two TCP long-lived sources acted as background traffic and so on until a number of short-lived sources were eventually introduced, so that all ten TCP sources were on as background traffic for the tenth experiment. Figure 10 summarizes the results, showing the correspondence between the mean over time of the bandwidth occupied by the video source (plotted against the left-hand vertical axis) and the mean available bandwidth. The data points are not evenly spread out because of the larger bandwidth contribution of the long-lived sources compared to the short-term sources. Some data points, at low available bandwidth, are partially superimposed in the plot. For convergence, the data points represent the mean of twenty independent runs. From the plot, the sending rate of the video is never above the available bandwidth, but tracks the available bandwidth throughout the experiments, leading to low packet losses. Consequently, the video quality (mean PSNR over time plotted on the right-hand vertical axis) follows the rate trend and does not suffer any degradation in quality.

Statistical Multiplexing over a Satellite Link

The second study (Ahmadi Aliabad, et al., 2010) concerns dynamically adapting the rate of multiplexed CBR video streams, so that the individual video stream rates more closely match their time-wise coding complexity. In statistical multiplexing, a constant bitrate, for example that available to a satellite's transponder's bandwidth, is allocated according to the coding complexity of the constituent video streams. Efficient statistical multiplexing can improve received video quality at the receiver, and may even increase the number of TV channels carried (Tanberg Television, 2008) by a transponder. In business terms, it is acknowledged that the revenue that can be potentially generated from combining video streams within a multimedia channel (Seeling & Reisslein, 2005) is related to the quality of the video delivered to end users. Statistical multiplexing can reduce deep quality fades (Kuhn & Antkowiak, 2000), thus increasing the quality of experience. Though there may be times at which the content of a majority of mul-

Figure 8. Tracking a varying available bandwidth

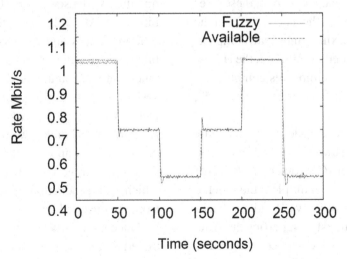

Figure 9. FLC congestion control of the 'news' sequence with changing available bandwidth and including the CTRL output from the FLC

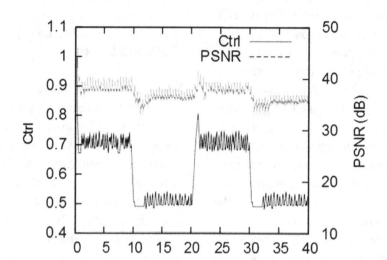

tiplexed TV channels demands a large bandwidth allocation, the essence of well-managed statistical multiplexing is that the duration of these intervals is short.

In statistical multiplexing with H.264/AVC, the same MPEG-2 TS (Wenger, 2003) as previously employed by the MPEG-2 codec itself can be used. The MPEG-2 TS (Fischer, 2008) can assemble up to 6 or 10 or even 20 independent television programs, depending on compression ratios. Programs bitrates can be constant or variable. In the case of variable data rate, these rates can be controlled based on the requirements of the system prior to multiplexing (statistical multiplexing). However, compared to MPEG-2, the H.264/AVC codec significantly improves compression ratios (Ghanbari, et al., 2006) by as much as 50%, especially for SDTV.

Assuming live or pre-encoded CBR video, the basic system of this case study is based on finding the Number of Non-Zero Coefficients (NNZC) of the input sequences to the multiplex (the scheme can be modified to work for VBR video). These coefficients are those transform coefficients in the encoded bitstream that have not been reduced to zero by the quantization process. The NNZC allow the spatial complexity of the video sequence to be judged, and by implication, the coding complexity required to achieve a given quality. Notice that coding complexity is a measure of the coding bits required for compression and not the computational complexity. Specifically the NNZC of an individual macroblock was found to be logarithmically proportional to the coding complexity of that macroblock. Notice that because the case study system is intended for broadcast quality TV, the spatial coefficients dominate the bitstream. At low bit rates, the data given over to motion vectors and headers in the compressed bitstream (Seeling & Reisslein, 2005) must be taken into account. As this is a no-reference system, a method is required to estimate the objective video quality (*Peak Signal-to-Noise Ratio [PSNR]*) based on knowledge of the NNZC. To do so requires an estimate of the video quality from the average Quantization Parameter (QP), which, in a production system, can be extracted from the encoded bitstream without full decode.

Once the relative coding complexity is determined across the sequences, the video quality is equalized across the input sequences. This operation is performed at each GOP boundary, though a refined version could also include scene change detection. The CBR rates are subsequently adjusted on a GOP-by-GOP basis to produce what might be called 'semi-CBR-VBR' streams. Research in Böröczy et al. (1999) also presents a CBR multiplex of streams previously stored at a high quality for the MPEG-2 codec. In implemented systems, such as that from Scopus (Scopus Video Networking, 2006) for the MPEG-2 codec, VBR video can be smoothed (Zhang, et al., 1997) prior to complexity analysis. However, it is important to notice that H.264/AVC video bitstreams have been found to be significantly more variable (Auwera, et al., 2008) than even MPEG-4 part 2 streams, due to the variety of coding modes available in H.264/AVC. It is also reported (Auwera, et al., 2008a) that, after H.264/AVC frame size smoothing, the output remained significantly more variable than unsmoothed MPEG- 4 part 2 output for the same films. CBR encoding allows planning of storage capacity and in video-on-demand schemes, it allows the bandwidth from a server to be tightly controlled. If the CBR video is not pre-encoded

at a high rate (prior to transcoding) then image 'dissolves,' fast 'action' and scenes with camera motion (pans, zooms, tilts, ...) all suffer. However, scenes with limited motion such as head-and-shoulder news sequences are not much affected by CBR encoding.

The envisaged system is shown in Figure 11 in which a bitrate transcoder bank modifies the input after NNZC statistics have been extracted in the compressed domain. For an example of a commercial transcoder bank for a different purpose refer to Kasai et al. (2002). In Figure 11, the statistical multiplexor receives n compressed bitstreams, which pass through a bank of bit-rate transcoders to adjust the combined bitrate according to the output channel constraint. The bandwidth share is defined by the statistical bandwidth manager, which receives content complexity measures (parameters) from each transcoder and returns the appropriate bandwidth share (α).

Because the system is GOP-based it is probably easily integrated into GOP-based call-admission control for bandwidth allocation systems for satellite channels (De Rango, et al., 2008), replacing H.264/AVC for MPEG-2 streams. In a *Digital*

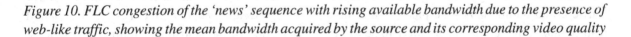

Figure 10. FLC congestion of the 'news' sequence with rising available bandwidth due to the presence of web-like traffic, showing the mean bandwidth acquired by the source and its corresponding video quality

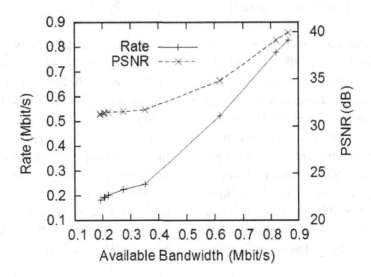

Figure 11. Statistical multiplexor architecture

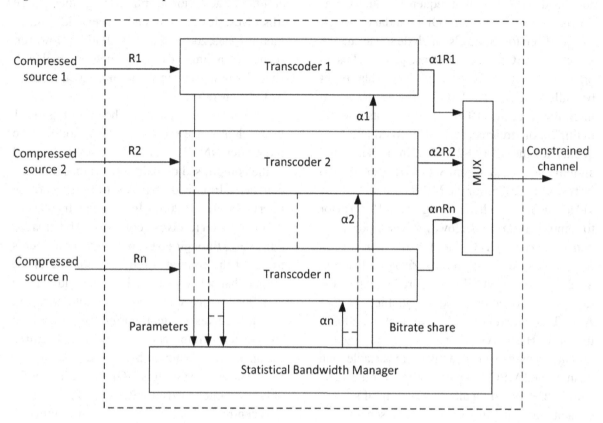

Video Broadcasting (DVB)-Return Channel via Satellite (DVB-RCS) system (Neal, et al., 2001), statistical multiplexing of aggregated user video traffic could additionally take place at the terrestrial hub, if the return channel were to be used for user video applications such as remote learning.

The scheme was evaluated through comparison of the system output quality with conventional equal allocation of bitrate multiplexing. In order to test the scheme, combined sequences of a set of reference video sequences were CBR encoded at 3 Mbit/s. Each of three combined sequences contained 900 frames consisting of FNS (*Foreman + News + Stefan*), NMF (*News + Mobile + Foreman*) and WHB (*Flower + Highway + Bus*). The average PSNR over all frames was calculated for both statistical multiplexing and constant CBR mode, as recorded in Table 1. The overall quality

difference is 0.42 (dB) gained by the statistical multiplexing scheme. The standard deviation of the PSNR is also calculated as a measure of variation of the PSNR around its average. The PSNR standard deviation of the statistical multiplexing scheme is lower by a considerable extent. This result illustrates the success of the scheme in equalizing the quality across the three sequences. Because of the quality fade for the observable constant CBR and because of the low overall quality of the WHB sequence (see Table 1), it would be necessary to send just two sequences at a rate of 1.5 Mbps each for fixed rate transport. Therefore, for this example, the number of programs that can be sent at 40 dB has been increased by a factor of 1.5.

A further gain from using the scheme was found to be a reduction in bitrate fluctuations arising from the H.264/AVC reference *Joint Model (JM)* codec software behavior in CBR

Table 1. Per sequence average PSNR and standard deviation of the sequences

	Statistical multiplexing		CBR	
Sequence	Average PSNR (dB)	PSNR standard deviation (dB)	Average PSNR (dB)	PSNR standard deviation (dB)
FNS	40.57	3.00	41.70	5.38
NMF	40.39	3.47	40.21	7.39
WHB	40.04	2.10	38.68	3.68

mode. The codec selects I- and B-frame QPs intelligently to avoid too rapid transitions in quality. Unfortunately, if at scene changes the new scene is more complex, then a large fluctuation in bitrate occurs. It was observed that the proposed scheme results in smoother resulting bitrate fluctuations than when requests are made for a uniformly constant bitrate. For 40 dB PSNR video streams, the scheme significantly reduces PSNR fluctuations. True VBR video may result in too frequent bitrate oscillations and in this case study, a GOP-by-GOP semi-VBR scheme has been investigated. This implies that the latency of the scheme is the time taken to process a GOP, normally about 0.5s.

Adaptive FEC for Broadband Wireless

IPTV generally supports two services: multicast delivery of popular TV channels and a video-on-demand, added-value service employing unicast delivery of individual streams. In this case study, we consider an adaptive unicast service, which adapts the extent of FEC protection across a mobile access network, and the effect of removing the adaptation for the multicast service. Thus, two services are provided for the price of one (Al-Jobouri, et al., 2012a).

If IPTV is to support a mobile service it must address the much increased risk of wireless channel errors. Clearly, compared to unicast, multicast improves bandwidth efficiency by sharing video

packets delivered through a network. However, unicast streaming permits individualized error protection for each stream, thus avoiding the principal detraction of multicasting. As previously mentioned, with multicast streaming there is a risk of a feedback implosion at the server, an overload of network resources due to the attempts of many receivers trying to send repair requests for a single packet.

Instead, this study asks whether the same protection scheme can be employed for unicast and multicast by turning *Automatic Repeat Requests (ARQs)* for repair packets on and off according to the casting mode. By testing the same adaptive, *Application Layer (AL)* FEC scheme for both unicast and multicast one can judge what the impact on video-quality penalty is from turning off repair packets in multicast mode. A related question is what level of fixed AL-FEC is necessary to improve video quality, compared to an adaptive scheme that responds to wireless channel conditions. Of course, in a short study one is not able to show all possible objective video qualities (PSNR) but it is still possible to say that the fixed-rate AL-FEC datarate overhead is about 5% more compared to an adaptive FEC scheme, as simulated results for IEEE 802.16e (mobile WiMAX) demonstrate.

Both the unicast and multicast modes of the scheme use adaptive rateless channel coding for AL-FEC. In the scheme, the probability of channel Byte Loss (BL), based on continuous serves to predict the amount of redundant data to be added to the payload. In an implementation, BL is found through measurement of channel conditions. As channel measurement is known to be inaccurate, in simulations, 5% normally distributed noise was added to the estimate of BL. If the original packet length is L, then the redundant data is given simply by

$$R = L \times BL + (L \times BL^2) + (L \times BL^3)...$$
$$= L / (1 - BL) - L.$$

$$(1)$$

The following statistical model (Luby, et al., 2007) was used to model the repair probability of the Raptor variety (Shokorallahi, 2006) of rateless channel coding:

$$P_f(m,k) = 1 \text{ if } m < k$$
$$= 0.85 \times 0.567^{m-k} \text{ if } m \geq k \qquad (2)$$

where is the decode failure probability of the code with k source symbols if m symbols have been successfully received. If a packet cannot be decoded despite the provision of redundant data then in the unicast mode of the scheme (not in the multicast mode) extra redundant data are added to the next packet. It is implied from (2) that if less than k symbols (bytes) in the payload are successfully received then a further $k - m + e$ redundant bytes can be sent, 'piggybacked' on the next packet, to reduce the risk of failure. In simulations, e was set to four, which if the extra data in the repair packet successfully arrive, reduces the risk of packet loss to 8.7%, because of the exponential decay of the risk that is evident from equation (2).

The tests were performed on the reference sequences *Paris* (moderate spatial-coding complexity) and *Football* (high temporal coding-complexity) video sequences encoded in CIF resolution at 30 Hz. In the H.264/AVC codec used, data partitioning was enabled, such that every slice was divided into three separate partitions according to their importance in reconstruction at the decoder. The GOP structure was IPPP..., i.e. one initial intra-coded I-picture and all predictively-coded P-pictures. By default, 2% intra-refresh macroblocks were randomly inserted within each P-picture to restrict temporal error propagation. *Constrained Intra Coding (CIP)* was turned on to ensure independent decoding of partition-B (containing inter-coded residual transform coefficients). For comparison, CBR streaming was tested at two different rates.

The video stream was simulated to be transmitted from a WiMAX base station to an IEEE

802.16e Mobile Station (MS), and, to provide sources of traffic congestion, a permanently available FTP source was introduced with TCP transport to a second MS. Likewise, a CBR source with packet size of 1000 B and inter-packet gap of 0.03s was also downloaded to a third MS. The simulations adopted the mandatory settings for a 10.67 Mbps *Downlink (DL)* rate with 3:1 DL/UL sub-frame ratio with WiMAX frame size of 5 ms, 16-QAM ½ modulation over a 10 MHz channel with IEEE 802.16e recommended antenna heights and transmit/receive powers. The channel model was deliberately kept simple for comparison purposes, with a Uniform distribution, *BLmean* = 0.008. This model can still potentially result (if not corrected) in packet losses of up to 10%.

Table 2 is the result of adaptive Raptor channel coding of the two video streams at two different streaming rates. The packet drop rates represent those packets that were corrupted by channel noise but could not be repaired. 'Corrupted packets' represent those packets that were affected by channel noise but that were repaired. This has resulted in no more than 20 ms delay before a packet is declared successfully received, compared to no more that 10 ms for uncorrupted packets, both delays being acceptable over normal path lengths. The objective video quality in all cases was good (over 30 dB), with the higher motion video clip benefitting most from the adaptive FEC scheme. The effect of withdrawing repair packets

Table 2. Unicast CBR streaming with adaptive raptor coding and repair packets

CBR datarate:	Football 500 kbps	Football 1 Mbps	Paris 500 kbps	Paris 1 Mbps
Packet drops (%)	0.6	0.88	1.28	1.66
Packet delay (s)	0.0068	0.0083	0.0067	0.0084
Mean PSNR (dB)	31.21	36.26	33.16	35.48
CPs (%)	2.94	8.07	4.11	8.55
CPs delay (s)	0.0163	0.0188	0.0163	0.0175

is demonstrated in Table 3 as a representation of multicast delivery. The main implication is a decrease in video quality. This is most likely to impact on clips with higher motion. Objective video quality less than 25 dB is judged as poor. Thus, again content type is critical.

We subsequently used Raptor coding at a fixed rate and raised the rate until all clips were received at greater than 30 dB PSNR. As recorded in Table 4, to achieve this in the given channel conditions implies a redundant data rate of 13.5%. In comparison, Table 5 gives the mean overhead from the adaptive scheme resulting in the performance of Tables 2 and 3. Notice that content type remains important: the overhead is higher when streaming the more complex *Football*. In respect to multicast delivery, the adaptive scheme cannot completely anticipate the impact of the channel but a fixed-rate overhead of 13.5% is high. Therefore, the desirable redundant rate probably lies between the purely adaptive rate and the given fixed rate.

FUTURE RESEARCH DEVELOPMENTS

For cost reasons rather than technical reasons, there is a swing towards adaptive simulcast video streaming. However, specifications such as DASH (Stockhammer, 2011), which is supported by the source coding community in the form of *Motion Picture Experts Group (MPEG)* and the cellular wireless community in the form of the *3rd Generation Partnership Project (3GPP)*, do not prescribe how source coding can be best performed. Thus, content-aware coding (Adzic, et al., 2012) which takes into account scene changes and content type is a potential area of investigation. However, there are other 'parameters' that are unclear in this form of video streaming such as the duration of each video chunk and the number of video streams to include. The rate of quality transition once a client request has been made is also worthy of investigation.

Table 3. Multicast CBR streaming with adaptive raptor coding without repair packets

CBR datarate:	*Football* 500 kps	*Football* 1 Mbps	*Paris* 500 kbps	*Paris* 1 Mbps
Packet drops (%)	5.11	7.55	5.25	11.28
Packet delay (s)	0.0068	0.0083	0.0068	0.0084
PSNR (dB)	21.97	25.61	29.26	28.69
CPs (%)	0	0	0	0
CP delay (s)	0	0	0	0

Table 4. Multicast CBR streaming with raptor coding at fixed-rate (13.5%) without repair packets

CBR datarate:	*Football* 500 kbps	*Football* 1 Mbps	*Paris* 500 kbps	*Paris* 1 Mbps
Packet drops (%)	0.25	2.30	1.55	2
Packet delay (s)	0.00698	0.00877	0.06921	0.00915
Mean PSNR (dB)	31.77	30.49	33.39	32.97
CPs (%)	0	0	0	0
CPs delay (s)	0	0	0	0

Table 5. Mean overhead from adaptive channel coding for multicast CBR streaming

Video trace/CBR datarate:	Average datarate overhead % for multicast, adaptive Raptor code
Football 500 kbps	8.05
Football 1 Mbps	8.79
Paris 500 kbps	7.53
Paris 1 Mbps	7.78

Another issue that has come to prominence is how to reduce energy consumption in video streaming, given the growing dominance of this form of streaming within the Internet. To reduce energy at mobile receivers, burst mode transmission has been specified in the *Digital Video Broadcasting-Handheld (DVB-H)* (Kornfeld & Reimers, 2005) standard for terrestrial broadcasting. The same concept has been applied to the wired Internet with the IEEE 802.3az standard for energy efficient Ethernet, which allows Ethernet receivers to enter a sleep mode, just as with their mobile counterparts. Whereas scheduling of video stream packets at a video server for entry onto a network has sometimes been considered a problem, it could be that this form of burst scheduling should be encouraged to improve energy consumption. Thus, video streaming should adapt itself to energy conservation (Razavi, et al., 2008).

The successor codec to H.264/AVC, known as High Efficiency Video Codec (HEVC) (Sullivan & Ohm, 2010), is under active development. It promises significant gains, perhaps eventually over 50%, in coding efficiency over H.264/AVC. At the time of writing, high-definition (1024×768 pixel/frame progressive at 24 frame/s) broadcast quality video, i.e. with intra refresh, is reported (Li, et al., 2011) as having an average coding gain of 44%, low-delay coding gain for the same resolution is reported to be on average 48% better, while all-Intra frame coding gain is 26%. However, as yet HEVC development has concentrated on compression efficiency and not network aspects, which will no doubt be added by the target finalization date of November 2013. In Nightingale et al. (2012), a prototype, the Test Model under Consideration (TMuC HM4.0), was tested with some additional resiliency features and simple error concealment at the decoder in a streaming environment in order to establish a benchmarking framework. Improved coding gain increases the need to protect the video stream during streaming if some form of adaptive HTTP streaming is not used. For applications such as video call and video conferencing it is likely that true streaming will be preferred, which increases the need for effective adaptive streaming. For stored video, the increased coding efficiency of HEVC implies that higher quality video can be transported directly by TCP. This is because if TCP needs approximately twice the available bandwidth as its maximum rate to be delivered at video rate, then improved coding efficiency implies high-quality video can be transported in a satisfactory manner, without delays.

Finally, the multimedia cloud (Zhu, et al., 2011) is being promoted by large organizations, not just as a form of media storage but also as a way that thin clients, typically mobile devices, could offload processing to a virtual processing facility. For example, in user-to-user video streaming it is unrealistic to expect a source mobile device to be able to convert to the format that can be handled by the receiving device. However, a cloud transcoding bank facility represents a way in which that could be achieved.

CONCLUSION

Adaptive video streaming remains a vibrant area of investigation. One indication of that is the shift from push-oriented streaming in which the server makes decisions about the receiving rate of the client to pull-oriented streaming in which the download logic is moved to the client. Thus, a new set of considerations needs to be considered such as the likely effect on network congestion of this arrangement. Moving away from client-server architectures, there is also an alternative architecture in peer-to-peer streaming. Pure peer-to-peer streaming for practical reasons is rare but hybrid forms with some privileged peers (or super-peers) are certainly around and have been very successful for low-quality TV streaming. In fact, peer-to-peer streaming has in some cases been integrated with content-delivery networks, though this may be easier to achieve in semi-command driven economies. Thus, peer-to-peer systems and their variants are certainly an area in which adaptive video streaming should be considered. 3D video streaming in particular may be well suited to peer-to-peer streaming.

REFERENCES

Abhari, A., & Soraya, M. (2010). Workload generation for YouTube. *Multimedia Tools and Applications*, *46*, 91–118. doi:10.1007/s11042-009-0309-5

Adobe. (2010). *HTTP dynamic streaming on the Adobe flash platform*. Retrieved from http://www.adobe.com/products/httpdynamicstreaming/pdfs/httpdynamicstreaming_wp_ue.pdf

Adzic, V., Kalva, H., & Furht, D. (2012). Content aware video encoding for adaptive HTTP streaming. In *Proceedings of the IEEE International Conference on Consumer Electronics*, (pp. 94-95). IEEE Press.

Ahmad, I., Wei, X., Sun, Y., & Zhang, Y. W. (2005). Video transcoding: An overview of various techniques and research issues. *IEEE Transactions on Multimedia*, *7*(5), 793–804. doi:10.1109/TMM.2005.854472

Ahmadi Aliabad, H., Moiron, S., Fleury, M., & Ghanbari, M. (2010). No-reference H.264/AVC statistical multiplexing for DVB-RCS. In *Proceedings of the 2nd International ICST Conference on Personal Satellite Services,* (pp. 163-178). ICST.

Al-Jobouri, L., Fleury, M., & Ghanbari, M. (2012). Comprehensive protection of data-partitioned video for broadband wireless IPTV streaming. *Mobile Information Systems*, *8*(2), 1–23.

Al-Jobouri, L., Fleury, M., & Ghanbari, M. (2012a). Versatile IPTV for broadband wireless with adaptive channel coding. In *Proceedings of the IEEE International Conference on Consumer Electronics*, (pp. 344-345). IEEE Press.

Alcock, S., & Nelson, R. (2011). Application flow control in YouTube video streams. *ACM SIGCOMM Computer Communication Review*, *41*(2), 25–30. doi:10.1145/1971162.1971166

Assunção, P. A. A., & Ghanbari, M. (1997). Transcoding of single-layer MPEG video into lower rates. *Proceedings on the Institute of Electronics Engineers: Vision, Image, and Signal Processing*, *144*(6), 377–383. doi:10.1049/ip-vis:19971558

Assunção, P. A. A., & Ghanbari, M. (1998). A frequency domain video transcoder for dynamic bit-rate reduction of MPEG-2 bit streams. *IEEE Transactions on Circuits and Systems for Video Technology*, *8*(8), 953–967. doi:10.1109/76.736724

Bansal, D., & Balakrishnan, H. (2001). Binomial congestion control algorithms. In *Proceedings of IEEE INFOCOM* (pp. 631–640). IEEE Press.

Bertone, F., Menkovski, V., & Liotta, A. (2012). Adaptive P2P streaming. In M. Fleury & N. Qadri (Eds.), *Streaming Media with Peer-to-Peer Networks: Wireless Perspectives*. Hershey, PA: IGI Global.

Bing, B. (2010). *3D and HD broadband video networking*. Norwood, MA: Artech House.

Böröczy, L., Ngai, A. Y., & Westermann, E. F. (1999). Statistical multiplexing using MPEG-2 video encoders. *IBM Journal of Research and Development*, *43*(4), 511–520. doi:10.1147/rd.434.0511

Braden, R., Zhang, L., Berson, S., Herzog, S., & Jamin, S. (1997). *Resource reservation protocol (RSVP)*. IETF RFC 2205. Retrieved from http://www.ietf.org/rfc/rfc2205.txt

Broadband Forum. (2007). *ADSL2/ADSL2plus performance test plan*. Technical Report, TR-100. Broadband Forum.

Cai, L., Shen, X., Pan, J., & Mark, J. W. (2005). Performance analysis of TCP-friendly AIMD algorithms for multimedia applications. *IEEE Transactions on Multimedia*, *7*(2), 339–355. doi:10.1109/TMM.2005.843360

Chang, S.-F., & Eleftheriadis, A. (1994). Error accumulation of repetitive image coding. In *Proceedings of the IEEE Symposium on Circuits and Systems,* (Vol. 3, pp. 201-204). IEEE Press.

Chang, S. F., & Messerschmidt, D. G. (1995). Manipulation and compositing of MC-DCT compressed video. *IEEE Journal on Selected Areas in Communications*, *13*(1), 1–11. doi:10.1109/49.363151

Chow, C.-O., & Ishii, H. (2007). Enhancing real-time video streaming over mobile ad hoc networks using multipoint-to-point communication. *Computer Communications*, *30*(8), 1754–1764. doi:10.1016/j.comcom.2007.02.004

Cicco, L. D., Mascolo, S., Bari, P., & Orabona, V. (2010). An experimental investigation of the Akamai adaptive video streaming. In *Proceedings of the 6ᵗʰ International Conference on HCI in Work and Learning, Life and Leisure,* (pp. 447-464). HCI.

Conklin, G., Greenbaum, G., Lillevold, K., Lippman, A., & Reznik, Y. (2001). Video coding for streaming media delivery over the Internet. *IEEE Transactions on Circuits and Systems for Video Technology*, *11*(3), 269–281. doi:10.1109/76.911155

de Prycker, M. (1993). *Asynchronous transfer mode: Solutions for broadband ISDN*. Upper Saddle River, NJ: Prentice Hall.

De Rango, F., Tropea, M., Fazio, P., & Marano, S. (2008). Call admission control for aggregate MPEG-2 traffic over multimedia geo-satellite networks. *IEEE Transactions on Broadcasting*, *54*(3), 612–622. doi:10.1109/TBC.2008.2002716

Dovrolis, C., Ramanathan, P., & Moore, D. (2004). Packet dispersion techniques and a capacity-estimation methodology. *IEEE/ACM Transactions on Networking*, *12*(6), 963–977. doi:10.1109/TNET.2004.838606

Du, Q., Shang, S., Lu, H., & Tang, X. (2008). A fast MPEG-2 to H.264 downscaling transcoder. In *Proceedings of the 8th Conference on Signal Processing, Computational Geometry, and Artificial Vision*, (pp. 230-233). IEEE.

Eleftheriadis, A., & Anastassiou, D. (1995). Constrained and general dynamic rate shaping of compressed digital video. In *Proceedings of the IEEE International Conference on Image Processing,* (vol. 3, pp. 396–399). IEEE Press.

Eleftheriadis, A., Civanlar, R., & Shapiro, O. (2006). Multipoint videoconferencing with scalable video coding. *Journal of Zhejiang University – Science A, 7*(5), 696-705.

Fischer, W. (2008). *Digital video and audio broadcasting technology* (2nd ed.). Berlin, Germany: Springer-Verlag.

Frossard, P., & Vercheure, O. (2001). Joint source/FEC rate selection for quality optimal MPEG-2 video delivery. *IEEE Transactions on Image Processing*, *10*(12), 1815–1825. doi:10.1109/83.974566

Ghanbari, M., Fleury, M., Khan, E., et al. (2006). *Future performance of video codecs*. London, UK: Office of Communications (Ofcom).

Gill, P., Arlitt, M., Li, Z., & Mahant, A. (2007). YouTube traffic characterization: A view from the edge. In *Proceedings of the 7th ACM SIGCOMM Conference on Internet Measurement,* (pp. 15-28). ACM Press.

Goyal, V. K. (2001). Multiple description coding: Compression meets the network. *IEEE Signal Processing Magazine*, *18*(5), 74–93. doi:10.1109/79.952806

Gürler, C. G., Görkemli, B., Saygili, G., & Tekalp, M. (2011). Flexible transport of 3D video over networks. *Proceedings of the IEEE*, *99*(4), 694–707. doi:10.1109/JPROC.2010.2100010

Handley, M., Floyd, S., Padhye, J., & Widmer, J. (2003). TCP friendly rate control (TFRC) protocol specification. RFC 3448. Retrieved from http://www.ietf.org/rfc/rfc3448.txt

Hartung, F., & Girod, B. (1998). Watermarking of uncompressed and compressed video. *IEEE Transactions on Signal Processing, 66*(3), 283–301.

Hwang, J.-N. (2009). *Multimedia networking: From theory to practice*. Cambridge, UK: Cambridge University Press. doi:10.1017/CBO9780511626654

ISO/IEC. (2000). International standard 13818-2 information technology — Generic coding of moving pictures and associated audio information: Video (mpeg-2/h.262) video buffering verifier. In *Annex C* (2nd ed.). Washington, DC: ISO.

Issariyakul, T., & Hossain, E. (2009). *Introduction to the ns2 simulator*. Berlin, Germany: Springer Verlag. doi:10.1007/978-0-387-71760-9

ITU-T. (1997). *Hypothetical reference decoder: Video coding for low bit rate communication*. Annex B. Retrieved from http://www.itut.int

Jammeh, E., Fleury, M., & Ghanbari, M. (2008). Fuzzy logic congestion control of transcoded video streaming without packet loss feedback. *IEEE Transactions on Circuits and Systems for Video Technology, 18*(3), 387–393. doi:10.1109/TCSVT.2008.918459

Jarnikov, D., & Özçelebi, T. (2011). Client intelligence for adaptive streaming solutions. *Signal Processing Image Communication, 26*(7), 378–389. doi:10.1016/j.image.2011.03.003

Kalva, H., Adzic, V., & Furht, B. (2012). Comparing MPEG AVC and SVC for adaptive HTTP streaming. In *Proceedings of the IEEE International Conference on Consumer Electronics*, (pp. 160-161). IEEE Press.

Karczewicz, M., & Kurceren, R. (2003). The SP-and SI-frames design for H.264/AVC. *IEEE Transactions on Circuits and Systems for Video Technology, 13*(7), 637–644. doi:10.1109/TCSVT.2003.814969

Kasai, H., Nilsson, M., Jebb, T., Whybray, M., & Tominaga, H. (2002). The development of a multimedia transcoding system for mobile access to video conferencing. *IEICE Transactions on Communications, 10*(2), 2171–2181.

Kim, J., & Kang, S. (2011). An ontology-based personalized target advertisement system on interactive TV. In *Proceedings of the IEEE International Conference on Consumer Electronics*, (pp. 895-896). IEEE Press.

Kornfeld, M., & Reimers, U. (2005). DVB-H — The emerging standard for mobile data communication. *EBU Technical Review*. Received from http://tech.ebu.ch/docs/techreview/trev_301-dvb-h.pdf

Kuhn, M., & Antkowiak, J. (2000). *Statistical multiplex what does it mean for DVB-T?* Berlin, Germany: FKT Fachzeitschrift für Fernsehen, Film und Elektronische Medien.

Kumar, A. (2007). *Mobile TV: DVB-H, DMB, 3G systems and rich media applications*. Amsterdam, The Netherlands: Focal Press.

Kurutepe, E., Aksay, A., Bilen, C., Gürler, C. G., Sikora, T., Akar, G. B., & Tekalp, A. M. (2007). A standards-based, flexibile, end-to-end multiview streaming architecture. In *Proceedings of the International Packet Video Workshop*, (pp. 302-307). IEEE.

Kurutepe, E., Civanlar, M. R., & Tekalp, A. M. (2007a). Client-driven selective streaming of multiview video for interactive 3DTV. *IEEE Transactions on Circuits and Systems for Video Technology, 17*(11), 1558–1565. doi:10.1109/TCSVT.2007.903664

Lakshman, T. V., Ortega, A., & Reibman, A. R. (1998). VBR video: Trade-offs and potentials. *Proceedings of the IEEE, 86*(5), 952–973. doi:10.1109/5.664282

Lefol, D., Bull, D., & Canagarajah, N. (2006). Performance evaluation of transcoding algorithms for H.264. *IEEE Transactions on Consumer Electronics, 25*(1), 215–222.

Li, B., Sullivan, G. J., & Xu, J. (2011). Comparison of compression performance of HEVC working draft 4 with AVC high profile. In *Proceedings of 7th Meeting of Joint Collaborative Team on Video Coding (JCT-VC).* JCT-VC.

Liu, Z., Wu, Z., Liu, H., & Stein, A. (2007). A layered hybrid-ARQ scheme for scalable video multicast over wireless networks. In *Proceedings of the Asilomar Conference on Signals, Systems and Computers,* (pp. 914-919). Asilomar.

Luby, M., Gasiba, T., Stockhammer, T., & Watson, M. (2007). Reliable multimedia download delivery in cellular broadcast networks. *IEEE Transactions on Broadcasting, 53*(1), 235–246. doi:10.1109/TBC.2007.891703

Ma, S., Gao, W., Wu, F., & Lu, Y. (2003). Rate control for JVT video coding scheme with HRD considerations. In *Proceedings of the IEEE International Conference on Multimedia and Expo,* (pp. 793-796). IEEE Press.

Meesters, L. M. J., Ijsselsteijn, W. A., & Seuntiems, P. J. H. (2004). A survey of perceptual evaluations and requirements of three-dimensional TV. *IEEE Transactions on Circuits and Systems for Video Technology, 14*(3), 381–391. doi:10.1109/TCSVT.2004.823398

Meng, J., & Chang, S. F. (1998). Embedding visible video watermarks in the compressed domain. In *Proceedings of the IEEE International Conference on Image Processing,* (Vol. 1, pp. 474-477). IEEE Press.

Merkle, P., Müller, K., Smolic, A., & Wiegand, T. (2007). Efficient compression of multiview depth data based on MVC. In *Proceedings of the 3DTV Conference,* (pp. 1-4). 3DTV.

Merkle, P., Smolic, A., Müller, K., & Wiegend, T. (2007a). Efficient prediction structures for multiview video coding. *IEEE Transactions on Circuits and Systems for Video Technology, 17*(11), 1461–1473. doi:10.1109/TCSVT.2007.903665

Nakajima, Y., Hori, H., & Kanoh, T. (1995). Rate conversion of MPEG coded video by requantization process. In *Proceedings of the IEEE International Conference on Image Processing,* (pp. 408-411). IEEE Press.

Neal, J., Green, R., & Landovskis, J. (2001). Interactive channel for multimedia satellite networks. *IEEE Communications Magazine, 39*(3), 192–198. doi:10.1109/35.910607

Ni, P., Eichhorn, A., Griwodz, C., & Halvorsen, P. (2010). Frequent layer switching for perceived video quality improvements of coarse-grained scalable video. *Multimedia Systems, 16*(3), 171–182. doi:10.1007/s00530-010-0186-9

Nightingale, J. M., Wang, Q., & Grecos, C. (2012). Benchmarking real-time HEVC streaming. In *Proceeding of SPIE Conference on Real-Time Imaging.* SPIE.

Open Mobile Alliance. (2011). *MMS architecture. Technical specification, OMA-AD-MMS-V1_3-20110913-A.* Washington, DC: Open Mobil Alliance.

Ortega, A., & Wang, H. (2007). Mechanisms for adapting compressed multimedia to varying bandwidth conditions. In van der Schaar, M., & Chou, P. A. (Eds.), *Multimedia over IP and Wireless Networks* (pp. 81–116). Amsterdam, The Netherlands: Academic Press. doi:10.1016/B978-012088480-3/50005-9

Padhye, J., Firoiu, V., Towsley, D., & Kurose, J. (1998). Modeling TCP throughput: A simple model and its empirical validation. In *Proceedings of ACM SIGCOMM*, (pp. 303–314). ACM Press.

Pantos, R., & May, W. (2011). *HTTP live streaming*. IETF Draft. Retrieved from http://tools.ietf.org/html/draft-pantos-http-live-streaming-06

Pao, I. M., & Sun, M. T. (2001). Encoding stored video for streaming applications. *IEEE Transactions on Circuits and Systems for Video Technology, 11*(2), 199–209. doi:10.1109/76.905985

Park, S., Yoon, H., & Kim, J. (2006). Network-adaptive HD MPEG-2 video streaming with cross-layered channel monitoring in WLAN. *Journal of Zhejiang University Science A, 7*(5), 885–893. doi:10.1631/jzus.2006.A0885

Perkins, C. (2003). *RTP: Audio and video for the internet*. Boston, MA: Addison Wesley.

Plissonneau, L., En-Najjary, T., & Urvoy-Keller, G. (2008). Revisiting web traffic from a DSL provider perspective: The case of YouTube. In *Proceedings of the 19th ITC Specialist Seminar on Network Usage and Traffic*. IEEE.

Rahda, H. M., van der Schaar, M., & Chen, Y. (2001). The MPEG-4 fine-grained scalable video coding method for multimedia streaming over IP. *IEEE Transactions on Multimedia, 3*(1), 53–68. doi:10.1109/6046.909594

Razavi, R., Fleury, M., & Ghanbari, M. (2008). Energy efficient video streaming over Bluetooth using rateless coding. *Electronics Letters, 44*(22), 1309–1310. doi:10.1049/el:20080851

Rejaie, R., Handley, M., & Estrin, D. (1999). RAP: An end-to-end rate-based congestion control mechanism for realtime streams in the internet. In *Proceedings of the IEEE INFOCOM*, (pp. 1337–1345). IEEE Press.

Rhee, I., Ozdemir, V., & Yi, T. (2000). *TEAR: TCP emulation at receivers. Technical Report*. Raleigh, NC: North Carolina State University.

Rubenstein, D., Kurose, J., & Towsley, D. (1998). *Real-time reliable multicast using proactive forward error correction*. Technical Report 98-19. Amherst, MA: University of Massachusetts.

Rupp, M. (Ed.). (2009). *Video and multimedia transmissions over cellular networks: Analysis, modelling and optimization in live 3G mobile networks*. Chichester, UK: John Wiley & Sons.

Savas, S. S., Tekalp, A. M., & Gürler, C. G. (2011). Adaptive multi-view video streaming over P2P networks considering quality of experience. In *Proceedings of the 2011 ACM Workshop on Social and Behavioural Networked Media Access*, (pp. 53-58). ACM Press.

Schierl, T., & Narasimhan, S. (2011). Transport and storage systems for 3-D video using MPEG-2 systems, RTP, and ISO file format. *Proceedings of the IEEE, 99*(4), 671–683. doi:10.1109/JPROC.2010.2091370

Schwarz, H., Marpe, D., & Wiegand, T. (2007). Overview of the scalable video coding extension of the H.264/AVC standard. *IEEE Transactions on Circuits and Systems for Video Technology, 17*(9), 1103–1120. doi:10.1109/TCSVT.2007.905532

Schwarz, H., & Wiegand, T. (2007). R-D optimized multilayer encoder control for SVC. In *Proceedings of the IEEE International Conference on Image Processing*, (pp. 281-284). IEEE Press.

Scopus Video Networking. (2006). *Advanced encoding mechanism and statistical multiplexing*. White Paper. Sunnyvale, CA: Scopus Video Networking.

Seeling, P., & Reisslein, M. (2005). The rate variability-distortion (VD) curve of encoded video and its impact on statistical multiplexing. *IEEE Transactions on Broadcasting, 51*(4), 473–492. doi:10.1109/TBC.2005.851121

Segall, C. A., & Sullivan, G. J. (2007). Spatial scalability within the H.264/AVC scalable video coding extension. *IEEE Transactions on Circuits and Systems for Video Technology, 17*(9), 1121–1131. doi:10.1109/TCSVT.2007.906824

Shade, J., Gortler, S., He, L. W., & Szeliski, R. (1998). Layered depth images. [ACM Press.]. *Proceedings of SIGGRAPH, 1998*, 231–242.

Shanableh, T., & Ghanbari, M. (2000). Heterogeneous video transcoding to lower spatio-temporal resolutions and different encoding formats. *IEEE Transactions on Multimedia, 2*(2), 101–110. doi:10.1109/6046.845014

Shokorallahi, A. (2006). Raptor codes. *IEEE Transactions on Information Theory, 52*(6), 2551–2567. doi:10.1109/TIT.2006.874390

Sisalem, D., & Wolisz, A. (2000). LDA+ TCP-friendly adaptation: A measurement and comparison study. In *Proceedings of the 10th International Workshop on Network and Operating Systems Support for Digital Audio and Video*, (pp. 25-28). IEEE.

Stockhammer, T. (2011). Dynamic adaptive streaming over HTTP – Design principles and standards. In *Proceedings of the Second Annual ACM Conference on Multimedia Systems*, (pp. 133-144). ACM Press.

Stockhammer, T., Fröjdh, P., Sodagar, I., & Rhyu, S. (Eds.). (2011). *Information technology — MPEG systems technologies — Part 6: Dynamic adaptive streaming over HTTP (DASH).* ISO/IEC MPEG Draft International Standard. Retrieved from developer.longtailvideo.com/trac/export/1509/.../adaptive/.../dash.pdf

Sullivan, G. J., & Ohm, J.-R. (2010). Recent developments in standardization of high efficiency video coding (HEVC). In *Proceedings of SPIE Applications of Digital Image Processing 23*. SPIE.

Sun, H., Chen, X., & Chiang, T. (2005). *Digital video transcoders for transmission and storage.* Boca Raton, FL: CRC Press.

Sun, H., Kwok, W., & Zdepski, J. (1996). Architectures for MPEG compressed bitstream scaling. *IEEE Transactions on Circuits and Systems for Video Technology, 6*(2), 191–199. doi:10.1109/76.488826

Sun, H., Vetro, A., Bao, J., & Poon, T. (1997). A new approach for memory-efficient ATV decoding. *IEEE Transactions on Consumer Electronics, 43*(3), 517–525. doi:10.1109/30.628667

Tan, W., & Zakhor, A. (1999). Multicast transmission of scalable video using layered FEC and scalable compression. *IEEE Transactions on Circuits and Systems for Video Technology, 11*(3), 373–386. doi:10.1109/76.911162

Tan, Y.-P., Liang, Y.-Q., & Yu, J. (2002). Video transcoding for fast forward/ reverse video playback. In *Proceedings of the IEEE International Conference Image Processing,* (vol. 1, pp. 713–716). IEEE Press.

Tandberg Television. (2008). *Reflex and data reflex statistical multiplexing system.* White Paper. Slough, UK: Tandberg Television.

Tanimoto, M., Tehrani, M. P., Fujii, T., & Yendo, T. (2011). Free-viewpoint TV. *IEEE Signal Processing Magazine, 28*(1), 67–76. doi:10.1109/MSP.2010.939077

Taubman, D. S. (2000). High performance scalable image compression with EBCOT. *IEEE Transactions on Image Processing, 9*(7), 1158–1170. doi:10.1109/83.847830 van der Auwera, G., David, P. T., & Reisslein, M. (2008b). Traffic and quality characterization of single-layer video streams encoded with the H.264/MPEG-4 advanced video coding standard and scalable video coding extension. *IEEE Transactions on Broadcasting*, *54*(3), 698–718. doi:10.1109/TBC.2008.2000422

van der Auwera, G., David, P. T., Reisslein, M., & Karam, L. J. (2008a). Traffic and quality characterization of the H.264/AVC scalable video coding extension. *Advances in Multimedia*, *2*, 1–27. doi:10.1155/2008/164027

van der Schaar, M., & Chou, P. A. (Eds.). (2007). *Multimedia over IP and wireless networks*. Amsterdam, The Netherlands: Academic Press.

Varsa, V., & Karzcewicz, M. (2001). Long window rate control for video streaming. In *Proceedings of the 11th International Packet Video Workshop*. IEEE.

Vetro, A., Christopoulos, C., & Sun, H. (2003). Video transcoding architectures and techniques: An overview. *IEEE Signal Processing Magazine*, *20*(2), 18–29. doi:10.1109/MSP.2003.1184336

Vetro, A., Wiegand, T., & Sullivan, G. J. (2011). Overview of the stereo and multiview video coding extensions of the H.264/MPEG-4 AVC extension. *Proceedings of the IEEE*, *99*(4), 626–642. doi:10.1109/JPROC.2010.2098830

Wang, B., Kurose, J. F., Shenoy, P. J., & Towsley, D. F. (2004). Multimedia streaming via TCP: An analytic performance study. In *Proceedings of ACM Multimedia Conference*, (pp. 908-915). ACM Press.

Wang, Y., Reibman, A. R., & Lin, S. (2005). Multiple description coding for video delivery. *Proceedings of the IEEE*, *93*(1), 57–70. doi:10.1109/JPROC.2004.839618

Watson, A. B. (1994). Image compression using the discrete cosine transform. *Mathematica Journal*, *4*(1), 81–88.

Wei, D. X., Jin, C., Low, S. H., & Hedge, S. (2006). FAST TCP: Motivation, architecture, algorithm and performance. *IEEE/ACM Transactions on Networking*, *4*(6), 1246–1259. doi:10.1109/TNET.2006.886335

Wein, M., Schwarz, H., & Oelbaum, T. (2007). Performance analysis of SVC. *IEEE Transactions on Circuits and Systems for Video Technology*, *17*(9), 1194–1203. doi:10.1109/TCSVT.2007.905530

Wenger, S. (2003). H.264/AVC over IP. *IEEE Transactions on Circuits and Systems for Video Technology*, *13*(7), 645–656. doi:10.1109/TCSVT.2003.814966

Widmer, J., Denda, R., & Mauve, M. (2001). A survey on TCP-friendly congestion control. *IEEE Network*, *15*(3), 28–37. doi:10.1109/65.923938

Wiegand, T., Noblet, L., & Rovati, F. (2009). Scalable video coding for IPTV services. *IEEE Transactions on Broadcasting*, *55*(2), 527–538. doi:10.1109/TBC.2009.2020954

Wien, M., Cazoulat, R., Graffunder, A., Hutter, A., & Amon, P. (2007). Real-time system for adaptive video streaming based on SVC. *IEEE Transactions on Circuits and Systems for Video Technology*, *17*(9), 1227–1237. doi:10.1109/TCSVT.2007.905519

Wittmann, R., & Zitterbart, M. (2001). *Multicast communication: Protocols and applications*. San Francisco, CA: Morgan Kaufmann.

Wu, D., Hu, Y. T., & Zhang, Y.-Q. (2000). Transporting real-time video over the Internet: Challenges and approaches. *Proceedings of the IEEE*, *88*(12), 1855–1875. doi:10.1109/5.899055

Xin, J., Sun, M. T., & Chun, K. (2002). Motion re-estimation for MPEG-2 to MPEG-4 simple profile transcoding. In *Proceedings of the International Workshop on Packet Video*. IEEE.

Zambelli, A. (2009). IIS smooth streaming technical overview. *Microsoft Corporation*. Retrieved from http://users.atw.hu/dvb-crew/applications/documents/IIS_Smooth_Streaming_Technical_Overview.pdf

Zhang, Y., Mao, S., Yang, L. T., & Chen, T. M. (Eds.). (2008). *Broadband mobile multimedia*. Boca Raton, FL: Auerbach Publications.

Zhang, Z.-L., Kurose, J., Salehi, J. D., & Towsley, D. (1997). Smoothing, statistical multiplexing and call admission control for stored video. *IEEE Journal on Selected Areas in Communications*, *15*(6), 1148–1166. doi:10.1109/49.611165

Zhao, L., & Kuo, C.-C. J. (2003). Buffer-constrained R-D optimized rate-control for video coding. In *Proceedings of the IEEE International Conference on Multimedia and Expo*, (pp. 377-380). IEEE Press.

Zhu, W., Luo, C., Wang, J., & Li, S. (2011). Multimedia cloud computing. *IEEE Signal Processing Magazine*, *28*(3), 59–69. doi:10.1109/MSP.2011.940269

Zhu, X., Agrawal, P., Singh, J. P., Alpcan, T., & Girod, B. (2007). Rate allocation for multi-user streaming over heterogeneous access networks. In *Proceedings of ACM Multimedia*, (pp. 37-46). ACM Press.

Zink, M., Künzel, O., Scmitt, J., & Steinmetz, R. (2003). Subjective impression of variations in layer encoded videos. In *Proceedings of the 11ᵗʰ International Workshop on Quality of Service*, (pp. 137-154). IEEE.

Zitnick, C. L., Kang, S. B., Uyttendaele, M., Winder, S., & Szeliski, R. (2004). High-quality video view interpolation using a layered representation. *ACM Transactions on Graphics*, *23*(3), 600–608. doi:10.1145/1015706.1015766

ADDITIONAL READING

Austerberry, D. (2004). *The technology of video and audio streaming* (2nd ed.). Amsterdam, The Netherlands: Focal Press.

Chen, C. W., Li, Z., & Lian, S. (Eds.). (2010). *Intelligent multimedia communication: Techniques and applications*. Berlin, Germany: Springer Verlag. doi:10.1007/978-3-642-11686-5

Fleury, M., & Qadri, N. (Eds.). (2012). *Streaming media with peer-to-peer networks: Wireless perspectives*. Hershey, PA: IGI Global. doi:10.4018/978-1-4666-1613-4

Follansbee, J. (2006). *Hands-on guide to streaming media: An introduction to delivering on-demand media*. Amsterdam, The Netherlands: Focal Press.

Gürler, C. G., Bagci, K., & Tekalp, A. M. (2010). Adaptive stereoscopic 3D video streaming. In *Proceedings of IEEE International Conference on Image Processing*, (2409-2412). IEEE Press.

Lee, J. Y. B. (2005). *Scalable continuous media streaming systems*. Chichester, UK: John Wiley & Sons Ltd. doi:10.1002/047001539X

Richter, S., & Ozer, J. (2006). *Hands-on guide to flash video: Web video and flash media server*. Amsterdam, The Netherlands: Focal Press.

Schwarz, H., Marpe, D., & Wiegand, T. (2006). An analysis of hierarchical B-pictures and MCTF. In *Proceedings of the IEEE International Conference on Multimedia and Expo*, (pp. 1929-1932). IEEE Press.

Shanableh, T., & Ghanbari, M. (2005). Multilayer transcoding with format portability for multicasting of single-layered video. *IEEE Transactions on Multimedia*, 7(1), 1–15. doi:10.1109/TMM.2004.840602

Simpson, W. (2008). *Video over IP*. Amsterdam, The Netherlands: Focal Press.

Zink, M. (2005). *Scalable video on demand*. Chichester, UK: Wiley-Blackwell Press.

KEY TERMS AND DEFINITIONS

Adaptive Streaming: A streaming service that adapts the bitrate of the sent stream according to the state of the transmission channel (mainly the available bandwidth in a wired network and error conditions in a wireless network) and the client state (screen resolution, available memory, processor capability, CPU load, battery level, and so on).

Content Distribution Network: An overlay network of localized servers. An overlay network provides application layer routing. These servers cache popular content at their location. The content is provided to them from a centralized source. When the central server is contacted by a streaming client, it transfers streaming control to the local server, which is nearest in some sense.

Multi-View Coding: Provides two or more views that can be selected by the viewer for 3D display. Two view video or TV is normally referred to as 3D video or TV. However, it is a sub-set of the 3D display possibilities. Free-viewpoint TV is an extension of multi-view coding, as now the viewer is no longer restricted to pre-chosen fixed 3D views. Current research and development has concentrated on efficient compression of multi-view video.

Multiple: Description Coding: Is a form of encoding that separates the compressed video into multiple representations (*descriptions*) of lower quality. The separation can take place in the spatial, temporal or frequency domain. When several descriptions are combined together, the resulting quality is enhanced compared to the quality of a single description.

Scalable Video Coding: Consists of encoding video in a set of hierarchical layers. The hierarchy can be arranged according to spatial resolution, temporal frame rate, video quality (also known as SNR scalability), or a combination of these. Early scalable codecs separated out the scalable layers, whereas codecs such as the JPEG2000 still image codec introduced embedded scalability within a single bitstream. In respect to Signal-to-Noise SNR scalability, starting from a base layer, that provides the lowest level of quality, each subsequent layer improves the quality of the reconstructed video. However, including more layers naturally increases the bitrate. Notice that higher layers are dependent on lower ones in order to be correctly decoded. The Scalable Video Coding extension to the H.264 codec is a standardized scalable codec with low loss of coding efficiency compared to the non-scalable version of the codec.

Stream Switching: Allows a stream's bitrate to be changed, usually at pre-determined synchronization points. Multiple versions of the same video are pre-encoded and stored at different bitrates. A variety of switching methods exist such as switching at periodic intra-coded I- or IDR-pictures, or at recovery points after gradual decoding refresh, or using the H.264/AVC primary and secondary SP-frames.

Transcoding: Is the process of converting a video stream to a different format from the original. In the sense used in this chapter, transcoding refers to changes that cause a bitrate change. (It is also possible to transcode from one codec format to another such as from MPEG-2 to H264/AVC format.) One way to change the bitrate is to change the quantization parameter. It is clearly not possible to increase the video quality beyond that of the original. Transcoding can also cause some delay and may cause drift error.

Chapter 5
Intra–Refresh Techniques for Mobile Video Streaming

Martin Fleury
University of Essex, UK

Ismail A. Ali
University of Essex, UK

Nadia Qadri
COMSATS Institute of Information & Technology, Pakistan

Mohammed Ghanbari
University of Essex, UK

ABSTRACT

Mobile devices are replacing the desktop computer in most spheres outside the workplace. This development brings a problem to video streaming services, as wireless channels are fundamentally error-prone, whereas video compression depends for most of its gains on predictive coding. The H.264 codec family has included a good number of error resilience facilities to counter-act the spatio-temporal error propagation brought on by packet loss. This chapter outlines these facilities before examining ways in which predictive coding can be temporally restrained. In particular, intra-refresh techniques are the focus, as these bring additional utility to the video stream. For example, the chapter compares periodic and gradual intra-refresh, each of which provides recovery points for the decoder and also allow stream switching or joining at these points. Thus, in intra-coding, the more normal temporal prediction is temporally replaced by spatial prediction, at a cost in coding efficiency but allowing a decoder in a mobile device to reset itself. After a review of research into this area, the chapter provides a case study in non-periodic intra-refresh before considering possible future research directions.

DOI: 10.4018/978-1-4666-2833-5.ch005

INTRODUCTION

As fixed and wireless networks converge (Watson, 2009; Ahson & Ilyas, 2010; Paul, 2011), there is a need to bring the user's video experience within the mobile wireless network closer to that of the fixed network. Real-time delivery of video is required for streaming applications such as Web TV, *Internet Protocol TV (IPTV)* in its various forms (Simpson, 2008), and the class of interactive video applications such as video conferencing and soft videophone. However, wireless access continues to present a bandwidth constriction, especially as high-definition video (720p, i.e. 1280×720 pixels/frame in progressive display at 30 frames/s) is extended (Bing, 2010) to displays on laptops and other mobile devices such as smart-phones. In terms of streaming to mobiles, the trend (e.g. Apple's FaceTime) is towards full VGA resolution (640 × 480 pixels/frame at 30 frames/s). This implies that compression is still very necessary and, because of the predictive nature of that compression, there is an ever present risk of spatio-temporal error propagation. The main differentiating feature of wireless networks is the various channel impairments that can occur. Therefore, ways are sought to arrest error propagation and intra-refresh techniques provided by the video codec itself are a way to do so, along with other forms of error resilience (Stockhammer & Zia, 2007).

Intra-coded video data within I-pictures relies on spatial reference within the video picture and, hence, is unaffected by the corruption of previous pictures. This is in contrast to inter-coded video data, which takes reference from past (P-pictures) or even future video pictures (or both temporal directions) (B-pictures). In both these forms of coding, it is the difference image (or residual data) that is processed in subsequent stages of a hybrid video encoder (Ghanbari, 2011). For networked TV, I-pictures can serve as a point at which the TV channel can be switched or zapped. For streaming of live video, periodic I-pictures can also act as the point in time of joining the broadcast stream. Pseudo-Video Cassette Recorder (VCR) functions (otherwise known as trick modes) such as fast forward, rewind and so on, can also be based on I-pictures. Periodic I-pictures are, in fact, the usual way to provide intra-refresh in those ways but it is also possible to provide *Gradual Decoding Refresh (GDR)* (Hannuksela, et al., 2004) by including intra-refresh *Macroblocks (MBs)* (the compression building blocks) within inter-coded pictures, provided the refreshment pattern is carefully considered.

Related to the issue of intra-refresh is the question of the Group-of-Picture (GoP) structure and size, as this can be varied statically and dynamically. It is also dependent on codec profile. For streaming to mobile devices it may actually be advisable not to use traditional intra-refresh I-pictures (or slices), provided distributed intra-refresh MBs can be used in some way to provide the functionality once served by I-pictures (see the previous remarks on GDR). It should also be borne in mind that some intra-coding is naturally inserted even on nominally inter-coded pictures and that in some circumstances it will be necessary to constrain intra-coding reference to avoid temporal error corruption arising from spatial reference to inter-coded MBs that themselves are corrupted. The authors of this chapter (Ali, et al., 2012) have shown that the insertion of a cyclic intra-coded line of MBs on a per-video picture basis as a convenient way to mitigate error propagation, if less-active video sequences are transmitted over wireless or other 'lossy' links. For more-active sequences (ones with substantial inter-picture motion), periodic insertion of I-pictures is preferable. Randomized insertion of intra-coded MBs (Haskell & Messerschmitt, 1992; Côté & Kossentini, 1999) is an alternative to a cyclic intra-coded line of MBS or some other more visually pleasing pattern. However, a random pattern may result in duplication of intra-coded MBs in successive pictures. However, there are some subtleties involved in this, as the randomized insertion of intra-coded MBs in the *Joint Model (JM)* reference software for the H.264/Advanced Video Coding (AVC) codec (Wiegand, et al., 2003) is not 'completely random' in the sense of earlier writers (Haskell & Messerschmitt, 1992; Côté & Kossentini, 1999). In fact, previous MBs are not duplicated and, once all MBs have been replaced with intra-coded MBs, the random cycle repeats, exactly as before.

The rest of this chapter is organized as follows. The following section by reference to the H.264/ AVC codec standard outlines the error protection and error resilience facilities currently available, before looking at commercial video streaming modalities. The next Section reviews the available possibilities for intra-refresh before reviewing how these have been exploited in the research literature. Before concluding, the chapter includes a case study with research results to show the trade-offs in employing one form of intra-refresh.

MOBILE VIDEO STREAMING

Mobile video streaming and video distribution in general is beginning to dominate Internet traffic, as is illustrated by a few facts and figures. Over one third of the top 50 sites by volume are now based on video distribution (Cisco, 2011). Netflix in 2011, a provider of on-demand Internet streaming video, alone accounted for almost 30% of peak period downstream traffic (Sandvine, 2011). The sum of all forms of video is expected to exceed 90% of global user traffic by 2014. The Web is becoming one of the main broadcasting platforms, due to the recent availability of the so-called 'three screens' (digital TV, PC, and smart-phones) capable of accessing it, and the increase of the number of faster connections to the network. Mobile Internet data traffic stands at 237 Peta Bytes (PB) per month in 2010 but is set to rise to 6,254 PB in 2015 and could exceed that of the wired Internet by 2015 according to Cisco's published estimates. Mobile Internet is by far the fastest growing new service

with a forecast increase in data traffic over the next five years of 92%. In fact, some markets such as South America according to Sandvine have largely replaced fixed or wired access with wireless or mobile access. Overall, according to Cisco, unmanaged IP consumer traffic is predicted to be 53.3 Exabytes per month by the end of 2015, as opposed to 11.8 for managed networks, and 4.9 over mobile networks.

Source Coding with H.264/AVC

A key tool in tackling the challenges of video streaming is the development of more efficient video codecs. In 2001, the ITU-T *Video Coding Experts Group (VCEG)* together with the ISO/IEC *Moving Picture Experts Group (MPEG)* formed the *Joint Video Team (JVT)* to develop a new video coding standard the name given ITU H.264/ AVC also known as ISO MPEG-4 Part 10, with an increased compression efficiency almost twice as that of the MPEG-2 (Ozbek & Tumnali, 2005). Usually an increase in compression efficiency may lead to a substantial increase in complexity; this was recorded to be approximately four times and nine times for the H.264/AVC decoder and encoder as compared to MPEG-2. It is important to note that this increase in complexity also depends on the selection of the different features and profiles described in the standard. The emerging *High Efficiency Video Coding (HEVC)* standard considers ways to improve implementation efficiency, particularly for *high-definition (HD)* video but as it is at a development stage, it would be premature to discuss it herein.

Apart from better coding efficiency, another important feature of the standard is enhanced error resiliency and the adaptability to various networks (Kumar, et al., 2006), with video representation ranging from 'conversational' (video telephony and video conferencing) to 'non-conversational' (storage, broadcast, or streaming) application (Narkhede & Kant, 2009). Video streaming can be further sub-divided into the streaming of pre-encoded material and the streaming of live material. In the latter, the actions that can be taken at the encoder are more limited though both the streaming modes are more relaxed in their delay constraints than the 'conversational' applications. In order to increase the flexibility and adaptability, H.264/AVC has adopted a two-layer structure (Sullivan & Wiegand, 2005), the *Video Coding Layer (VCL)*, which is designed to efficiently represent the video content, and a *Network Abstraction Layer (NAL)*, which formats the VCL representation of the video and provides header information for ready transmission over the network. The NAL can provide compressed video data in two media container formats, for the stream-based protocols with H.320, H.324 or MPEG-2 *Transport Stream (TS)* and for the packet-based protocols with *Real-Time Transport Protocol (RTP)*/UDP/ IP and TCP/IP. For the stream-based protocols, the data is provided with start codes such that the transport layers and the decoder can easily identify the structure of the bitstream, while for the packet-based protocols the data are provided without these start codes. In the H.264/AVC, data from the VCL are packetized into a *Network Abstraction Layer Unit (NALU)*. Each NALU is encapsulated in an RTP packet, with subsequent addition of UDP/IP headers.

Video compression or video coding refers to the reduction of quantity of data used to represent video sequences. In the standard video method, compression is based on spatial and temporal redundancy reduction along with entropy coding (static or adaptive Huffman and Arithmetic). (Entropy coding is a way to compress (and losslessly recover) digital data by exploiting statistical redundancy). Spatial redundancy reduction results in intra-pictures, in which each block of the picture is predicted from its neighboring coded block without reference to any other picture. The difference between the two blocks is transformed using a *Discrete Cosine Transform (DCT)* a type of orthonormal transformation matrix, followed by quantization and entropy coding. (The DCT

is now replaced by a reversible integer transform in H.264 to avoid drift during inverse transform). As the intra pictures do not rely on other pictures, they are used for a number of operations, typically for random access and confining drift errors due to losses during transmission.

Inter pictures exploit the temporal redundancy between pictures. In this process, the movements of objects in the neighboring pictures are modeled using motion vectors. The resulting picture is subtracted from the original picture to be encoded, and the residue is transformed using the DCT followed by entropy coding. Inter pictures can be further divided into Predictive (P) pictures and Bi-predictive (B) pictures. In P pictures the prediction is made from earlier P or I pictures, while in B pictures the prediction can be made from an earlier and/or later I and/or P picture, but not a B picture. In H.264/AVC this restriction is removed and a B-picture can be used as reference for predicting other pictures. A sequence of pictures grouped between two I pictures is referred to as *Group of Pictures (GOP)* as shown in Figure 1.

Error Protection for Mobile Streaming

This Section outlines the methods available to protect vulnerable compressed video data, before specializing in the following Section to one form of protection, namely source-coded error resilience. Wireless technologies generally provide physical-layer *Forward Error Correction (FEC)*. The advantage of FEC provision at this layer is that real-time applications may be spared the need to implement their own FEC schemes. However, FEC can only protect a limited range of errors, as otherwise the FEC overhead might become intolerable. Due to the data dependencies within a compressed video bitstream, errors not only affect the immediate vicinity of the error but also may propagate in time until an intra-refresh reset can take place. The problem of error propagation has become more severe with the

Figure 1. A general diagram of a GOP

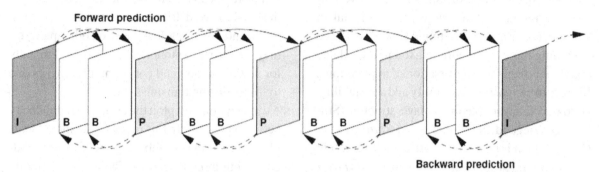

increased efficiency of recent codecs, especially H.264/AVC. Therefore, additional packet-level *application-layer (AP)*-FEC (Luby, et al., 2008) has generally been adopted for multicast and broadcast video services over wireless cellular networks. The *Digital Video Broadcasting (DVB)*, *3rd Generation Partnership Project (3GPP)*, and ITU *Focus Group (FG)* on IPTV include erasure codes in their recommendations for download and streaming video services. AP-FEC is intended to correct packet erasures. However, if the video is first transmitted over a managed wired network or during benign channel conditions then AP-FEC is a burden. However, notice (Luby, et al., 2008) that burst errors leading to a packet being declared lost are also a problem in *Asymmetric Digital Subscriber Link (ADSL)* access.

Error control through one of the varieties of Automatic Repeat Request (ARQ) (Girod & Färber, 1999) is an alternative to ARQ, which unlike FEC is only applied when it is needed. However, ARQ may be unsuitable for interactive or conversational video services such as video conferencing and videophone, because of the additional latency introduced by the need to wait for the ARQ reply. However, if the journey to the source server is relatively short in duration, then ARQ can be effective. However, simple ARQ and even hybrid ARQ (e.g. for IEEE 802.16e (Andrews, et al., 2007)) may be available at the datalink layer. The main issue then is how many repeat ARQs to configure before a repeated packet is abandoned.

End-to-end latency is the key restriction. For example, IPTV standards recommend (Agilent Technologies, 2008) no more than 50 ms for the complete network path.

One way to compensate for burst errors is to interleave packets so that an error burst does not affect a contiguous set of packets. However, because of latency constraints, packet-interleaving at the application layer (Razavi, et al., 2009) may result in an unacceptable packet delay resulting in buffer underflow at the video decoder. Interleaving can also be applied as a way of countering error bursts and random errors within entropy-encoded data. *Error-Resilient Entropy Coding (EREC)* (Redmill & Kingsbury, 1996) used this form of interleaving to counter errors within H.263 data-partitioned packets.

The previous discussion has considered Equal Error Protection (EEP). However, there are many forms of Unequal Error Protection (UEP) based on applying additional protection to the more important types of video (such as intra-coded data) or to more important content (such as that content which has the most effect if lost on the resulting picture distortion after decoding). UEP FEC or *Unequal Packet Loss Protection (UPLP)* techniques can be particularly useful for the streaming of scalable (Schwarz, et al., 2007) and multiview video (Merkle, et al., 2009). This is because some parts of the video, such as enhancement layers in scalable video, may not be decidable if other data is not received. In Hellge et al. (2011), *Layer-Aware (LA)*-FEC is proposed in which the base layer remains protected by the original FEC provision but the base layer and further enhancement layers are also protected with additional redundant data. Thus, if the base layer is labeled layer 0, then repair data for layer 0 and enhancement layer 1 is generated, implying that layer 0 is protected with its original allocation of FEC and is additionally protected by data generated to protect the base layer combined with the first enhancement layer. This process continues for layers 0 through to layer 2, and so on. To generate repair data in this flexible manner rateless coding is availed of in the form of Raptor coding (Shokrollahi, 2006). In Hellge et al. (2011), the effectiveness of LA-FEC is demonstrated. Wen et al. (2007) is an earlier paper that anticipated the approach taken by Hellge et al. (2011).

Error Resilience for Mobile Streaming

H.264/AVC along with its efficient compression has strong error resiliency features. Error resilience is a form of error protection through source coding. Due to the growing importance of multimedia communication over wireless, the range of these techniques has been expanded in the H.264/AVC codec (Wenger, 2003). Error resilience introduces limited delay and as such is suitable for real-time, interactive video streaming, especially video-telephony, and video conferencing. It is also suitable for one-way streaming over cellular wireless networks and broadband wireless access networks to the home. As physical-layer *Forward Error Correction (FEC)* is normally already present at the wireless physical layer, application-layer FEC (rather than error resilience) may duplicate its role. The exception is if application-layer FEC can be designed to act as an outer code after inner coding at the physical layer, in the manner of concatenated channel coding. Various forms of ARQ are also possible

as a means of error control. For example, in Mao et al. (2001), if packets from the base layer of a layered video stream (sent over multiple paths) failed to arrive then a single ARQ is permitted. More ARQs can lead to a deterioration in overall packet latency.

H.264 has combined the error resiliency schemes of the previous codecs with some techniques introduced newly or implemented differently such as *IntraMB (Intra MacroBlock), DP (Data Partitioning),* and slicing borrowed from H.261, H.262, and MPEG-1 and -2. Some of the newly introduced techniques by H.264 are Redundant Slices or Pictures and *Flexible Macroblock Ordering (FMO).* All these techniques are discussed below.

As mentioned in the Introduction, Intra pictures or intra MB placements in a slice are used to confine drifting errors (Ostermann, et al., 2004). H.264/AVC supports two types of intra pictures, the *Intra Decode Refresh (IDR)* and the intra coded picture or I-picture. Furthermore it also support intra coded MBs inside the inter-coded pictures. Both the I- and the IDR picture are intra coded pictures without reference to any other pictures. The difference between the two is that the IDR picture invalidates all the short term reference memory buffers and, thus, can completely confines the drift errors due to previous pictures. The -intra picture, on the other hand, does not invalidate the reference memory buffers and, thus, can confine drift error at that picture position only. If future pictures refer to any picture older than the intra picture the drift error can occur again. The insertion of intra-coded MBs into pictures normally encoded through motion-compensated prediction allows temporal error propagation to be arrested if matching MBs in a previous picture are lost. IDR through periodic insertion of pictures with all MBs encoded through spatial reference (intra-coded) is the usual way of catching error propagation. However, such pictures cause periodic increases in the data rate when encoding at a variable bit rate.

Slicing

H.264/AVC can divide a picture into slices, whose size can be as small as a MB and as large as one complete picture. MBs are assigned to slices in raster scan order, unless FMO (explained later) is used (Wenger, 2003). Intra prediction across the slice boundaries is not allowed, making (Ostermann, et al., 2004) slices self-contained units that can be decoded without referring to other slices of the picture. Thus, they can prevent error propagation (Son & Jeong, 2008). However, the non-availability of intra prediction across slice boundaries reduces the compression efficiency, which can be further decreased with an increase in the number of slices per picture. Slices can be decoded independently. However, some information from other slices may be needed for applying the H.264 de-blocking filter (Sullivan & Wiegand, 2005).

Data Partitioning

In the video bitstream, some syntax elements are more important than the others and, thus, the error robustness can be enhanced by separating these data from one another and protecting it unequally based on their importance (Sullivan & Wiegand, 2005). Data partitioning in H.264/AVC, Figure 2, separates the compressed bitstream into: a) configuration data (IDR pictures) and motion vectors; b) intra-coded transform coefficients;

and c) inter-coded coefficients (Wenger, 2003). This data forms A, B, and C partitions, which are packetized as separate NALUs. The arrangement allows a picture to be reconstructed even if the inter-coded MBs (or rather their residual transform coefficients) in partition C are lost, provided that the motion vectors in partition A survive. Partition A is normally strongly FEC-protected at the application layer or physical-layer protection may be provided such as the hierarchical modulation scheme in Barmada et al. (2005) for broadcast TV. Notice that in codecs prior to H.264/AVC, data partitioning was also applied but no separation into NALUs occurred. The advantage of integral partitioning is that additional resynchronization markers are available that reset entropy decoding. This mode of data partitioning is still available in H.264/AVC and is applied to I-pictures.

Redundant Slices/Pictures

To enhance error robustness in H.264/AVC, the encoder may create reduced fidelity some or all parts of a picture. Redundant pictures (Richardson, 2003) (or strictly redundant slices making up a picture) are coarsely quantized pictures that can avoid sudden drops in quality marked by freeze picture effects if a complete picture (or slice) is lost. To improve error resilience redundant pictures intended for error resilience in H.264/AVC, can also serve to better reconstruct pictures received in error. The main weakness of the redundant picture

Figure 2. H264/AVC data partitioning in which a single slice is split into three NAL units (types 2 to 4). The relative size of the C partition will depend on the quantization parameter (QP), with a lower QP leading to higher quality and a larger C partition.

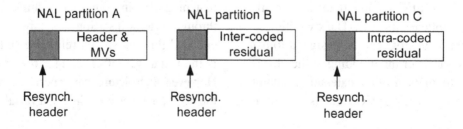

solution is that these pictures are discarded if not required. However, they are still a more efficient solution than including extra I-picture synchronization, as redundant pictures are predictively coded and require less bits as compared to I-pictures. A subsidiary weakness is the delay in encoding and transmitting redundant pictures, making it more suitable for one-way communication. If the redundant picture/slice replaces the loss of the original picture/slice there will still be some mismatch between encoder and decoder. This is because the encoder will assume the original picture/slice was used. However, the effect will be much less than if no substitution took place.

FMO (Flexible Macro-Block Ordering)

FMO allows different arrangements of MBs in a slice by utilizing the concept of slice groups. The MBs are arranged in a slice in different order than the scan order, enhancing the error resilience. In each slice group, the MBs are arranged according to MB to slice group map. Notice that slice groups can be employed independently of FMO. A further facility that also may be employed independently of FMO is *Arbitrary Slice Ordering (ASO)* by which slices within the same picture can be decoded in order of arrival at the decoder, rather than by their geometrical position within a picture.

In H.264/AVC (Thomos, et al., 2005), by varying the way in which the MBs are assigned to a slice (or rather group of slices), FMO gives a way of reconstructing a picture even if one or more slices are lost. Within a picture up to eight slice groups are possible. H.264/AVC provides different MB classification patterns. Assignment of MBs to a slice group can be general (type 6) but the other six types utilize a pre-defined assignment formula, thus reducing the coding overhead from providing a full assignment map. Pre-defined types are interleaved, checkerboard, foreground, Box out, Raster scan, and wipe.

The checkerboard type stands apart from other types, as it does not employ adjacent MBs as cod-

ing references, which decreases its compression efficiency and the relative video quality after decode. However, if there are safely decoded MBs in the vicinity of a lost packet, error concealment can be applied. Consequently, the rate of decrease in video quality with an increase in loss rate is lower than for the other pre-set types.

Arranging MBs in multiple slice groups increases error resilience. For example if one of the slices in the dispersed map is 'lost,' the missing slice can be concealed by interpolation from the available slices. Experiments show (Son & Jeong, 2005) that at a loss rate of 10% in case of video conferencing, the impairments due to losses can be kept so small that it is very difficult to observe the effect of the loss.

In FMO the possible errors are scattered to the whole picture to avoid its accumulation in a limited area. In this way, the distance between the correctly recovered block and the erroneous block is reduced. As the distance between correct and recovered block is reduced, the distortion in the recovered block is less and vice versa. Therefore, it is easier to conceal scattered errors as compared to the errors concentrated in a region.

Approaches to Video Streaming

There are two main commercial approaches to distributing video, depending on whether communication is one-way or two-way, that have hitherto been adapted for reasons of deployment expediency. Firstly, in the case of one-way distribution of video-on-demand-type streaming, a video chunk or segment-based approach has been used, which has recently been employed in an adaptive form. Because delays are higher in respect to chunk-based encoding (Lohmar, et al., 2011) it is debatable whether this approach can also be successfully employed for live streaming of video, such as for sports events. Secondly, in the case of two-way communication, true frame-by-frame streaming has been employed rather than encapsulating each set of frames or group of pictures inside a video chunk. Unfortunately, when it comes to wireless streaming the existing deficiencies of both approaches are likely to be magnified.

Within the wired Internet commercial operators have generally opted for a form of pseudo-streaming or progressive download (Yetgin & Secking, 2008) in which video chunks are sent as files in overlapped fashion. The initial chunk contains metadata necessary to decode chunks. *Download and Play (D&P)* is the main alternative to streaming but downloading (say) a two-hour movie results in a long start-up-time and excessive storage requirements. As the complete content is stored on the user machine it is commercially unwelcome for reasons of protection of intellectual property. Progressive download is an evolution of D&P that preserves content confidentiality, as only a small portion of a video is buffered on the user machine before it is overwritten. This is not to say that methods of de-streaming do not exist, but they have to be actively applied by the user. However, before the download starts the viewer needs to choose the most appropriate version if there are multiple offerings with different resolutions and/or qualities of the same content. If there is not enough available bandwidth for the selected version, the viewer may experience frequent picture freezes and re-buffering. Therefore, the industry solution is unsatisfactory and if complete video chunks are lost, as is likely in a wireless environment, then the only alternative could be to resend the chunk with between 2 and 15 s delay at a minimum, according to the default chunk size.

To answer these problems one approach is to use an adaptive form of progressive download (Färber, et al., 2006) in which the user device may at each reconnection time with the video server request a different lower resolution chunk according to an estimate of the available bandwidth. Unfortunately, either each chunk is short in duration, resulting in loss of compression efficiency, or it is long in which case the adaption rate is slow. One way that compression efficiency can

be improved within longer chunks is by means of long GOPs. To allow gradual quality adaptation sometimes lengthy transition periods are required (Cicco, et al., 2010) which contrasts with a wireless channel, which is subject to volatile variations in conditions. Notice that D&P, progressive download, and adaptive progressive download all employ the *Transmission Control Protocol (TCP)* as the underlying transport protocol, which is a reliable, connection oriented transport protocol (For streams targeting the currently popular browser-based Adobe Flash Player, the *Real-Time Messaging Protocol (RTMP)* is employed. RTMP itself is a complex multi-channeled protocol, which employs TCP as its underlying transport protocol.). Of course, if TCP is used there is no need for error resilience or other forms of error control. However, while TCP has some value in penetrating security firewalls and coping with *Network Address Translation (NAT)* the need to maintain reliability through re-transmissions introduces additional latency. There is a further problem in that a feedback channel is required to service a continuous stream of *Acknowledgments (ACKs)*. To improve the response either a reliable stream of ACKs must be maintained or a specialist form of TCP (Bing, 2010) should be used to reduce the number of ACKs. Further issues exist with TCP's interpretation of most packets losses as losses due to congestion. Whereas this is an acceptable assumption in a wired environment, in a wireless environment packet loss can occur through temporary channel impairment. In which case, it is preferable to ignore the loss rather than staunch the flow. A good number of methods also exist to cope with this problem (Balkrishnan, et al., 2007) but they can involve implementation of video-specific units at the boundary of the wired and wireless network.

Despite the problems that using adaptive progressive download may encounter, commercial operators have pressed ahead with the approach (Cicco, et al., 2010), which has been standardized as Dynamic Adaptive HTTP Streaming (DASH) for the *3GPP* and the *Motion Picture Experts Group (MPEG)* (Stockhammer, et al., 2011). This in turn has resulted in recent publications that survey mobile video streaming with the *Hyper-Text Transfer Protocol (HTTP)* (Ma, et al., 2011; Lohmar, et al., 2011) this development. In Ma et al. (2011), it is predominantly the business case for employing HTTP that is made. This paper (Ma, et al., 2011) also provides a useful overview of current commercial practices in delivering video to mobile devices, especially smartphones. HTTP is a ubiquitous protocol available within Web browsers on all devices. Because HTTP is layered over TCP, there is no degradation of quality, which consequently means that the "brand-image" of content providers is not compromised, though of course the streaming service provider would be compromised if unacceptable delays were to occur through this form of delivery. On the other hand, using the true-streaming approach, outlined below in the context of two-way video streaming, may not gracefully degrade the video quality, especially if selection of data to be dropped is left to network routers, which are generally not multimedia aware.

DASH itself requires each video chunk to commence with a random access point, typically an I-picture. The remainder of the chunk is likely to be encoded as a set of P-pictures, with no other I-picture to achieve good coding gain. Management of stream switching for bitrate adaptation is the responsibility of the streaming destination or client. Generally this is achieved by dynamic calculation of the next chunk's address, which is specified as a *Uniform Resource Locator (URL)*. The client may also use HTTP range requests (allowing only part of an HTTP message to be selected), if these are supported, to request the next chunk. Because DASH works by switching between streams encoded at pre-agreed bitrates, it is necessary to store a set of chunks for each of these streams (typically up to six streams) at the server. This arrangement adds considerably to the management of the chunk store at the server

and to the chunk selection process. The size of the chunks determines the potential adaptation rate but it needs to be borne in mind that smaller chunks impose an overhead at the server.

Interactive forms of networked streaming are unable or unlikely to use the chunk-based approach. Instead, operators seem to have opted for true picture-by-picture streaming. For example, the *Internet Streaming Media Alliance (ISMA)* (Jack, 2007) was founded by Apple Computer, Cisco Systems, IBM, Kasenna, Philips an Sun Microsystems. ISMA has issued a set of profiles that adopt the following standardized protocols. These forms of streaming in a packet switched network usually employ the *User Datagram Protocol (UDP)* as the underlying transport protocol. To avoid congestion at buffers on the traditional network, a congestion controller is employed (Widmer, et al., 2001). Basically, such a controller, just as in the TCP case previously described, in a wireless environment is unable to distinguish between packet loss by congestion and packet loss due to volatile wireless conditions. One attempt to rescue the traditional congestion controller is to employ multiple connections (Chen & Zakhor, 2005), so that if one or more of the connections streaming rate is staunched then the others can compensate.

UDP (and TCP) clearly lacks timing synchronization information, which must then be provided. A convenient way to do this is through RTP (Perkins, 2003) (RFC 3550). However, notice that unlike TCP and UDP, RTP is not a protocol in the sense that there is additional controlling software as part of a protocol stack. RTP provides headers but not congestion or flow control. RTP has timing and packet sequence numbers. It also provides media information to allow synchronization of playback (for audio and in particular for lip synchronization) and for buffer management. RTP has standardized packet formats for MPEG-4 (RFC 3640), H.264/AVC (RFC 3984), and Microsoft's Video Codec-1 (VC-1) (Lee & Kalva, 2008) codec (RFC 4425). The *Real-time Transport Control Protocol (RTCP)*

(RFC 3550) is a companion protocol to RTP that can return performance information such as lost packets, delay jitter, and roundtrip time. RTCP can also be employed for adaptive management of jitter buffers. However, the combination of RTP and RTCP is most useful for synchronization of video and audio streams contained in multiple RTP streams.

Media containers are employed when more than one video stream with associated audio stream is carried on the same multimedia stream, as, for example, might occur when streaming a set of IPTV channels. The MPEG-2 TS allows packets from a number (typically seven, each 188 B in size) of video and audio channels to be multiplexed into a single RTP packet. It is also possible to encapsulate packets bearing an H.264/AVC bitstream into an MPEG-2 TS. The MPEG-2 TS is widely adopted, most likely because of the previous experience of broadcasters with it. However, other containers such as MPEG-4 may well result in less overhead. A remaining facility lacking in RTP is the ability to set up a streaming connection. Because the underlying transport protocol is UDP, many Internet firewalls will block the video stream. The *Real-Time Transport Streaming Protocol (RTSP)* (RFC 2326) can provide this facility, though its role is more general than that, as it works as a signaling protocol. It allows choice of transport protocol and feedback of pseudo-VCR commands such as rewind, pause and so on. RTSP may require separate RTP/UDP ports if separate video and RTP streams are maintained. However, if audio is included within the and MPEG-2 TS, itself encapsulated within an RTP stream, multiple UDP ports may not be necessary.

INTRA-REFRESH TECHNIQUES

The need for intra-refresh arises from the predictive nature of video coding, which exploits redundancies in the spatial and temporal domains, amongst other sources of redundancy (Ghanbari, 2011).

Intra-refresh can be viewed as a way of restricting the prediction range, and as such, it has various applications some of which have been alluded to in the Introduction.

Certainly, in mobile transmission, intra refresh can provide error resilience. Lost packets, when not replaced by erasure coding, or corrupted data (when errors go undetected and do not halt the decoder) can both create insecure prediction bases. Error concealment of affected areas is a partial remedy but can still lead to a difference between the encoder's view of the decoding process and that actually implemented by the decoder. Any areas that are unsafe can then be predicted from. This in turn leads to a corrupted area growing in spatial extent, at a maximum rate governed by the range of motion estimation.

The ability to randomly access a video sequence from either the point of a periodic intra-coded picture (I-picture) or gradually to perform the refresh process is, perhaps, the main use of intra-refresh. However, it is possible that intra-refresh could aid in the parallel processing of multi-resolution video (as now specified in the High profile of H.264/AVC) by restricting the prediction range. The same might apply to scalable video. However, notice that there are other methods of restricting the prediction range. The sub-picture slice structure of H.264/AVC can be configured without cross slice inter prediction and intra prediction is not permitted across slice boundaries. A complicating factor is that limited interference across slice boundaries if the H.264/AVC deblocking loop filter is not restricted to slice boundaries. The search range for motion vectors can also be restricted. B-pictures can also be readily dropped in codecs prior to H.264/AVC, as they do not act as references for other pictures.

There are various forms of intra-refresh. Historically, periodic insertion of I-pictures is the most common form, as it has been used in early digital broadcasting with the MPEG-1/2 codecs (Ghanbari, 2011). In this case, the sequence is periodically refreshed with a fully intra-coded picture

(I-picture) at regular intervals. In the presence of a feedback channel, it is possible dynamically to send an I-picture whenever it is requested by the receiver in order to recover from transmission error or for channel zapping purposes. It is also possible (Färber, et al., 1996; Wand & Chang, 1999) to send individual intra-refresh MBs on request through a feedback channel. Furthermore, in Liao and Villasenor (1996), a record of each MB is kept and MBs that might have a significant effect are sent if lost are intra-coded. The latter method avoids the need for a feedback channel but there is an obvious computational impact from maintaining the impact metric for every MB. As an example of a suitable feedback channel, in the ITU-T H.232 and H.324 video conferencing standards, a decoder at the receiver can request through the H.245 control protocol an intra-update of an entire video picture or of certain MBs.

GDR techniques take a different approach to periodic refresh. The video sequence is gradually updated with intra-coded MBs with a given number of MBs per picture. The selection of MBs to be intra coded and their number and pattern depends on the specific technique used. To improve the performance of this method, *Constrained Intra Prediction (CIP)* should be forced to prevent intra-coded MBs needing to use inter-coded samples for prediction. The category of gradual picture refresh can be further sub-divided into non-adaptive intra-refresh and adaptive intra-refresh techniques. The main types of non-adaptive intra-refresh are circular intra refresh and random intra refresh, Isolated region refresh (Hannuksela, et al., 2004) is somewhat different, as an entire picture region is initially intra-coded, with no need to set CIP, as the region is isolated in a coding sense from the rest of the picture.

In circular intra refresh, the video sequence is updated with intra-coded MBs using a pre-defined scan order and a selected number of MBs per picture. In an implementation of this method that will be called Intra-Refresh line, the sequence is refreshed with a line of intra coded MBs moving from top-to-bottom. In random intra refresh, the video sequence is updated with intra-coded MBs using a random pattern and a selected number of MBs per picture taking into account the estimated loss rate.

Turning to adaptive intra-refresh techniques, these are distinguished by whether a feedback channel is used or not. Clearly if a feedback channel is used, there are implications in terms of the need to provide that channel and the possible increased latency that may arise. Channel conditions can be taken into account in the selection of the coding mode of the MBs. Receiver statistics are used to indicate the channel conditions by means of feedback messages. On the other hand, non-adaptive techniques select the coding mode of each MB based on the combined optimization of both the rate and distortion.

In Jiang et al. (2008), the encoder was required to keep track of which parts of the image area were recently refreshed. The encoder would then refresh those MBs, which had more of an impact on error propagation. Alternatively, Tan and Pearmain (2010) proposed a scheme in which FMO was combined with adaptive MB grouping. Because such schemes are at an individual MB level, as previously remarked, they significantly increase the computational complexity arising from the required video content analysis. Moreover, methods using 'explicit' FMO also increase the bitrate and the degree of inter-packet dependency due to the need to include additional packets with the updated MB maps for every picture. In Wand and Chang (1999), once the decoder detects an error, it informs the encoder, which transmits intra-coded MBs to halt any error propagation. However, this procedure is unsuitable for conversational video services such as videophone or mobile teleconferencing.

Other schemes (Schreier & Rothermel, 2006; Krause, et al., 1991) improve upon the deterministic application of the cyclic refresh line method by resolving a problem that exists at the boundary between a cleansed area and an area yet to be cleansed by intra-refresh. Suppose the direction of motion within the sequence is from a potentially corrupted region to a cleansed region. Then motion compensated prediction could predict a cleansed region from a suspect region. In that case, the cycle needs to revisit those predicted areas in order to undo the new corruption. It is possible to restrict the range of prediction (Krause, et al., 1991) in these circumstances, but this will reduce coding efficiency. Alternatively, the direction of motion can be estimated (Schreier & Rothermel, 2006) by observation of motion vectors in the border regions between the clean and yet to be cleansed regions. Based on this a refresh pattern is found. However, this method does not lend itself easily to hardware implementation and depends on estimates. The main gain from this technique (Schreier & Rothermel, 2006) seems to be an improvement in data rate of about 1% rather than improved video quality.

In Haskell and Messerschmitt (1992) randomized placement of MBs was proposed to be up to a percentage determined by the average lifetime of errors. Random insertion may result in the duplication of the error propagation property if the same MB is selected in successive pictures. To improve on random placement in Côté and Kossentini (1999), based on knowledge of the form of error concealment at the decoder, rate distortion analysis is performed on error-concealed pictures of an H.263 codec. Thus, well-concealed MBs are not considered for intra-refresh MB replacement. However, as error concealment is now a non-normative feature of an H.264/AVC decoder, meaning that the form of error concealment is not standardized.

In Hannuksela et al. (2004), evolving isolated regions were employed to achieve GDR. The isolated region takes advantage of H.264/AVC's slice group mechanism. An isolated region consists of one or more slices that do not reference other slices within a picture. For GDR, the MBs within an initial seed region are intra-coded. The region is then evolved, meaning that it grows in successive

pictures. In an evolved isolated region, references must all derive from the seeded region. Thus, provided there is no error, a GDR recovery point is established when the isolated region evolves to cover a picture. Tests reported in Hannuksela et al. (2004) report that compared to periodic refresh with H.264/AVC IDR pictures there was between 11 and 17% bitrate loss. The overhead may well result from the reduction in coding efficiency by the restriction of the reference range when using an isolated region combined with the need to send a slice group map. Of course, refresh from IDR pictures may occur at a cost in a sudden increase in data rate if using VBR and a decrease in quality if using CBR. Moreover, in error-prone environments with packet loss rates of 3, 5, 10, and 20%, GDR results in higher objective quality video compared to periodic IDR pictures (tests were conducted streaming the *Paris* sequence (see the Case Study) at the 3G wireless rate of 384 kbps). This may well be due to the fact that a significant cause of quality degradation is the loss of packets from the IDR pictures themselves, as in periodic intra-refresh, their loss cannot be compensated for until the next IDR picture.

Forced intra-refresh with a cyclic MB line is a classical way to achieve GDR. If there are N lines per picture then the worst-case GDR should take place within $2N$-1 pictures (Schreier & Rothermel, 2006). The refresh rate can be increased by cycling more than one line at a time. However, this will increase the data rate, as intra-coding is markedly less efficient than intra-coding. For example, in a comparison between the datarates for I-pictures and P-pictures an overhead of as much as ten times was found in Schreier and Rothermel (2006) for coarsely quantized frames. Circular MB line refresh was one of the GDR options simulated in Hannuksela et al. (2004) for low bitrate streaming over a 3G wireless link. Unfortunately, the bitrates considered meant that the video quality never reached 30 dB even at zero packet loss. Nevertheless, the tests indicated that objective video quality (Peak Signal-to-Noise

Ratio (PSNR)) was only better for circular MB line refresh at low packet loss rates (3%). At higher loss rates (5, 10, and 20%), the use of an isolated region that evolved in a circular fashion was better in terms of equivalent video quality.

It was also found that an H.264/AVC algorithm that adjusted rate-distortion according to the expectation of lost MBs (loss-adjusted macroblock mode rate-distortion analysis) was superior at packet loss rates of 10% or more. However, that algorithm or a combination of that algorithm with the isolated region algorithm is generally thought (Wang, et al., 2003) to be computationally prohibitive for current deployments. Nevertheless, anticipating a future growth in computational capability at the end device, both Côté et al. (2000) and Zhang et al. (2000) consider adaptive selection of MB mode (intra- or inter-coding) depending on feedback from the receiver of packet loss patterns and distortion analysis.

CASE STUDY EXPLORATION OF INTRA-REFRESH WITH CYCLIC LINE

The scheme presented in this Section exploits the periodic insertion of intra-coded MB lines in a cyclic pattern within successive temporally predicted video pictures. The objective of inserting intra-coded MB lines is to mitigate error propagation at the cost of lower coding efficiency than purely predictive inter coding. Using a horizontal (or vertical) sliding intra-coded line, Figure 1, reduces the error drift arising from packet loss. However, it should be noted that an intra-coded MB line within a temporally predicted picture represents a significant percentage of the bits devoted to compressing the whole picture. Nonetheless, a packet containing data from a line of intra-coded MBs represents a small portion of the image area. Therefore, only a small potential quality penalty arises from the loss of a packet containing intra-refresh MBs due to the small image area affected (see Figure 3).

In evaluation tests, two versions of a Constant Bit Rate (CBR) stream with Common Intermediate Format (CIF) resolution (352 × 288 pixels/picture) at 30 picture/s were used, the one with intra-refresh MB line and the other without. In the Baseline Profile of H.264/AVC intended for mobile devices, only predictively coded P-pictures are permitted, as bi-predictive B-pictures result in an increased computational cost resulting from the identification of matching MBs in more than one picture. Therefore, this study employed a GoP structure of either IPPP pictures, i.e. a potentially infinite sequence of P-pictures after an initial intra-coded picture. Packets were limited at the codec to 1 kB. Motion estimation search range was set to eight.

Packet sizes were limited to a maximum of 1 kB using H.264/AVC's RTP packetization mode. The size of 1 kB is a characteristic *Maximum Transport Unit (MTU)* that avoids network segmentation. A single reference picture was used, as this limits the error introduced when using multiple reference pictures in conjunction with intra-refresh MBs (Moiron, et al., 2010). By setting the periodic refresh rate to eighteen the date rates for the two streams under comparison could be made equal (as there are 18 MB lines in one CIF picture). That CBR was 500 kbps.

Figure 4 shows the objective video quality for the semi-active video sequence *Paris*. The two studio presenters in this reference clip are seated and though they make some motions (see Figure 1), the main coding complexity is spatial arising from the bookcase in the background. From Figure 4, there is no difference in quality at zero error, indicating that there is no difference in the overall overhead from the two forms of intra refresh placement. However, as the Uniformly distributed data drop rate increases, then there is a marked decrease in the mean *Peak Signal-to-Noise Ratio (PSNR)* for the periodically refreshed version. In fact, the quality becomes poor after around 5% data drops, when the objective quality measure falls below 25 dB.

Figure 3. Cyclic intra-refresh MB line technique for the Paris public-domain test sequence, showing successive MB lines in lighter shading, with some slice boundaries also shown

Figures 5 and 6 show the effect on more active sequences, when it is immediately apparent that the gain is less. This is because motion (either object or camera motion) can disrupt the cleansing effect of the cyclic MB line. However, it is the direction of motion in the border area between cleansed and about to be cleansed regions that is important. If predictions are made from a yet to be cleansed area of a previous picture to an already cleansed area, then cleansing must recover the newly corrupted area.

Table 1 summarizes CBR results in which for comparison purposes all sequences were limited to 300 pictures. The table shows the PSNR gain (dB) from employing cycle intra-refresh line.

Gains for very static sequences such as *Akiyo* can be high while negative 'gains' do occur for sequences with adverse patterns of motion but the loss is usually less than 1 dB. Active sequences can be detected by inspection of the number of non-zero motion vectors.

FUTURE RESEARCH DEVELOPMENTS

The ability of the various means of intra-refresh has been studied in terms of error resilience and the provision of stream entry or swapping positions. However, a thoroughgoing comparison of

Figure 4. PSNR versus video data loss rate for Paris CBR streaming

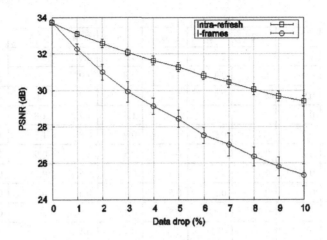

Figure 5. PSNR versus video data loss rate for Stefan CBR streaming

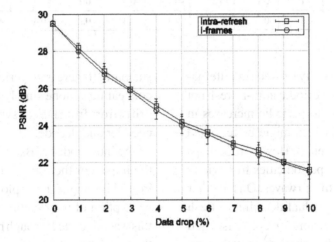

Figure 6. PSNR versus video data loss rate for Mobile CBR streaming

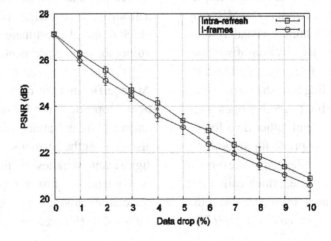

Table 1. Mean PSNR gain (decibels) when using intra-refresh MBs compared to using periodic I-pictures in CBR streaming

Data loss rate (%):	0	1	2	3	4	5	6	7	8	9	10
Akiyo	0.1	1.1	2.1	2.7	3.5	4.3	4.5	5	5.1	6	5.8
Paris	0	0.9	1.6	2.1	2.6	2.8	3.2	3.5	3.7	3.9	4
News	-0.2	0.9	2	2.3	2.5	3.3	3.2	3.6	3.5	3.5	4.2
Hall	-0.1	0.1	0.6	0.8	1	1.4	1.7	1.8	1.9	2.3	2.8
Tempete	0	0.2	0.4	0.6	0.7	0.9	0.8	1	1.2	1.1	1.6
Flower	0	0.2	0.3	0.4	0.4	0.5	1	0.9	1.1	1.1	1.1
Tennis	0	0.1	0	0.3	0.5	0.7	0.5	0.7	0.8	0.7	0.6
Mobile	0	0.3	0.5	0.3	0.5	0.4	0.5	0.4	0.4	0.3	0.4
Stefan	0	0.3	0.5	0.3	0.5	0.4	0.5	0.4	0.4	0.3	0.4
Football	-0.1	-0.1	-0.1	-0.1	-0.1	0	0.1	0	-0.1	-0.2	0.1
Bridge-close	-0.1	-0.1	-0.1	-0.1	-0.1	-0.1	-0.1	-0.1	-0.1	-0.2	-0.1
Highway	-0.2	-0.2	-0.2	-0.2	-0.3	-0.2	-0.2	-0.2	-0.3	-0.3	-0.1
Bus	-0.1	-0.1	-0.1	-0.3	-0.3	-0.3	-0.1	-0.2	-0.2	-0.1	-0.3
Soccer	0.1	-0.4	-0.3	-0.5	-0.4	-0.4	-0.2	-0.4	-0.4	-0.6	

the impact upon video quality between the alternative methods is lacking. Periodic intra-refresh not unexpectedly gives rise to periodic increases in the per-picture PSNR, which might be reflected in the subjective experience. However, perhaps we are all used to these periodicities from viewing TV broadcast over the airwaves. On the other hand, it could be that gradual decoding refresh through cyclic line insertion (or cyclic isolated region) produces more disturbing artifacts in some subjective sense. Randomized insertion (or the pseudo-randomized insertion of the JM implementation of H.264/AVC) of intra-coded MBs may be more appealing to the human visual system. There has also been an investigation of whether an isolated region that evolves in a spiral fashion to achieve gradual decoding refresh (Wang, et al., 2003) may be less disturbing. Another open issue is whether error concealment rather than forced intra-refresh is more appropriate for smooth regions in the image. This will depend on the type(s) of error concealment in place, which calls for a standardization of this area, even perhaps in the emerging *High Efficiency Video Coding (HEVC)*

standard. If there is no periodic refresh, does error concealment alone simply lead to the build-up of drift errors from lack of synchronization between encoder and decoder?

As intra-coded MBs act as prediction anchors, the impact of their loss is often visually disturbing. This 'might' be exploited through selective encryption of their coding data. In Tang (1996), this was attempted through random permutation of the transform AC coefficients instead of a zigzag scan. The zero-frequency DC coefficient's value was changed and its position altered. Optionally, a group of DC coefficients can be encrypted by a block cipher. Unfortunately, that attempt proved vulnerable to replacement attacks in which a DC coefficient is replaced by a fixed value and the AC coefficients are recovered from their statistical properties (for low-resolution images). A known plaintext attack (Qiao, et al., 1997) can also recover the permutation order. An additional significant weakness is that permuting the AC coefficients prior to entropy coding removes the correlations amongst the coefficients that the coder can exploit (for example as much as 46% in the

MPEG-2 codec). As a result, there is an increase in the bitrate. It should also be pointed out that encrypting the whole of each periodic I-picture is also vulnerable (Agi & Gong, 1996), because the blocks predicted from the I-picture can help reconstruct the I-picture. Besides as we have seen, P-pictures can contain intra-coded MBs, not just those forcibly inserted but also those naturally inserted by an encoder when no convenient reference MB exists. Thus the naturally inserted MBs (in MPEG-2 codecs) make portions of the image 'visible' irrespective of the difficulty of reconstructing from the predictive residuals. Selective encryption attention has now switched to the tables or output of entropy coders. However, it could still be that some security means exists that can use the properties of intra-coding.

One should also mention that in simulcast streaming systems I-pictures provided switching points to move to an alternative stream. More recently, the same need to switch streams has arisen in multi-view video. In the H.264/AVC codec's Extended profile special switching frames are specified (Karczewicz & Kurceren, 2003) to avoid the need for switching through I-pictures. SP frames within the stream take the place of I-pictures, with some gain in bitrate (and some decrease in quality). When a stream switch is decided upon either an SI frame (with no backward reference) or a secondary SP frame is substituted for the SP frame. However, this method has the disadvantage that switching frames for transitioning between streams, possibly in either direction, must be pre-coded and stored at the server. It appears that the bitstream's size and smoothness is traded for an increase in storage. Some progress has been made in reducing storage space but it may still possible to improve upon this. As an alternative for switching between streams at different rates, Psannis and Ishibashi (2008) proposed marionette frames (P(M)-frames). Primary SP-frames, though superior in coding efficiency that I-pictures, still introduce a drop in efficiency compared to P-pictures. P(M)-frames encode the difference between the same frame encoded at one Quantization Parameter (QP) and another target QP (the QP of the stream to be switched to). When a stream switch is required a pre-stored or transcoded P(M)-frame is transmitted before switching to the target stream. This introduces an additional frame into the stream but avoids the need to include SP-frames (or I-pictures) within a stream. However, some form of error-resilience will still be needed if an all P-picture sequence were to be transmitted over wireless.

CONCLUSION

The field of video coding is very rich and, consequently, numerous means exist of enhancing that coding. Error-resilience provision has become important within source coding, because of the extra dimension of mobility that has been added to consumer devices. It is also the case that FEC is already provided at the physical and link layers of wireless protocol stacks and that application-layer FEC may duplicate or be incompatible with FEC at lower layers (though packet erasure coding is often applied at the application layer). However, rather than examine all types of error resilience, this chapter has outlined the principle measures and then concentrated on intra-refresh methods. Linked into this is the possibility of gradual decoding refresh, which on mobile devices avoids the effect on bandwidth of large periodic intra-coded pictures, which can fill up output buffers and stretch the constrained bandwidth capacity available.

REFERENCES

Agi, I., & Gong, L. (1996). An empirical study of MPEG video transmissions. In *Proceedings of the Internet Society Symposium on Network and Distributed System Security*, (pp. 137-144). IEEE.

Agilent-Technologies. (2008). *Validating IPTV service quality under multiplay network conditions.* White Paper. Agilent-Technologies.

Ahson, S. A., & Ilyas, M. (2010). *Fixed-mobile convergence handbook.* Boca Raton, FL: CRC Press. doi:10.1201/EBK1420091700

Ali, I., Fleury, M., & Ghanbari, M. (2012). Content-aware intra-refresh for video streaming over lossy links. In *Proceedings of the IEEE International Conference on Consumer Electronics.* IEEE Press.

Andrews, J. G., Ghosh, A., & Muhamed, R. (2007). *Fundamentals of WiMAX: Understanding broadband wireless networking.* Upper Saddle River, NJ: Prentice Hall.

Balkrishnan, H., Padmanabhan, V., Seshan, S., & Katz, R. (2007). A comparison of mechanisms for improving TCP performance over wireless links. *IEEE/ACM Transactions on Networking, 5*(6), 756–769. doi:10.1109/90.650137

Barmada, B., Ghandi, M., Jones, E., & Ghanbari, M. (2005). Prioritized transmission of data partitioned H. 264 video with hierarchical QAM. *IEEE Signal Processing Letters, 12*(8), 577–580. doi:10.1109/LSP.2005.851261

Bing, B. (2010). *3D and HD broadband video networking.* Norwood, MA: Artech House.

Chen, M., & Zakhor, A. (2005). Rate control for streaming video over wireless. *IEEE Wireless Communications, 12*(4), 32–41. doi:10.1109/MWC.2005.1497856

Cicco, L. D., Mascolo, S., Bari, P., & Orabona, V. (2010). An experimental investigation of the Akamai adaptive video streaming. In *Proceedings of the 6th International Conference on HCI in Work and Learning, Life and Leisure: Workgroup Human-Computer Interaction and Usability Engineering,* (pp. 447-464). HCI.

Cisco. (2011). *Cisco visual networking index: Forecast and methodology, 2010-2011.* White Paper. San Jose, CA: Cisco.

Côté, G., & Kossentini, F. (1999). Optimal intra coding of blocks for robust video communication over the Internet. *EUROSIP Journal of Image Communication.* Retrieved from http://wftp3.itu.int/av-arch/video-site/9811_Seo/q15f38attach.pdf

Färber, N., Döhla, S., & Issing, J. (2006). Adaptive progressive download based on the MPEG-4 file format. *Journal of Zhejiang University Science A, 7*(1), 106–111. doi:10.1631/jzus.2006.AS0106

Färber, N., Steinbach, E., & Girod, B. (1996). Robust H.263 compatible transmission over wireless channels. In *Proceedings of the International Coding Picture Symposium,* (pp. 575-578). IEEE.

Ghanbari, M. (2011). *Standard codecs: Image compression to advanced video coding* (3rd ed.). Stevenage, UK: IET Press. doi:10.1049/PBTE054E

Girod, B., & Färber, N. (1999). Feedback-based error control for mobile video transmission. *Proceedings of the IEEE, 87*(10), 1707–1723. doi:10.1109/5.790632

Hannuksela, M. M., Wang, Y. K., & Gabbouj, M. (2004). Isolated regions in video coding. *IEEE Transactions on Multimedia, 6*(2), 259–267. doi:10.1109/TMM.2003.822784

Haskell, P., & Messerschmitt, D. (1992). Resynchronization of motion compensated video affected by ATM cell loss. In *Proceedings of IEEE International Conference on Acoustics, Speech, and Signal Processing,* (pp. 545-548). IEEE Press.

Hellge, C., Gomez-Barquero, D., Schierl, T., & Wiegand, T. (2011). Layer-aware forward error correction for mobile broadcast of layered media. *IEEE Transactions on Multimedia, 13*(3), 551–562. doi:10.1109/TMM.2011.2129499

Jack, K. (2007). *Video demystified* (5th ed.). Amsterdam, The Netherlands: Newnes.

Jiang, J., Guo, B., & Mo, W. (2008). Efficient intra refresh using motion affected region tracking for surveillance video over error prone networks. In *Proceedings of the International Conference on Intelligent System Design and Applications*, (pp. 242 –246). IEEE.

Karczewicz, M., & Kurceren, R. (2003). The SP- and SI-frames design for H.264/AVC. *IEEE Transactions on Circuits and Systems for Video Technology*, *13*(7), 637–644. doi:10.1109/TC-SVT.2003.814969

Krause, E., et al. (1991). *Method and apparatus for refreshing motion compensated sequential video images*. US 5,057,916. Washington, DC: United States Patent Office.

Kumar, S., Xu, L., Mandal, M., & Panchanathan, S. (2006). Error resiliency schemes in H. 264/AVC standard. *Elsevier Journal of Visual Communication and Image Representation*, *17*, 425–450. doi:10.1016/j.jvcir.2005.04.006

Lee, J.-B., & Kalva, H. (2008). *The VC-1 and H.264 video compression standards for broadband video services*. New York, NY: Springer. doi:10.1007/978-0-387-71043-3

Liao, J., & Villasenor, J. (1996). Adaptive intra update for video coding over noisy channels. In *Proceedings of IEEE International Conference on Image Processing (ICIP)*, (vol. 3, pp. 763-766). Lausanne, Switzerland: IEEE Press.

Lohmar, T., Einarsson, T., Fröjdh, P., Gabin, F., & Kampmann, M. (2011). Dynamic adaptive HTTP streaming of live content. In *Proceedings of IEEE Symposium on the World of Wireless, Mobile and Multimedia Networks*, (pp. 1-8). IEEE Press.

Luby, M., Stockhammer, T., & Watson, M. (2008). Application layer FEC in IPTV services. *IEEE Communications Magazine*, *45*(5), 95–101.

Ma, K. J., Bartos, R., Bhatia, S., & Nair, R. (2011). Mobile video delivery with HTTP. *IEEE Communications Magazine*, *49*(4), 166–175. doi:10.1109/MCOM.2011.5741161

Mao, S., Lin, S., Panwar, S., & Wang, Y. (2001). Reliable transmission of video over ad-hoc networks using automatic repeat request and multi-path transport. In *Proceedings of IEEE Vehicular Technology Conference*, (pp. 615-619). IEEE Press.

Merkle, P., Morvan, Y., Smolic, A., Farin, D., Müller, K., de With, P. H. N., & Wiegand, T. (2009). The effects of multiview depth video compression on multiviewrendering. *Signal Processing Image Communication*, *24*(1–2), 73–88. doi:10.1016/j.image.2008.10.010

Moiron, S., Ali, I., Ghanbari, M., & Fleury, M. (2010). Limitations of multiple reference frames with cyclic intra-refresh line for H.264/AVC. *Electronics Letters*, *47*(2), 103–104. doi:10.1049/el.2010.3018

Narkhede, N. S., & Kant, N. (2009). The emerging H.264/AVC advanced video coding standard and its applications. In *Proceedings of International Conference on Advances in Computing, Communication and Control*, (pp. 300-305). IEEE.

Ostermann, J., Bormans, J., List, P., Marpe, D., Narroschke, M., & Pereira, F. (2004). Video coding with H.264/AVC: Tools, performance and complexity. *IEEE Circuits and Systems Magazine*, *4*(1), 7–28. doi:10.1109/MCAS.2004.1286980

Ozbek, N., & Tumnali, T. (2005). A survey on the H. 264/AVC standard. *Turk Journal of Electrical Engineering*, *13*, 287–302.

Paul, S. (2011). *Digital video distribution in broadband, television, mobile and converged networks*. Chichester, UK: John Wiley & Sons Ltd. doi:10.1002/9780470972915

Perkins, C. (2003). *RTP: Audio and video for the internet*. Boston, MA: Addison Wesley.

Psannis, K., & Ishibashi, Y. (2008). Enhanced H.264/AVC stream switching over varying bandwidth networks. *IEICE Electronics Express, 5*(19), 827–832. doi:10.1587/elex.5.827

Qiao, L., Nahrstedt, K., & Tam, M.-C. (1997). Is MPEG encryption by using random list instead of zigzag order secure? In *Proceedings of the IEEE International Symposium on Consumer Electronics*, (pp. 226-229). IEEE Press.

Razavi, R., Fleury, M., & Ghanbari, M. (2009). Adaptive packet-level interleaved FEC for wireless priority-encoded video streaming. *Advances in Multimedia*. Retrieved from http://www.hindawi.com/journals/am/2009/982867/

Redmill, D. W., & Kingsbury, N. G. (1996). The EREC: An error resilient technique for coding variable-length blocks of data. *IEEE Transactions on Image Processing, 5*(4), 565–574. doi:10.1109/83.491333

Richardson, I. (2003). *H.264 and MPEG-4 video compression: Video coding for next-generation multimedia*. New York, NY: John Wiley & Sons Inc. doi:10.1002/0470869615

Sandvine. (2011). *Sandvine's global internet phenomena report*. White Paper. New York, NY: Sandvine.

Schreier, R. M., & Rothermel, A. (2006). Motion adaptive intra refresh for low-delay video coding. In *Proceedings of the IEEE International Conference on Consumer Electronics*, (pp. 453-454). IEEE Press.

Schwarz, H., Marpe, D., & Wiegand, T. (2007). Overview of the scalable video coding extension of the H.264/AVC standard. *IEEE Transactions on Circuits and Systems for Video Technology, 17*(9), 1103–1120. doi:10.1109/TCSVT.2007.905532

Shokrollahi, A. (2006). Raptor codes. *IEEE Transactions on Information Theory, 52*(6), 2551–2567. doi:10.1109/TIT.2006.874390

Simpson, W. (2008). *Video over wireless*. Burlington, MA: Focal Press.

Son, N., & Jeong, J. (2008). An effective error concealment for H.264/AVC. In *Proceedings of IEEE 8th International Conference on Computer and Information Technology Workshops*, (pp. 385-390). IEEE Press.

Stockhammer, T., Fröjdh, P., Sodagar, I., & Rhyu, S. (Eds.). (2011). *Information technology — MPEG systems technologies — Part 6: Dynamic adaptive streaming over HTTP (DASH)*. ISO/IEC MPEG Draft International Standard. Retrieved from http://developer.longtailvideo.com/trac/export/1509/.../adaptive/.../dash.pdf

Stockhammer, T., & Zia, W. (2007). Error resilient coding and decoding strategies for video communication. In van der Schaar, M., & Chou, P. A. (Eds.), *Multimedia over IP and Wireless Networks* (pp. 13–58). Burlington, MA: Academic Press. doi:10.1016/B978-012088480-3/50003-5

Sullivan, G., & Wiegand, T. (2005). Video compression—From concepts to the H. 264/AVC standard. *Proceedings of the IEEE, 93*(1), 18–31. doi:10.1109/JPROC.2004.839617

Tan, K., & Pearmain, A. (2010). An FMO-based error resilience method in H.264/AVC and its UEP application in DVB-H link layer. In *Proceedings of the IEEE International Conference on Multimedia and Expo*, (pp. 214 –219). IEEE Press.

Tang, L. (1996). Methods for encrypting and decrypting MPEG video data efficiently. In *Proceedings of the 4th ACM International Multimedia Conference and Exhibition*, (pp. 219-229). ACM Press.

Thomos, N., Argyropoulos, S., Boulgouris, N., & Strintzis, M. (2005). Error-resilient transmission of H.264/AVC streams using flexible macroblock ordering. In *Proceedings of Second European Workshop on the Integration of Knowledge, Semantic, and Digital Media Techniques*, (pp. 183-189). IEEE.

Wand, J.-T., & Chang, P.-C. (1999). Error-propagation prevention technique for real-time video transmission over ATM networks. *IEEE Transactions on Circuits and Systems for Video Technology*, *9*(3), 513–523. doi:10.1109/76.754780

Wang, Y.-K., Hannuksela, M. M., & Gabbouj, M. (2003). Error-robust inter/intra macroblock mode selection using isolated regions. In *Proceedings of the 13th Packet Video Workshop*. IEEE.

Watson, R. (2009). *Fixed-mobile convergence and beyond*. Amsterdam, The Netherlands: Elsevier (Newnes).

Wen, W.-C., Hsiao, H.-F., & Yu, J.-Y. (2007). Dynamic FEC-distortion optimization for H.264 scalable video streaming. In *Proceedings of IEEE Workshop on Multimedia Signal Processing*, (pp. 147-150). IEEE Press.

Wenger, S. (2003). H. 264/AVC over IP. *IEEE Transactions on Circuits and Systems for Video Technology*, *13*(7), 645–656. doi:10.1109/TCSVT.2003.814966

Widmer, J., Denda, R., & Mauve, M. (2001). A survey on TCP-friendly congestion control. *IEEE Network*, *15*(3), 28–37. doi:10.1109/65.923938

Wiegand, T., Sullivan, G. J., Bjøntegaard, G., & Luthra, A. (2003). Overview of the H.264/AVC video coding standard. *IEEE Transactions on Circuits and Systems for Video Technology*, *13*(7), 560–576. doi:10.1109/TCSVT.2003.815165

Yetgin, Z., & Seckin, G. (2008). Progressive download for 3G wireless multicasting. *International Journal of Hybrid Information Technology*, *2*(2), 67–82.

Zhang, R., Regunthan, S. L., & Rose, K. (2000). Video coding with optimal inter/intra-mode switching for packet loss resilience. *IEEE Journal on Selected Areas in Communications*, *18*(6), 966–976. doi:10.1109/49.848250

ADDITIONAL READING

Côté, G., Kossentini, F., & Wenger, F. (2001). Error resilience coding. In Sun, M.-T., & Reibman, A. R. (Eds.), *Compressed Video over Networks* (pp. 309–342). New York, NY: Marcel Dekker, Inc.

Côté, G., Shirani, S., & Kossentini, F. (2000). Optimal mode selection and synchronization for robust video communications over error-prone networks. *IEEE Journal on Selected Areas in Communications*, *18*(6), 952–965. doi:10.1109/49.848249

Girod, B., & Färber, N. (2001). Wireless video. In Sun, M.-T., & Reibman, A. R. (Eds.), *Compressed Video over Networks* (pp. 465–511). New York, NY: Marcel Dekker, Inc.

Massoudi, A., Lefebrve, F., de Vleeschouwer, C., Macq, B., & Quisquater, J.-J. (2008). Overview on selective encryption of image and video: Challenges and perspectives. *EURASIP Journal on Information Security*. Retrieved from http://jis.eurasipjournals.com/content/2008/1/179290

Ngan, K. N., Yap, C. W., & Tan, K. T. (2001). *Video coding for wireless communications systems*. New York, NY: Marcel Dekker, Inc.

Richardson, I. E. G. (2003). *H.264 and MPEG-4 video compression: Video coding for next-generation multimedia*. Chichester, UK: John Wiley & Sons Ltd. doi:10.1002/0470869615

Rupp, M. (Ed.). (2009). *Video and multimedia transmissions over cellular networks*. Chichester, UK: John Wiley & Sons Ltd. doi:10.1002/9780470747773

Sadka, A. G. (2002). *Compressed video communications*. Chichester, UK: John Wiley & Sons Ltd. doi:10.1002/0470846712

Schulzrinne, H. (2001). IP networks. In Sun, M.-T., & Reibman, A. R. (Eds.), *Compressed Video over Networks* (pp. 38–138). New York, NY: Marcel Dekker, Inc.

Yang, C., Li, Y., & Chen, J. (2011). A new mobile streaming system base-on http live streaming protocol. In *Proceedings of the International Conference on Wireless Communications, Networking and Mobile Computing*, (pp. 1-4). IEEE.

Zimmermann, M., & Seilheimer, M. (2012). Reviewing HTTP and RTSP work in two actual commercial media delivery platforms for multimedia services and mobile devices. In Sarmiento, A. S., & Lopez, E. M. (Eds.), *Multimedia Services and Streaming for Mobile Devices: Challenges and Innovations* (pp. 91–110). Hershey, PA: IGI Global. doi:10.4018/978-1-61350-144-3.ch005

KEY TERMS AND DEFINITIONS

Error Resilience: The protection of compressed video bitstreams by source-coded measures such as ASO, FMO, redundant slices, and flexible frame referencing. Alternatively or additionally, forward error correction employs channel coding to contain errors.

Fixed-Mobile Convergence: The industry trend towards merging the network provision across wired and wireless networks in such a way that a consumer will be unable to distinguish the service available on a mobile and static device.

Gradual Decoding Refresh: Is the ability to commence decoder recovery at a point other than at a periodic intra coded picture (in the H.264/AVC codec at an Instantaneous Decoding Refresh (IDR) picture).

High Efficiency Video Coding (HEVC): HEVC is billed as the successor to H.264/AVC and has even been called H.265. However, its developers suggest that it will mainly improve on computational performance through parallel processing. It is also reported to be conceptually similar in structure to H.264/AVC.

Instantaneous Decoding Refresh: Due to the possibility of multi-reference in H.264/AVC and Instantaneous Decoding Refresh (IDR) picture is required to prevent prediction reference to a point prior to that IDR picture. Intra-coded I-pictures do not necessarily have the IDR property in an H.264?AVC codec. However, in standard codecs prior to H.264/AVC they do have the IDR property.

Intra-Refresh: The use of intra-coded macroblocks to provide a recovery point from which a decoder can recover part or the whole of a picture in the event of data loss or corruption.

Isolated Region: A region within a picture is coded completely independently of other parts of the picture. In an evolving isolated region, the isolated region changes its shape in successive pictures but may not reference parts of other pictures outside that isolated region.

Quantization Parameter: The Quantization Parameter (QP) controls the amount of detail present in a video stream. In an H.264/AVC codec it varies between 0 and 51. A high QP results in low-quality video and a low bitrate. Conversely, a high QP results in high-quality video but a high bitrate. By setting the QP at a fixed value, open loop coding results in a Variable Bitrate (VBR). However, VBR may strain the bandwidth capacity and can overwhelm the decoder input buffer. However, in closed loop coding according to a target bitrate, a Constant Bitrate (CBR) results. Unfortunately, this can result in fluctuations in video quality and unnecessary consumption of bandwidth. A compromise possibility is truncated VBR in which the rate is prevented from exceeding a threshold. However, this possibility does not appear to be supported in the JM reference software for H.264/AVC. Both CBR encoding and truncated VBR require a more complex encoder, which can introduce encoding delay.

Rate-Distortion Analysis: The relation between source coding means and data rate is not a linear one. Consequently, rate-distortion seeks to optimize that relationship, typically through the method of Lagrangian multipliers. For example, the relationship between quantization parameter and objective video quality (PSNR) can be systematically tested to find an optimal point. Rate distortion analysis can explore other relationships such as herein between intra- and inter-coding of macroblocks.

Chapter 6
Region–of–Interest:
Processing and Coding Techniques
Overview of Recent Trends and Directions

Dan Grois
Ben-Gurion University of the Negev, Israel

Ofer Hadar
Ben-Gurion University of the Negev, Israel

ABSTRACT

This chapter comprehensively covers the topic of the Region-of-Interest (ROI) processing and coding for multimedia applications. The variety of end-user devices with different capabilities, ranging from cell phones with small screens and restricted processing power to high-end PCs with high-definition displays, have stimulated significant interest in effective technologies for video adaptation. Therefore, the authors make a special emphasis on the ROI processing and coding with regard to the relatively new H.264/SVC (Scalable Video Coding) standard, which have introduced various scalability domains, such as spatial, temporal, and fidelity (SNR/quality) domains. The authors' observations and conclusions are supported by a variety of experimental results, which are compared to the conventional Joint Scalable Video Model (JSVM).

INTRODUCTION

The number of video applications has been dramatically increased in the last decade, due to many reasons, such as rapid changes in the video coding standardization process driven by the increase of the computing power and significant developments of network infrastructures (Grois & Hadar, 2012). Nowadays, the most common video applications include wireless and wired Internet video streaming, high-quality video conferencing, High-Definition (HD) TV broadcasting, HD DVD storage and Blu-ray storage, while employing a variety of video transmission and storage systems

DOI: 10.4018/978-1-4666-2833-5.ch006

(e.g., MPEG-2 for broadcasting services over satellite, cable, and terrestrial transmission channels, or H.320 for conversational video conferencing services (Schwarz, et al., 2007). Also, most access networks are usually characterized by a wide range of connection qualities, and a wide range of end-user devices with different capabilities, starting from cell phones/mobile devices with relatively small displays and limited computational resources to powerful Personal Computers (PCs) with high-resolution displays (Schwarz, et al., 2007).

As a result, due to the continuous need for scalability, much of the attention in the field of video processing and coding is currently directed to the Scalable Video Coding (SVC), which was standardized in 2007 as an extension of H.264/AVC (Schwarz, et al., 2007), since the bit-stream scalability for video is currently a very desirable feature for many multimedia applications (e.g., video conferencing, video surveillance, telemedical applications, etc.). The need for the scalability arises from the need for spatial formats, bit-rates or power (Wiegand & Sullivan, 2003; Grois & Hadar, 2011a, 2011b). To fulfill these requirements, it would be beneficial to simultaneously transmit or store video in a variety of spatial/temporal resolutions and qualities, leading to video bit-stream scalability. Major requirements for the Scalable Video Coding are to enable encoding of a high-quality video bitstream that contains one or more subset bitstreams, each of which can be transmitted and decoded to provide video services with lower temporal or spatial resolutions, or to provide reduced reliability, while retaining reconstruction quality that is highly relative to the rate of the subset bitstreams. Therefore, Scalable Video Coding provides important functionalities, such as the spatial, temporal and SNR (quality) scalability, thereby enabling power adaptation. In turn, these functionalities lead to enhancements of video transmission and storage applications.

SVC has achieved significant improvements in coding efficiency compared to the scalable profiles of prior video coding standards due to the largely increased flexibility and adaptability (e.g., SVC enables to provide the graceful degradation in lossy transmission environments as well as the bit-rate, format, and power adaptation). In addition to the temporal, spatial, and quality scalabilities, the SVC supports the Region-of-Interest (ROI) scalability. The ROI is a desirable feature in many future scalable video coding applications, such as mobile device applications, which have to be adapted to be displayed on a relatively small screen (thus, a mobile device user may require to extract and track only a predefined Region-of-Interest within the displayed video). At the same time, other users having a larger mobile device screen may wish to extract other ROI(s) to receive greater video stream resolution (Grois & Hadar, 2011c, 2011d). Therefore, to fulfill these requirements, it would be beneficial to simultaneously transmit or store a video stream in a variety of Regions-of-Interest (e.g., each Region-of-Interest having different spatial resolution, as presented in Figure 1), as well to enable efficiently tracking the predefined Region-of-Interest.

This chapter is organized as follows: first, recent Region-of-Interest (ROI) detection and tracking techniques are described in detail, while presenting the pixel-domain approach and compressed-domain approach, and further providing various models and techniques, such as the visual attention model, object detection, face detection, skin detection. Secondly, ROI coding for the H.264/SVC (Schwarz, et al., 2007) standard is presented, including the ROI scalability by performing cropping and the ROI scalability by using Flexible Macroblock Ordering (FMO) technique (Lambert, et al., 2006). Then, the bit-rate control techniques for the ROI coding are discussed. After that, the complexity-aware ROI scalable video coding by performing pre-processing is discussed in detail, while focusing on the complexity-aware SVC bit allocation. Finally, this chapter is concluded by presenting future research directions.

Figure 1. Defining ROIs with different spatial resolutions (e.g., QCIF, CIF, SD/4CIF resolutions) to be provided within a scalable video coding stream

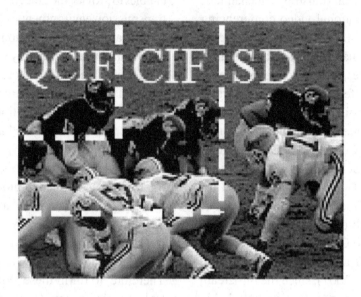

REGION-OF-INTEREST DETECTION AND TRACKING

In order to successfully perform the ROI coding, it is important to accurately detect, and then correctly track, the desired Region-of-Interest. There are mainly two methods for the ROI detection and tracking: (a) the pixel-domain approach; and (b) the compressed-domain approach. The pixel-domain approach is more accurate compared to the compressed-domain approach, but it requires relatively high computational resources. On the other hand, the compressed-domain approach does not consume many resources since it exploits the encoded information (such as DCT coefficients, motion vectors, macroblock types that are extracted in a compressed bitstream, etc.) (Manerba, et al., 2008; Kas & Nicolas, 2009; Hanfeng, et al., 2001; Zeng, et al., 2005). However, the compressed-domain approach results in a relatively poor performance due to many reasons, such as unreliability of the encoded information, sparse assignment of the block-based data; also, it is mainly applicable to simple scenarios and is poorly applicable for abrupt appearance changes and occlusions. In addition,

for the same reason, the compressed-domain approach has significantly fast processing time and is adaptive to compressed videos.

Both the pixel-domain and compressed-domain approaches are explained in detail in the following sections.

Pixel-Domain Approach

Generally, the main research on object detection and tracking has been focused on the pixel domain approach since it can provide a powerful capability of object tracking by using various technologies. The pixel-domain detection can be classified into the following types:

- **Region-Based Methods:** *"According to these methods, the object detection is performed according to ROI features, such as motion distribution and color histogram. The information with regard to the object colors can be especially useful when these colors are distinguishable from the image background or from other objects within the image"* (Vezhnevets, 2002).

- **Feature-Based Methods:** (Shokurov, et al., 2003). *"According to these methods, various motion parameters of feature points are calculated (the motion parameters are related to affine transformation information, which in turn contains rotation and 2D translation data)."*
- **Contour-Based Methods:** *"According to these methods, the shape and position of objects are detected by modeling the contour data"* (Wang, et al., 2002).
- **Template-Based Methods:** *"According to these methods, the objects (such as faces) are detected by using predetermined templates"* (Schoepflin, et al., 2001).

As mentioned above, the pixel-domain approach is, generally, more accurate than the compressed-domain approach, but has relatively high computational complexity and requires further additional computational resources for decoding compressed video streams. Therefore, the desired ROI can be predicted in a relatively accurate manner by defining various pixel-domain models, such as visual attention models, object detection models, face detection and skin detection models, as presented in detail in the following sub-sections.

Visual Attention

The visual attention models refer to the ability of a human user to concentrate his/her attention on a specific region of an image/video. This involves selection of the sensory information by the primary visual cortex in the brain by using a number of characteristic, such as intensity, color, size, orientation in space, and the like (Hu, et al., 2008). Actually, the visual attention models simulate the behavior of the Human Visual System (HVS), and in turn enable detection of the Region-of-Interest within the image/video, such as presented in Figure 2.

Several research investigations have been conducted with this regard in order to achieve better ROI detection performance, and in turn improve the ROI visual presentation quality. Thus, for example Cheng et al. (2005) present a framework for automatic video Region-of-Interest determination based on a user attention model, while considering the three types of visual attention features, i.e. intensity, color and motion. The contrast-based intensity model is based on the fact that particular color pairs, such as red-green and blue-yellow possess high spatial and chromatic opposition; the same characteristics exist in high difference lighting or intensity pairs. Thus, according to Cheng et al. (2005), the intensity,

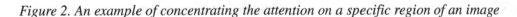

Figure 2. An example of concentrating the attention on a specific region of an image

red-green color and blue-yellow color constant models should be included into the user attention representation module. In addition, when there is more than one ROI within the frame (e.g., a number of football players), then a saliency map is used which shows the ability to characterize the visual attraction of the image/video. The saliency map, which represents visual saliency of a corresponding visual scene, is divided into *n* regions, and ROI is declared for each such region, thereby enabling to dynamically and automatically determine ROI for each frame-segment.

Also, Sun et al. (2010) propose a visual attention based approach to extract texts from a complicated background in camera-based images. Firstly, the approach applies the simplified visual attention model to highlight the Region of Interest (ROI) in an input image and to yield a map consisting of the ROIs. Secondly, an edge map of image containing the edge information of four directions is obtained by Sobel operators; character areas are detected by connected component analysis and merged into candidate text regions. Finally, the map consisting of the ROIs is employed to confirm the candidate text regions.

Further, other visual attention models have been recently proposed to improve the ROI visual presentation quality, such as Engelke et al. (2009), which discusses two ways of obtaining subjective visual attention data that can be subsequently used to develop visual attention models based on the selective region-of-interest and visual fixation patterns. Chen et al. (2010) disclose a model of the focus of attention for detecting the attended regions in video sequences by using the similarity between the adjacent frames, establishing the gray histogram, selecting the maximum similarity as a predicable model, and finally obtaining a position of the focus of attention in the next fame. Li et al. (2010) present a three-stage method that combines the visual attention model with target detection by using the saliency map, covering the region of interest with blocks and measuring the similarity between the blocks and the template. Kwon et

al. (2010) show a ROI based video preprocessor method that deals with the perceptual quality in a low-bit rate communication environment, further proposing three separated processes: the ROI detection, the image enhancement, and the boundary reduction in order to deliver better video quality at the video-conferencing application for use in a fixed camera and to be compatible as a preprocessor for the conventional video coding standards. Duchowski (2000) presents a wavelet-based multiresolution image representation method for matching the Human Visual System (HVS) spatial acuity within multiple Regions of Interest (ROIs). ROIs are maintained at high (original) resolution, while peripheral areas are gracefully degraded. As seen from the above, the visual attention approach has recently become quite popular among researchers, and many improved techniques have been lately presented.

Object Detection

Automatic object detection is one of the important steps in image processing and computer vision (Bhanu, et al., 1997; Lin & Bhanu, 2005). The major task of object detection is to locate objects in images and extract the regions containing them (the extracted regions are ROIs). The quality of object detection is highly dependent on the effectiveness of the features used in the detection. Finding or designing appropriate features to capture the characteristics of objects and building the feature-based representation of objects are the key to the success of detection. Usually, it is not easy for human experts to figure out a set of features to characterize complex objects, and sometimes, simple features directly extracted from images may not be effective in object detection.

The ROI detection is especially useful for medical applications (Liu, et al., 2006). Automatic detection of an ROI in a complex image or video like endoscopic neurosurgery video is an important task in many image and video processing applications such as an image-guided surgery system, real-time patient monitoring system, and object-based video compression. In telemedical applications, object-based video coding is highly useful because it produces a good perceptual quality in a specified region, i.e., a region of interest (ROI), without requiring an excessive bandwidth. By using a dedicated video encoder, the ROI can be coded with more bits to obtain a much higher quality than that of the non-ROI, which is coded with fewer bits.

In the last decade, various object detection techniques have been proposed. For example, Han and Vasconcelos (2008) present a fully automated architecture for object-based ROI detection, based on the principle of discriminant saliency, which defines as salient the image regions of strongest response to a set of features that optimally discriminate the object class of interest from all the others. It consists of two stages, saliency detection, and saliency validation. The first detects salient points; the second verifies the consistency of their geometric configuration with that of training examples. Both the saliency detector and the configuration model can be learned from cluttered images downloaded from the Web.

In addition, Wang et al. (2008) describe a simple and novel algorithm for detecting foreground objects in video sequences using just two consecutive frames. The method is divided into three layers: sensory layer, perceptual layer, and memory layer (short-term memory in conceptual layer). In the sensory layer, successive images are obtained from one fixed camera, and some early computer vision processing techniques are applied here to extract the image information, which are edges and visually inconsistent spliced regions. In the perceptual layer, moving objects are extracted based on the information from the sensory layer, and may request the sensory layer to support with more detail. The detected results are stored in the memory layer, and help the perceptual layer to detect the temporal static objects.

In addition, Jeong et al. (2006) proposes an object image detection system based on the ROI.

The system proposed by Jeong et al. (2006) excels in that ROI detection method is specialized in object image detection. In addition, a novel feature consisting of weighted SCD based on an ROI and skin color structure descriptor is presented for classifying objects in an image. Using the ROI detection method, Jeong et al. (2006) can reduce the noisy information in an image and extract more accurate features for classifying objects in an image.

Further, Lin and Bhanu (2005) use Genetic Programming (GP) to synthesize composite operators and composite features from combinations of primitive operations and primitive features for object detection. The motivation for using GP is to overcome the human experts' limitations of focusing only on conventional combinations of primitive image processing operations in the feature synthesis. GP attempts many unconventional combinations that in some cases yield exceptionally good results. Compared to a traditional region-of-interest extraction algorithm, the composite operators learned by GP are more effective and efficient for object detection. Still further, Kim and Wang (2009) propose a method for smoke detection in outdoor video sequences, which contains three steps. The first step is to decide whether the camera is moving or not. While the camera is moving, the authors skip the ensuing steps. Otherwise, the second step is to detect the areas of change in the current input frame against the background image and to locate Regions of Interest (ROIs) by connected component analysis. In the final step, the authors decide whether the detected ROI is smoke by using the k-temporal information of its color and shape extracted from the ROI.

Face Detection

Face detection can be regarded as a specific case of object-class detection. In object-class detection, the task is to find the locations and sizes of all objects in an image that belong to a given class (such as pedestrians, cars, and the like). Also, face detection can be regarded as a more general case of face localization. In face localization, the task is to find the locations and sizes of a known number of faces (usually one). In face detection, one does not have this additional information.

Early face-detection algorithms focused on the detection of frontal human faces, whereas recent face detection methods aim to solve the more general and difficult problem of multi-view face detection. Face detection from an image video is considered to be a relatively difficult task due to a plurality of possible visual representations of the same face: the face scale, pose, location, orientation in space, varying lighting conditions, face emotional expression, and many others (e.g., as presented in Figure 3). Therefore, in spite of the recent technological progress, this field still has many challenges and problems to be resolved.

Generally, the challenges associated with face detection can be attributed to the following factors (Yang, et al., 2002):

- **Facial Expression:** The appearance of faces is directly affected by a person's facial expression.
- **Pose:** The images of a face vary due to the relative camera-face pose (frontal, 45 degree, profile, upside down), and some facial features such as an eye or the nose may become partially or wholly occluded.
- **Occlusion:** Faces may be partially occluded by other objects. In an image with a group of people, some faces may partially occlude other faces.
- **Image Orientation:** Face images directly vary for different rotations about the camera's optical axis.
- **Imaging Conditions:** When the image is formed, factors such as lighting (spectra, source distribution and intensity) and camera characteristics (sensor response, lenses) affect the appearance of a face.
- **Presence or absence of structural components:** Facial features such as beards, mustaches, and glasses may or may not be present and there is a great deal of variability among these components including shape, color, and size.

During the last decade, many researchers around the world tried to improve the face detection and develop an efficient and accurate detection system. For example, Mustafah et al. (2009) propose a design of a face detection system for a real-time high resolution smart camera, while making an emphasis on the problem of crowd surveillance where the static color camera is used to monitor a wide area of interest, and utilizing a

Figure 3. An example of a plurality of possible visual representations of the same face, which has an influence on the accurate face detection. Although the accuracy of face detection systems has dramatically increased during the last decade, such systems still have many challenges and problems to be resolved, such as varying lighting conditions, facial expression, presence, or absence of structural components, etc.

background subtraction method to reduce the Region-of-Interest (ROI) to areas where the moving objects are located. Another work was performed by Zhang et al. (2009), in which was presented a ROI based H.264 encoder for videophone with a hardware macroblock level face detector. The ROI definition module operates as a face detector in a videophone, and it is embedded into the encoder to define the currently processed and encoded ROI macroblocks, while the encoding process is dynamically controlled according to the ROI (the encoding parameters vary according to ROI).

Further, other face detection techniques have been recently proposed to improve the face detection, such as: Micheloni et al. (2005) present an integrated surveillance system for the outdoor security; Qayyum and Javed (2006) disclose a notch based face detection, tracking and facial feature localization system, which contains two phases: visual guidance and face/non-face classification; and Sadykhov and Lamovsky (2008) disclose a method for real-time face detection in 3D space.

Skin Detection

The successful recognition of the skin ROI simplifies the further processing of such ROI. The main aim of traditional skin ROI detection schemes is to detect skin pixels in images, thereby generating skin areas. According to Abdullah-Al-Wadud and Oksam (2007), if the ROI detection process misses a skin region or provides regions having lots of holes in it, then the reliability of applications significantly decreases. Therefore, it is important to maintain the efficiency of the Human-Computer Interaction (HCI)-based systems. In turn, Abdullah-Al-Wadud and Oksam (2007) present an improved region-of-interest selection method for skin detection applications. This method can be applied in any explicit skin cluster classifier in any color space, while not requiring any learning or training procedure. The proposed algorithm mainly operates on a grayscale image (DM), but the processing is based on color information. The

scalar distance map contains the information of the vector image, thereby making this method relatively simple to implement.

In addition, Yuan and Mu (2007) present an ear detection method, which is based on skin-color and contour information, while introducing a modified Continuously Adaptive Mean Shift (CAMSHAFT) algorithm for rough and fast profile tracking. The aim of profile tracking is to locate the main skin-color regions, such as the ROI that contains the ear. The CAMSHIFT algorithm is based on a robust non-parameter technique for climbing density gradients to find the peak of probability distribution called the mean shift algorithm. The mean shift algorithm operates on a probability distribution, so in order to track colored objects in video sequence, the color image data has to be first represented as the color distribution. According to Yuan and Mu (2007), the modified CAMSHIFT method is performed as follows:

- Generating the skin-color histogram on training set skin images.
- Setting the initial location of the 2D mean shift search window at a fixed position in the first frame such as the center of the frame.
- Using the generated skin-color histogram to calculate the skin-color probability distribution of the 2D region centered at the area slightly larger than the mean shift window size.
- Calculating the zeroth moment (area of size) and mean location (the centroid).
- For the next frame, centering the search window at the calculated mean location and setting the window size using a function of the zeroth moment. Then the previous two steps are repeated.

In addition, Chen et al. (2003) present a video coding H.263 based technique for robust skin-color detection, which is suitable for real time videoconferencing. According to Chen et al.

(2003), the ROIs are automatically selected by a robust skin-color detection which utilizes the Cr and RGB variance instead of the traditional skin color models, such as YCbCr, HSI, etc. The skin color model defined by Cr and RGB variance can choose the skin color region more accurately than other methods. The distortion weight parameter and variance at the macroblock layer are adjusted to control the qualities at different regions. As a result, the quality at the ROI can significantly improved.

In the following two sub-sections, the recent advances in the compressed-domain detection and region-of-interest tracking are presented.

Compressed-Domain Detection

The conventional compressed domain algorithms exploit motion vectors or Discrete Cosine Transform (DCT) coefficients instead of original pixel data as resources in order to reduce computational complexity of object detection and tracking (You, 2010).

In general, the compressed domain algorithms can be categorized as follows: the clustering-based methods and the filtering-based methods.

The clustering-based methods (Benzougar, et al., 2001; Babu, et al., 2004; Ji & Park, 2000; Jamrozik & Hayes, 2002) attempt to perform grouping and merging all blocks into several regions according to their spatial or temporal similarity. Then, these regions are merged with each other or classified as background or foreground. The most advanced clustering-based method, which handles the H.264/AVC standard, is the region growing approach, in which several seed fragments grow spatially and temporally by merging similar neighboring fragments.

On the other hand, the filtering-based methods (Aggarwal, et al., 2006; You, et al., 2007, 2009) extract foreground regions by filtering blocks, which are expected to belong to background or by classifying all blocks into foreground and background. Then, the foreground region is split into several object parts through a clustering procedure.

Region-of-Interest Tracking

Object tracking based on a video sequence plays an important role in many modern vision applications such as intelligent surveillance, video compression, human-computer interfaces, sports analysis (Haritaoglu, et al., 2000). When an object is tracked with an active camera, traditional methods such as background subtraction, temporal differencing, and optical flow may not work well due to the motion of camera, tremor of camera and disturbance from the background (Xiang, 2009).

Some researchers propose methods of tracking a moving target with an active camera, yet most of their algorithms are too computationally complex due to their dependence on accurate mathematical model and motion model, and cannot be applied to real-time tracking in the presence of fast motion from the object or the active camera, irregular motion and un-calibrated camera. Xiang (2009) makes a great effort to find a fast, computationally efficient algorithm, which can handle fast motion, and can smoothly track a moving target with an active camera, by proposing a method for real-time follow-up tracking fast moving object with an active camera. Xiang (2009) focused on the color-based Mean Shift algorithm, which shows excellent performance both on computationally complexity and robustness.

Wei and Zhou (2010) present a novel algorithm that uses the selective visual attention mechanisms to develop a reliable algorithm for object tracking that can effectively deal with the relatively big influence of external interference in a-priori approaches. To extract the ROI, it makes use of the "local statistic" of the object. By integrating the image feature with state feature, the synergistic benefits can bring following obvious advantages:

- It doesn't use any a-priori knowledge about blobs and no heuristic assumptions must be provided;
- The computation of the model for a generic blob does not take a long processing time.

According to Wei and Zhou (2010), during the detection phase, there are some false-alarms in tracking an object in any actual image. To reduce the fictitious targets as much as possible, a method needs to identify the extracted ROI, while the tracing target can be defined by the following characteristics:

- The length of boundary of the tracing target in the ROI.
- **Aspect Ratio:** The length and the width of the target can be expressed by the two orthogonal axes of minimum enclosing rectangle. The ratio between them is the aspect ratio.
- **Shape Complexity:** The ratio between the length of the boundary and the area.

The ROI, whose parameters accord with the above three features, can be considered as the ROI including the real- target.

Further, there are many other recent tracking methods, such as: Mehmood (2009) implements kernel tracking of density-based appearance models for real-time object tracking applications; Wang et al. (2009) disclose a wireless, embedded smart camera system for cooperative object tracking and event detection; Sun and Sun (2008) present an approach for detecting and tracking dynamic objects with complex topology from image sequences based on intensive restraint topology adaptive snake mode; Wang and Zhu (2008) present a sensor platform with multi-modalities, consisting of a dual-panoramic peripheral vision system and a narrow field-of-view hyperspectral fovea; thus, only hyperspectal images in the ROI should be captured; Liu et al. (2006) present a new method that addresses several challenges in

automatic detection of ROI of neurosurgical video for ROI coding, which is used for neurophysiological intraoperative monitoring (IOM) system. According to Liu et al. (2006), the method is based on an object tracking technique with multivariate density estimation theory, combined with the shape information of the object, thereby by defining the ROIs for neurosurgical video, this method produces a smooth and convex emphasis region, within which surgical procedures are performed. Abousleman (2009) present an automated region-of-interest-based video coding system for use in ultra-low-bandwidth applications.

In the following section, the recent research directions in the SVC Region-of-Interest coding are discussed in detail.

REGION-OF-INTEREST CODING IN H.264/SVC STANDARD

Region-of-Interest coding is a desirable feature in the future SVC applications, especially in applications for the wireless networks, which have a limited bandwidth. However, the H.264 standard does not explicitly teach as how to perform the ROI coding.

The ROI coding is supported by various techniques in the H.264/AVC standard (Wiegand & Sullivan, 2003) and the SVC (Schwarz, et al., 2007) extension. Some of these techniques include quantization step size control at the slice and macroblock levels, and are related to the concept of slice grouping, also known as Flexible Macroblock Ordering (FMO). For example, Lu et al. (2005a) handle the ROI-based Fine Granular Scalability (FGS) coding, in which a user at the decoder side requires to receive better decoded quality ROIs, while the pre-encoded scalable bit-stream is truncated. Lu et al. (2005a) present a number of ROI enhancement quality layers to provide fine granular scalability. In addition, Thang et al. (2005) present a ROI-based spatial scalability scheme, concerning two main issues:

overlapped regions between ROIs and providing different ROIs resolutions. However, Thang et al. (2005) follow the concept of slice grouping of H.264/AVC, considering the following two solutions to improve the coding efficiency:

- Supporting different spatial resolutions for various ROIs by introducing a concept of virtual layers; and
- Enabling the avoidance of the duplicate coding of overlapped regions in multiple ROIs by encoding the overlapped regions such that the corresponding encoded regions can be independently decoded.

Further, Lu et al. (2005b) present an ROI-based Coarse Granular Scalability (CGS), using a perceptual ROI technique to generate a number of quality profiles, and in turn, to realize the CGS. According to Lu et al. (2005b), the proposed ROI based compression achieves better perceptual quality and improves coding efficiency. Moreover, Lambert et al. (2006) relate to extracting the ROIs (i.e., of an original bit-stream) by introducing a description-driven content adaptation framework. According to Lambert et al. (2006), two methods for ROI extraction are implemented:

1. The removal of the non-ROI portions of a bit-stream; and
2. The replacement of coded background with corresponding placeholder slices.

In turn, bit-streams that are adapted by this ROI extraction process have a significantly lower bit-rate than their original versions. While this has, in general, a profound impact on the quality of the decoded video sequence, this impact is marginal in the case of a fixed camera and static background. This observation may lead to new opportunities in the domain of video surveillance or video conferencing. According to Lambert et al. (2006), in addition to the bandwidth decrease, the adaptation process has a positive effect on the

decoder due to the relatively easy processing of placeholder slices, thereby increasing the decoding speed.

Below a novel dynamically adjustable and scalable ROI video coding scheme is presented (Grois, et al., 2010a, 2010b). The technique enables adaptively and efficiently setting the desirable ROI location, size, resolution and bit-rate, according to the network bandwidth (especially, if it is a wireless network in which the bandwidth is limited), power constraints of resource-limited systems (such as mobile devices/servers) where low power consumption is required, and according to end-user resource-limited devices (such as mobile devices, PDAs, and the like), thereby effectively selecting best encoding scenarios suitable for most heterogeneous and time-invariant end-user terminals (i.e., different users can be connected each time) and network bandwidths.

In the following two sub-sections, different types of ROI scalability are presented: the ROI scalability by performing cropping, and ROI scalability by employing the Flexible Macroblock Ordering (FMO) technique, respectively.

ROI Scalability by Performing Cropping

According to the first method for the ROI video coding, and in order to enable obtaining a high-quality ROI on resource-limited devices (such as mobile devices), the ROI is cropped from the original image, and it is used as a base layer (or other low enhancement layers, such as *Layer 1* or *2*), as schematically illustrated in Figure 4 (Grois, et al., 2010a). Then, an Inter-layer prediction is performed within similar sections of the image, i.e., in the cropping areas. As a result, for example (see Figure 4), by using Inter-layer prediction for the three-layer (QCIF-CIF-SD) coding (with similar Quantization Parameter (QP) settings at each layer), a significantly low bit-rate overhead is achieved. Prior to cropping the image, the location of a cropping area in the successive layers of the image is determined (in *Layer 1*, and then in *Layer 2*, as shown on Figure 4). For this, an ESS (Extended Spatial Scalability) method can be employed (Shoaib & Anni, 2010). In addition, a GOP for the SVC is defined as a group between two I/P frames, or any combination thereof (Grois, et al., 2010a, 2010b). Thus, as shown for example in Table 1, for the *"SOCCER"* video sequence (30 f/sec; 300 frames; GOP size 16; QPs varying from 22 to 34) a bit-rate overhead of only 4.7% to 7.9% is achieved compared to the conventional single layer coding.

In addition, Table 1 presents R-D (Rate-Distortion) experimental results for the variable-layer coding with different cropping spatial resolutions, while using the Inter/Intra-layer prediction. As is clearly seen from this Table, there is significantly low bit–rate overhead, which is especially important for transmitting over limited-bandwidth networks (such as wireless networks). Particularly, the Table 1 presents the R-D (Rate-Distortion) experimental results for the three-layer coding (QCIF-CIF-SD) with the QCIF-CIF cropping versus the single layer coding.

It is clearly seen from the above experimental results that when using the Inter/Intra-layer pre-diction, the bit-rate overhead is very small and is much less than 10%.

ROI Scalability by Using Flexible Macroblock Ordering

The second method refers to ROI dynamic adjustment and scalability (Grois, et al., 2010a) by using the FMO in the scalable baseline profile (not for *Layer 0*, which is similar to the H.264/AVC baseline profile without the FMO).

One of the basic elements of an H.264 video sequence is a slice, which contains a group of macroblocks. Each picture can be subdivided into one or more slices and each slice can be provided with increased importance as the basic spatial segment, which can be encoded independently from its neighbors (the slice coding is one of the techniques used in H.264 for transmission) (Chen, et al., 2008; Liu, et al., 2005; Ndili & Ogunfunmi, 2006; Kodikara, et al., 2006). Usually, slices are provided in a raster scan order with continuously ascending addresses; on the other hand, FMO is an advanced tool of H.264 that defines the information of slice groups and enables to allocate different macroblocks to slice groups according to mapping patterns.

Figure 4. An example of ROI dynamic adjustment and scalability (e.g., for mobile devices with different spatial resolutions) by using a cropping method

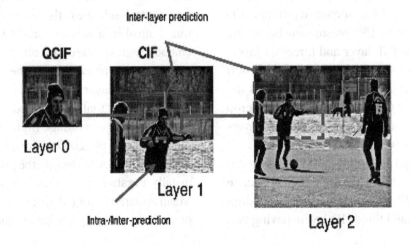

Table 1. Three-layer (QCIF-CIF-SD) spatial scalability coding vs. single layer coding ("SOCCER" video sequence, 30 fp/s, 300 frames, GOP size 16)

Quantization Parameters	Single Layer		QCIF-CIF-SD		Bit–Rate Overhead (%)
	PSNR (dB)	Bit-Rate (k/sec)	PSNR (dB)	Bit-Rate (k/sec)	
22	41.0	5663.3	41.0	5940.6	4.7
26	38.8	3054.9	38.8	3248.1	6.0
30	36.8	1770.2	36.8	1894.9	6.6
34	34.8	1071.3	34.8	1163.6	7.9

Each slice of each picture/frame is independently intra predicted, and the macroblock order within a slice must be in the ascending order. In the H.264 standard, FMO consists of seven slice group map types (*Type 0* to *Type 6*), six of them are predefined fixed macroblock mapping types (interleaved, dispersed, foreground, box-out, raster scan and wipe-out), which can be specified through picture parameter setting (PPS), and the last one is a custom type, which allows the full flexibility of assigning macroblock to any slice group. In this regard, the ROI can be defined as a separate slice in the FMO *Type 2*, which enables defining slices of rectangular regions, and then the whole sequence can be encoded accordingly, while making it possible to define more than one ROI (these definitions should be made in the SVC configuration files, according to the JSVM 9.19 reference software manual (JSVM, 2009).

Table 2 presents experimental results for four layers spatial scalability coding versus six layers coding of the "*SOCCER*" sequence (30 fp/s; 300 frames; GOP size is 16), where four layers are presented by one CIF layer and three SD layers having the CIF-resolution ROI in an upper-left corner of the image. In turn, the six layers are presented by three CIF layers (each layer is a crop from the SD resolution) and three 4CIF/SD layers (Grois, et al., 2010a, 2010b).

Further, Table 3 presents R-D (Rate-Distortion) experimental results for the HD video sequence "*STOCKHOLM*" by using four-layer coding (640×360 layer and three HD layers having two

ROIs (CIF and SD resolution, respectively) in the upper-left corner of the image) versus six-layer coding (three CIF and three SD layers).

As it is clearly observed from Tables 2 and 3, there are very significant bit-rate savings—up to 39%, when using the FMO techniques.

In the following section, recent advances in the ROI bit-rate control techniques are presented.

BIT-RATE CONTROL FOR REGION-OF-INTEREST CODING

Bit-rate control is crucial in providing desired compression bit-rates for H264/AVC video applications, and especially for Scalable Video Coding, which is an extension of H264/AVC.

Bit-rate control has been intensively studied in existing single-layer coding standards, such as MPEG 2, MPEG 4, and H.264/AVC (Li, et al., 2003). According to the existing single-layer rate-control schemes, the encoder employs the rate control as a way to control varying bit-rate characteristics of the coded bit-stream. Generally, there are two objectives of the bit-rate control for the single-layer video coding: one is to meet the bandwidth that is provided by the network, and another is to produce high-quality decoded pictures (Li, et al., 2007). Thus, the inputs of the bit-rate control scheme are: the given bandwidth; usually, the statistics of video sequence including Mean Squared Error (MSE); and a header of each predefined unit (e.g., a basic unit, macroblock,

Table 2. FMO: four-layer spatial scalability coding vs. six-layer coding ("SOCCER" video sequence, 30 fp/s, 300 frames, GOP size 16)

Quantization Parameters	4 Layers (CIF and three SD layers)		Six Layers (three CIF layers and three SD layers)		Bit-Rate Savings (%)
	PSNR (dB)	Bit-Rate (k/sec)	PSNR (dB)	Bit-Rate (k/sec)	
32	36.0	2140.1	36.0	2290.1	6.6
34	35.1	1549.4	35.1	1680.1	7.8
36	34.0	1140.1	34.0	1279.4	10.9

frame, slice). In turn, the outputs are a Quantization Parameter (QP) for the quantization process and another QP for the Rate-Distortion Optimization (RDO) process of each basic unit, while these two quantization parameters, in the single layer video coding, are usually equal in order to maximize the coding efficiency.

In the current JSVM reference software (JSVM, 2009) there is no rate control mechanism, besides the base-layer rate control, which does not consider enhancement layers. The target bit-rate for each SVC layer is achieved by coding each layer with a fixed QP, which is determined by a logarithmic search (JSVM, 2009; Liu, et al., 2008). Of course, this is very inefficient and time-consuming. To solve this problem, only a few works have been published during the last few years, trying to provide an efficient rate control mechanism for the SVC. However, none of them handles scalable bit-rate control for the Region-of-Interest (ROI) coding. Thus, in Xu et al. (2005), the Rate Distortion Optimization (RDO) involved in the step of encoding temporal subband pictures is only implemented on low-pass subband pictures, and

rate control is independently applied to each spatial layer. Furthermore, for the temporal subband pictures obtained from the Motion Compensation Temporal Filtering (MCTF), the target bit allocation and quantization parameter selection inside a GOP make a full use of the hierarchical relations inheritance from the MCTF. In addition, Liu et al. (2008) proposes a switched model to predict the Mean Absolute Difference (MAD) of the residual texture from the available MAD information of the previous frame in the same layer and the same frame in its "Base-Layer." Further, Anselmo and Alfonso (2010) describe a constant quality Variable Bit-Rate (VBR) control algorithm for multiple layer coding. According to Anselmo and Alfonso (2010), the algorithm allows the achievement of a target quality by specifying the memory capabilities and the bit-rate limitations of the storage device. In the more recent work of Roodaki et al. (2010), the joint optimization of layers in the layered video coding is investigated. The authors show that spatial scalability, like the SNR scalability, does benefit from joint optimiza-

Table 3. FMO: four-layer coding vs. six-layer coding ("STOCKHOLM," 30 fp/s, 96 frames, GOP size 8)

Quantization Parameters	Four Layers (640×360, and three HD layers)		Six Layers (three CIF layers and three SD layers)		Bit–Rate Savings (%)
	PSNR (dB)	Bit-Rate (k/sec)	PSNR (dB)	Bit-Rate (k/sec)	
32	34.5	2566.2	34.5	3237.0	19.3
34	33.9	1730.2	33.9	2359.1	29.7
36	33.3	1170.0	33.3	1759.0	39.9

tion, though not being able to exploit the relation between the quantizer step sizes.

Further, Grois et al. (2010b) present a method and system for the efficient ROI Scalable Video Coding, according to which a bit-rate that is very close to the target bit-rate is achieved, while being able to define the desirable ROI quality (in term of QP or Peak Signal-To-Noise Ratio (PSNR)) and while adaptively changing the background region quality (the background region excludes the ROI), according to the overall bit-rate. In order to provide the different visual presentation quality to at least one ROI and to the background (or other less important region of the frame), Grois et al. (2010b) divide each frame into at least two slices, while one slice is used for defining the ROI and at least one additional slice is used for defining the background region, for which fewer bits should be allocated.

In the following section, a novel complexity-aware ROI pre-processing scheme for the Scalable Video Coding is presented and discussed in detail.

COMPLEXITY-AWARE REGION-OF-INTEREST SCALABLE VIDEO CODING BY PERFORMING PRE-PROCESSING

Below, is presented a novel method and system for an efficient adaptive spatial ROI pre-processing/pre-filtering for the Scalable Video Coding (Grois & Hadar, 2011c, 2011d). This scheme is based on an adaptive SVC Computational Complexity-Rate-Distortion (SVC C-R-D) model, which is discussed in detail in the following section (the C-R-D model was extended from the authors' previous research (Kaminsky, et al., 2008; Grois, et al., 2010c) to support SVC systems).

The SVC spatial pre-filtering scheme enables to decrease the motion activity of the background region (or other less important regions within the video sequence), thereby decreasing overall quantization fluctuations. As a result, smaller

quantization parameters can be used for encoding the ROIs, which leads to obtaining the high-quality visual presentation quality of the ROIs, while keeping the overall bandwidth substantially the same. Thus, the device computational resources are employed in an optimal manner, leading to the optimal video presentation quality even for devices with very limited computational resources (e.g., mobile/cellular devices).

The pre-filter, which should be preferably used in the above adaptive SVC system (see Figure 5), is a Gaussian filter due to its relatively low computational complexity, in term of the CPU processing time (i.e. the CPU clocks). According to the simulation results, which are further presented in Table 4, the pre-processing of the Gaussian filter wastes relatively few computational complexity/power resources in term of the CPU processing time (CPU clocks), and also it lasts for a relatively short period of time, comparing to the Wiener or Wavelet filters. The pre-filtering scheme of Grois and Hadar (2011c, 2011d) is adaptively implemented for each layer of the Scalable Video Coding, according to the layer scalability (e.g., according to a particular spatial scalability such as the CIF or HD resolution). As a result, the pre-filtering process for each layer varies according to the above particular layer scalability; the pre-filter standard deviation σ and other parameters (e.g., the kernel matrix size) are adaptively adjusted by the Adaptive SVC Pre-Filter C-R-D Controller (as illustrated in Figure 5), according to the C-R-D analysis, which is discussed in the following section.

Table 4 presents the simulation comparison results for pre-processing the "*SUNFLOWER*" and "*SHIELDS*" video sequences, having various spatial resolutions: 1080p, 720p, and 4CIF (SD), by means of Gaussian, Wiener and Wavelet filters. In addition, the values of the Gaussian standard deviation parameter (σ) are varied from 2 to 8 (the test platform is: Intel® Core 2 Duo CPU, 2.33 GHz, 2GB RAM with Windows® XP Profes-

Table 4. Comparison results for pre-processing of "SUNFLOWER" video sequences having 1080p, 720p, and 4CIF spatial resolutions (80 frames, 30 fp/sec)

Resolution	Pre-Filter Type	Processing Time)sec(	Average PSNR-Y) dB(	SSIM-Y
720p	Gaussian (σ=2)	59.2	36.6	0.99
	Gaussian (σ=5)	48.2	30.4	0.93
	Gaussian (σ=8)	48.0	29.4	0.90
	Weiner	84.9	33.4	0.93
	Wavelet	145.2	41.8	0.99
4CIF	Gaussian (σ=2)	30.7	32.8	0.96
	Gaussian (σ=5)	25.5	27.1	0.82
	Gaussian (σ=8)	25.6	26.2	0.78
	Weiner	40.6	29.7	0.90
	Wavelet	63.9	38.4	0.99

sional® operating system, version 2002, Service Pack 3).

As clearly seen from Table 4, the pre-processing performed by the Gaussian filter wastes much less computational resources, when compared to the Weiner or Wavelet filters. Also, the PSNR and Structural Similarity (SSIM) values of the pre-processed video sequences remain relatively high, which is sufficient for pre-processing the background (or other predefined less important) regions. In such a way, it is ensured that the computational complexity for the pre-processing stage remains relatively low, and the CPU resources are not wasted.

In addition, Table 5 shows the influence of the varying kernel matrix size (varying from 3×3 to 15×15) on the computational complexity of the pre-filtering process in term of the CPU processing time. The tested sequence is *"FOREMAN,"* QCIF - *Layer 0*, CIF - *Layer 1*, 300 frames, 30 fp/s. Further, Table 6 presents, sample 3×3 Gaussian filter kernel matrices for different standard deviation (σ) values (varying from σ=0.5 to σ=1.5).

As is observed from Table 5, the Gaussian kernel matrices having larger sizes lead to obtaining larger computational complexity values. Therefore, in order to save computational re-

sources, small-size Gaussian matrices should be used. In addition, as seen from Table 6, the greater standard deviation values of the Gaussian filter lead to obtaining a smaller value at the center of the matrix, which means that the filtered image becomes smoother as the standard deviation value increases.

According to the pre-processing scheme of Figure 5, the Adaptive SVC Pre-Filter C-R-D Controller performs the C-R-D analysis of the input scalable video sequence, and adaptively determines the Gaussian filter standard deviation for pre-filtering the background region (or other predefined less important region) of each input sequence layer in order to optimally encode the video sequence in term of visual presentation quality. The bits that are removed from the background region due to the pre-filtering process are reallocated according to the C-R-D model to the ROI, and as a result, the high-quality ROI visual presentation is achieved. The ROI can be defined either as a separate SVC layer (e.g., the Base-Layer: *Layer 0*) or as a portion of any SVC layer (either the Base or Enhancement Layer).

As already noted, the above-mentioned adaptive pre-filtering scheme is provided for each layer of the Scalable Video Coding and it adaptively

Figure 5. Block-diagram of the proposed adaptive pre-filtering scheme for the ROI scalable video coding. For simplicity, only two layers (base-layer/layer 0 and enhancement layer/layer 1) are presented.

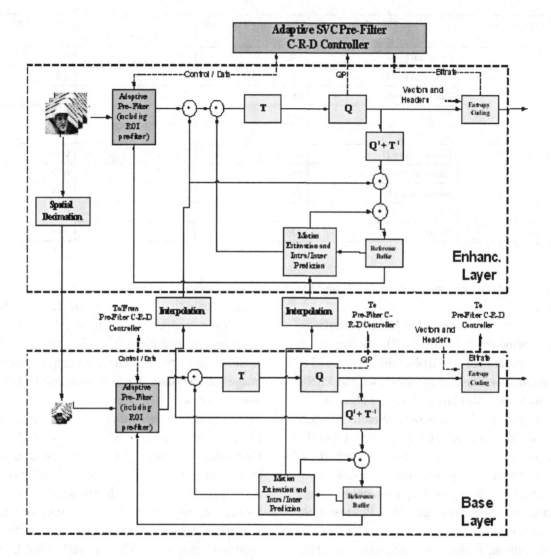

updates various parameters, such as a particular SVC layer scalability, filter standard deviation σ, filter kernel matrix size, a number of filters for the dynamic pre-processing of a transition region between the ROI and background (see Figure 6), etc.

The Gaussian filters are used for the low-pass filtering of the background and transition regions of each frame within each SVC layer (except for *Layer 0*, where the ROI is provided as a whole frame). The bits which are removed due to the

pre-filtering process, are reallocated according to the C-R-D analysis to the ROI, and as a result, the high-quality ROI visual presentation is achieved. The ROI can be defined either as a separate SVC layer (e.g., the *Base-Layer: Layer 0*) or as a portion of any SVC layer (either the *Base or Enhancement Layer*). It should be noted that automatically determining and tracking of the exact ROI region can be performed according to many conventional techniques and algorithms (You et al., 2007), and therefore this issue is out

Table 5. Gaussian kernel matrix size vs. computational complexity of the "foreman" video sequence (QCIF - layer 0, CIF - layer 1)

Standard Deviation (σ)	Gaussian Kernel Matrix Sizes/Computational Complexity in Terms of CPU Processing Time (sec.)			
	3×3	7×7	11×11	15×15
σ=0.5	132.5	241.3	264.9	382.9
σ=1	129.1	250.0	302.2	342.0
σ=1.5	142.2	233.7	312.2	333.9
σ=2.0	117.9	230.8	301.4	305.1
σ=2.5	135.6	217.7	308.7	362.3

of the scope of this chapter (it is supposed that the ROI region is appropriately determined).

In order to ensure a smooth transition from the ROI to the background, a transition region around the ROI is *adaptively* and *dynamically* defined for each SVC spatial layer, as schematically illustrated in Figure 6. The dynamic transition region of each SVC layer contains a variable number of filtered straps (e.g., from 5 to 20 filtered straps, where each strap has a width of a predefined number of pixels) around the ROI, while the closer the strap is located with respect to the ROI, then the higher visual quality it should have. For varying the visual quality of the straps, each strap can be pre-filtered by means of a Gaussian filter with a different standard deviation σ.

It should be noted that a gradual transition of the visual quality from the ROI toward the background is discussed in Karlsson and Sjostrom (2005) and Gopalan (2009). However, Karlsson and Sjostrom (2005) and Gopalan (2009) do not provide a solution for the scalable video coding, where different adaptive gradual transitions are required for each layer, according to a particular layer scalability (e.g., according to a particular spatial resolution, such as the CIF, HD or a particular temporal resolution, and at 15, 30, or 60 frames per second).

The corresponding computational complexity for providing the pre-filtered transition region with a varying number of Gaussian straps is summarized in Table 7, which refers to the pre-filtering of *Layer 1* of the following video sequences: "*FOREMAN*," "*CREW*," and "*MOBILE*" (the standard deviation range is 1-1.5; 100 frames; 30 fp/sec; Gaussian kernel matrix size is 3×3; *Layer 0* is the ROI (a

Table 6. 3x3 Gaussian filter matrices with varying standard deviation values

Standard Deviation (σ)	Matrices		
σ=0.5	0.0113	0.0838	0.0113
	0.0838	**0.6192**	0.0838
	0113.	0.0838	0.0113
σ=1	0.0751	0.1238	0.751
	0.1238	**0.2041**	0.1238
	0.0751	0.1238	0.0751
σ=1.5	0.0947	0.183	0.0947
	0.1183	**0.1477**	0.1183
	0.0947	0.1183	0.0947

Figure 6. Providing a dynamic transition region for gradual transition from ROI to the background of an SVC video sequence

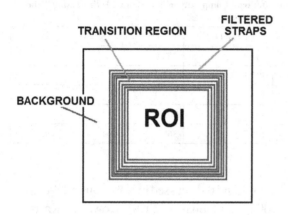

crop of *Layer 1*)). As seen from Table 7, the more Gaussian filters are used, the greater is the CPU processing time of the pre-filtering process. Also, according to the experimental results of Grois et al.(2011b, 2011c), in order to save computational resources, the small-size Gaussian kernel matrices should be used (e.g., the kernel matrix size of 3X3).

In the following section, the recent advances in the complexity-aware bit allocation for the ROI Scalable Video Coding are further discussed.

Complexity-Aware Bit Allocation for the Region-of-Interest Scalable Video Coding

There are several attempts to provide a solution for the complexity issues with regard to the Scalable Video Coding. Lin et al. (2010) present a computation control motion estimation method, which can perform motion estimation adaptively under different computation or power budgets, further evaluating the impact of different tools (such as key picture, etc.) on the performance of an SVC decoder. In addition, Tan et al. (2009) propose a singularly parameterized complexity-scalable scheme for designing power-aware H.264 video encoders. However, traditional solutions, which consider the computational complexity con-

straints, are not related to the Region-of-Interest (ROI) scalable video coding for providing the ROI high-quality visual presentation.

According to the novel complexity-aware adaptive ROI SVC pre-filtering scheme of Grois et al. (2011b, 2011c), in addition to calculating the target bit-rate, the authors calculate the target encoding computational complexity that is required for encoding the current frame (and in turn, for encoding each macroblock within the frame) at each layer of the Scalable Video Coding. The target encoding computational complexity $C_t^{Layer}(i)$ (Kaminsky, et al., 2008; Grois, et al., 2009, 2010c) for the frame i in each SVC layer should be a weighted combination of the remaining computational complexity for encoding the remaining frames, and the target encoding computational complexity allocated for frame i, which is formulated as:

$$C_t^{Layer}(i) = \alpha \cdot \widetilde{C}_r^{Layer}(i) - (1-\alpha) \cdot \widehat{C}_t^{Layer}(i) \quad (1)$$

where α is a weight coefficient; $\widetilde{C}_r^{Layer}(i)$ is the remaining computational complexity for encoding the current frame i; and $\widehat{C}_t^{Layer}(i)$ is the target encoding computational complexity allocated for frame i in the current GOP. $\widetilde{C}_r^{Layer}(i)$ and $\widehat{C}_t^{Layer}(i)$ are represented by the following expressions:

$$\widetilde{C}_r^{Layer}(i) = \frac{C_r^{Layer}(i)}{N_r}, and \quad (2)$$

$$\widehat{C}_t^{Layer}(i) = \frac{\frac{C_r^{Layer}(i)}{N_r} \cdot \Theta^2(i)}{\frac{1}{N-N_r} \sum_{j=1}^{N-N_r} \Theta^2(j)} \quad (3)$$

where, N_r is the number of remaining frames; and N is the total number of frames in the current GOP; $C_r^{Layer}(i)$ is the remaining computational complexity for encoding the remaining frames;

Table 7. Computational complexity for pre-filtering the transition region with a different number of Gaussian filters (100 frames, 30 fp/sec.)

Video Sequences			No. of Pre-Filters
MOBILE (sec)	**CREW (sec)**	**FOREMAN (sec)**	
1.19	1.19	1.24	Pre-Processing only the Background
1.73	1.76	1.90	Pre-Processing also the Transition Region with 5 Gaussian Pre-Filters (5 straps)
2.92	2.91	3.17	Pre-Processing also the Transition Region with 10 Gaussian Pre-Filters (10 straps)
4.13	4.07	4.24	Pre-Processing also the Transition Region with 15 Gaussian Pre-Filters (15 straps)

$\Theta(i)$ is the predicted Mean Absolute Deviation (MAD) of the current frame i; and $\Theta(j)$ is the actual MAD of the previous frame. By using a linear prediction model, the target encoding complexity allocated for the i frame in the current GOP can be determined based on the predicted MAD of the current frame and the actual MADs of the previous frames. In turn, the bit-rate control C-S-R and R-Q-C models (Kaminsky, et al., 2008; Grois, et al., 2010c), for each SVC layer are as follows:

$$C^{SVC\_Layer}(S,R) = A_{C_1} \cdot S^{-1} + A_{C_2} \cdot S^{-2} + A_{C_3}$$
(4)

and

$$R^{SVC\_Layer}(Q,C) = A_{R_1} \cdot Q^{-1} + A_{R_2} \cdot Q^{-2} + A_{R_3} \cdot C^{SVC\_Layer}$$
(5)

where A_{C_1}, A_{C_2}, A_{C_3} and A_{R_1}, A_{R_2}, A_{R_3} are corresponding coefficients that are calculated for each SVC layer regressively (according to a linear regression method (Kaminsky, et al., 2008; Grois, et al., 2010c); S is the complexity step for selecting a corresponding group of coding modes for each SVC layer (e.g., Inter-Search16X8, Inter-Search8X16, Inter-Search8X8, Inter-Search8X4

modes); C^{SVC_Layer} is a computational complexity of encoding each SVC layer; R^{SVC_Layer} is the bit-rate for each SVC layer; and Q is the corresponding quantization step-size for performing the Rate-Distortion-Optimization (RDO) for each macroblock in the current frame.

As a result, the modified SVC encoder is capable of allocating computational complexity and bit-rate resources for each SVC layer for obtaining much better coding efficiency with improved processing speed in term of processing time. It should be noted that the computational complexity is mainly presented in term of the CPU processing time since the processing time can be easily tracked for updating the bit-rate control models and for the parameter adjustments on any computing platform.

For evaluating and testing the presented complexity-aware bit-rate control for the ROI

Table 8. Bit-rate and PSNR values for encoding SVC video sequences on JSVM 9.19

Average PSNR of the ROI	Bit-Rate (K/s)	Tested Video Sequences (on JSVM 9.19)	
39.5	2695.5	Layer 1	FOREMAN
38.6	2689.3	Layer 1	CREW
33.0	6801.0	Layer 1	MOBILE

Table 9. Comparison of the proposed pre-filtering scheme

Average PSNR of the ROI in Layer 1 (dB)	Average PSNR (dB)	Bit-Rate (Kb/s)	Tested Video Sequences (According to the SVC Pre-Filtering Approach of Figure 5)	
41.8	37.9	2719.5	Layer 1	FOREMAN (0.5<σ<1)
42.3	37.4	2674.4	Layer 1	FOREMAN (1<σ<1.5)
38.8	37.6	2772.7	Layer 1	CREW (0.5<σ<1)
39.1	37.3	2621.3	Layer 1	CREW (1<σ<1.5)
39.0	33.8	7207.0	Layer 1	MOBILE (0.5<σ<1)
40.4	34.0	7258.1	Layer 1	MOBILE (1<σ<1.5)

scalable video coding, the following test platform was used: Intel® Core 2 Duo CPU, 2.33 GHz, 2GB RAM with Windows® XP Professional® operating system, version 2002, Service Pack 3. The general conditions for evaluating the presented adaptive ROI SVC pre-processing scheme are as follows: QP for *Layer 0* is 30; QP for *Layer 1* varies from 20 to 30; Spatial Resolution for *Layer 0* is QCIF or CIF, and for *Layer 0* is CIF or 4CIF/SD; 30fp/sec for either *Layer 0* or *Layer 1*; Motion Vector (MV) search range is 16; Number of coded frames is from 100 to 300; Fast Search is ON.

Table 8 presents the PSNR and bit-rate values for the SVC video sequences ("*FOREMAN*," "*CREW*," and "*MOBILE*") encoded by a conventional JSVM reference encoder (JSVM 9.19).

On the other hand, Table 9 presents the encoding of the same SVC video sequences ("*FORE-MAN*," "*CREW*," and "*MOBILE*") by using the novel adaptive ROI SVC pre-filtering scheme of Grois et al. (2011b, 2011c) with varying standard deviation σ and with varying quantization parameters (QPs). The size of the Region-of-Interest in *Layer 1* (which has the 4CIF/SD resolution) is about 25% of the frame size. The *Layer 0* (CIF resolution) is not shown in Tables 3 and 4, since it represents only the Region-of-Interest (without the background).

As is clearly seen from Table 9, there is a significant improvement (compared to the results of Table 8—according to JSVM 9.19), for the same bit-rate, in the PSNR values (*approximately, up to 7 dB*) for the ROI region in *Layer 1*, while

Figure 7. a) The original frame of the "CREW" SVC sequence (CIF resolution, 30fp/sec); b) the decoded frame (PSNR of ROI = 39.1), which was pre-processed and encoded by the adaptive ROI SVC pre-processing system (Figure 5)

varying the Gaussian filter standard deviation σ values from 0.5 to 1.5.

In addition, Figure 7 is an example of using the adaptive ROI SVC pre-processing scheme of Figure 5. The ROI is defined as a rectangular region including upper parts of the crew member bodies.

This chapter is concluded by presenting the following future research directions.

FUTURE RESEARCH DIRECTIONS AND CONCLUSION

The research in 3D video coding is limited, and the issues, such as the 3D ROI detection and tracking, 3D object/face detection and recognition, and the like are very challenging, especially when used in surveillance systems.

In addition, both in 2D and 3D, the ROI temporal filtering for scalable video coding should be handled in a more appropriate manner in order to improve the overall visual presentation quality (which is reduced due to the filtering of the background in a temporal domain). A possible solution for this can be performing "intelligent" post-filtering at the decoder end in order to at least partially recover the lost temporal information.

Also, there is still a need to reduce the computational complexity for processing and coding videos with one or more regions-of-interest, especially for real-time/live systems, and in turn, to decrease the end-to-end delay in order to improve the user visual experience.

Further, techniques mentioned in this chapter (such as the ROI complexity-aware bit allocation, ROI cropping, ROI coding by using the FMO, bit-rate control for the ROI coding, etc.) should be evaluated and tested in detail with regard to the emerging High Efficiency Video Coding (HEVC) standard, which is expected to be officially declared during the next year, in turn, which

is expected to widely penetrate the multimedia market due to its significantly improved coding efficiency (e.g., currently provides up to 50% bit-rate savings).

This chapter covered the Region-of-Interest (ROI) processing and coding for video applications, making a special emphasis on the new H.264/SVC (Scalable Video Coding) standard, which introduced various scalability domains, such as spatial, temporal, and SNR/quality domains. First of all, the authors presented the Region-of-Interest detection and tracking techniques, while discussing various approaches, such as the visual attention model, object detection, face detection, and skin detection. Secondly, the ROI coding techniques for the H.264/SVC standard were presented, including the ROI scalability by performing cropping and the ROI scalability by using the Flexible Macroblock Ordering (FMO) technique. Then, the bit-rate control techniques for the ROI coding were discussed, followed by the detailed discussion of the complexity-aware ROI scalable video coding, while focusing on the complexity-aware SVC bit allocation; all observations and conclusions were supported by a variety of experimental results, which were further compared to the conventional SVC reference software (JSVM 9.19).

ACKNOWLEDGMENT

This work was supported by the NEGEV consortium, MAGNET Program of the Israeli Chief Scientist, Israeli Ministry of Trade and Industry under Grant 85265610. The authors thank Guy Azulay, Adir Atias, Ronen Varfman, Idan Ori, Ron Heiman, Igor Medvetsky, Evgeny Kaminsky, and Alexander Samochin for their assistance in evaluation and testing.

In addition, we thank anonymous reviewers for their valuable comments and suggestions.

REFERENCES

Abdullah-Al-Wadud, M., & Oksam, C. (2007). Region-of-interest selection for skin detection based applications. In *Proceedings of the International Conference on Convergence Information Technology, 2007*, (pp. 1999-2004). IEEE.

Abousleman, G. P. (2009). Target-tracking-based ultra-low-bit-rate video coding. In *Proceedings of Military Communications Conference, 2009*, (pp. 1-6). IEEE Press.

Aggarwal, A., Biswas, S., Singh, S., Sural, S., & Majumdar, A. K. (2006). Object tracking using background subtraction and motion estimation in MPEG videos. *Lecture Notes in Computer Science, 3852*, 121–130. doi:10.1007/11612704_13

Anselmo, T., & Alfonso, D. (2010). Constant quality variable bit-rate control for SVC. In *Proceedings of the 2010 11th International Workshop on Image Analysis for Multimedia Interactive Services (WIAMIS)*, (pp. 1-4). WIAMIS.

Babu, R. V., Ramakrishnan, K. R., & Srinivasan, S. H. (2004). Video object segmentation: A compressed domain approach. *IEEE Transactions on Circuits and Systems for Video Technology, 14*(4), 462–474. doi:10.1109/TCSVT.2004.825536

Benzougar, A., Bouthemy, P., & Fablet, R. (2001). MRF-based moving object detection from MPEG coded video. In *Proceedings of the IEEE International Conference on Image Processing*, (vol. 3, pp. 402-405). IEEE Press.

Bhanu, B., Dudgeon, D. E., Zelnio, E. G., Rosenfeld, A., Casasent, D., & Reed, I. S. (1997). Guest editorial introduction to the special issue on automatic target detection and recognition. *IEEE Transactions on Image Processing, 6*(1), 1–6. doi:10.1109/TIP.1997.552076

Chen, H., Han, Z., Hu, R., & Ruan, R. (2008). Adaptive FMO selection strategy for error resilient H.264 coding. In *Proceedings of the International Conference on Audio, Lang. and Image Processing, ICALIP 2008*, (pp. 868-872). Shanghai, China: ICALIP.

Chen, M.-J., Chi, M.-C., Hsu, C.-T., & Chen, J.-W. (2003). ROI video coding based on H.263+ with robust skin-color detection technique. In *Proceedings of the 2003 IEEE International Conference on Consumer Electronics, ICCE 2003*, (pp. 44-45). IEEE Press.

Chen, Q.-H., Xie, X.-F., Guo, T.-J., Shi, L., & Wang, X.-F. (2010). The study of ROI detection based on visual attention mechanism. In *Proceedings of the 2010 6th International Conference on Wireless Communications Networking and Mobile Computing (WiCOM)*, (pp. 1-4). IEEE.

Cheng, W.-H., Chu, W.-T., Kuo, J.-H., & Wu, J.-L. (2005). Automatic video region-of-interest determination based on user attention model. In *Proceedings of IEEE International Symposium on Circuits and Systems (ISCAS 2005)*, (pp. 3219-3222). IEEE Press.

Duchowski, A. T. (2000). Acuity-matching resolution degradation through wavelet coefficient scaling. *IEEE Transactions on Image Processing, 9*(8), 1437–1440. doi:10.1109/83.855439

Engelke, U., Zepernick, H.-J., & Maeder, A. (2009). Visual attention modeling: Region-of-interest versus fixation patterns. In *Proceedings of the Picture Coding Symposium, 2009, PCS 2009*, (pp. 1-4). PCS.

Gopalan, R. (2009). *Exploiting region-of-interest for improved video coding*. (PhD Thesis). The Ohio State University. Columbus, OH.

Grois, D., & Hadar, O. (2011a). Efficient adaptive bit-rate control for scalable video coding by using computational complexity-rate-distortion analysis. In *Proceedings of the 2011 IEEE International Symposium on Broadband Multimedia Systems and Broadcasting (BMSB),* (pp. 1-6). Nuremberg, Germany: IEEE Press.

Grois, D., & Hadar, O. (2011b). Recent advances in region-of-interest coding. In J. del ser Lorente (Ed.), *Recent Advances on Video Coding,* (pp. 49-76). Intech.

Grois, D., & Hadar, O. (2011c). Complexity-aware adaptive bit-rate control with dynamic ROI pre-processing for scalable video coding. In *Proceedings of the 2011 IEEE International Conference on Multimedia and Expo (ICME),* (pp. 1-4). Barcelona, Spain: IEEE Press.

Grois, D., & Hadar, O. (2011d). Complexity-aware adaptive spatial pre-processing for ROI scalable video coding with dynamic transition region. In *Proceedings of the International Conference on Image Processing (ICIP 2011).* Brussels, Belgium: ICIP.

Grois, D. & Hadar, O. (2012). Advances in Region-of-Interest Video and Image Processing. In Multimedia Networking and Coding, ed. Reuben Farrugia and Carl James Debono, pp. 76-123, IGI Global.

Grois, D., Kaminsky, E., & Hadar, O. (2009). Buffer control in H.264/AVC applications by implementing dynamic complexity-rate-distortion analysis. In *Proceedings of the International Symposium on Broadband Multimedia Systems and Broadcasting, 2009,* (pp. 1-7). IEEE Press.

Grois, D., Kaminsky, E., & Hadar, O. (2010a). ROI adaptive scalable video coding for limited bandwidth wireless networks. In *Proceedings of Wireless Days (WD), 2010 IFIP,* (pp. 1-5). IFIP.

Grois, D., Kaminsky, E., & Hadar, O. (2010b). Adaptive bit-rate control for region-of-interest scalable video coding. In *Proceedings of the 26th Convention on Electrical and Electronics Engineers in Israel (IEEEI),* (pp. 761-765). IEEE Press.

Grois, D., Kaminsky, E., & Hadar, O. (2010c). Optimization methods for H.264/AVC video coding. In Angelides, M. C., & Agius, H. (Eds.), *The Handbook of MPEG Applications: Standards in Practice.* Chichester, UK: John Wiley & Sons, Ltd. doi:10.1002/9780470974582.ch7

Han, S., & Vasconcelos, N. (2008). Object-based regions of interest for image compression. In *Proceedings of the Data Compression Conference, 2008,* (pp. 132-141). IEEE.

Hanfeng, C., Yiqiang, Z., & Feihu, Q. (2001). Rapid object tracking on compressed video. In *Proceedings of the 2nd IEEE Pacific Rim Conference on Multimedia,* (pp. 1066-1071). IEEE Press.

Haritaoglu, I., Harwood, D., & Davis, L. S. (2000). W⁴: Real-time surveillance of people and their activities. *IEEE Transactions on Pattern Analysis and Machine Intelligence, 22*(8), 809–830. doi:10.1109/34.868683

Hu, Y., Rajan, D., & Chia, L. (2008). Detection of visual attention regions in images using robust subspace analysis. *Journal of Visual Communication and Image Representation, 19*(3), 199–216. doi:10.1016/j.jvcir.2007.11.001

Jamrozik, M. L., & Hayes, M. H. (2002). A compressed domain video object segmentation system. In *Proceedings of the IEEE International Conference on Image Processing,* (vol. 1, pp. 113-116). IEEE Press.

Jeong, C. Y., Han, S. W., Choi, S. G., & Nam, T. Y. (2006). An objectionable image detection system based on region of interest. In *Proceedings of the 2006 IEEE International Conference on Image Processing,* (pp. 1477-1480). IEEE Press.

Ji, S., & Park, H. W. (2000). Moving object segmentation in DCT-based compressed video. *Electronics Letters, 36*(21). doi:10.1049/el:20001279

JSVM. (2009). *JSVM software manual, ver. JSVM 9.19 (CVS tag: JSVM_9_19), Nov. 2009.* JSVM.

Kaminsky, E., Grois, D., & Hadar, O. (2008). Dynamic computational complexity and bit allocation for optimizing H.264/AVC video compression. *Journal of Visual Communication and Image Representation, 19*(1), 56–74. doi:10.1016/j.jvcir.2007.05.002

Karlsson, L. S., & Sjostrom, M. (2005). Improved ROI video coding using variable Gaussian pre-filters and variance in intensity. In *Proceedings of the IEEE International Conference on Image Processing, 2005, ICIP 2005,* (vol. 2, pp. 313-316). IEEE Press.

Kas, C., & Nicolas, H. (2009). Compressed domain indexing of scalable H.264/SVC streams. *Signal Processing Image Communication,* 484–498. doi:10.1016/j.image.2009.02.007

Kim, D.-K., & Wang, Y.-F. (2009). Smoke detection in video. In *Proceedings of the 2009 WRI World Congress on Computer Science and Information Engineering,* (vol. 5, pp. 759-763). WRI.

Kodikara Arachchi, H., Fernando, W. A. C., Panchadcharam, S., & Weerakkody, W. A. R. J. (2006). Unequal error protection technique for ROI based H.264 video coding. In *Proceedings of the Canadian Conference on Electrical and Computer Engineering,* (pp. 2033-2036). Ottawa, Canada: IEEE.

Kwon, H., Han, H., Lee, S., Choi, W., & Kang, B. (2010). New video enhancement preprocessor using the region-of-interest for the videoconferencing. *IEEE Transactions on Consumer Electronics, 56*(4), 2644–2651. doi:10.1109/TCE.2010.5681152

Lambert, P., Schrijver, D. D., Van Deursen, D., De Neve, W., Dhondt, Y., & Van de Walle, R. (2006). A real-time content adaptation framework for exploiting ROI scalability in H.264/AVC. In *Proceedings of Advanced Concepts for Intelligent Vision Systems* (pp. 442–453). IEEE. doi:10.1007/11864349_40

Li, Z., Pan, F., Lim, K. P., Feng, G., Lin, X., & Rahardja, S. (2003). Adaptive basic unit layer rate control for JVT. In *Joint Video Team (JVT) of ISO/IEC MPEG and ITU-T VCEG (ISO/IEC JTC1/SC29/WG11 and ITU-T SG16 Q.6), Doc. JVT-G012.* Pattaya, Thailand: JVT.

Li, Z., Zhang, X., Zou, F., & Hu, D. (2010). Study of target detection based on top-down visual attention. In *Proceedings of the 2010 3rd International Congress on Image and Signal Processing (CISP),* (vol. 1, pp. 377-380). CISP.

Li, Z. G., Yao, W., Rahardja, S., & Xie, S. (2007). New framework for encoder optimization of scalable video coding. In *Proceedings of the 2007 IEEE Workshop on Signal Processing Systems,* (pp. 527-532). IEEE Press.

Lin, W., Panusopone, K., Baylon, D. M., & Sun, M.-T. (2010). A computation control motion estimation method for complexity-scalable video coding. *IEEE Transactions on Circuits and Systems for Video Technology, 20*(11), 1533–1543. doi:10.1109/TCSVT.2010.2077773

Lin, Y., & Bhanu, B. (2005). Object detection via feature synthesis using MDL-based genetic programming. *IEEE Transactions on Systems, Man, and Cybernetics. Part B, Cybernetics, 35*(3), 538–547. doi:10.1109/TSMCB.2005.846656

Liu, B., Sun, M., Liu, Q., Kassam, A., Li, C.-C., & Sclabassi, R. J. (2006). Automatic detection of region of interest based on object tracking in neurosurgical video. In *Proceedings of the 27th International Conference of the Engineering in Medicine and Biology Society, 2005,* (pp. 6273-6276). IEEE Press.

Liu, L., Zhang, S., Ye, X., & Zhang, Y. (2005). Error resilience schemes of H.264/AVC for 3G conversational video services. In *Proceedings of the Fifth International Conference on Computer and Information Technology*, (pp. 657- 661). Binghamton, NY: IEEE.

Liu, Y., Li, Z. G., & Soh, Y. C. (2008). Rate control of H.264/AVC scalable extension. *IEEE Transactions on Circuits and Systems for Video Technology, 18*(1), 116–121. doi:10.1109/TC-SVT.2007.903325

Lu, Z., Lin, W., Li, Z., Pang Lim, K., Lin, X., Rahardja, S., Ping Ong, E., & Yao, S. (2005b). *Perceptual region-of-interest (ROI) based scalable video coding.* JVT-O056, Busan, KR.

Lu, Z., Peng, W.-H., Choi, H., Thang, T. C., & Shengmei, S. (2005a). *CE8: ROI-based scalable video coding.* JVT-O308, Busan, KR.

Manerba, F., Benois-Pineau, J., Leonardi, R., & Mansencal, B. (2008). Multiple object extraction from compressed video. *EURASIP Journal on Advances in Signal Processing.* Retrieved from http://asp.eurasipjournals.com/content/2008/1/231930

Mehmood, M. O. (2009). Study and implementation of color-based object tracking in monocular image sequences. In *Proceedings of the 2009 IEEE Student Conference on Research and Development (SCOReD),* (pp. 109-111). IEEE Press.

Micheloni, C., Salvador, E., Bigaran, F., & Foresti, G. L. (2005). An integrated surveillance system for outdoor security. In *Proceedings of the IEEE Conference on Advanced Video and Signal Based Surveillance, 2005, AVSS 2005,* (pp. 480-485). IEEE Press.

Mustafah, Y. M., Bigdeli, A., Azman, A. W., & Lovell, B. C. (2009). Face detection system design for real time high resolution smart camera. In *Proceedings of the Third ACM/IEEE International Conference on Distributed Smart Cameras, 2009, ICDSC 2009,* (pp. 1-6). ACM/IEEE.

Ndili, O., & Ogunfunmi, T. (2006). On the performance of a 3D flexible macroblock ordering for H.264/AVC. *Digest of Technical Papers International Conference on Consumer Electronics,* 37-38.

Qayyum, U., & Javed, M. Y. (2006). Real time notch based face detection, tracking and facial feature localization. In *Proceedings of the International Conference on Emerging Technologies, 2006, ICET 2006,* (pp. 70-75). ICET.

Roodaki, H., Rabiee, H. R., & Ghanbari, M. (2010). Rate-distortion optimization of scalable video codecs. *Signal Processing Image Communication, 25*(4), 276–286. doi:10.1016/j.image.2010.01.004

Sadykhov, R. K., & Lamovsky, D. V. (2008). Algorithm for real time faces detection in 3D space. In *Proceedings of the International Multiconference on Computer Science and Information Technology, 2008, IMCSIT 2008,* (pp. 727-732). IMCSIT.

Schoepflin, T., Chalana, V., Haynor, D. R., & Kim, Y. (2001). Video object tracking with a sequential hierarchy of template deformations. *IEEE Transactions on Circuits and Systems for Video Technology, 11*(11), 1171–1182. doi:10.1109/76.964784

Schwarz, H., Marpe, D., & Wiegand, T. (2007). Overview of the scalable video coding extension of the H.264/AVC standard. *IEEE Transactions on Circuits and Systems for Video Technology, 17*(9), 1103–1120. doi:10.1109/TCSVT.2007.905532

Shoaib, M., & Anni, C. (2010). Efficient residual prediction with error concealment in extended spatial scalability. In *Proceedings of the Wireless Telecommunications Symposium (WTS)*, (pp. 1-6). WTS.

Shokurov, A., Khropov, A., & Ivanov, D. (2003). Feature tracking in images and video. In *Proceedings of the International Conference on Computer Graphics between Europe and Asia (GraphiCon-2003)*, (pp. 177-179). GraphiCon.

Sun, Q., Lu, Y., & Sun, S. (2010). A visual attention based approach to text extraction. In *Proceedings of the 2010 20th International Conference on Pattern Recognition (ICPR)*, (pp. 3991-3995). ICPR.

Sun, Z., & Sun, J. (2008). Tracking of dynamic image sequence based on intensive restraint topology adaptive snake. In *Proceedings of the 2008 International Conference on Computer Science and Software Engineering*, (vol. 6, pp. 217-220). IEEE.

Tan, Y. H., Lee, W. S., Tham, J. Y., Rahardja, S., & Lye, K. M. (2009). Complexity control and computational resource allocation during H.264/SVC encoding. In *Proceedings of the Seventeenth ACM International Conference on Multimedia*, (pp. 897-900). Beijing, China: ACM Press.

Thang, T. C., Bae, T. M., Jung, Y. J., Ro, Y. M., Kim, J.-G., Choi, H., & Hong, J.-W. (2005). *Spatial scalability of multiple ROIs in surveillance video*. JVT-O037, Busan, KR.

Vezhnevets, M. (2002). Face and facial feature tracking for natural human-computer interface. In *Proceedings of the International Conference on Computer Graphics between Europe and Asia (GraphiCon-2002)*, (pp. 86-90). GraphiCon.

Wang, H., Leng, J., & Guo, Z. M. (2002). Adaptive dynamic contour for real-time object tracking. In *Proceedings of the Image and Vision Computing New Zealand (IVCNZ 2002)*. IVCNZ.

Wang, J.-M., Cherng, S., Fuh, C.-S., & Chen, S.-W. (2008). Foreground object detection using two successive images. In *Proceedings of the IEEE Fifth International Conference on Advanced Video and Signal Based Surveillance, 2008, AVSS 2008*, (pp. 301-306). IEEE Press.

Wang, Y., Casares, M., & Velipasalar, S. (2009). Cooperative object tracking and event detection with wireless smart cameras. In *Proceedings of the 2009 Sixth IEEE International Conference on Advanced Video and Signal Based Surveillance (AVSS 2009)*, (pp. 394-399). Washington, DC: IEEE Press.

Wei, Z., & Zhou, Z. (2010). An adaptive statistical features modeling tracking algorithm based on locally statistical ROI. In *Proceedings of the 2010 International Conference on Educational and Information Technology (ICEIT)*, (vol. 1, pp. 433-437). ICEIT.

Wiegand, T., & Sullivan, G. (2003). *Final draft ITU-T recommendation and final draft international standard of joint video specification (ITU-T Rec. H.264 ISO/IEC 14 496-10 AVC)*. In Joint Video Team (JVT) of ITU-T SG16/Q15 (VCEG) and ISO/IEC JTC1/SC29/WG1, Annex C. Pattaya, Thailand, Doc. JVT-G050.

Xiang, G. (2009). Real-time follow-up tracking fast moving object with an active camera. In *Proceedings of the 2nd International Congress on Image and Signal Processing, 2009, CISP 2009*, (pp. 1-4). IEEE.

Xu, L., Ma, S., Zhao, D., & Gao, W. (2005). Rate control for scalable video model. *Proceedings of SPIE: Visual Communication and Image Processing, 5960*, 525.

Yang, M.-H., Kriegman, D. J., & Ahuja, N. (2002). Detecting faces in images: A survey. *IEEE Transactions on Pattern Analysis and Machine Intelligence, 24*(1), 34–58. doi:10.1109/34.982883

You, W. (2010). *Object detection and tracking in compresses domain.* Retrieved from http://knol. google.com/k/wonsang-you/object-detection-and-tracking-in/3e2si9juvje7y/7#

Yuan, L., & Mu, Z.-C. (2007). Ear detection based on skin-color and contour information. In *Proceedings of the 2007 International Conference on Machine Learning and Cybernetics,* (vol. 4, pp. 2213-2217). IEEE.

You, W., Sabirin, M. S. H., & Kim, M. (2007). Moving object tracking in H.264/AVC bitstream. *Lecture Notes in Computer Science, 4577,* 483–492. doi:10.1007/978-3-540-73417-8_57

You, W., Sabirin, M. S. H., & Kim, M. (2009). Real-time detection and tracking of multiple objects with partial decoding in H.264/AVC bitstream domain. In N. Kehtarnavaz & M. F. Carlsohn (Eds.), *Proceedings of SPIE.* San Jose, CA: SPIE.

Zeng, W., Du, J., Gao, W., & Huang, Q. (2005). Robust moving object segmentation on H.264/AVC compressed video using the block-based MRF model. *Real-Time Imaging, 11*(4), 290–299. doi:10.1016/j.rti.2005.04.008

Zhang, T., Liu, C., Wang, M., & Goto, S. (2009). Region-of-interest based H.264 encoder for videophone with a hardware macroblock level face detector. In *Proceedings of the IEEE International Workshop on Multimedia Signal Processing, 2009,* (pp. 1-6). IEEE Press.

ADDITIONAL READING

Bae, T. M., Thang, T. C., Kim, D. Y., Ro, Y. M., Kang, J. W., & Kim, J. G. (2006). Multiple region-of-interest support in scalable video coding. *ETRI Journal, 28*(2), 239–242. doi:10.4218/etrij.06.0205.0126

Bing, L., Mingui, S., Qiang, L., Kassam, A., Li, C.-C., & Sclabassi, R. J. (2006). Automatic detection of region of interest based on object tracking in neurosurgical video. In *Proceedings of the 2005 27th Annual International Conference of the Engineering in Medicine and Biology Society,* (pp. 6273-6276). IEEE Press.

Chiang, T., & Zhang, Y.-Q. (1997). A new rate control scheme using quadratic rate distortion model. *IEEE Transactions on Circuits and Systems for Video Technology, 7*(1), 246–250. doi:10.1109/76.554439

Dai, W., Liu, L., & Tran, T. D. (2005). Adaptive block-based image coding with pre-/post-filtering. In *Proceedings of DCC 2005,* (pp. 73-82). DCC.

Foo, B., Andreopoulos, Y., & Van der Schaar, M. (2008). Analytical rate-distortion-complexity modeling of wavelet-based video coders. *IEEE Transactions on Signal Processing, 56*(2), 797–815. doi:10.1109/TSP.2007.906685

Hadar, O., Stern, A., & Koresh, R. (2001). Enhancement of an image compression algorithm by pre- and post-filtering. *Optical Engineering (Redondo Beach, Calif.), 40*(2), 193–199. doi:10.1117/1.1339203

Hannuksela, M. M. (2001). Syntax for supplemental enhancement information. In *Proceedings of the VCEG 14th Meeting,* (pp. 24-27). Santa Barbara, CA: VCEG.

He, Z., Cheng, W., & Chen, X. (2008). Energy minimization of portable video communication devices based on power-rate-distortion optimization. *IEEE Transactions on Circuits and Systems for Video Technology, 18*(5), 596–608. doi:10.1109/TCSVT.2008.918802

He, Z., Liang, Y., Chen, L., Ahmad, I., & Wu, D. (2005). Power-rate-distortion analysis for wireless video communication under energy constraints. *IEEE Transactions on Circuits and Systems for Video Technology, 15*(5), 645–658. doi:10.1109/TCSVT.2005.846433

Itti, L. (2004). Automatic foveation for video compression using a neurobiological model of visual attention. *IEEE Transactions on Image Processing*, *13*(10), 1304–1318. doi:10.1109/TIP.2004.834657

Jiang, M., & Ling, N. (2006). Lagrange multiplier and quantizer adjustment for H.264 frame-layer video rate control. *IEEE Transactions on Circuits and Systems for Video Technology*, *16*(5), 663–668. doi:10.1109/TCSVT.2006.873159

Kannangara, C. S., Richardson, I. E. G., & Miller, A. J. (2008). Computational complexity management of a real-time H.264/AVC encoder. *IEEE Transactions on Circuits and Systems for Video Technology*, *18*(9), 1191–1200. doi:10.1109/TCSVT.2008.928881

Kim, S. D., & Ra, J. B. (2003). Efficient block-based video encoder embedding a Wiener filter for noisy video sequences. *Journal of Visual Communication and Image Representation*, *14*(1), 22–40.

Kwon, D.-K., Shen, M.-Y., & Kuo, C.-C. J. (2007). Rate control for H.264 video with enhanced rate and distortion models. *IEEE Transactions on Circuits and Systems for Video Technology*, *17*(5), 517–529. doi:10.1109/TCSVT.2007.894053

Richardson, I. E. G. (2010). *The H.264 advanced video compression standard* (2nd ed.). Chichester, UK: John Wiley & Sons. doi:10.1002/9780470989418

Schaar, M., & Andreopoulos, Y. (2005). Rate-distortion-complexity modeling for network and receiver aware adaptation. *IEEE Transactions on Multimedia*, *7*(3), 471–479. doi:10.1109/TMM.2005.846790

Song, B. C., Kim, N. H., & Chun, K. W. (2005). Transform-domain wiener filtering for H.264/AVC video encoding and its implementation. In *Proceedings of IEEE ICIP 2005*, (vol. 3, pp. 529-532). IEEE Press.

Su, L., Lu, Y., Wu, F., Li, S., & Gao, W. (2009). Complexity-constrained H.264 video encoding. *IEEE Transactions on Circuits and Systems for Video Technology*, *19*(4), 477–490. doi:10.1109/TCSVT.2009.2014017

Tran, T. D., Liang, J., & Tu, C. (2003). Lapped transform via time-domain pre- and post-filtering. *IEEE Transactions on Signal Processing*, *51*(6), 1557–1571. doi:10.1109/TSP.2003.811222

Vanam, R., Riskin, E. A., & Ladner, R. E. (2009). H.264/MPEG-4 AVC encoder parameter selection algorithms for complexity distortion tradeoff. In *Proceedings Data Compression Conference*, (pp. 372-381). IEEE.

Wiegand, T., & Girod, B. (2001). Parameter selection in Lagrangian hybrid video coder control. In *Proceedings International Conference on Image Processing*, (vol. 3, pp. 542-545). Thessaloniki, Greece: IEEE.

Wiegand, T., Schwarz, H., Joch, A., Kossentini, F., & Sullivan, G. J. (2003). Rate-constrained coder control and comparison of video coding standards. *IEEE Transactions on Circuits and Systems for Video Technology*, *13*(7), 688–703. doi:10.1109/TCSVT.2003.815168

Wiegand, T., Sullivan, G., Reichel, J., Schwarz, H., & Wien, M. (2006). *Joint draft 8 of SVC amendment*. Paper presented at the ISO/IEC JTC1/SC29/WG11 and ITU-T SG16 Q.6 9 (JVT-U201) 21st Meeting. Hangzhou, China.

Wu, S., Huang, Y., & Ikenaga, T. (2009). A macroblock-level rate control algorithm for H.264/AVC video coding with context-adaptive MAD prediction model. In *Proceedings* of the *International Conference on Computer Model Simulation*, (pp. 124-128). IEEE.

Zeng, H., Cai, C., & Ma, K.-K. (2009). Fast mode decision for H.264/AVC based on macroblock motion activity. *IEEE Transactions on Circuits and Systems for Video Technology*, *19*(4), 1–10.

Zhan, C. Q., & Karam, L. J. (2003). Wavelet-based adaptive image denoising with edge preservation. In *Proceedings of the ICIP 2003*, (vol. 1, pp. 97-100). ICIP.

KEY TERMS AND DEFINITIONS

Flexible Macroblock Ordering (FMO): One of the basic elements of the H.264 video sequence is a slice, which contains a group of macroblocks. Each picture can be subdivided into one or more slices and each slice can be provided with increased importance as the basic spatial segment, which can be encoded independently from its neighbors. FMO is an advanced tool of H.264 that defines the information of slice groups and enables to employ different macroblocks to slice groups of mapping patterns.

H.264/AVC (Advanced Video Coding): H.264/AVC is a video coding standard of the ITU-T Video Coding Experts Group and the ISO/IEC Moving Picture Experts Group, which was officially issued in 2003. H.264/AVC has achieved a significant improvement in rate-distortion efficiency relative to prior standards.

H.264/SVC (Scalable Video Coding/SVC): SVC is an extension of H.264/AVC video coding standard; SVC enables the transmission and decoding of partial bit streams to provide video services with lower temporal or spatial resolutions or reduced fidelity, while retaining a reconstruction quality that is high relative to the rate of the partial bit streams.

HEVC (High Efficiency Video Coding): Is a high-definition video compression standard, which is developed by a Joint Collaborative Team on Video Coding (JCT-VC) formed by ITU-T VCEG and ISO/IEC MPEG. The HEVC standard is intended to provide significantly better compression capability than the existing AVC (ITU-T H.264 | ISO/IEC MPEG-4 Part 10) standard.

MAD: Is an abbreviation for the Mean Absolute Difference, and is one of the most popular block matching criterions, according to which corresponding pixels from each block are compared and their absolute differences are summed.

Motion Compensation: Is a technique employed for video compression, which describes a picture in terms of the transformation of a reference picture to the current picture. The reference picture may be a previous or future picture.

Motion Estimation: Is a process of determining motion vectors that describe the transformation from one image to another; usually, from adjacent frames in a video sequence. The motion vectors may relate to the whole image or to particular image portions, such as rectangular blocks, etc.

PSNR: Is an abbreviation for the Peak Signal-to-Noise Ratio, which is the ratio between the maximum possible power of a signal and the power of corrupting noise that affects the fidelity of the signal representation.

Scalability: In video coding, the term scalability relates to providing diffierent qualities in various domains, such as in the spatial domain (by varying the video resolution), temporal domain (by varying the frame rate), and fidelity domain (by varying the SNR/quality), which are embedded into a single SVC bit stream.

SSIM: Is an abbreviation for the Structural Similarity index, which relates to a method for measuring the similarity between two images. The SSIM index is a full reference metric, in other words, the measuring of image quality based on an initial uncompressed or distortion-free image as a reference.

Chapter 7
Recent Advances in Computational Complexity Techniques for Video Coding Applications

Dan Grois
Ben-Gurion University of the Negev, Israel

Ofer Hadar
Ben-Gurion University of the Negev, Israel

ABSTRACT

The computational complexity issue is critical for present and future video applications implemented by relatively new video coding standards, such as the H.264/AVC (Advanced Video Coding), which has a large number of coding modes. One of the main reasons for the importance of providing an efficient complexity control in video coding applications is a strong need to decrease the encoding/decoding computational complexity, especially when the encoding and/or decoding devices are resource-limited, such as portable devices. In turn, efficient complexity control enables reducing the video coding processing time and enables saving power resources during the encoding and/or decoding process. Since the recent dramatic progress in the development of multimedia technologies has made portable devices widespread everywhere, especially in order to provide or receive real-time video contents, the need to enhance the computational complexity control in video coding applications is expected to be further significantly increased as a function of the dramatic increase in the mobile/portable device penetration into the every-day life environment. In this chapter, the authors perform a comprehensive review of the recent advances in computational complexity techniques for video coding applications. This chapter will not only summarize the recent advances in this field, but will also provide explicit directions for the design of the future complexity-aware video coding applications.

DOI: 10.4018/978-1-4666-2833-5.ch007

BACKGROUND

The H.264/AVC (ISO/IEC MPEG-4 Part 10) video coding standard (Wiegand & Sullivan, 2003; Wiegand, et al., 2003a, 2003b), which was officially issued in 2003, has become a challenge for real-time video applications. Compared to others standards, it gains about 50% in bit-rate, while providing the same visual quality. In addition to having all the advantages of MPEG-2, H.263 and MPEG-4, the H.264 video coding standard possesses a number of improvements, such as the Content-Adaptive-Based Arithmetic Coder (CABAC), enhanced transform and quantization, prediction of Intra macroblocks (spatial prediction), and others. H.264 is designed for both Constant Bit-Rate (CBR) and Variable Bit-Rate (VBR) video coding, useful for transmitting video sequences over statistically multiplexed networks (e.g., the CBR statistical multiplexing is common in terrestrial broadcast applications, whereas researchers usually prefer the VBR multiplexing). This video coding standard can also be used at any bit-rate range for various applications, varying from wireless video phones to High Definition Television (HDTV) and Digital Video Broadcasting (DVB). In addition, H.264 provides significantly improved coding efficiency and greater functionality, such as rate scalability, Intra-prediction and error resilience in comparison with its predecessors, MPEG-2 and H.263. However, H.264/AVC is much more complex in comparison to other coding standards and to achieve maximum quality encoding, high computational resources are required (Grois, et al., 2010c, 2009; Kaminsky, et al., 2008).

One of the important concepts of H.264/AVC is the separation of the system into two layers, as schematically illustrated in Figure 1: a Video Coding Layer (VCL), providing the high-compression representation of data, and a Network Adaptation Layer (NAL), packaging the coded data in an appropriate manner based on the characteristics of the transmission network

(Saponara, et al., 2003; Wiegand, et al., 2003a, 2003b). The basic coding framework defined by the H.264/AVC is similar to the one of previous video coding standards: translational block-based motion estimation and compensation, residual coding in a transformed domain and entropy coding of quantized transform coefficients. Additional tools improve the compression efficiency at *an increase of computational complexity cost*: e.g., the motion compensation scheme supports multiple previous reference pictures and a large number of different block sizes, from *16x16* up to 7 modes, including *16x8, 8x16, 8x8, 8x4, 4x8,* and *4x4* pixel blocks; also, the motion vector field can be specified with a higher spatial accuracy (quarter or eighth-pixel resolution instead of half pixel) and a Rate-Distortion (RD) Lagrangian technique (Sullivan & Wiegand, 1998) optimizes both motion estimation and coding mode decisions (Saponara, et al., 2003).

The computational complexity issue is critical for the present and future real-time video applications using the H.264/AVC standard, which has a large number of coding modes. In conventional advanced video coding applications, these coding modes are not fully selected at the time of video sequence encoding, since selecting all possible coding modes leads to a significant increase of the overall computational complexity. The greater computational complexity is, the larger processing (power) resources are required; the power issue becomes critical for wireless applications. On the other hand, not selecting all possible coding modes leads to an increase of the encoded video sequence distortion, and in turn to a decrease of the overall video quality.

It should be noted that in order to change the overall computational constraints, the method needs to be robust. The difference in the overall video sequence quality, related to an optimal and constant computational complexity and bit allocations, defines the level of robustness (Grois, et al., 2010c).

Figure 1. Structure of H.264/AVC video encoder (Wiegand, et al., 2003b)

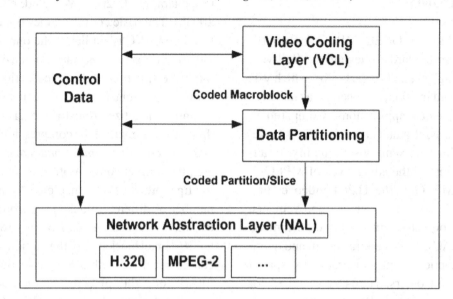

With this regard, Figure 2 presents the need for a generic coding control methodology for any given coded bit-rate and computational complexity in a video encoder, according to which computation can no longer be considered to be fixed. Generally, the video encoder performance (mainly, from the software point of view) can be considered as a function of three variables: computational complexity, coded bit-rate, and video presentation quality (Grois, et al., 2010c). However, the maximum achievable video quality depends on the available coded bit-rate and processing resources. Optimizing the complexity, rate and distortion performance of a video encoder requires flexible allocation of complexity and bits for each coded element (i.e., a macroblock, slice, or frame), leading to the development of effective rate-complexity control algorithms and models for video encoding (Grois, et al., 2010a, 2010b, 2010c; Vanam, et al., 2010; Foo, et al., 2008).

The rapid development of multimedia technologies makes portable multimedia devices increasingly found everywhere, and the prevalence of user-created content has created unsurpassed demand of real-time video compression on portable multimedia devices (Lee, et al., 2009).

However, the energy capacity of the battery in portable device is only being improved at a much slower pace (Powers, 1995), and it becomes the major restriction on video applications for the portable devices. A popular technique to achieve low energy consumption in portable multimedia device is to have multiple power modes and to choose a mode best suited for a user preference or battery status (Lian, et al., 2007). Dynamic Voltage Scaling (DVS) is another well-known power control technique widely used on portable multimedia devices. Based on the fact that power consumption is determined mainly by supply voltage or clock frequency, the dynamic voltage scaling changes them to control power consumption in mobile devices.

The H.264/AVC standard entails high computational complexity with many advanced coding tools such as variable block size motion compensation, multiple reference frames, quarter-pel motion vector accuracy, multi-directional intra-prediction, and context adaptive entropy coding (Lee, et al., 2009). Figure 3 presents sample complexity profiling results of the H.264/AVC encoder (Baseline profile). As seen from Figure 3, most of the computational complexity resources are allocated for

Figure 2. Available computational resources

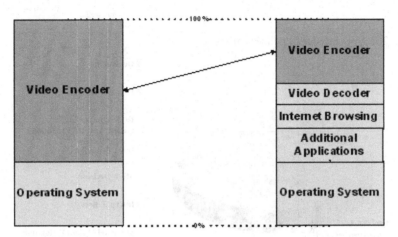

the Motion Estimation (ME) portion. The H.264/AVC standard uses the motion compensated transform coding, similarly to the previous standards (Huang, et al., 2006). The improvement in coding performance comes mainly from the inter/intra-prediction part, while the motion estimation at quarter-pixel accuracy with variable block sizes and multiple reference frames greatly reduces the prediction errors (Huang, et al., 2006).

Generally, it should be noted that the computational complexity of an application can be estimated by its overall execution time, which can be easily measured (Alfonso, 2010). However, this kind of measure is coarse, since the total elapsed time relates to the time spent by the CPU for a particular process of interest and also for all other processes, which run on behalf of other users or on behalf of the Operating System (OS). Even in the ideal case when a single process is running, the OS can put the CPU in one of the following states (Alfonso, 2010):

- **Executing in the "User Mode":** The CPU is executing the machine code of a process that accesses its own data space in memory.
- **Executing in the "System Mode" (i.e. the "Kernel Mode"):** The CPU is executing a system call, which is made by the process to request a Kernel service.

- **Idle waiting for I/O:** Processes are sleeping, while waiting for the completion of I/O to disk or other block devices.
- **Idle:** No processes are ready-to-run on the CPU or are sleeping, while waiting for the block I/O, or the keyboard input, or the network I/O.

Based on this observations, the Elapsed Time can be defined as a total time elapsed from the initiation of a process to its termination; the User Time can be defined as the overall CPU time that the process uses in the "User Mode"; and the System Time is the overall CPU time used by the OS on behalf of the process (in the "System Mode"). The CPU load, defined in the following equation, expresses the ratio of the total time that the CPU spends during the execution operation (Alfonso, 2010):

$$CPU\_Load\_Ratio = \frac{User\ Time + System\ Time}{Elapsed\ Time}$$

(1)

Of course, there is no interest in measuring the performance of the hard disk or the time spent by the CPU in executing the Kernel routines, because these factors are not related to the intrinsic complexity of the video coding system. To improve

Figure 3. Computational complexity usage (in percents) for the H.264/AVC Baseline encoder (Huang, et al., 2006)

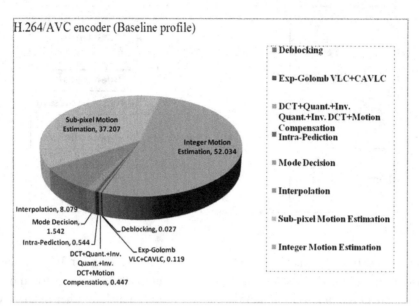

the accuracy of the measure, the evaluation of the Elapsed Time can be rather replaced by the evaluation of the User Time of the process of interest (Alfonso, 2010).

Also, it should be noted that the H.264/AVC standard defines twenty-one sets of capabilities, which are referred to as profiles, targeting specific classes of applications and resulting in consuming different computational complexity resources. These profiles include, for example, the Constrained Baseline Profile (CBP), which is primarily used for low-cost applications, such as videoconferencing and mobile applications; Baseline Profile (BP), which is mainly used for low-cost applications that require additional data loss robustness (the BP profile includes all features that are supported in the CBP, plus three additional features that can be used for loss robustness or for other purposes such as low-delay multi-point video stream compositing); Main Profile (MP) that is used for standard-definition digital TV broadcasts that use the MPEG-4 format, as defined in the Digital Video Broadcasting (DVB) standard; and many other profiles (Wiegand, et al., 2003b). Based on the above observations,

in order to provide efficient complexity control schemes, many complexity reduction algorithms have been developed, as is described in the following sections of this chapter.

This chapter is organized as follows. First, the recent *computational complexity allocation* techniques in the H.264/AVC are discussed in detail. Then, the computational complexity management for the *real-time H.264/AVC–based video encoders* is presented. After that, the computational complexity reduction by performing an efficient *block mode selection and efficient intra-prediction mode selection*, respectively, is reviewed. Further, low computational complexity techniques for *multi-reference motion estimation* are discussed; and finally, chapter conclusions are provided.

COMPUTATIONAL COMPLEXITY ALLOCATION TECHNIQUES IN H.264/AVC

There are two main issues related to the computational complexity and bit allocation of conventional video encoding and decoding systems

(Grois, et al., 2010c; Kaminsky, et al., 2008), which do not consider the computational complexity constraints. One issue is that according to the conventional R-D (Rate-Distortion) analysis, a user/system is unable to define the desired computational complexity (such as a number of CPU clocks, memory bandwidth usage, and power consumption) for the overall encoding process (e.g., for encoding a number of various video sequences on a server for a predetermined period of time). This issue becomes critical for video applications used on mobile/portable devices due to the waste of power recourses. The second issue is that the computational complexity resources, such as the CPU processing resources are not fully used during real-time encoding and transmission, since the CPU processing usage depends on encoder buffer occupancy, which is usually limited by the buffer's memory size. If the CPU has high processing resources, then the encoder buffer is loaded too quickly under a constant quantization. One of the conventional solutions for this problem in real-time video transmission is to pause the CPU processing by sending the CPU a "wait cycle" command. Still another conventional solution is to decrease the CPU clock frequency (if the CPU is implemented to operate with variable clock cycles). However, by sending the "wait cycle" command or decreasing the CPU clock frequency, the available CPU resources are wasted. Still another conventional solution is to increase the quantization settings. However, an increase of the quantization settings leads to a decrease of video quality. As a result, providing an efficient computational complexity control scheme is a very desirable and challenging task.

In the H.264/AVC *baseline* profile encoder, the computational complexity can be modeled as follows (Lee, et al., 2009):

$$C_{mb}(i) = C_{ME} + C_{Ipred} + C_{Text} + C_{Debl} + C_{VLC} \quad (2)$$

and

$$C_{frame} = \sum_{k=0}^{M-1} C_{mb}(i) \quad (3)$$

where $C_{mb}(i)$ is the computational complexity of the *i-th* macroblock and C_{frame} is the whole computational complexity of a frame C_{ME} is the computational complexity for the Motion Estimation (ME) and mode decision. C_{Ipred} is the computational complexity of the Intra-prediction step; C_{Text} is the computational complexity of the DCT, quantization and reconstruction; C_{Debl}, and C_{VLC} are the computational complexity values of the deblocking filter and CAVLC (context adaptive VLC (Variable Length Coding)), respectively. It should be noted that the H.264/AVC encoder performs the Motion Estimation/Motion Compensation steps for the seven different block sizes: *16×16, 16×8, 8×16, 8×8, 8×4, 4×8*, and *4×4* (Lee, et al., 2009). Since most of the computational complexity resources are allocated to the motion estimation part, as shown for example in Figure 3, Lee et al. (2009) propose an approximation of each macroblock's computational complexity:

$$C_{mb}(i) \cong C_{ME} + C_{rest} \quad (4)$$

where C_{rest} represents the computational complexity of the Intra-prediction, texture, deblocking filter, and VLC coding, which it is assumed to have a substantially constant value. Based on the result of complexity profiling of baseline H.264/AVC encoding, Lee et al. (2009) selects two major parameters to control the computational complexity of encoding: a set of possible candidate macroblock modes for Inter-mode decision and the number of iterations in the sub-pel Motion Estimation.

Also with this regard, Grois et al. (2010c) and Kaminsky et al. (2008) present a technique for providing complexity-aware high quality video coding for H.264/AVC applications, thereby suggesting an approach for optimizing the H.264/AVC video compression by dynamically allocating computational complexity (such as a number of CPU clocks) and bits of each basic unit within a video sequence, according to its predicted MAD (Mean Absolute Difference). As a result, the video sequence distortion is minimized and better video quality is achieved.

According to Grois et al. (2010c) and Kaminsky et al. (2008), in addition to calculating the target bit-rate, it is necessary to calculate the target encoding computational complexity, which is required for encoding the current frame (and in turn, for encoding each macroblock within the frame). The target encoding computational complexity $C_t(i)$ for the frame i should be a weighted combination of the remaining computational complexity for encoding the remaining frames, and the target encoding computational complexity allocated for frame i, which is formulated as:

$$C_t(i) = \alpha \cdot \widetilde{C}_r(i) - (1-\alpha) \cdot \widehat{C}_t(i) \qquad (5)$$

where α is a weight coefficient; $\widetilde{C}_r(i)$ is the remaining computational complexity for encoding the current frame i; and $\widehat{C}_t(i)$ is the target encoding computational complexity allocated for frame i in the current GOP. $\widetilde{C}_r(i)$ and $\widehat{C}_t(i)$ are represented by the following expressions:

$$\widetilde{C}_r(i) = \frac{C_r(i)}{N_r}, and \qquad (6)$$

$$\widehat{C}_t(i) = \frac{\frac{C_r(i)}{N_r} \cdot \Theta^2(i)}{\frac{1}{N-N_r} \sum_{j=1}^{N-N_r} \Theta^2(j)} \qquad (7)$$

where, N_r is the number of remaining frames; and N is the total number of frames in the current GOP; $C_r(i)$ is the remaining computational complexity for encoding the remaining frames; $\Theta(i)$ is the predicted MAD of the current frame

i; and $\Theta(j)$ is the actual MAD of the previous frame j.

By using a linear prediction model, the target encoding complexity allocated for the i frame in the current GOP can be determined based on the predicted MAD of the current frame and the actual MADs of the previous frames. Also, according to Grois et al. (2010c) and Kaminsky et al. (2008), most Distortion-Complexity-Rate (DCR) surfaces are actually Exponential Distortion-Complexity-Rate (EDCR) surfaces (the average DCR surface of the video sequence is an EDCR surface). As a result, Grois et al. (2010c) and Kaminsky et al. (2008) propose representing the distortion as a function of computational complexity (C) and bit-rate (R):

$$D(C,R) = A \cdot \alpha^{-R} \cdot \beta^{-T}, \text{for } C > 0, R > 0 \qquad (8)$$

where $A > 0$, $\beta > 0$ and $\alpha > 1$. From (8), the corresponding average complexity-rate-distortion (CRD) and Rate-Complexity-Distortion (RCD) surfaces are given by:

$$C(R,D) = \log_\beta(A \cdot \alpha^{-R} / D), \text{ for } A \cdot \alpha^{-R} > D > 0 \qquad (9)$$

and

$$R(C,D) = \log_\alpha(A \cdot \beta^{-C} / D), \text{ for } A \cdot \beta^{-C} > D > 0 \qquad (10)$$

Figure 4 shows experimental and fitted EDCR surfaces of different video sequences: "Silent," "News," "Foreman" and "Container." The average errors between the experimental and fitted surfaces are 2.86%, 3.76%, 3.82% and 3.27%, respectively, which means that the equation (8) accurately represents the Computational Complexity-Rate-Distortion (C-R-D) relationship.

In turn, according to Grois et al. (2010c) and Kaminsky et al. (2008), the bit-rate control C-S-R and R-Q-C models, are as follows:

Figure 4. Experimental and fitted EDCR surfaces (the fitted surfaces are based on Equation [8]) of: a) the "SILENT" video sequence, an average error between the experimental and fitted surfaces is 2.86%; b) the "NEWS" video sequence, an average error between the experimental and fitted surfaces is 3.76%; c) the "FOREMAN" video sequence, an average error between the experimental and fitted surfaces is 3.82%; d) the "CONTAINER" video sequence, an average error between the experimental and fitted surfaces is 3.27%

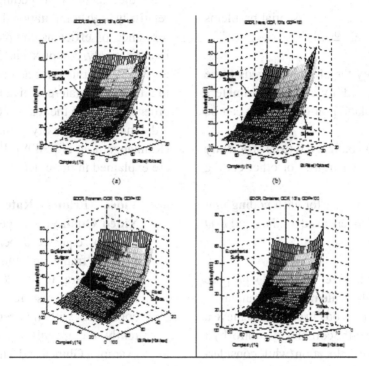

$$C(S,R) = A_{C_1} \cdot S^{-1} + A_{C_2} \cdot S^{-2} + A_{C_3} \cdot R \quad (11)$$

and

$$R(Q,C) = A_{R_1} \cdot Q^{-1} + A_{R_2} \cdot Q^{-2} + A_{R_3} \cdot C \quad (12)$$

where A_{C_1}, A_{C_2}, A_{C_3} and A_{R_1}, A_{R_2}, A_{R_3} are corresponding coefficients that are calculated regressively (according to a linear regression method); S is the complexity step for selecting a corresponding group of coding modes (e.g., Inter-Search16X8, Inter-Search8X16, Inter-Search8X8, Inter-Search8X4 modes); C is the encoding computational complexity; R is the bit-rate; and Q is the corresponding quantization step-size for perform-

ing the Rate-Distortion-Optimization (RDO) for each macroblock in the current frame.

As a result, the modified encoder is capable of allocating computational complexity and bit-rate resources for obtaining much better coding efficiency with improved processing speed in term of processing time. It should be noted that the computational complexity is mainly presented in term of the CPU processing time since the processing time can be easily tracked for updating the bit-rate control models and for the parameter adjustments on any computing platform.

Based on the above observation and due to the strong need for providing complexity-aware coding techniques, it is clearly seen that the rate-complexity-distortion optimization (RCDO) at the encoder side has drawn much attention in the recent years (Li, et al., 2011). The RCDO is actually an extension of the RDO, which was mentioned above. In real-time video applications, the computational complexity-rate-distortion information is not available, which makes the RCDO quite difficult.

Generally, there are three essential problems in the RCDO (Li, et al., 2011):

1. The complexity target budget needs to be properly mapped to coding parameters such that regulation on complexity can be obtained;
2. The complexity budget has to be efficiently distributed among frames or other coding units; and
3. The allocated budget for each coding unit should be effectively used for the good R-D performance.

With this regard, Li et al. (2011) propose single-scale and multi-scale medium-granularity complexity control methods in order to achieve a large dynamic range in the complexity control by a frame level complexity allocation (while considering the inter-frame dependency) and by performing an adaptive mode and reference searching, which is designed to efficiently utilize the complexity budget at the macroblock level. Further, Foo et al. (2008) presents a rate-distortion-complexity (R-D-C) analysis for the state-of-the-art wavelet video coding methods by modeling several aspects found in operational coders, i.e., embedded quantization, quadtree decompositions of block significance maps and context-adaptive entropy coding of subband blocks.

The following section discusses in detail recent advances in computational complexity management for real-time video encoders.

COMPUTATIONAL COMPLEXITY MANAGEMENT FOR REAL-TIME H.264/AVC-BASED VIDEO ENCODERS

Since the H.264/AVC encoding process consumes relatively high computational resources, the performance of efficient complexity management is a relatively complex challenge. The typical approaches to this issue are as follows: 1) reducing the frame-rate/perceived video quality in order to maintain the real-time coding; and/or 2) reducing the complexity of the encoding process (Kannangara, et al., 2008). Below, these issues are explained in more details:

- **Video Frame Rate and Perceived Video Quality:** The perceived quality of a video sequence depends on the quality of each frame and the video frame rate (Kannangara, et al., 2008). However, although much research has been recently conducted with this regard, there is currently no unequivocally conclusion. For example, Ghinea and Thomas (1998) show that a significant loss in frame rate does not proportionately reduce the perceptual quality or the acceptability of the encoded video and that the ability to understand the information content of the sequence does not degrade significantly with a reduction in frame rate. Also, McCarthy et al. (2004) claim that when watching sports events by using small displays (such as amobile phone display), the picture quality is more important than the frame rate. Further, Apteker et al. (1995) indicate that users perceive the effect of reduced frame rate differently depending on the video content: the users find that reduced frame rate leads to progressively lower acceptability ratings, particularly for scenes with low activity and movement and less so for high-activity scenes.

- **Reduced Complexity Video Coding:**
Many low-complexity algorithms have been recently proposed. For example, low-complexity motion estimation algorithms for H.264 are described in Chung et al. (2003) and Stottrup-Andersen et al. (2004) and a significant amount of work has been carried out to reduce the complexity of the mode selection process (Jiao, et al., 2011; Paul, et al., 2009; Bystrom, et al., 2008; Yu & Martin, 2004; Kim & Altunbasak, 2004; Dai, et al., 2004; Chen, et al., 2004). In addition, significant research has been carried out with regard to the power-distortion optimization (Kim, et al., 2011; Solak & Labeau, 2010; He, et al., 2008; Parlak, et al., 2008). As a result, the common characteristics of these researches are (Kannangara, et al., 2008):
  - Significant reductions in coding complexity may reduce rate-distortion performance; and
  - The saving in complexity is typically content-dependent.

With this regard, Kannangara et al. (2008) propose to manage complexity in order to maximize perceptual video quality in a real-time computational constrained scenario, as presented in Figure 5. The assumption of Kannangara et al. (2008) is that higher perceptual video quality can be achieved by maintaining a smooth frame rate even with some loss in frame quality as opposed to allowing a drop in frame rate.

Ivanov and Bleakley (2010) propose a fast Mode Decision (MD) algorithm that contains two parts: a) a class decision; and b) a mode search with the Early Termination (ET). From the preceding analysis, a class decision algorithm is based on the three metrics: Frame Difference (FD), J_{prev}, and $SAD_{8\times8}$, as presented in Figure 6. The FD metric is used to identify the MBs with the high motion activity. MBs for which the FD exceeds a threshold are considered to be active, and are allocated to Class A. For inactive macroblocks, the Rate-Distortion (RD) cost of the same MB in the previous frame is compared to a predefined threshold. MBs with the high J_{prev} are considered as being coded with poor efficiency in the previous frame, and are allocated to Class B. The RD cost in the case of a SKIP decision is predicted according to:

$$J_{SKIP\_predicted} = \alpha \cdot SAD_{8\times8} + C_1 \qquad (13)$$

where α and C_1 are constants. If the predicted RD cost of a SKIP decision is less than the mean RD cost in the previous frame, i.e. $J_{SKIP\ predicted} < J_{mean}$, then a SKIP decision is made, and the MB is allocated to Class E. If a SKIP decision is not made, then the partial Sum of Absolute Differences (SAD) result is compared to a threshold. If it exceeds the threshold, the MB is considered to be static and it is encoded with the low quality in the previous frame, and is allocated to Class D. Otherwise the MB is considered to be static and encoded with the high quality in the previous frame, and is allocated to Class C (Ivanov & Bleakley, 2010).

Further, Akyol et al. (2007) present a macroblock-level fast motion estimation mode search algorithm, where the complexity is jointly adapted by parameters that determine: the "aggressiveness" of an early stop criteria, a number of the modes searched, and the accuracy of the Inter-mode motion estimation. In addition, in order to optimize the H.264/AVC encoder Rate-Distortion (R-D) curve, Kim et al. (2010) analyze the encoding structure and establish an analytic model for computational complexity. Based on this analysis, Kim et al. (2010) propose an improved real-time encoder, which maintains both frame rate and video quality by analyzing the results of P16x16 Motion Estimation (ME) in the current frame and deciding about the number of skipped Macroblocks (MBs) in order to encode the frame

Figure 5. Managing the encoder computational complexity (Kannangara, et al., 2008)

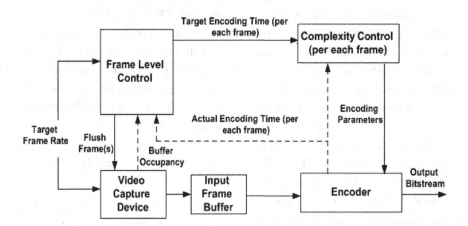

within the allocated target encoding time. Figure 7 presents a scheme proposed by Kim et al. (2010).

The complexity control encoding scheme of Kim et al. (2010) consists of two frame-level stages. At the first stage, a new frame is encoded by using the P16X16 mode to estimate the encoding time and parameters for the second stage (the current frame information is used for obtaining better estimation). At the second stage, a frame is encoded according to the target encoding time, which was determined at the first stage; the coding complexity of each frame is controlled by determining whether or not to use other modes for each MB (all possible modes except for the P16X16 mode). In order to minimize the redundant operations, the motion vector (of P16X16 mode), which was determined at the first stage, is used at the second stage. Since the P16X16 mode is always executed at the first stage, then the computational complexity is limited. However, the encoder structure proposed by Kim et al. (2010) leads to the minimal coding efficiency.

It should be noted that the emergence of high-definition video into mainstream applications has imposed a challenge to researchers and engineers in the field: achieving optimal balance between programmability and hardware efficiency (Wang & Hua, 2008). With this regard, Rhee et al. (2010) present a processing time control algorithm for

a hardware-based H.264/AVC encoder, which employs three complexity-scaling methods: a) partial cost evaluation for Fractional Motion Estimation (FME), block size adjustment for FME, and search range adjustment for Integer Motion Estimation (IME). With these methods, 12 complexity levels are defined to support tradeoffs between the processing time and compression efficiency. Also, Rhee et al. (2010) proposes a speed control algorithm for selecting a complexity level that leads to the most efficient compression (among complexity levels which meet the target time budget); the time budget is allocated to each macroblock based on the complexity of the macroblock and on the execution time of other macroblocks in the frame.

According to Rhee et al. (2010), in an H.264/AVC main profile encoder, ME is performed in three directions: forward, backward, and bidirectional. To achieve the best compression efficiency, the encoder needs to perform all three directions and choose the direction with the best compression efficiency. Evaluating the cost of these three directions significantly increases the computational complexity. Therefore, Rhee et al. (2010) proposes a simplified ME scheme called direction filtering to reduce the computational burden of evaluating all three directions, as shown in Figure 8. For the Integer Motion Estimation

Figure 6. Schematic flow of the MB class decision algorithm (Ivanov & Bleakley, 2010)

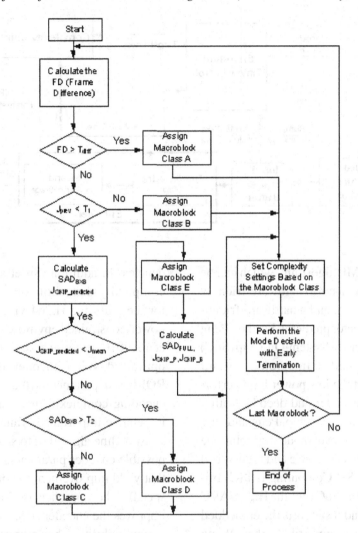

(IME), both the forward and backward directions are performed, whereas the direction(s) of FME is chosen by comparing the costs from the two IMEs. If the difference between the two cost estimates is between the upper threshold, TH_{max}, and the lower threshold, TH_{min}, both forward and backward FMEs are performed. In addition, the bi-directional prediction is executed and the results of three predictions are compared to obtain the best Motion Vector (MV).

In addition, Qi et al. (2011) present a five-stage pipelined architecture of H.264/AVC real-time encoder by separating off-chip data read from the IME and using it as a substantive stage, which in turn shortens the critical path and improves the performance of encoder. In addition, Qi et al. (2011) adopt an optimized Motion Estimation (ME) algorithm for removing data dependencies and shared storage policies to save hardware resources. In addition, Adibelli et al. (2011) propose pixel equality and pixel similarity based techniques for reducing the amount of computations performed by the H.264 Deblocking Filter (DBF), and in turn reducing the energy consumption of the H.264 DBF hardware for the real-time usage, for example. In addition, Wen et al. (2011) propose

Figure 7. The block-diagram of the improved real-time encoder, which is proposed by Akyol et al. (2007)

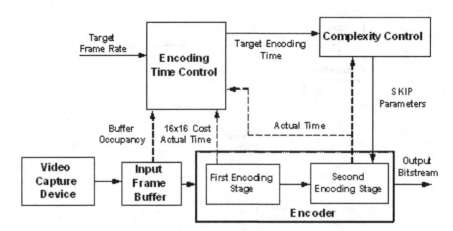

a hardware-friendly ME algorithm, which can be used in real-time, with a novel hardware-friendly (R-D)-like cost function, and a hardware-friendly modified motion vector predictor. Further, Peng and Jing (2003) designed a System-On-Chip (SoC) and Application-Specific Integrated Circuit (ASIC) to implement the low power high performance H.264/AVC encoder and decoder with a 32-bit RISC CPU on a single chip. Peng and Jing (2003) used the system-level modeling technique to develop the H.264/AVC codec and high-speed Reduced Instruction Set Computing (RISC) microprocessor cores. The SoC runs the H.264/AVC codec in hardware, and it supports the embedded real-time operating system. Still further, Wang and Hua (2008) investigate how to efficiently implement H.264/AVC high-definition encoder and decoder based on a C64+ DSP (Digital Signal Processor) core accompanied by the on-chip co-processor. Wang and Hua (2008) provide real-time performance analysis based on such a system, in terms of system loading for high-definition encoders.

There are other recent real-time complexity-rate-distortion techniques, such as of Tan et al. (2010) that propose a complexity scalable video encoding technique for controlling the compelxity-coding performance by using a "wavefront" scheme to optimize the R-D curve of independent MBs, while considering the inter-MB depen-

dencies. In addition, Su et al. (2009) present a complexity-distortion optimization approach for the real-time H.264 video coding under the power-constrained environment. Further, Vanam et al. (2010) propose a Region-of-Interest (ROI)-optimized video encoder that includes three ROI-based parameters that allow variations in encoding complexity per-macroblock based on its relative importance. Vanam et al. (2010) use a fast offline algorithm to search the space of all possible encoder parameters, to find parameters that yield improvement in complexity (encoding speed). The approach of Vanam et al. (2010) improves the encoder speed with negligible loss in intelligibility, when compared to the real-time x264 encoder (an implementation of the H.264/AVC standard) default parameter settings.

In the following section, recent techniques for the computational complexity reduction by performing an efficient block mode selection are discussed.

COMPUTATIONAL COMPLEXITY REDUCTION BY PERFORMING EFFICIENT BLOCK MODE SELECTION

To select the best coding mode, the Rate-Distortion (R-D) costs of all possible H.264/AVC modes

Figure 8. The best motion vector selection scheme, as proposed by Rhee et al. (2010)

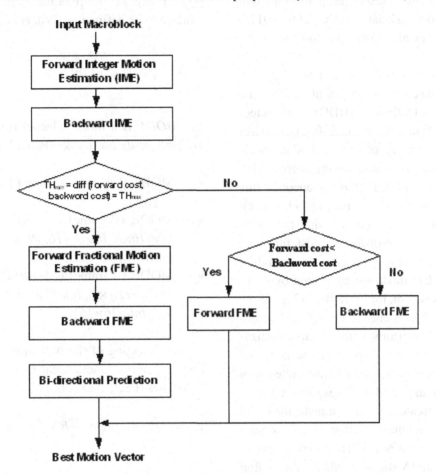

can be exhaustively calculated, since each mode selection has to be evaluated for its effect on distortion, which in turn can lead to a massive growth in computational complexity in case when every possible coding mode is assessed. Therefore, there is a need to perform the coding mode selection in an efficient manner without wasting valuable computational resources.

As specified in H.264/AVC, there are seven different block sizes *16x16, 16x8, 8x16, 8x8, 8x4, 4x8,* and *4x4*, which can be used in the Inter-frame motion estimation/compensation (Wu, et al., 2005). In addition, the SKIP/Direct modes can be used (the SKIP/Direct modes are referred to when a block is coded without sending the residual error or motion vectors; in turn, the decoder deduces

the motion vector of the Direct/Skip mode coded block from other already decoded blocks). These different block sizes form a two-level hierarchy inside a MB. The first level comprises block sizes of *16x16, 16x8, or 8x16*. In the second level, the MB is specified as *P8x8* type, of which each 8x8 block can be one of the subtypes such as *8x8, 8x4, 4x8,* or *4x4*. The relationship between these different block sizes is shown in Figure 9 (Wu, et al., 2005).

While selecting the minimal encoding computational complexity, an optimal set of coding modes for encoding each basic unit cannot be obtained, and as a result, this usually leads to the maximal distortion. In other words, the minimal encoding computational complexity relates to a single coding mode, and as a result, to a maximal distortion (Grois & Hadar, 2011b, 2011c, 2011d, 2011e; Grois, et al., 2010c; Kaminsky, et al., 2008).

In order to achieve the good Rate-Distortion (RD) performance (Bystrom, et al., 2008), the Rate Distortion Optimized (RDO) mode selection process (Wiegand, et al., 2003a) evaluates the distortion and rate of each candidate mode prior to selecting the mode for the current MB. In the Joint Model (JM) reference encoder, this is carried out by coding the macroblock in each of the possible modes and selecting the mode that minimizes a rate-distortion cost function. The rate (R) and distortion (D) corresponding to each candidate mode are calculated by using the process as shown, for example, in Figure 10.

The source MB is encoded by using the Intra or Inter-prediction, a forward transform, quantization, and source coding, to produce a sequence of R bits, where R indicates the rate associated with a particular candidate mode. The quantized coefficients are rescaled, inverse transformed and reconstructed, and the distortion, D, i.e. the Sum of Squared Differences (SSD) between the source and reconstructed MBs, is calculated. According to Lim et al. (2005a, 2005b) and Wiegand et al. (2003a), the RDO for each macroblock is performed for selecting an optimal coding mode for each mackroblock by minimizing the Lagrangian function as shown as follows.

$$
\begin{aligned}
& J(orig, rec, MODE \,|\, \lambda_{MODE}) \\
& = D(orig, rec, MODE \,|\, QP) \\
& + \lambda_{MODE} \cdot R(orig, rec, MODE \,|\, QP)
\end{aligned}
\tag{14}
$$

where the distortion $D(orig, rec, MODE \,|\, QP)$ can be the sum of squared differences (SSD) or the sum of absolute differences (SAD) between the original block (*orig*) and the reconstructed block (*rec*); QP is the macroblock quantization parameter; $R(orig, rec, MODE \,|\, QP)$ is the number of bits associated with selecting *MODE*; λ_{MODE}

is a Lagrangian multiplier for the mode decision, which can be defined as (Wu, et al., 2005):

$$
J_{MODEP} = \begin{cases} 0.85 \cdot 2^{QP/3}, & for\,a\,P - frame \\ \max(2, \min(4, QP\,/\,6)) \cdot \lambda_{MODEP}, & for\,a\,B - frame \end{cases}
\tag{15}
$$

and *MODE* is a mode selected from the set of available prediction modes (Pan & Kwong, 2011):

- MODE (I-frame) ∈ {*Intra4 × 4, Intra16 × 16*};
- MODE (P-frame) ∈ {*SKIP, Inter16×16, Inter16×8, Inter8×16, P8 × 8, Intra4 × 4, Intra16 × 16*}; and
- MODE (B-frame) ∈ {*DIRECT, Inter16 × 16, Inter16 × 8, Inter8 × 16, P8 × 8, Intra4 × 4, Intra16×16*}.

The selecting of the best mode m_b from the Lagrangian cost function, can be also represented as follows (Paul, et al., 2011):

$$
m_n = \arg \min(J(m_i)) \,\big|\, R(m_i) \leq R^T, \forall m_i
\tag{16}
$$

where R^T is the target bet rate.

However, the best mode selection process (e.g., as shown in Figure 11) of exhaustively evaluating each mode requires significant computational resources.

As noted above, for the *inter*-frame MB coding, there are eleven candidate modes (Zeng, et al., 2009): *SKIP, 16x16, 16x8, 8x16, 8x8, 8x4, 4x8, 4x4, Intra4x4, Intra8x8*, and *Intra16x16*, while the last three modes are denoted as *I4MB, I8MB*, and *I16MB*, respectively. On the other hand, for the intra-frame MB coding, only *I4MB, I8MB*, and *I16MB* are applicable. As a result, the exhaustive mode decision algorithm implemented in the H.264 reference software (Richardson, 2003b, 2003c) can be generally described as follows (Zeng, et al., 2009):

Figure 9. Dividing each macroblock (MB) into different partitions: a) MB partitions and b) MB sub-partitions (Wu, et al., 2005)

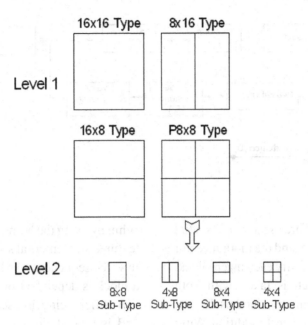

1. Check whether the current MB can be considered as an intra MB, which is always the case if the MB comes from an intra-frame. If the answer is "yes," then go to step (f); otherwise, proceed to the next step.
2. Compute the cost of the SKIP mode.
3. Perform the motion estimation process and compute the costs of three large block-sized modes (namely, 16x16, 16x8, and 8x16) for the current MB, respectively.
4. Select one of the four 8x8 blocks within the current MB, perform the motion estimation process and compute the costs of four smaller block-sized modes (namely, 8x8, 8x4, 4x8, and 4x4), respectively.
5. Repeat step (d) likewise for the other three 8x8 blocks, individually.
6. Perform the intra prediction procedure and compute the costs of I4MB and I16MB (i.e., Intra4x4 and Intra16x16 modes), respectively.
7. Among all the modes that have been checked in the previous steps, select the one that yields the minimum cost as the best mode.

Therefore, by automatically estimating the outcome of the process, it may be possible to eliminate much of the complexity, while maintaining high compression performance and good video quality (Bystrom, et al., 2008). It is noted that when not enabling all available modes after performing the high-complexly RDO and using full search motion estimation, the computational complexity allocation required for encoding each macroblock within each frame type (*I, P* or *B*) is *constant* and *not optimal*. In addition, it is noted that when implementing the Fast Motion Estimation (FME), the computational complexity allocation is *variable*, but is still *not optimal* (Grois, et al., 2010c; Kaminsky, et al., 2008).

With this regard, Bystrom et al. (2008) address the challenge of early estimation of coder mode selection, and sets the mode decision process in

Figure 10. Processing of a candidate mode for a macroblock in order to determine the corresponding rate and distortion (Bystrom, et al., 2008)

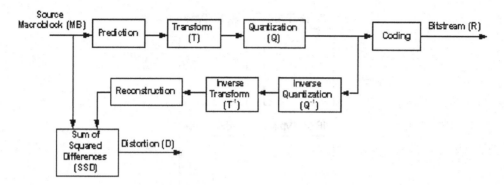

a Bayesian framework. Three features, a spatial feature, a temporal feature, and the motion vector magnitude, are used to classify each macroblock, and a Bayesian approach is taken with both parametric and non-parametric models for the examined mode cost difference. In addition, Wu et al. (2005) present a fast inter-mode decision algo-

rithm by using the homogeneity of video object's textures and temporal stationarity characteristics in video sequences (the homogeneity decision of a block is dependent on edge information), and MB differencing is used to judge whether the MB is time-stationary. According to Wu et al. (2005), the fast inter-mode decision algorithm is

Figure 11. Schematic flowchart of the H.264/AVC full-mode decision scheme (Pan & Kwong, 2011)

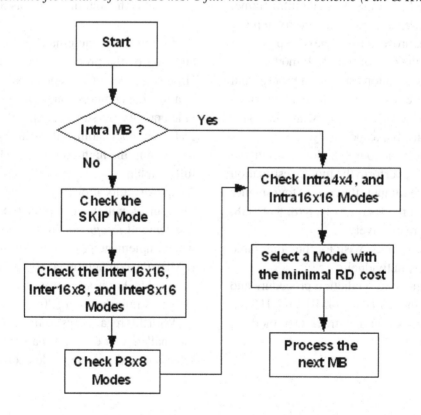

able to achieve a reduction of 30% in encoding time on average.

Further, Paul et al. (2011) propose reducing a significant amount of computational time by the direct mode decision and exploitation of available motion vector information from phase correlation. The phase correlation technique, which is based on the Discrete Fourier Transform (DFT), can be used for estimation of the relative displacement between co-located (i.e., with zero-motion vector) blocks. The magnitude of the transformed blocks tells "how much" of a frequency component is present and the phase tells "where" the frequency component is in the blocks. According to Paul et al. (2011), the phase information is used for prediction of the initial motion vector (i.e., PCDMV), and the phase-matched error, PME, has been adopted for the mode selection. The schematic block-diagram of generating the PME is presented in Figure 12. The PCDMV (*dx, dy*) is considered as the distance between the largest peak position and the middle position of the block, i.e., the

PCDMV (0, 0) means that the largest peak in is in the exact middle.

In addition, Wu and Lin (2009) propose an efficient inter/intra mode decision for H.264/AVC inter-frame transcoding. First, a zero-block decision scheme is first employed to select candidate inter and intra modes. Then, a fast multi-reference frame motion estimation is performed with an adaptive search window. For the intra mode decision, Wu and Lin (2009) suggest a mode refinement scheme to eliminate undesirable modes during the RDO process. Further, Pan and Kwong (2011) propose a fast inter-mode decision scheme based on luminance difference. The potential modes are divided into three sub-sets, while one of the three sub-sets is selected according to Macroblock (MB) motion activity, which is measured by the luminance difference between the current MB and its co-located MB in the previous frame.

Figure 12. The schematic block-diagram of generating a phase-matched error (Paul, et al., 2011; To, et al., 2004)

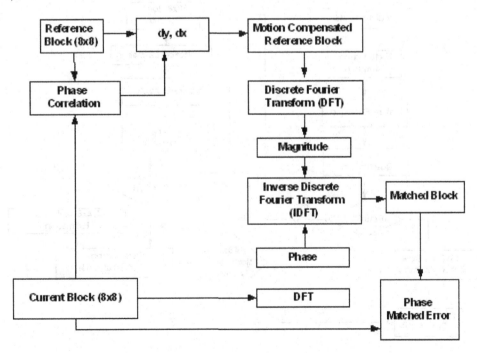

Further, Chiang et al. (2011) propose fast algorithms based on statistical learning to reduce the computational cost involved in the:

- Inter-mode decision;
- Intra-prediction mode selection; and
- Multi-reference motion estimation (ME).

According to Chiang et al. (2011), representative features are extracted to build the learning models. Then, an offline pre-classification approach is used to determine the best results from the extracted features; thus, a significant amount of computational resources is reduced. The flowchart of the inter-mode decision method is presented in Figure 13.

According to Chiang et al. (2011), the offline pre-classification approach is used to minimize the computation time involved in the classification procedure. The idea is to generate all possible combinations of the quantized feature vectors and pre-classify them with SVM (Support Vector

Machines) (Chang & Lin, 2003). To reduce the total number of possible combinations, the Lloyd-Max quantizer (Cover, 1991), which has an adaptive step size, is applied on the training samples because it can optimize the quantization of a feature for a specific distribution. The results are stored as a look-up table. Hence, the computation time can be significantly reduced by using this offline pre-classification approach (Chiang, et al., 2011). In addition, Lee and Lin (2006) propose to perform a fast inter-mode decision algorithm in which the variance of the sum of absolute difference (SAD) between the 4x4 small blocks is employed to describe the stationary and homogeneous characteristics of a macroblock.

In addition, Kannangara et al. (2009) propose a computational complexity control algorithm, which is based on a macroblock mode prediction algorithm that employs a Bayesian framework for the accurate early SKIP decision. The complexity control is achieved by relaxing the Bayesian Maximum-Likelihood (ML) criterion in order

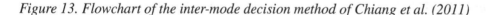

Figure 13. Flowchart of the inter-mode decision method of Chiang et al. (2011)

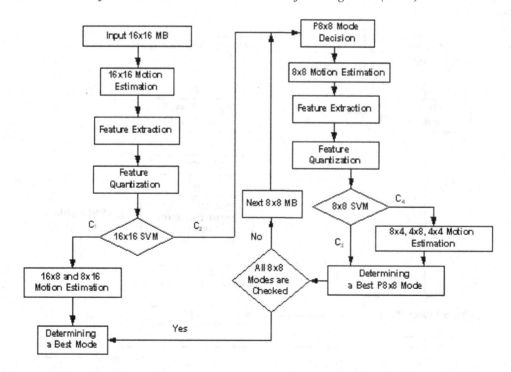

to match the mode decision threshold to a target complexity level. In addition, a feedback algorithm is used to maintain the performance of the algorithm with respect to achieving an average target complexity level, reducing frame-by-frame complexity variance and optimizing rate-distortion performance (Kannangara, et al., 2009).

Further, Paul et al. (2009) propose a simplified Lagrange multipliers approach, in which the mode selection criteria consider distortion along with the amount of data required for the motion vectors and the variable block type. Still further, Pang et al. (2011) address the problem of optimal Lagrange determination for intra-frames in a signal dependent environment, and a novel frame complexity guided Lagrange multiplier selection method is introduced. Firstly, Pang et al. (2011) propose a novel signal dependent – model, which uses the average input frame gradient to measure the signal complexity. Secondly, based on the

proposed signal dependent R-D model, Pang et al. (2011) derives the closed-form solution of the optimal Lagrange multiplier.

Also, Kim et al. (2011) propose an efficient relativelly simple block size decision method. According to Kim et al. (2011), only 2 block sizes are decided as candidate block sizes, and further the prediction modes of each block size are reduced using an improved prediction mode selection algorithm. Figure 14 presents a flow of the block size decision method proposed by Kim et al. (2011), which is based on the following steps:

1. Calculating the variance of a MB to be coded and calculating the corresponding threshold.
2. If variance of the MB is greater than the threshold, then I4MB and I8MB modes are selected as the candidate MB modes; otherwise, I8MB and I16MB are selected.

Figure 14. A flow of the improved macroblock mode decision method (Kim, et al., 2011)

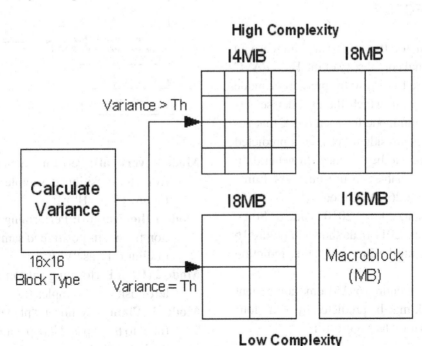

Further, Lee and Park (2011) present a fast mode decision method for inter-picture macroblocks, which significantly reduces the number of candidate modes for the R-D optimization process by detecting spatially and temporally homogeneous regions and analyzing motion costs for inter-modes and intra-prediction costs for intra-modes, as illustrated in Figure 15. It is noted that the 3rd minimum cost mode is eliminated according to Table 1.

Further, there are many additional low-complexity mode selection techniques, such as Zhu et al. (2010), who present an adaptive fast block mode selection method, in which the threshold information is used to effectively select the SKIP mode, and then to skip unnecessary modes based on the coded frames /block information. In the following section, the computational complexity reduction by performing the efficient intra-prediction mode selection is discussed in detail.

COMPUTATIONAL COMPLEXITY REDUCTION BY PERFORMING EFFICIENT INTRA-PREDICTION MODE SELECTION

Another reason for the significant increase in the computational complexity of the H.264/AVC encoder is related to the intra-prediction mode selection process, in which the encoder selects the best prediction mode for each block that actually minimizes a residual between a predicted block and a block to be encoded. In general, in the H.264/AVC enables to use nine *4x4* Luma Intra-prediction H.264/AVC modes (H.264/AVC, 2003; Richardson, 2003a, 2010; Park & Song, 2006; Grois, et al., 2011a), as shown in Figure 16 (the arrows indicate the direction of the prediction in each mode).

In addition, the entire 16x16 Luma component of a macroblock may be predicted by using four modes, as shown in the Figure 17:

Figure 15. The flowchart of the fast inter-mode decision method (Lee & Park, 2011)

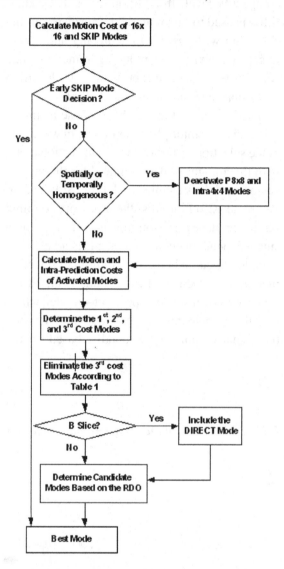

Mode 0 (vertical): Performing an extrapolation from top-positioned samples (denoted in Figure 17 as "H");

Mode 1 (horizontal): Performing an extrapolation from left- positioned samples (denoted in Figure 17 as "V");

Mode 2 (DC): Performing a mean of the top and left-positioned samples ("H+V");

Mode 3 (Plane): A linear "plane" function is fitted to the top and left-positioned samples ("H" and "V").

Table 1. Elimination of the third minimum cost mode (Lee & Park, 2011)

Sub-Optimal Mode	Third Minimum Cost Mode, which should be eliminated
SKIP	*16x8, 8x16, P8X8, Inra16x16, Inra8x8, Inra4x4*
16x16	*SKIP, 16x8, 8x16, P8X8, Inra16x16, Inra8x8, Inra4x4*
16x8	*SKIP, 8x16, P8X8, Inra16x16, Inra8x8, Inra4x4*
8x16	*SKIP, 16x8, P8X8, Inra16x16, Inra8x8, Inra4x4*
P8X8	*SKIP, Inra16x16, Inra8x8, Inra4x4*
Intra16x16	*P8X8*
Intra8x8	*P8X8*
Intra4x4	*SKIP, 16X16, 16x8, 8x16, P8X8, Inra16x16*

Therefore, for determining the best intra-prediction mode, usually high computational complexity resources are required. As a result, in order to reduce the encoder complexity, the extensive research was recently conducted. For example,

Jiao et al. (2011) propose an efficient pipelining method for the 4x4 blocks intra-prediction mode selection. In particular, Jiao et al. (2011) exploit the Graphics Processing Unit (GPU) streaming architecture at the 4x4 intra-prediction mode selection in the H.264/AVC encoder and develops a special strategy, including instruction optimization, while taking full advantage of shared memory in order to further exploit the fine-grained parallelism of GPUs. Also, Zhang et al. (2010) combine a fast algorithm and efficient fast mode decision for intra-prediction, while first, the intra-prediction modes are determined according to the Sum of Absolute Difference (SAD) values.

In addition, Kim and Kim (2011) present an intra-mode search algorithm to reduce the computational complexity of inter-frames for the H.264/AVC video encoding system. To decrease the computational burden due to the intra-mode search, Kim and Kim (2011) propose an adaptive algorithm based on distribution characteristics of the SAD of the best inter-mode when the intra-

Figure 16. 4x4 Luma intra-prediction modes

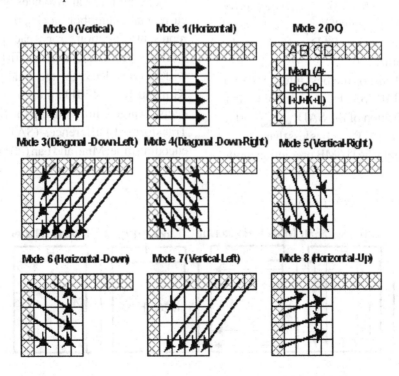

mode is the final coding mode. It is assumed that there is a relatively large SAD cost difference between the I16MB and I4MB intra-modes. The SAD cost of the motion estimation is calculated before the R-D calculation and the intra-mode decision process. Thus, a proper intra-mode can be determined based on the SAD cost (the I4MB type is within the nine prediction modes and the I16MB type is within the four prediction modes for luminance blocks in the H.264/AVC Main profile).

For a macroblock the SAD cost is determined as shown as follows:

$$SAD(macroblock 16 \times 16) = \sum_{k=0}^{15} \sum_{l=0}^{15} \left| f_N(k,l) - f_{N-m}(k+v_y, l+v_x) \right|$$

(17)

where $f_N(k, l)$ is an intensity value at pixel (k, l). v_x and v_y are the components of the best motion vector after motion estimation, and $(N-m)$ is a time index of the reference frame. The algorithm's flow is presented in Figure 18, and it can be summarized as follows:

Step 1: For the current MB ($MB^N_{(i,j)}$), the adaptive thresholds T^N_{I16MB} and T^N_{I4MB} are calculated by using the average SAD costs from the previous frame.

Step 2: The SAD cost of the best inter-mode for the current MB ($SAD^N_{Interbest}$) are obtained. Then, the position of the $SAD^N_{Inter-best}$ cost is determined. If $SAD^N_{Inter-best}$ is less than T^N_{I16MB} value, then only the I16MB prediction mode

is considered. If the cost of $SAD^N_{Interbest}$ is larger than T^N_{I4MB}, then only the I4MB prediction mode search is performed.

Step 3: For an MB with a $SAD^N_{Inter-best}$ value between T^N_{I16MB} and T^N_{I4MB}, the best mode of the neighbouring MBs ($MB^{N-1}_{(i,j)}$, $MB^N_{(i-1,j)}$ and $MB^N_{(i,j-1)}$) is determined. If $MB^{N-1}_{(i,j)}$, $MB^N_{(i-1,j)}$, or $MB^N_{(i,j-1)}$ are encoded as an intra-mode, the full intra-mode search is performed for the current block MBN(i, j).Otherwise, the intra-mode search is skipped.

Step 4: After finishing the mode search for all MBs in the current frame, the adaptive thresholds T^N_{I16MB} and T^N_{I4MB} for the next frame.

Further, there are many additional low-complexity mode selection techniques, such as that of Yang et al. (2010), which propose a low-complexity Fractional Motion Estimation (FME) design, which supports the adaptive mode selection; Meng et al. (2010) present a fast algorithm based on the quantization parameter, and the macroblock motion characteristics are used to reduce the encoder complexity; Lei et al. (2010) propose to calculate overlapping interval distances with regard to the current and neighbor blocks, thereby enabling to represent a direction error of each prediction mode, while the prediction modes with the smaller error will be considered for the final mode selection; Kim and Jeong (2011) present a fast intra-mode decision algorithm by using the Sum of Absolute Transformed Differences (SATD) to reduce the encoding time; Miao and Fan (2011) presents an ef-

Figure 17. 16x16 Luma intra-prediction modes

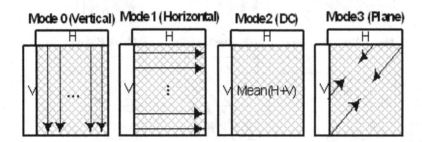

ficient fast algorithm, to predict the best directional mode except for the DC mode based on the edge detection. At the first step, Miao and Fan (2011) use a pre-processing mode selection algorithm to find the primary modes, and at the second step, the selected candidate modes are finally applied for calculating the R-D cost; Zeng et al. (2010) present a fast intra mode decision algorithm, called the Hierarchical Intra Mode Decision (HIMD), to speed up the mode decision process by reducing the number of modes required to be checked for each macroblock. According to Zeng et al. (2010), the candidate modes are selected according to their Hadamard distances and prediction directions, and only one of the hierarchical paths is chosen to compute the smallest R-D cost; Quyang et al. (2010) presents a simplified directional mode selection method based on the edge-detection algorithm; Ren and Dong (2010) presents a fast intra mode selection algorithm based on the grayscale; Kim et al. (2011) proposes a fast intra mode decision method, which enable to predict the edge direction of the current block by using sub-sampled pixels; Wang et al. (2011) selects the optimal intra mode by extracting macroblock spatial features by using a two-dimensional histogram.

In the following section, the recent low-complexity techniques for multi-reference motion estimation are discussed in detail.

LOW COMPUTATIONAL COMPLEXITY TECHNIQUES FOR MULTI-REFERENCE MOTION ESTIMATION

In order to improve the encoding efficiency, H.264/AVC adopts motion estimation and compensation algorithm in multi-reference frames (Xiao & Cheng, 2011). Although the multi-reference frame selection algorithm can find the best matching block by exhaustively testing all the candidate blocks within the multi-reference frames, its computation complexity is extremely

great: the processing time increases linearly with the number of allowed reference frames. In order to speed up the processing of block-matching motion estimation in multi-reference frame, many researchers have been working hard to propose fast multi-reference frame selection algorithms.

In H.264/AVC standard, among adopted Block Matching Algorithms (BMAs), the Full Search Block Matching Algorithm (FSBMA) is used (Chuang, et al., 2011) for determining the best reference frame (the best motion vector can be found in no more than five reference frames), according to the minimal rate-distortion cost, which is determined by the following function $J(ref / \lambda_{Motion})$, as shown in the following.

$$
\begin{aligned}
&J(ref / \lambda_{Motion}) \\
&= SAD(s, c(ref, m(ref))) \\
&+ \lambda_{Motion} \cdot (R(m(ref) - p(ref)) \\
&+ R(ref))
\end{aligned}
\tag{18}
$$

and

$$
\lambda_{Motion} = \sqrt{0.85 \cdot 2^{QP/3}}
\tag{19}
$$

where *ref* is a number of the particular reference frame; λ_{Motion}, which is defined in (11), is the Lagrange multiplier for the motion estimation; $SAD(s, c(ref, m(ref)))$ is the Sum of Absolute Difference (or the Hadamard transformation) between the original block *s* and its reconstruction *c; m(ref)* is the current block's motion vector in the reference frame *ref; p(ref)* is the current block predictive motion vector in the reference frame *ref; $R(m(ref) - p(ref)) + R(ref)$* represents a number of bits, which are required for encoding the motion vector difference (i.e. the difference between the actual and predicted motion vector); *R(ref)* represents a number of bits, which are required for encoding the particular reference frame *ref;* and QP is the quantization parameter. First, for the P-frame, the ME rate-distortion cost in the

Figure 18. A flow of the intra-mode decision algorithm (Kim & Kim, 2011)

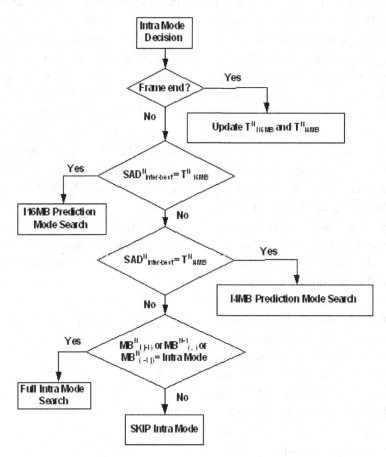

first reference frame, i.e. $J(0 / \lambda_{Motion})$ is computed (Xiao & Cheng, 2011). Then, the ME rate-distortion costs in the second, third, fourth and fifth frames $J(1 / \lambda_{Motion})$, $J(2 / \lambda_{Motion})$, $J(3 / \lambda_{Motion})$, and $J(4 / \lambda_{Motion})$ are computed, respectively. The reference frame (*ref*) that minimizes the cost function $J(ref / \lambda_{Motion})$ is selected as the best reference frame. It is noted that instead of determining the SAD value, as mentioned above, the SATD (Sum of Absolute Transform Differences), or SSD (Sum of Square Differences) values can be determined; for example, in the H.264/AVC reference software (JM), the SAD is used for the integer pixel motion estimation, while SATD is used for the sub-pel motion estimation if the Hadamard coding option is enabled (Lim, et al., 2005a, 2005b; Kuo & Lu, 2008).

The following Figure 19 schematically illustrates the multiple reference frame motion search (up to 5 reference frames, as defined in the H.264/AVC standard).

Therefore, although this method improves the encoder coding efficiency, the computation complexity is significantly increased. Therefore, there is a need to develop a fast multi-reference frame selection algorithm in order to save available computational resource and to speed up the performance of the H.264/AVC encoder. With this regard, Xiao and Cheng (2011) propose a fast multi-reference frame selection algorithm, which is based on the correlation between the multi-reference frame selection and the size of the motion vector and the monotonicity of the different reference frame's rate-distortion. The method of

Xiao and Cheng (2011) enable one to determine the number of reference frame according to the size of the motion vector, and enables to skip the reference frame selection for the undesirable reference frames or block modes.

In addition, Kuo and Lu (2008) present an effective technique (see Figure 20) to select proper reference frames for the H.264/AVC motion estimation, which enables working with any existing motion search algorithm. According to Kuo and Lu, (2008), in the first stage, the target 16X16 macroblock is split into four blocks of 8x8 to perform a motion search on the previous frame (*ref-frame t-1*), thereby obtaining the four motion vectors $V^{t-1} = \{V_i^{t-1} | i = 0, 1, 2, 3\}$ and their corresponding minimal R-D costs (e.g., determined by equation [10]), where i denotes an index of the 8x8 block. Next, the value of V^{t-1} is determined. If both the variances of x and y components of V^{t-1} are not greater than a small threshold *TH*, then this indicates that the macroblock has a greater probability of having the static content. In such a case, the encoder operation is terminated early by designating only a previous frame

as valid, without checking the remaining reference frames.

On the other hand, if the flow goes to the second stage, the 8x8 block motion search should be tested on all of the remaining reference frames to obtain their motion vectors $V_i^{t-2}, V_i^{t-3}, ..., V_i^{t-N}$ and their corresponding R-D costs $\{C(V_i^{t-k}) | i = 0, 1, 2, 3; k = 2, 3, ..., N\}$, where N indicates a maximal number of reference frames. For a given block index i, the k is set as $l = \arg\min\{C(V_i^{n-k})\} | k = 1, 2, ..., N$, which means that block i has the best motion vector with the lowest cost, by referring to *ref-frame n-k* rather than any other frame. As a result, the *ref-frame n-k* is set as the valid qualified reference frame (upon checking all four blocks, i.e. from *i=0* to *i=3*), and the unqualified frames are dropped.

In addition, Jun and Park (2010) present a priority-based reference frame selection algorithm by computing the priorities of all reference frames by using spatial and temporal correlation of the reference frame index and motion vectors, and then selecting candidate reference frames for reducing

Figure 19. Schematic illustration of the multi-frame motion compensation (with regard to the multiple reference frame motion search): up to 5 reference frames, as defined in the H.264/AVC standard (Wiegand, et al., 2003b)

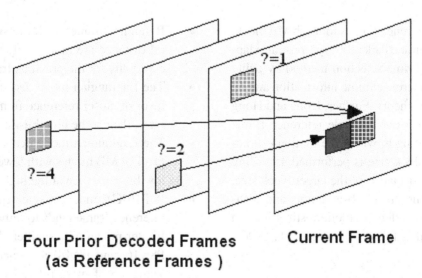

Four Prior Decoded Frames   Current Frame
(as Reference Frames)

Figure 20. Improved reference frame selection technique, according to Kuo and Lu (2008)

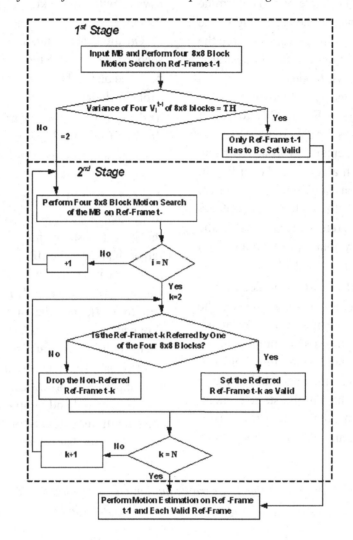

a number of reference frames for each ME block. In addition, Jeon and Park (2009) propose an adaptive reference frame selection method by using the neighbor reference frame information and by using the Bayes theory. Further, Jeong and Hong (2008) present a fast multiple reference frame method, according to which the motion estimation of small block size is performed according to the motion estimation of the larger block size.

Based on the above observations and techniques, the multi-reference motion estimation can be summarized as follows (Huang, et al., 2006):

- Reference frames with the smaller temporal distance are more likely to be optimal, especially for the previous frame.
- The Lagrangian mode decision is more in favor of closer reference frames at low bit-rates, than at the high bit-rates.
- The Lagrangian mode decision is more in favor of MB modes with fewer partitions at low bit-rates than at the high bit-rates.
- If P16X16 mode is selected, the optimal reference frame tends to remain the same.
- If inter-modes with smaller blocks (P16X8, P8X16, and P8X8) are selected, then the searching of more reference frames tends to be helpful.

- If texture is not significant and if the Quantization Parameter (QP) value is large, then the SKIP mode is often selected as the best MB mode.
- I4MB cost value can be used as a criterion to measure the complexity of the texture.
- If MVs of larger blocks are very similar to those of smaller blocks, it is likely that no occlusion or uncovering occurs in the MB, so one reference frame may be enough.
- If MVs of larger blocks are very different from those of smaller blocks, the MB may cross object boundaries where the motion field is inhomogeneous; thereby more reference frames can be required.
- Fractional-pixel ME can improve the prediction for sampling defects, but for highly textured areas, multiple reference frames are more helpful.
- If the object undergoes an integer motion, the prediction gain of multiple reference frames may not be significant.

There are also other methods, such as that of Cheong and Ortega (2007), who present a quantizer, which is optimized to minimize the impact of quantization on indentifying the nearest neighbor, and which can be used for the motion estimation applications. The main advantage of this scheme is: a) it can significantly reduce a number and complexity of required arithmetic operations; b) the complexity is not increased along with the input bit-size; and c) the decrease of the quality due to the complexity reduction is very small. In addition, Wang et al. (2007) propose a fast multiple reference frame selection method, according to which the Lagrangian cost of neighboring blocks is determined for selecting the best reference frame. In addition, Tsui et al. (2010) suggest a *selective* multiple reference frame motion estimation technique, which characterizes the multiple reference frame motion estimation as a stationary Gaussian random process. In addition, Lee et al. (2009) propose a fast reference frame selection algorithm

based on the information from a reference region, which is used in the ME process of the previous frame. Further, Chiang and Lai (2009) propose to use a statistical learning approach to reduce the computational complexity, which is involved in the multi-reference motion estimation.

FUTURE RESEARCH DIRECTIONS AND CONCLUSION

High Efficiency Video Coding (HEVC) is an emerging video compression standard, which is currently under development of the Joint Collaborative Team on Video Coding (JCT-VC) group (which is a group of video coding experts from the ITU-T Study Group 16 Visual Coding Experts Group [VCEG]) and the ISO/IEC JTC1/ SC29/ WG11 Moving Picture Experts Group [MPEG]). HEVC is expected to significanty improve coding efficiency of the H.264/AVC standard. Currently, it has been reported that in many cases HEVC can provide comparable subjective visual quality with the H.264/AVC High profile at only half bit-rate, which can be especially useful for high-definition video applications. However, HEVC is also expected to consume more computational resources compared to H.264/AVC, particularly, with computational complexity ranging from 0.5 to 3 times that of the H.264/AVC High profile.

Therefore, one of the important issues is to efficiently reduce the computational complexity of this emerging standard without decreasing the visual presentation quality (Ugur, et al., 2010).

Additional standard that should be noted is the Audio Video Coding Standard of China (AVS) that provides a streamlined, highly efficient video coder employing the latest video coding tools and dedicated to coding of the high-definition TV content. Compared to the H.264/AVC, the AVS has been designed to provide near optimum performance and a considerable reduction in complexity (Gao, et al., 2004).

As a result, possible research directions can focus on further decrease of the AVS computational complexity for enabling to provide ultra low-cost video coder implementations.

In addition, there is a strong need to address complexity issues with regard to hardware codec implementations, which is especially vital for mobile/cellular devices. As noted in this chapter, the H.264/AVC significantly outperforms previous video coding standards (such as MPEG-2), but the extraordinary huge computation complexity and memory access requirements make the hardware codec solution to be a tough job. Also, due to the complex, sequential, and highly data-depended characteristics of all essential algorithms in H.264/AVC, not only the pipeline structure but also efficient memory hierarchy can be required (Chen, et al., 2006). Finally, it should be noted that various H.264/AVC features, such as data partitioning, can be considered to be supported in more H.264/AVC profiles, especially in Baseline and Main Profiles.

As was shown in this chapter, the computational complexity issue is critical for the H.264/AVC-based systems and applications. With this regard, the authors conducted a comprehensive review of the recent advances in computational complexity techniques for video coding applications. However, due to the increasing need for the high visual quality video presentation (e.g., for displaying high/ultra-high definition resolution videos, such as of 1080p, 4K, 8K, and the like), these issues require further research and development.

ACKNOWLEDGMENT

The authors thank anonymous reviewers for their valuable comments and suggestions.

REFERENCES

Adibelli, Y., Parlak, M., & Hamzaoglu, I. (2011). Energy reduction techniques for H.264 deblocking filter hardware. *IEEE Transactions on Consumer Electronics*, *57*(3), 1399–1407. doi:10.1109/TCE.2011.6018900

Akyol, E., Mukherjee, D., & Liu, Y. (2007). Complexity control for real-time video coding. In *Proceedings of the 2007 IEEE International Conference on Image Processing, ICIP 2007*, (vol. 1, pp. 77-80). IEEE Press.

Alfonso, D. (2010). *Proposals for video coding complexity assessment*. Paper presented at the JCTVC-A026, Joint Collaborative Team on Video Coding (JCT-VC) of ITU-T SG16 WP3 and ISO/IEC JTC1/SC29/WG11 1st Meeting. Dresden, Germany.

Apteker, R. T., Fisher, J. A., Kisimov, V. S., & Neishlos, H. (1995). Video acceptability and frame rate. *IEEE MultiMedia*, *2*(3), 32–40. doi:10.1109/93.410510

Bystrom, M., Richardson, I., & Zhao, Y. (2008). Efficient mode selection for H.264 complexity reduction in a Bayesian framework. *Signal Processing Image Communication*, *23*(2), 71–86. doi:10.1016/j.image.2007.11.001

Chang, C.-C., & Lin, C.-J. (2003). *LIBSVM: A library for support vector machines*. Retrieved from http://www.csie.ntu.edu.tw/~cjlin/libsvmtools

Chen, J., Qu, Y., & He, Y. (2004). A fast mode decision algorithm in H.264. In *Proceedings of the Picture Coding Symposium*, (pp. 46-49). San Francisco, CA: IEEE.

Chen, T.-C., Lian, C.-J., & Chen, L.-G. (2006). Hardware architecture design of an H.264/AVC video codec. In *Proceedings of the 2006 Asia and South Pacific Design Automation Conference (ASP-DAC 2006)*, (pp. 750-757). Piscataway, NJ: IEEE Press.

Cheong, H.-Y., & Ortega, A. (2007). Distance quantization method for fast nearest neighbor search computations with applications to motion estimation, signals, systems and computers, 2007. In *Proceedings of the Conference Record of the Forty-First Asilomar Conference*, (pp. 909-913). ACSSC.

Chiang, C.-K., & Lai, S.-H. (2009). Fast multi-reference motion estimation via statistical learning for H.264/AVC. In *Proceedings of the IEEE International Conference on Multimedia and Expo, 2009, ICME 2009*, (pp. 61-64). IEEE Press.

Chiang, C.-K., Pan, W.-H., Hwang, C., Zhuang, S.-S., & Lai, S.-H. (2011). Fast H.264 encoding based on statistical learning. *IEEE Transactions on Circuits and Systems for Video Technology, 21*(9), 1304–1315. doi:10.1109/TCSVT.2011.2147250

Chuang, C.-H., Chen, B.-N., Chang, C.-C., & Cheng, S.-C. (2011). Computation-aware fast motion estimation for H.264/AVC using image indexing. *Journal of Visual Communication and Image Representation, 22*(6), 451–464. doi:10.1016/j.jvcir.2011.05.003

Chung, H., Romacho, D., & Ortega, A. (2003). Fast long-term motion estimation for H.264 using multiresolution search. In *Proceedings of the International Conference on Image Processing*, (pp. 905-908). Barcelona, Spain: IEEE.

Cover, T. M. (1991). *Information theory*. New York, NY: Wiley-Interscience.

Dai, Q., Zhu, D., & Ding, R. (2004). Fast mode decision for inter prediction in H.264. In *Proceedings of the IEEE International Conference on Image Processing*, (pp. 119-122). Singapore, Singapore: IEEE Press.

Foo, B., Andreopoulos, Y., & Van der Schaar, M. (2008). Analytical rate-distortion-complexity modeling of wavelet-based video coders. *IEEE Transactions on Signal Processing, 56*, 797–815. doi:10.1109/TSP.2007.906685

Gao, W., Reader, C., Wu, F., He, Y., Yu, L., & Lu, H. ... Pan, X. (2004). AVS—The Chinese next-generation video coding standard. In *Proceedings of the National Association of Broadcasters (NAB) Conference*. Las Vegas, NV: NAB.

Ghinea, G., & Thomas, J. P. (1998). QoS impact on user perception and understanding of multimedia video clips. In *Proceedings of the 6th ACM International Conference on Multimedia*, (pp. 49-54). Bristol, UK: ACM Press.

Grois, D., & Hadar, O. (2011b). Efficient adaptive bit-rate control for scalable video coding by using computational complexity-rate-distortion analysis. In *Proceedings of the 2011 IEEE International Symposium on Broadband Multimedia Systems and Broadcasting (BMSB)*, (pp. 1-6). Nuremberg, Germany: IEEE Press.

Grois, D., & Hadar, O. (2011c). Recent advances in region-of-interest coding. In J. del ser Lorente (Ed.), *Recent Advances on Video Coding*, (pp. 49-76). Intech.

Grois, D., & Hadar, O. (2011d). Complexity-aware adaptive bit-rate control with dynamic ROI pre-processing for scalable video coding. In *Proceedings of the 2011 IEEE International Conference on Multimedia and Expo (ICME)*, (pp. 1-4). Barcelona, Spain: IEEE Press.

Grois, D., & Hadar, O. (2011e). Complexity-aware adaptive spatial pre-processing for ROI scalable video coding with dynamic transition region. In *Proceedings of the International Conference on Image Processing (ICIP 2011)*. Brussels, Belgium: ICIP.

Grois, D., Kaminsky, E., & Hadar, O. (2009). Buffer control in H.264/AVC applications by implementing dynamic complexity-rate-distortion analysis. In *Proceedings of the 2009 IEEE International Symposium on Broadband Multimedia Systems and Broadcasting, BMSB 2009*, (pp. 1-7). IEEE Press.

Grois, D., Kaminsky, E., & Hadar, O. (2010a). ROI adaptive scalable video coding for limited bandwidth wireless networks. In *Proceedings of Wireless Days (WD), 2010 IFIP*, (pp. 1-5). IFIP.

Grois, D., Kaminsky, E., & Hadar, O. (2010b). Adaptive bit-rate control for region-of-interest scalable video coding. In *Proceedings of the 2010 IEEE 26th Convention of Electrical and Electronics Engineers in Israel (IEEEI)*, (pp. 761-765). IEEE Press.

Grois, D., Kaminsky, E., & Hadar, O. (2010c). Optimization methods for H.264/AVC video coding . In Angelides, M. C., & Agius, H. (Eds.), *The Handbook of MPEG Applications: Standards in Practice*. Chichester, UK: John Wiley & Sons, Ltd. doi:10.1002/9780470974582.ch7

Grois, D., Kaminsky, E., & Hadar, O. (2011a). Efficient real-time video-in-video insertion into a pre-encoded video stream. *ISRN Signal Processing*, *2011*, 1–11. doi:10.5402/2011/975462

He, Z., Cheng, W., & Chen, X. (2008). Energy minimization of portable video communication devices based on power-rate-distortion optimization. *IEEE Transactions on Circuits and Systems for Video Technology*, *18*(5), 596–608. doi:10.1109/TCSVT.2008.918802

Huang, Y.-W., Hsieh, B.-Y., Chien, S.-Y., Ma, S.-Y., & Chen, L.-G. (2006). Analysis and complexity reduction of multiple reference frames motion estimation in H.264/AVC. *IEEE Transactions on Circuits and Systems for Video Technology*, *16*(4), 507–522. doi:10.1109/TCSVT.2006.872783

Ivanov, Y. V., & Bleakley, C. J. (2010). Real-time H.264 video encoding in software with fast mode decision and dynamic complexity control. *ACM Transactions on Multimedia Computer Communication Applications*, *6*(1).

Jeon, D.-S., & Park, H.-W. (2009). An adaptive reference frame selection method for multiple reference frame motion estimation in the H.264/AVC, In *Proceedings of the 2009 16th IEEE International Conference on Image Processing (ICIP)*, (pp. 629-632). IEEE Press.

Jeong, C.-Y., & Hong, M.-C. (2008). Fast multiple reference frame selection method using inter-mode correlation. In *Proceedings of the 14th Asia-Pacific Conference on Communications, 2008, APCC 2008*, (pp. 1-4). APCC.

Jiao, L., Zhou, J., & Chen, R. (2011). Efficient parallel intra-prediction mode selection scheme for 4x4 blocks In *Proceedings of the 2011 International Conference on Intelligent Computation Technology and Automation (ICICTA)*, (vol. 2, pp. 527-530). ICICTA.

Jun, D.-S., & Park, H.-W. (2010). An efficient priority-based reference frame selection method for fast motion estimation in H.264/AVC. *IEEE Transactions on Circuits and Systems for Video Technology*, *20*(8), 1156–1161. doi:10.1109/TCSVT.2010.2057016

Kaminsky, E., Grois, D., & Hadar, O. (2008). Dynamic computational complexity and bit allocation for optimizing H.264/AVC video compression. *Journal of Visual Communication and Image Representation*, *19*(1), 56–74. doi:10.1016/j.jvcir.2007.05.002

Kannangara, C. S., Richardson, I. E., Bystrom, M., & Zhao, Y. (2009). Complexity control of H.264/AVC based on mode-conditional cost probability distributions. *IEEE Transactions on Multimedia*, *11*(3), 433–442. doi:10.1109/TMM.2009.2012937

Kannangara, C. S., Richardson, I. E., & Miller, A. J. (2008). Computational complexity management of a real-time H.264/AVC encoder. *IEEE Transactions on Circuits and Systems for Video Technology*, *18*(9), 1191–1200. doi:10.1109/TCSVT.2008.928881

Kim, H., & Altunbasak, Y. (2004). Low-complexity macroblock mode selection for H.264/AVC encoders. In *Proceedings of the International Conference on Image Processing*, (pp. 765-768). Singapore, Singapore: IEEE.

Kim, J., & Jeong, J. (2011). Fast intra mode decision algorithm using the sum of absolute transformed differences. In *Proceedings of the 2011 International Conference on Digital Image Computing Techniques and Applications (DICTA)*, (pp. 655-659). DICTA.

Kim, J.-H., & Kim, B.-G. (2011). Efficient intra-mode decision algorithm for inter-frames in H.264/AVC video coding. *IET Image Processing*, *5*(3), 286–295. doi:10.1049/iet-ipr.2009.0097

Kim, T., Hwang, U., & Jeong, J. (2011). Efficient block mode decision and prediction mode selection for intra prediction in H.264/AVC high profile. In *Proceedings of the 2011 International Conference on Digital Image Computing Techniques and Applications (DICTA)*, (pp. 645-649). DICTA.

Kim, W., You, J., & Jeong, J. (2010). Complexity control strategy for real-time H.264/AVC encoder. *IEEE Transactions on Consumer Electronics*, *56*(2), 1137–1143. doi:10.1109/TCE.2010.5506050

Kim, Y., Kim, W., & Jeong, J. (2011). Fast intra mode decision algorithm using sub-sampled pixels. In *Proceedings of the 2011 IEEE 15th International Symposium on Consumer Electronics (ISCE)*, (pp. 290-293). IEEE Press.

Kuo, T.-Y., & Lu, H.-J. (2008). Efficient reference frame selector for H.264. *IEEE Transactions on Circuits and Systems for Video Technology*, *18*(3), 400–405. doi:10.1109/TCSVT.2008.918111

Lee, H., Jung, B., Jung, J., & Jeon, B. (2009). Computational complexity scalable scheme for power-aware H.264/AVC encoding. In *Proceedings of the IEEE International Workshop on Multimedia Signal Processing, 2009, MMSP 2009*, (pp. 1-6). IEEE Press.

Lee, J., & Park, H. (2011). A fast mode decision method based on motion cost and intra prediction cost for H.264/AVC. *IEEE Transactions on Circuits and Systems for Video Technology*, *22*(3), 393–402. doi:10.1109/TCSVT.2011.2163460

Lee, K., Jeon, G., & Jeong, J. (2009). Fast reference frame selection algorithm for H.264/AVC. *IEEE Transactions on Consumer Electronics*, *55*(2), 773–779. doi:10.1109/TCE.2009.5174453

Lee, Y.-M., & Lin, Y.-Y. (2006). A fast intermode mode decision for H.264 video coding. In *Proceedings of the 2006 8th International Conference on Signal Processing*, (vol. 2). IEEE.

Lei, Y. Q., Wang, Y.-G., & Liang, F. (2010). Overlapping interval differences-based fast intra mode decision for H.264/AVC. In *Proceedings of the 2010 Sixth International Conference on Intelligent Information Hiding and Multimedia Signal Processing (IIH-MSP)*, (pp. 659-663). IIH-MSP.

Li, X., Wien, M., & Ohm, J.-R. (2011, Jul.). Rate-Complexity-Distortion optimization for hybrid video coding. *IEEE Transactions on Circuits and Systems for Video Technology*, *21*(7), 957–970. doi:10.1109/TCSVT.2011.2133750

Lian, S.-J., Chien, S.-Y., Lin, C.-P., Tseng, P.-C., & Chen, L.-G. (2007). Power-aware multimedia: Concepts and design perspectives. *IEEE Circuits and Systems Magazine*, *7*(2), 26–34. doi:10.1109/MCAS.2007.4299440

Lim, K.-P., Sullivan, G., & Wiegand, T. (2005a). Text description of joint model reference encoding methods and decoding concealment methods. *Study of ISO/IEC 14496-10 and ISO/IEC 14496-5/ AMD6 and Study of ITU-T Rec. H.264 and ITU-T Rec. H.2.64.2, in Joint Video Team JVT of ISO/IEC MPEG and ITU-T VCEG*, Hong Kong, JVT-N046.

Lim, K.-P., Sullivan, G., & Wiegand, T. (2005b). Text description of joint model reference encoding methods and decoding concealment methods. *Study of ISO/IEC 14496-10 and ISO/IEC 14496-5/ AMD6 and Study of ITU-T Rec. H.264 and ITU-T Rec. H.2.64.2, in Joint Video Team (JVT) of ISO/ IEC MPEG and ITU-T VCEG*, Busan, Korea, Doc. JVT-O079.

McCarthy, J. D., Sasse, M. A., & Miras, D. (2004). Sharp or smooth? Comparing the effects of quantization versus frame rate for streamed video. In *Proceedings of the SIGCHI Conference on Human Factors Computing Systems*, (pp. 535-542). Vienna, Austria: ACM Press.

Meng, B., Li, M.-Z., & Ren, Y.-H. (2010). Fast mode selection for H.264/AVC based on MB motion characteristics. In *Proceedings of the 2010 International Conference on Intelligent Computing and Integrated Systems (ICISS)*, (pp. 205-208). ICISS.

Miao, C.-H., & Fan, C.-P. (2011). Efficient mode selection with extreme value detection based pre-processing algorithm for H.264/AVC fast intra mode decision. In *Proceedings of TENCON 2011 - 2011 IEEE Region 10 Conference*, (pp. 316-320). IEEE Press.

Pan, Z., & Kwong, S. (2011). A fast inter-mode decision scheme based on luminance difference for H.264/AVC. In *Proceedings of the 2011 International Conference on System Science and Engineering (ICSSE)*, (pp. 260-263). ICSSE.

Pang, C., Au, O. C., Dai, J., & Zou, F. (2011). Frame complexity guided Lagrange multiplier selection for H.264 intra-frame coding. *IEEE Signal Processing Letters*, *18*(12), 733–736. doi:10.1109/LSP.2011.2172940

Park, J. S., & Song, H. J. (2006). Selective intra-prediction mode decision for H.264/AVC encoders. *Transactions on Engineering. Computing and Technology*, *13*, 51–55.

Parlak, M., Adibelli, Y., & Hamzaoglu, I. (2008). A novel computational complexity and power reduction technique for H.264 intra prediction. *IEEE Transactions on Consumer Electronics*, *54*(4), 2006–2014. doi:10.1109/TCE.2008.4711266

Paul, M., Frater, M. R., & Arnold, J. F. (2009). An efficient mode selection prior to the actual encoding for H.264/AVC encoder. *IEEE Transactions on Multimedia*, *11*(4), 581–588. doi:10.1109/ TMM.2009.2017610

Paul, M., Lin, W., Lau, C. T., & Lee, B.-S. (2011). Direct intermode selection for H.264 video coding using phase correlation. *IEEE Transactions on Image Processing*, *20*(2), 461–473. doi:10.1109/ TIP.2010.2063436

Peng, Q., & Jing, J. (2003). H.264 codec system-on-chip design and verification. In *Proceedings of the 5th International Conference*, (vol. 2, pp. 922-925). IEEE.

Powers, R. A. (1995). Batteries for low power electronics. *Proceedings of the IEEE*, *83*(4), 687–693. doi:10.1109/5.371974

Qi, B., Zhang, D., Song, Y., Du, G., & Zheng, Y. (2011). Design and implementation of a new pipelined H.264 encoder. In *Proceedings of the 2011 International Conference on Computer Science and Network Technology (ICCSNT)*, (vol. 1, pp. 130-133). ICCSNT.

Quyang, K., Chen, C., & Chen, J. (2010). Simplified directional mode selection for h.264/avc intra mode decision. In *Proceedings of the 2010 International Conference on Computational Intelligence and Software Engineering (CiSE)*, (pp. 1-4). CiSE.

Ren, F., & Dong, J. (2010). Fast and efficient intra mode selection for H.264/AVC. In *Proceedings of the Second International Conference on Computer Modeling and Simulation, 2010, ICCMS 2010*, (vol. 2, pp. 202-205). ICCMS.

Rhee, C. E., Jung, J.-S., & Lee, H.-J. (2010). A real-time H.264/AVC encoder with complexity-aware time allocation. *IEEE Transactions on Circuits and Systems for Video Technology, 20*(12), 1848–1862. doi:10.1109/TCSVT.2010.2087834

Richardson, I. E. (2003a). *H.264 and MPEG-4 video compression: Video coding for next-generation multimedia*. Chichester, UK: John Wiley & Sons, Ltd. doi:10.1002/0470869615

Richardson, I. E. (2003b). *H.264/MPEG-4 part 10 white paper: Intra prediction*. Retrieved from http://www.vcodex.com/h264.html

Richardson, I. E. (2003c). *H.264/MPEG-4 part 10 white paper: Inter prediction*. Retrieved from http://www.vcodex.com/h264.html

Richardson, I. E. (2010). *The H.264 advanced video compression standard* (2nd ed.). Chichester, UK: John Wiley & Sons, Ltd. doi:10.1002/9780470989418

Saponara, S., Blanch, C., Denolf, K., & Bormans, J. (2003). The JVT advanced video coding standard: Complexity and performance analysis on a tool-by-tool basis. In *Proceedings of the Packet Video Workshop (PV 2003)*. Nantes, France: PV.

Solak, S. B., & Labeau, F. (2010). Complexity scalable video encoding for power-aware applications. In *Proceedings of the 2010 International Green Computing Conference*, (pp. 443-449). IEEE.

Stottrup-Andersen, J., Forchhammer, S., & Aghito, S. M. (2004). Rate-distortion-complexity optimization of fast motion estimation in H.264/MPEG-4 AVC. In *Proceedings of the IEEE International Conference on Image Processing*, (pp. 119-122). Singapore, Singapore: IEEE Press.

Su, L., Lu, Y., Wu, F., Li, S., & Gao, W. (2009). Complexity-constrained H.264 video encoding. *IEEE Transactions on Circuits and Systems for Video Technology, 19*(4), 477–490. doi:10.1109/TCSVT.2009.2014017

Sullivan, G. J., & Wiegand, T. (1998). Rate-distortion optimization for video compression. *IEEE Signal Processing Magazine, 15*(6), 74–90. doi:10.1109/79.733497

Tan, Y. H., Lee, W. S., Tham, J. Y., Rahardja, S., & Lye, K. M. (2010). Complexity scalable H.264/AVC encoding. *IEEE Transactions on Circuits and Systems for Video Technology, 20*(9), 1271–1275. doi:10.1109/TCSVT.2010.2058480

To, L., Pickering, M., Frater, M., & Arnold, J. (2004). A motion confidence measure from phase information. In *Proceedings of the 2004 International Conference on Image Processing, 2004*, (vol. 4, pp. 2583-2586). IEEE.

Tsui, C.-C., Lee, Y.-M., & Lin, Y. (2010). Selective multiple reference frames motion estimation for H.264/AVC video coding. In *Proceedings of the 2010 International Symposium on Information Theory and its Applications (ISITA)*, (pp. 237-242). ISITA.

Ugur, K., Andersson, K., Fuldseth, A., Bjontegaard, G., Endresen, L. P., & Lainema, J. (2010). High performance, low complexity video coding and the emerging HEVC standard. *IEEE Transactions on Circuits and Systems for Video Technology, 20*(12), 1688–1697. doi:10.1109/TCSVT.2010.2092613

Vanam, R., Chon, J., Riskin, E. A., Ladner, R. E., Ciaramello, F. M., & Hemami, S. S. (2010). Rate-distortion-complexity optimization of an H.264/AVC encoder for real-time videoconferencing on a mobile device. In *Proceedings of the Fifth International Workshop on Video Processing and Quality Metrics for Consumer Electronics (VPQM 2010)*. Scottsdale, AZ: VPQM.

Wang, H.-J., Wang, L.-L., & Li, H. (2007). A fast multiple reference frame selection algorithm based on H.264/AVC. In *Proceedings of the Third International Conference on Intelligent Information Hiding and Multimedia Signal Processing, 2007,* (vol. 1, pp. 525-528). IEEE.

Wang, J., & Hua, G. (2008). Implementing high definition video codec on TI DM6467 SOC. In *Proceedings of the 2008 IEEE International SOC Conference,* (pp. 193-196). IEEE Press.

Wang, L.-L., Jia, K.-B., & Lu, Z.-Y. (2011). A multi-stage fast intra mode decision algorithm in H.264. In *Proceedings of the 2011 Seventh International Conference on Intelligent Information Hiding and Multimedia Signal Processing (IIH-MSP),* (pp. 236-239). IIH-MSP.

Wen, X., Au, O. C., Xu, J., Fang, L., Cha, R., & Li, J. (2011). Novel RD-optimized VBSME with matching highly data re-usable hardware architecture. *IEEE Transactions on Circuits and Systems for Video Technology*, *21*(2), 206–219. doi:10.1109/TCSVT.2011.2106274

Wiegand, T., Schwarz, H., Joch, A., Kossentini, F., & Sullivan, G. (2003a). Rate-constrained coder control and comparison of video coding standards. *IEEE Transactions on Circuits and Systems for Video Technology*, *13*(7), 688–703. doi:10.1109/TCSVT.2003.815168

Wiegand, T., & Sullivan, G. (2003). *Final draft ITU-T recommendation and final draft international standard of joint video specification (ITU-T Rec. H.264 ISO/IEC 14 496-10 AVC)*. In Joint Video Team (JVT) of ITU-T SG16/Q15 (VCEG) and ISO/IEC JTC1/SC29/WG1, Annex C. Pattaya, Thailand, Doc. JVT-G050.

Wiegand, T., Sullivan, G. J., Bjontegaard, G., & Luthra, A. (2003b). Overview of the H.264/AVC video coding standard. *IEEE Transactions on Circuits and Systems for Video Technology*, *13*(7), 560–576. doi:10.1109/TCSVT.2003.815165

Wu, C.-D., & Lin, Y. (2009). Efficient inter/intra mode decision for H.264/AVC inter frame transcoding. In *Proceedings of the 2009 16th IEEE International Conference on Image Processing (ICIP),* (pp. 3697-3700). IEEE Press.

Wu, D., Pan, F., Lim, K. P., Wu, S., Li, Z. G., & Lin, X. (2005). Fast intermode decision in H.264/AVC video coding. *IEEE Transactions on Circuits and Systems for Video Technology*, *15*(7), 953–958. doi:10.1109/TCSVT.2005.848304

Xiao, M., & Cheng, Y. (2011). A fast multi-reference frame selection algorithm for H.264/AVC. In *Proceedings of the 2011 IEEE International Conference on Computer Science and Automation Engineering (CSAE),* (vol. 4, pp. 615-619). IEEE Press.

Yang, C.-C., Tan, K.-J., Yang, Y.-C., & Guo, J.-I. (2010). Low complexity fractional motion estimation with adaptive mode selection for H.264/AVC. In *Proceedings of the 2010 IEEE International Conference on Multimedia and Expo (ICME),* (pp. 673-678). IEEE Press.

Yu, A. C., & Martin, G. R. (2004). Advanced block size selection algorithm for inter frame coding in H.264/MPEG-4 AVC. In *Proceedings of the International Conference on Image Processing,* (pp. 95-98). Singapore, Singapore: IEEE.

Zeng, H., Cai, C., & Ma, K.-K. (2009). Fast mode decision for H.264/AVC based on macroblock motion activity. *IEEE Transactions on Circuits and Systems for Video Technology, 19*(4), 491–499. doi:10.1109/TCSVT.2009.2014014

Zeng, H., Ma, K.-K., & Cai, C. (2010). Hierarchical intra mode decision for H.264/AVC. *IEEE Transactions on Circuits and Systems for Video Technology, 20*(6), 907–912. doi:10.1109/TCSVT.2010.2045802

Zhu, W., Ma, Y., & Wang, R. (2010). An adaptive fast algorithm for coding block mode selection. In *Proceedings of the 2010 Third International Symposium on Intelligent Information Technology and Security Informatics (IITSI)*, (pp. 402-406). IITSI.

ADDITIONAL READING

Aysu, A., Sayilar, G., & Hamzaoglu, I. (2011). A low energy adaptive hardware for H.264 multiple reference frame motion estimation. *IEEE Transactions on Consumer Electronics, 57*(3), 1377–1383. doi:10.1109/TCE.2011.6018897

Bu, J., Mo, L., Chen, C., Yang, Z., & Song, M. (2006). Multiple-reference-frame based fast motion estimation & mode decision for H.263-to-H.264 transcoder. In *Proceedings of the 2006 IEEE International Conference on Image Processing*, (pp. 849-852). IEEE Press.

Chang, K., Men, A., & Zhang, W. (2009). A novel fast multiple reference frames selection method for H.264/AVC. In *Proceedings of the Ninth IEEE International Conference on Computer and Information Technology, 2009, CIT 2009*, (vol. 1, pp. 156-160). IEEE Press.

Cheng, Y.-S., Chen, Z.-Y., & Chang, P.-C. (2009). An H.264 spatio-temporal hierarchical fast motion estimation algorithm for high-definition video. In *Proceedings of the 2009 IEEE International Symposium on Circuits and Systems, ISCAS 2009*, (pp. 880-883). IEEE Press.

Cho, C.-Y., Chang, S.-K., & Wang, J.-S. (2006). A multiframe motion estimation architecture for H.264/AVC. In *Proceedings of the 2006 IEEE International Conference on Image Processing*, (pp. 1357-1360). IEEE Press.

da Fonseca, T. A., & de Queiroz, R. L. (2008). Complexity reduction techniques applied to the compression of high definition sequences in digital TV. In *Proceedings of the 26th Simposio Brasileiro de Telecomunicacoes*, (pp. 43–47). IEEE.

da Fonseca, T. A., & de Queiroz, R. L. (2009). Complexity-constrained H.264 HD video coding through mode ranking. In *Proceedings of the Picture Coding Symposium, 2009, PCS 2009*, (pp. 1-4). PCS.

da Fonseca, T. A., & de Queiroz, R. L. (2009). Macroblock sampling and mode ranking for complexity scalability in mobile H.264 video coding. In *Proceedings of the 2009 16th IEEE International Conference on Image Processing (ICIP)*, (pp. 3753-3756). IEEE Press.

Facun, Z., Su, Q., & Duan, J. (2010). A new algorithm of fast intra prediction mode decision in H.264/AVC. In *Proceedings of the 2010 2nd International Conference on Information Engineering and Computer Science (ICIECS)*, (pp. 1-4). ICIECS.

Garcia, D. C., da Fonseca, T. A., & de Queiroz, R. L. (2011). Video compression complexity reduction with adaptive down-sampling. In *Proceedings of the 2011 18th IEEE International Conference on Image Processing (ICIP)*, (pp. 745-748). IEEE Press.

He, Z., Liang, Y., Chen, L., Ahmad, I., & Wu, D. (2005). Power-rate-distortion analysis for wireless video communication under energy constraints. *IEEE Transactions on Circuits and Systems for Video Technology, 15*(5), 645–658. doi:10.1109/TCSVT.2005.846433

Hong, D., Horowitz, M., Eleftheriadis, A., & Wiegand, T. (2010). H.264 hierarchical P coding in the context of ultra-low delay, low complexity applications. In *Proceedings of the Picture Coding Symposium (PCS), 2010,* (pp. 146-149). PCS.

Horowitz, M., Joch, A., Kossentini, F., & Hallapuro, A. (2003). H.264/AVC baseline profile decoder complexity analysis. *IEEE Transactions on Circuits and Systems for Video Technology, 13*(7), 704–716. doi:10.1109/TCSVT.2003.814967

Hsu, C.-L., & Cheng, C.-H. (2009). Reduction of discrete cosine transform/quantisation/inverse quantisation/inverse discrete cosine transform computational complexity in H.264 video encoding by using an efficient prediction algorithm. *IET Image Processing, 3*(4), 177–187. doi:10.1049/iet-ipr.2008.0213

Huo, C., Cao, C., & Gong, B. (2010). A fast intra-prediction mode selection algorithm for H.264/AVC. In *Proceedings of the 2010 29th Chinese Control Conference (CCC),* (pp. 4189-4192). CCC.

Jackson, M., Anderson, A. H., McEwan, R., & Mullin, J. (2000). Impact of video frame rate on communicative behaviour in two and four party groups. In *Proceedings of the 2000 ACM Conference on Computer Supported Cooperative Work (CSCW 2000)*. New York, NY: ACM Press.

Jiang, M., & Ling, N. (2005). On enhancing H.264/AVC video rate control by PSNR-based frame complexity estimation. *IEEE Transactions on Consumer Electronics, 51*(1), 281–286. doi:10.1109/TCE.2005.1405733

Kannangara, C. S., & Richardson, I. E. G. (2005). Computational control of an H.264 encoder through Lagrangian cost function estimation. *VLBV Workshop*. Retrieved from http://www.rgu.ac.uk/eng/cvc

Kannur, A. K., & Li, B. (2009). Power-aware content-adaptive H.264 video encoding. In *Proceedings fo the 2009 IEEE International Conference on Acoustics, Speech and Signal Processing, ICASSP 2009,* (pp. 925-928). IEEE Press.

Kim, J., Kim, J., Kim, G., & Kyung, C.-M. (2011). Power-rate-distortion modeling for energy minimization of portable video encoding devices. In *Proceedings of the 2011 IEEE 54th International Midwest Symposium on Circuits and Systems (MWSCAS),* (pp. 1-4). IEEE Press.

Kim, S.-E., Han, J.-K., & Kim, J.-G. (2006). An efficient scheme for motion estimation using multireference frames in H.264/AVC. *IEEE Transactions on Multimedia, 8*(3), 457–466. doi:10.1109/TMM.2006.870740

Lam, S.-Y., Au, O. C., & Wong, P. H. W. (2008). Complexity adaptive H.264 encoding using multiple reference frames. In *Proceedings of the 2008 IEEE International Conference on Acoustics, Speech and Signal Processing, ICASSP 2008,* (pp. 1025-1028). IEEE Press.

Lee, K., Jeon, G., Falcon, R., Ha, C., & Jeong, J. (2009). An adaptive fast multiple reference frame selection algorithm for H.264/AVC using reference region data. In *Proceedings of the 2009 IEEE Workshop on Signal Processing Systems, SiPS 2009,* (pp. 93-96). IEEE Press.

Lei, M., & Shu, Z. (2010). Research on fast mode selection and frame algorithm of H.264. In *Proceedings of the 2010 2nd IEEE International Conference on Information and Financial Engineering (ICIFE),* (pp. 844-847). IEEE Press.

Li, H.-J., Hsu, C.-T., & Chen, M.-J. (2004). Fast multiple reference frame selection method for motion estimation in JVT/H.264. In *Proceedings of the 2004 IEEE Asia-Pacific Conference on Circuits and Systems*, (vol. 1, pp. 605-608). IEEE Press.

Li, X., Hutter, A., & Kaup, A. (2009). One-pass frame level budget allocation in video coding using inter-frame dependency. In *Proceedings of the 2009 IEEE International Workshop on Multimedia Signal Processing, MMSP 2009*, (pp. 1-6). IEEE Press.

Li, X., Wien, M., & Ohm, J.-R. (2010). Rate-complexity-distortion evaluation for hybrid video coding. In *Proceedings of the 2010 IEEE International Conference on Multimedia and Expo (ICME)*, (pp. 685-690). IEEE Press.

Li, X., Wien, M., & Ohm, J.-R. (2010). Medium-granularity computational complexity control for H.264/AVC. In *Proceedings of the 2010 Picture Coding Symposium (PCS)*, (pp. 214-217). PCS.

Lin, S.-F., Chang, C.-Y., Su, C.-C., Lin, Y.-L., Pan, C.-H., & Chen, H. (2005). Fast multi-frame motion estimation and mode decision for H.264 encoders. In *Proceedings of the 2005 International Conference on Wireless Networks, Communications and Mobile Computing*, (vol. 2, pp. 1237-1242). IEEE.

Lin, S.-F., Lu, M.-T., Chen, H., & Pan, C.-H. (2005). Fast multi-frame motion estimation for H.264 and its applications to complexity-aware streaming. In *Proceedings of the 2005 IEEE International Symposium on Circuits and Systems, ISCAS 2005*, (vol. 2, pp. 1505-1508). IEEE Press.

Lin, W., Panusopone, K., Baylon, D., & Sun, M.-T. (2009). A new one-pass complexity-scalable computation-control method for video coding. In *Proceedings of the 2009 IEEE International Symposium on Circuits and Systems, ISCAS 2009*, (pp. 868-871). IEEE Press.

Lindroth, T., Avessta, N., Teuhola, J., & Seceleanu, T. (2006). Complexity analysis of H.264 decoder for FPGA design. In *Proceedings of the 2006 IEEE International Conference on Multimedia and Expo*, (pp. 1253-1256). IEEE Press.

Liu, Z., Li, L., Song, Y., Li, S., Goto, S., & Ikenaga, T. (2008). Motion feature and hadamard coefficient-based fast multiple reference frame motion estimation for H.264. *IEEE Transactions on Circuits and Systems for Video Technology, 18*(5), 620–632. doi:10.1109/TCSVT.2008.918844

Ma, L.-N., Shen, Y.-C., & Gao, Z. (2010). A fast motion estimation algorithm scheme based on multi-reference frame. In *Proceedings of the 2010 5th International Conference on Future Information Technology (FutureTech)*, (pp. 1-5). FutureTech.

Ma, Z., Hu, H., & Wang, Y. (2011). On complexity modeling of H.264/AVC video decoding and its application for energy efficient decoding. *IEEE Transactions on Multimedia, 13*(6), 1240–1255. doi:10.1109/TMM.2011.2165056

Mahajan, A. K., Kondayya, S., & Xiao, S. (2006). Exploiting reference frame history in h.264/avc motion estimation. In *Proceedings of the 2006 IEEE Asia Pacific Conference on Circuits and Systems, APCCAS 2006*, (pp. 410-413). IEEE Press.

Miao, C.-H., & Fan, C.-P. (2010). Extensive pixel-based fast direction detection algorithm for H.264/AVC intra mode decision. In *Proceedings of the TENCON 2010 - 2010 IEEE Region 10 Conference*, (pp. 1636-1640). IEEE Press.

Milani, S., Celetto, L., & Mian, G. A. (2008). An accurate low-complexity rate control algorithm based on (p, E_q)-domain. *IEEE Transactions on Circuits and Systems for Video Technology, 18*(2), 257–262. doi:10.1109/TCSVT.2007.913965

Oh, D., & Parhi, K. K. (2009). Low complexity decoder architecture for low-density parity-check codes. *Journal of Signal Processing Systems for Signal, Image, and Video Technology, 56,* 217–228. doi:10.1007/s11265-008-0231-5

Olmo, G., Cucco, C., Grangetto, M., & Magli, E. (2003). Few decoders in the encoder: A low complexity encoding strategy for H.26L. In *Proceedings if the 2003 IEEE International Conference on Acoustics, Speech, and Signal Processing, (ICASSP 2003),* (vol. 3, pp. 641-644). IEEE Press.

Pan, W., & Ortega, A. (2000). Complexity-scalable transform coding using variable complexity algorithms. In *Proceedings of the Data Compression Conference, DCC 2000,* (pp. 263-272). DCC.

Paul, M., Weisi, L., Chiew, T.-L., & Lee, B.-S. (2010). Video coding using the most common frame in scene. In *Proceedings of the 2010 IEEE International Conference on Acoustics Speech and Signal Processing (ICASSP),* (pp. 734-737). IEEE Press.

Porto, R., Bampi, S., & Agostini, L. (2010). Hardware design for fast intermode decision and for residues generaton in a variable block size motion estimation compliant with the H.264/AVC video coding standard. In *Proceedings of the 2010 VI Southern Programmable Logic Conference (SPL),* (pp. 183-186). SPL.

Rao, G. N., & Gupta, P. (2006). Temporal motion prediction for fast motion estimation in multiple reference frames. In *Proceedings of the 2006 IEEE International Symposium on Signal Processing and Information Technology,* (pp. 817-820). IEEE Press.

Sarwer, M. G., & Wu, Q. M. J. (2011). Enhanced low complex cost function for H.264/AVC intra mode decision. In *Proceedings of the 2011 International Conference on Multimedia and Signal Processing (CMSP),* (vol. 1, pp. 46-50). CMSP.

Shen, L., Liu, Z., Zhang, Z., & Shi, X. (2007). An adaptive and fast H.264 multi-frame selection algorithm based on information from previous searches. In *Proceedings of the 2007 IEEE International Conference on Multimedia and Expo,* (pp. 1591-1594). IEEE Press.

Shen, L., Liu, Z., Zhang, Z., & Wang, G. (2007). Video nature considerations for multi-frame selection algorithm in H.264. In *Proceedings of the 2007 IEEE/ACS International Conference on Computer Systems and Applications, AICCSA 2007,* (pp. 708-711). IEEE Press.

Slowack, J., Skorupa, J., Mys, S., Lambert, P., Grecos, C., & Van de Walle, R. (2010). Flexible distribution of complexity by hybrid predictive-distributed video coding. *Signal Processing Image Communication, 25*(2). doi:10.1016/j.image.2009.12.002

Tan, Y. H., Lee, W. S., & Tham, J. Y. (2009). Complexity control and computational resource allocation during H.264/SVC encoding. In *Proceedings of the 17th ACM International Conference on Multimedia (MM 2009),* (pp. 897-900). New York, NY: ACM Press.

Tan, Y. H., Lee, W. S., Tham, J. Y., & Rahardja, S. (2009). Complexity-rate-distortion optimization for real-time H.264/AVC encoding. In *Proceedings of 18th International Conference on Computer Communications and Networks, ICCCN 2009,* (pp. 1-6). ICCCN.

Tan, Y. H., Lee, W. S., Tham, J. Y., & Rahardja, S. (2009). Complexity scalable rate-distortion optimization for H.264/AVC. In *Proceedings of the 2009 16th IEEE International Conference on Image Processing (ICIP),* (pp. 3397-3400). IEEE Press.

Ting, C.-W., Po, L.-M., & Cheung, C.-H. (2003). Center-biased frame selection algorithms for fast multi-frame motion estimation in H.264. In *Proceedings of the 2003 International Conference on Neural Networks and Signal Processing*, (Vol. 2, pp. 1258-1261). IEEE.

van der Schaar, M., & Andreopoulos, Y. (2005). Rate-distortion-complexity modeling for network and receiver aware adaptation. *IEEE Transactions on Multimedia*, 7(3), 471–479. doi:10.1109/TMM.2005.846790

Vanam, R., Riskin, E. A., Hemami, S. S., & Ladner, R. E. (2007). Distortion-complexity optimization of the H.264/MPEG-4 AVC encoder using the GBFOS algorithm. In *Proceedings of the Data Compression Conference, 2007, DCC 2007*, (pp. 303-312). DCC.

Vanam, R., Riskin, E. A., & Ladner, R. E. (2009). H.264/MPEG-4 AVC encoder parameter selection algorithms for complexity distortion tradeoff. In *Proceedings of the Data Compression Conference, 2009*, (pp. 372-381). DCC.

Yang, J., & An, B. (2011). Low-complexity rate-distortion optimization for robust H.264 video coding. In *Proceedings of the 2011 IEEE International Conference on Multimedia and Expo (ICME)*, (pp. 1-4). IEEE Press.

KEY TERMS AND DEFINITIONS

CBR: Is an abbreviation for the Constant Bit-Rate, which means that the rate, at which the codec's output data should be consumed, is constant.

H.264/AVC (Advanced Video Coding): H.264/AVC is a video coding standard of the ITU-T Video Coding Experts Group and the ISO/IEC Moving Picture Experts Group, which was officially issued in 2003. H.264/AVC has achieved a significant improvement in rate-distortion efficiency relative to prior standards.

HEVC (High Efficiency Video Coding): Is a high-definition video compression standard, which is developed by a Joint Collaborative Team on Video Coding (JCT-VC) formed by ITU-T VCEG and ISO/IEC MPEG. The HEVC standard is intended to provide significantly better compression capability than the existing AVC (ITU-T H.264 | ISO/IEC MPEG-4 Part 10) standard.

MAD: Is an abbreviation for the Mean Absolute Difference, and is one of the most popular block matching criterions, according to which corresponding pixels from each block are compared and their absolute differences are summed.

Motion Compensation: Is a technique employed for video compression, which describes a picture in terms of the transformation of a reference picture to the current picture. The reference picture may be a previous or future picture.

Motion Estimation: Is a process of determining motion vectors that describe the transformation from one image to another; usually, from adjacent frames in a video sequence. The motion vectors may relate to the whole image or to particular image portions, such as rectangular blocks, etc.

SAD: Is an abbreviation for the Sum of Absolute Differences, which is a widely used algorithm for measuring the similarity between image blocks. It relates to taking the absolute difference between each pixel in the original block and the corresponding pixel in the block being used for comparison. These differences are summed to create a simple metric of block similarity.

VBR: Is an abbreviation for the Variable Bit-Rate; as opposed to the constant bit-rate coding, the VBR coding enables to vary the amount of output data per time segment. VBR allows a higher bit-rate (and therefore more storage space) to be allocated to the more complex segments of media files, while less space is allocated to less complex segments.

Section 3
Multimedia Content Adaptation

Chapter 8
Intelligent Approaches for Adaptation and Distribution of Personalized Multimedia Content

Artur Miguel Arsenio
Nokia Siemens Networks SA, Portugal & Instituto Superior Técnico, Portugal

ABSTRACT

Telecommunication operators need to deliver their clients not only new profitable services, but also good quality and interactive content. Some of this content, such as advertisements, generate revenues, while other contents generate revenues associated to a service, such as Video on Demand (VoD). One of the main concerns for current multimedia platforms is therefore the provisioning of content to end-users that generates revenue. Alternatives currently being explored include user-content generation as the content source (the prosumer model). However, a large source of revenue has pretty much been neglected, which corresponds to the capability of transforming, adapting content produced either by Content Providers (CPs) or by the end-user according to different categories, such as client location, personal settings, or business considerations, and to distribute such modified content. This chapter discusses and addresses this gap, proposing a content customization and distribution system for changing content consumption, by adapting content according to target end-user profiles (such as end-user personal tastes or its local social or geographic community). The aim is to give CPs ways to allow users and/or Service Providers (SPs) to configure contents according to different criteria, improving users' quality of experience and SPs' revenues generation, and to possibly charge users and SPs (e.g. advertisers) for such functionalities. The authors propose to employ artificial intelligence techniques, such as mixture of Gaussians, to learn the functional constraints faced by people, objects, or even scenes on a movie stream in order to support the content modification process. The solutions reported will allow SPs to provide the end-user with automatic ways to adapt and configure the (on-line, live) content to their tastes—and even more—to manipulate the content of live (or off-line) video streams (in the way that photo editing did for images or video editing, to a certain extent, did for off-line videos).

DOI: 10.4018/978-1-4666-2833-5.ch008

INTRODUCTION

There is a growing interest on delivering, efficiently, new interactive content to end-users. Such content must be adapted to end users' characteristics, according to personal tastes, group category, geographic location, or social community. This chapter presents work on content manipulation and distribution, especially the transformation of image and audio content on a video stream by users, service providers or advertisers, and its efficient distribution to a user multicast network. We exploit therefore the application of computer vision and machine learning techniques, traditionally employed on computer vision, into multimedia applications.

Computer Vision in Multimedia Applications

The adaptation of multimedia content can be achieved through the application of computer vision techniques. For instance, upon the detection and modeling of a face on a video stream, it becomes possible its replacement by the face of another person, or even transforming such face in order to make it appear larger, etc. There are currently several computer vision technologies, which may be especially useful, such as: object/face detection and recognition, 3D modeling, deformable contours, multi-feature tracking, head gaze inference estimation, eye detection and tracking, image perspective projections. Other algorithms that are employed to adapt video content are: leaps tracking and movement generation, emotional expression recognition and generation.

New Paradigms for Multimedia Content Adaptation Based on Machine Learning

Several statistical frameworks have been proposed in the literature to capture knowledge stored in the world, in order to learn the relative probability distribution of objects and people in a scene, which can then be employed for content adaptation, as described on the following sections. The goal is thus to enable individual object detection and prediction using the statistics of low-level features in real-world images, conditioned to the presence or absence of objects and their locations, sizes, depth and orientation. Therefore, machine learning offers powerful tools to infer image constraints (such as object locations) for objects, while computer vision provides techniques to transform and insert extra multimedia content representing such objects into the image.

Content Interactivity and Adaptation

Currently it is common for viewers/listeners to search for programs of their interest over an entire set of available/subscribed TV/Audio channels (e.g., news, sports, science, entertainment, music...) that they want to see/listen, especially at system power up or at the end of the program they were seeing/listening. This normally results in the viewer/listener having to manually switch (zap) possibly many channels in order to find what she prefers. And in general, all users receive the same content, and the same advertisements—the later usually in the form of extra content introduced between channel programs (or within scheduled programs). However, on one hand there is a need for advertisers to direct their content to target audiences. On the other hand, there is also the need for users to receive customized content:

- Advertisements should be added directly to channel content, such as a movie. Why not make James Bond drive a FIAT in Italy or a Volkswagen in Germany for the same movie?
- People should be able to customize the multimedia content objects in order to adapt these to their tastes.

Therefore, we will discuss techniques for advertisers to add personalized advertisements into high value content as if advertisements were part of the original content. Alternatively, we also detail techniques to allow content producers (or a prosumer) to adapt their content dynamically to end-users profiles including her tastes or trends/geographical location.

BACKGROUND

Various methods exist for processing of multimedia content (in particular audio and video content) in order to enrich the multimedia content e.g. with information or entertainment and thus make listening or viewing more attractive. For instance with current broadcast or multicast channels infotainment, content can be provided by a temporary interruption of the multimedia content, in particular a TV program, at an operator defined time for a subsequent provisioning (of e.g. an interview, a news report or an advertisement), that then is provided to all listeners or viewers of the channel. Furthermore, methods exist, which by means of usage of insertion techniques directly add infotainments into a multimedia content. An example for such a method is the insertion of news ticker information. Moreover, recognition and substitution algorithms may be used to automatically detect event triggers for delivering an infotainment and for convenient insertion of the infotainment into a multimedia content, in particular a TV program.

Individual adaptations of a multimedia content for a single viewer or listener are also possible. Disadvantages are the usage of unicast transmission and resulting high network load, when provided for a large number of users in parallel. Further disadvantages are high capital investments, to be available for real time or live programs, for which the individual adaptations have to be decided upon and executed on the fly with minimum delay implying high performance requests

for central processing platforms in the network. There is, therefore, currently a need to provide an improved efficient usage of network resources and network load distribution for a network operator or application service providers offering user group specific or user individual adaptation of a multimedia content, in particular when to be provided under real time conditions to a large user (i.e. listener and/or viewer) community. We are interested in solutions that are able to adapt video content to users, by merging other data to create new content or transforming existing one. This is a rather distinct approach than traditional solutions already in the market such as Content Management Systems, used either for simply managing content provided by external entities or to provide some removal operations such as advertisement removal for videos. These solutions have the disadvantages that they are not able to adapt video content to users, by merging other data to create new content or transforming existing one—their power rely mostly on the removal of features and on adding descriptive texts or images (e.g. biography of a soccer player on soccer games, which may be event triggered, such as the push of a button by a user). In resume, current solutions are appropriate for merging video segments from different sources (in slices), or by removing segments (such as advertisement segments), or even replacing some segments by others. The other main application of current solutions is to respond to events on a video stream (e.g. upon appearance of a certain object, to display information concerning this object). We will now review the main computer vision and machine learning techniques useful for content adaptation.

Personalized Multimedia Content Publishing and Distribution

Different techniques have been proposed for the distribution of personalized multimedia content. Indeed, Sebe and Tian (2007) present a deep overview on personalization for Multimedia Information Retrieval, analyzing different aspects of the challenges raised by such problem. The most popular techniques include content filtering employing client profiles, and adapting multimedia content to terminal devices, such as mobile phone resolution. An example of the former is the work by Ferman et al. (2002), which proposes to automatically determine a user's profile from his/her usage history (employing a profiling agent), and using such client profile for filtering the content (denoted the filtering agent). An example of the later is the distribution of content on P2P networks adapted to terminals by exploiting H.264 SVC feature (Cruz, et al., 2010). Another important aspect addressed by Sebe and Tian (2007) is the usage of artificial intelligence algorithms for multimedia personalization. However, there is a lack of previous works on allowing the user to use intelligent techniques to change streaming content. Perhaps one of the most relevant techniques is multimedia annotation, but which only addresses partially the problem. Hence, state-of-the-art technologies are still lagging behind consumers' needs and expectations. Dogan (2002) discusses the challenges brought by the customization of multimedia data delivery taking into account network heterogeneity and diversity of user terminal capabilities. He proposes to deliver multimedia content adaptation via transcoding at a central media gateway, acquiring from several distributed client-side gateways terminal capability and link characteristics data. A similar problem is addressed by Wei, Bhandarkar, and Li (2009), who presented a client-centered multimedia adaptation system to satisfy multiple client-side resource constraints, assuming knowledge of the client's preference(s) regarding the video content and the various client-side resource constraints. Inspired on speech recognition learning techniques, they propose a multi-level Hidden Markov Model (HMM)-based approach to automatically segment and index video streams in a single pass, without using any domain-dependent knowledge about the structure of video programs.

A technique recently proposed is Dynamic Adaptive Streaming over HTTP (DASH), which allows video clients to dynamically adapt the requested video quality for ongoing video flows, to match their current download rate as good as possible. Fuente et al. (2011) describe a solution based on Scalable Video Coding (SVC) for a DASH environment. The PENG project (Pasi & Villa, 2005) proposes an interactive and personalized tool for multimedia news composition and delivery, based on personalized filtering, retrieval and composition of multimedia news. Several architectures for personalized multimedia are also described by Ramanathan and Venkat Rangan (1994). However, none of the solutions described address the problem of transforming objects on video streams employing AI and computer vision techniques, and the creation of new multimedia content from such transformations, as proposed by this chapter.

Machine Learning for Location Inference

Holistic Approaches for Machine Learning

Agents, such as humans and objects, are situated in a dynamic world, full of information stored in its own structure. Given the image of an object, its meaning is often a function of the surrounding context (Torralba, et al., 2003). Context cues are useful to remove such ambiguity. Ideally, contextual features should incorporate the functional constraints faced by people or objects (e.g. people cannot fly and offices usually have doors). From a location point of view, contextual selection is very important both to constrain the search space for locating objects (optimizes computational resources) and also to determine common places on a scene to drop or store objects. Hence, the structure of real world scenes is most often constrained by configurational rules similar to those that apply to an object. The scene context puts a very important constraint on the type of places in which a certain object might be found. In Arsenio (2004), a statistical framework is used for a robot to learn the probable location of objects in images, together with estimates of their size and orientation. For instance, the probability of a chair being located in front of a table is much bigger than that of being located on the ceiling (see Figure 1). Such strategies can be very useful not only for robots, but also for a multimedia system to determine natural places on a video stream for the placement of extra multimedia content.

Torralba (2003) presents results for locating human heads (and their appropriate size on the images) on a diverse set of scenes. In other work, Torralba et al. (2003) presented results for the prediction of a diverse set of objects on scenes. These approaches require the existence of a database of tagged images, which is used to train machine-learning algorithms. After training, the algorithm estimates for instance, based on the contextual features such as the spatial distribution of the image spectral components, probable location, size, and orientation for an object. Several different contextual features can be found in the research literature. The approach presented by Fu et al. (1994) assumes prior knowledge about regularities of a reduced world where the system is situated. In another approach presented by Bobick and Pinhanez (1995), visual routines are

selected from contextual information. The context consists of a model of hand-written rules of a reduced world in which the vision system operates. Oliva and Torralba (2001) and Torralba (2003) apply Windowed Fourier Transforms (similar to STFTs) and Gabor Filters, respectively, as contextual features. The system proposed by Arsenio (2004) employs Wavelets Coefficients to infer contextual information. Let us consider Wavelets (Strang & Nguyen, 1996) as contextual features, and apply processing iteratively through the low frequency branch of the transform over T scales, while higher frequencies along the vertical, horizontal and diagonal orientations are stored (because of signal polarity, this corresponds to a compact representation of six orientations in three images). The input is thus represented by $v(x,y) = v(\vec{p}) = \{v_k(x,y), k = 1, ..., N\}$. With $h(x,y)$ a Gaussian window, each ith level wavelet component is down-sampled by 2^i:

$$\bar{v}(x,y) = \sum_{i,j} v(i,j)h(i-x,j-y)$$

The dimensionality problem should be reduced (Torralba, 2003; Arsenio, 2004) to become tractable by applying Principal Component Analysis (PCA). The image features $\bar{v}(\vec{p})$ are decomposed into basis functions provided by the PCA, encod-

Figure 1. Results for probable locations for a chair in a given environment, taken from Arsenio (2004). Even if the chair is not present on the image, the algorithm determines that the probable location for a chair to appear is in front of a table (its "natural place"—or most probable place—for the occurrence of a chair in nature).

 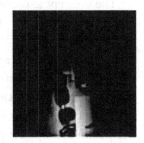  

ing the main spectral characteristics with a coarse description of its spatial arrangement:

$$v(\vec{p}) = \sum_{i=1}^{D} c_i \varphi_k^i(\vec{p})$$

where the functions $\varphi_k^i(\vec{p})$ are the eigenfunctions of the covariance operator given by $v_k(\vec{p})$. These functions incorporate both spatial and spectral information. The decomposition coefficients are obtained by projecting the image features $v_k(\vec{p})$ into the principal components:

$$c_i = \sum_{p,k} v_k(\vec{p}) \varphi_k^i(\vec{p})$$

computed using a database of annotated images. The vector $\vec{c} = \{c_i, i = 1,...,D\}$ denotes the resulting D-dimensional input vector, with $D = E_m$, $2 \leq D \leq T_h$, where m denotes a class, T_h an upper threshold and E_m denotes the number of eigenvalues within 5% of the maximum eigenvalue.

The coefficients c_i, used as input context features, can be viewed as a scene's holistic representation since all the regions of the image contribute to all the coefficients, as objects are not encoded individually.

Mixture models are applied to find interesting places to put a bounded number of local kernels that can model large neighborhoods. In D-dimensions a mixture model is denoted by density factorization over multivariate Gaussians (spherical Gaussians give faster processing times). The output space is defined by a n-dimensional vector $\vec{x}$, which may include a position vector, a size vector, orientation, or other positioning information. Therefore, given the context $\vec{c}$, one needs to evaluate the PDF $p(\vec{x}|o_n, \vec{c})$ from a mixture of (spherical) Gaussians (Gershenfeld, 1999):

$$p(\vec{x}, \vec{c}|o_n) = \sum_{m=1}^{M} b_{m,n} G(\vec{x}, \vec{\eta}_{m,n}, X_{m,n}) G(\vec{x}, \vec{\mu}_{m,n}, C_{m,n})$$

The mean of the new Gaussian $G(\vec{x}, \vec{\eta}_{m,n}, X_{m,n})$ is a function: $\vec{\eta} = f(\vec{c}, \beta_{m,n})$, that depends on $\vec{c}$ and on a set of parameters $\beta_{m,n}$. Choosing a locally afine model for f, results:

$$\{\beta_{m,n} = (\vec{a}_{m,n}, A_{i,n}) : \vec{\eta}_{m,n} = \vec{a}_{m,n} + A^T \vec{c}\}.$$

The learning equations are given by the EM algorithm (see Gershenfeld, 1999, for a detailed description):

- E-step for k-iteration: From the observed data $\vec{c}$ and $\vec{x}$, computes the posteriori probabilities $e_{m,n}^k(l) = p(\vec{c}_{m,n}|\vec{c}, \vec{x})$, where l indexes the number L of samples.
- M-step for k-iteration: estimation of cluster parameters.

The conditional probability follows then from the joint PDF of the presence of an object o_n, at the spatial location $\vec{x}$, given a set of contextual image measurements $\vec{c}$

$$p(\vec{x}|o_n, \vec{c}) = \frac{\sum_{m=1}^{M} b_{m,n}^k G(\vec{x}, \vec{\eta}_{m,n}^k, X_{m,n}^k) G(\vec{c}, \vec{\mu}_{m,n}^k, C_{m,n}^k)}{\sum_{m=1}^{M} b_{m,n}^k G(\vec{c}, \vec{\mu}_{m,n}^k, C_{m,n}^k)}$$

The concept associated to the determination of natural image places for objects offers significant potential for many multimedia applications, such as advertisement interactive placement on a multimedia stream. This is traditionally done in a static fashion, in which advertisements are placed for instance at the bottom of the video stream, not within appropriate places of the multimedia content to be transmitted.

Stationarity Estimation

Besides the natural places of objects, a multimedia system can determine stationary spatial regions on a video stream for the placement of extra multimedia content. Stationarity can be estimated using the optical flow, which is the apparent motion of image brightness (Horn, 1986). The optical flow constraint assumes that 1) brightness $I(x; y; t)$ smoothly depends on coordinates (x, y) on most of the image, 2) brightness at every point of an object does not change in time, and 3) higher order terms are discarded.

Some methods to compute the optical flow field are the Horn and Schunk algorithm (Horn, 1986) and Proesmans's algorithm—essentially a multiscale, anisotropic diffusion variant of Horn and Schunk's algorithm.

Computer Vision and Machine Learning Approaches

This chapter presents solutions that employ and integrates several well-known technologies, such as object/face detection and recognition, 3D modeling, deformable contours, multi-feature tracking, head gaze inference estimation, eye detection and tracking, or image perspective projections. Other algorithms that are employed to transform and adapt video content are leaps tracking and movement generation, emotional expression recognition and generation. A subset or all these techniques may be required to personalize a multimedia content, in order to transform the original content into new content according to the transmitted metadata information (and eventually extra multimedia content). Audio content adaptation also involves knowledge on speech synthesis and voice modulation, as well as emotional state estimation from speech signals. Therefore, a brief overview on these techniques will be presented, and it is discussed their role for multimedia content adaptation.

Face Detection

One interesting functionality for adaptive multimedia content is the detection and recognition of a face in a sequence of images, and the replacement of such face by another human face, or else a synthetic face (e.g. generated 3D model of an artificially character), or even to transform the detected face according some transformation (e.g. enlargement). Viola and Jones (2001) at MIT proposed a computationally efficient algorithm to detect faces in cluttered scenes, which is applied to each video frame. If a face is detected, the algorithm estimates a window containing that face. Other methods include tracking ovals, employed at the MIT Kismet and Cog robots to detect faces, so that the outline of the moving face is approximated as a deforming oval for tracking. The cropped image may then be sent to a face recognition, or gaze inference algorithms for further processing.

Face Recognition

The scale, pose, illumination, or facial expression can considerably change the geometry of a face. Attempts to model the latter based on a muscular point of view have been proposed (Sirovich & Kirby, 1987). Other methods include feature-based approaches in which geometric face features, such as eyebrow's thickness, face breadth, position, and width of eyes, nose, and mouth, or invariant moments, are extracted to represent a face (Chellappa, et al., 1995). However, feature extraction poses serious problems for such techniques (Chellappa, et al., 1995). Appearance-based approaches project face images into a linear subspace of reduced dimensions (Turk & Pentland, 1991). In order to efficiently describe a collection of face images, it is necessary to determine such a subspace—the set of directions corresponding to maximum face variability—using the standard procedure previously described of Principal Component Analysis (PCA) on a set of training images. The corresponding eigenvectors are called eigenfaces because they are face-like in appearance (Aryananda, 2002; Arsenio, 2004). Classification of a face image consists of projecting it into eigenface components, by correlating the eigenvectors with it, for obtaining the coefficients of this projection, which form a vector representing how well each eigenface describes the input face image. A face is then classified by selection of the minimal L_2 distance to each object's coefficients in the database. Eigenfeatures, such as eigeneyes or eigenmouth for the detection of facial features (Belhumeur, et al., 1997), is an alternative variant for the eigenfaces method.

Head Gaze Inference and Eye Tracking

The projection of a face into an image requires the identification of the head orientation on the image, as well as the estimation of the eyes' gazing direction, so that the new face is projected into the sequence of images according to similar postures. Head poses can be estimated by applying the same eigenobjects based strategy used for face recognition (and therefore the eigenvectors are really eigenposes for this problem). The single algorithmic difference is that the input space, instead of including faces labelled by identity, will consist of faces labelled by head gaze direction. Arsenio (2004) proposed to separate the classification problem into five category classes (although a finer resolution could be used to account for more classes), along two axes: left and right gaze, top and down gaze, and front gaze. Back propagation neural networks can also be employed as alternative to the eigenfunctions methods (Yucel & Ali Salah, 2009). There exists several approaches and diverse algorithmic implementations for training a multi-layer back-propagation neural network. A standard approach consists on updating weights by backpropagation of errors at the output of perceptron units (Yucel & Ali Salah, 2009).

Object Recognition

More generally, adaptive multimedia content requires the detection and recognition of an object in a sequence of images, and executing an operation on it, such as its removal, or its replacement by another object. The recognition of objects has to occur over a variety of scene contexts, and it can be applied on a multimedia stream in order to replace a scene object by another object belonging to the same category.

Previous object recognition work using exclusively local (intrinsic, not contextual) object features (computed from object image templates) for performing object recognition tasks are mostly based in object-centered representations obtained from these templates. Swain and Ballard (1991) proposed an object representation based on color histogram. Objects are then identified by matching two color histograms—one from a sample of the object, and the other from an image template. The fact that for many objects other properties besides color are important led (Schiele & Crowley, 2000) to generalize the color histogram approach to multidimensional receptive field histograms. These receptive fields are intended to capture local structure, shape or other local characteristics appropriate to describe the local appearance of an object. A kalman filter based approach for object recognition was presented by Rao and Ballard (1997), and applied to explain neural responses of a monkey while viewing a natural scene. Zhang and Malik 2003) described a shape context approach to learn a discriminative classifier. Papageorgiou and Poggio (2000) approach used instead Haar Wavelets to train a support vector machine classifier, while Freeman and Adelson (1991) applied steerable filters.

Another set of methodologies apply the Karhunen-Loeve Transform (KLT) (Fukunaga, 1990) or Principal Component Analysis (PCA) for the calculation of eigenpictures to recognize faces (Turk & Pentland, 1991) and objects (Murase & Nayar, 1995). The KLT approach yields a decomposition of a random signal by a set of orthogonal functions with uncorrelated coefficients.

Whenever there is a priori information concerning the locus of image points whether to look for an object of interest (from prior segmentation, for instance), a simpler template matching approach suffices to solve the recognition problem for a wide set of object categories (assuming they look different). Geometric hashing (Wolfson & Rigoutsos, 1997) is a rather useful technique for high-speed performance. In this method, invariants (or quasi-invariants) are computed from training data, and then stored in hash tables. Recognition consists of accessing and counting the contents of hash buckets.

Deformable Contours

Snakes (active contour models) are deformable contours that have been used in many image analysis applications, including the image-based tracking of rigid and non-rigid objects, such as leaps modeling and tracking, eye tracking, or the construction of 3D dynamic models such as that of the heart (Curwen & Blake, 1992; Kass, et al., 1988). This deformable contour moves under the influence of image forces. One can define a deformable contour by constructing a suitable deformation energy $P_i(z)$, where z represents the contour points. The external forces on the contour result from a potential $P_e(z)$. A snake is a deformable contour that minimizes the energy:

$P(z) = P_e(z) + P_i(z),\ z(s) = (x(s), y(s))$

For a simple (linear) snake, the internal deformation energy is

$$P_i = \int_0^T \frac{\alpha(s)}{2} |\dot{z}(s)|^2 + \frac{\beta(s)}{2} |\ddot{z}(s)|^2\ ds$$

where the differentiation is in respect to s. This energy function models the deformation of a stretchy, flexible contour $z(s)$ and includes two physical parameter functions: $\alpha(s)$ controls the tension and $\beta(s)$ the rigidity of the contour. To attract the snake to edge points one specifies the external potentials $P_e(z)$:

$$P_e(z) = \int_0^T P_{image}(z(s))ds$$

where $P_{image}(x,y)$ is a scalar potential function defined over the image plane and which is typically computed through image processing. The local minima of P_{image} can be viewed as snake attractors. Hence, the snake will have an affinity to intensity edges if $P_{image}(x,y) = -\left[G_\sigma * |\nabla I(x,y)| \right]$

P_{image} thus represents the convolution of the image gradient with a gaussian smoothing filter whose characteristic width, σ, controls the spatial extent of the attractive depression of P_{image}. This convolution is time expensive, but one can convolve the image with two one-dimensional gaussians, since the 2D gaussian is a separable filter.

Multi-Target Tracking

The replacement of an object by another usually will happen over a collection of video frames, requiring the detection and tracking of the object to be replaced on the video sequence over a specified time interval. This requires the usage of object tracking algorithms, initializing the moving object as follows.

A grid of points homogeneously sampled from the object are initialized in the moving region, and thereafter tracked over a time interval. At each frame, each tracker's velocity is computed together with the tracker's location in the next frame. The motion trajectory for each tracker over this time interval can be computed using different methods: some approaches are based on the computation of the image optical flow field as described previously, other algorithms rely on discrete point tracking, either by block matching (employing for instance cross-correlation of image pixels over time), or else by applying a pyramidal implementation of the Lucas-Kanade algorithm to track good feature points (Shi & Tomasi, 1994). The methods based on optical flow field are good for velocity estimation, but not so for position estimation due to dependence on a smoothing factor. B-splines have also been also proposed for tracking in Terzopoulos and Szeliski (1992). The authors also proposed the usage of a Kalman-filter in order to integrate the track estimation data with an a-priori model of motion.

Emotional Recognition and Generation

The replacement of a face on a video stream is dependent upon the face contours detection, the recognition of the person corresponding to that face, the head pose, the eyes gazing direction, the leaps positioning, as well the emotional state of the face. Having such descriptions, one is then able to replace a face by another face with a similar emotional state, head pose, eye gazing, or else to transform such face (for instance, changing the emotional state or the eye gaze direction). Emotional responses play a very important role for socially interacting multimedia content. Display of emotional states is useful to transmit one's motivational status. Velasquez (1997) one of the first advocates for modeling emotional processes in robots, proposed a computational model for the generation of emotions named Cathexis. The works of both Velasquez (1997) and Ferrell (1998) in social machines propose emotional mechanisms inspired by Damasio's somatic marker hypothesis (Damasio, 1994). This hypothesis argues that human learning and reasoning is biased by a nonconscious, covert mechanism, which tags decisions with (either negative or positive) somatic markers from previous experienced emotional responses. In another work (Arsenio, 2004), motivational drives were used to modulate behaviors for safe navigation, social interaction and object inspection for treacherous environments. In general,

the acoustic pattern of speech contains information concerning affective communicative intent. Breazeal and Aryananda (2002) presented an approach for recognizing four distinct prosodic patterns that communicate praise, prohibition, attention, and comfort to preverbal infants, and to generate emotional display patterns on a robot.

Image Perspective Projections

Once a replacement description is available for a given object on a video stream, the new object must be projected into a set of images according to the replacement metadata, which may consist of the same projection as the original object, or a transformation (such as a stretched face). Perspective transformations of a 3D model is a well studied subject (see for instance, Faugeras, 1993).

Virtual Personal Metadata Channels

CMS systems insert typically advertisements between frames, as well as additional text or images and other information. We describe, instead, the usage of an additional metadata channel in order to insert extra content directly into specific image locations and time instants on a video stream. A similar approach was proposed by Veiga (2009), who employs a special type of channel that unlike "normal" channels does not transport video/audio contents. Instead, it only transports the channel reference to be viewed/listened at any given moment for that virtual channel. On the virtual channel, only the channel references are sent, not the channel contents, i.e., such control channel only sends the reference—such as an IDentifier like number—to the channel that should be viewed/listened at a given moment. The proposal of Veiga (2009) differs from the solution here presented in the sense that he proposed virtual control channels with information to switch between channels, not to add extra information into within a specific channel (see Figure 2, which illustrates comparatively the two solutions, identifying the main differences).

We denote this virtual channel as personal metadata channel, that besides transporting the extra content, it may transport as well metadata information that describes how to insert the extra content into the original frames. This additional metadata information can be:

- Position coordinates to insert extra content in a frame (for instance, location *loc1* where to insert the multimedia content (marked on image as light grey), and loc2 (to insert dark grey multimedia content into frame t6).
- Size of extra content to be inserted, or perspective transformation parameters for projecting a 3D model of the extra content, and distortions to be applied to extra content.

- Definition variables specifying properties of image regions where the extra content should be placed (such as stationary image regions).
- Definition of the type of content, and/or specifying the insertion of the extra content on image regions where this type of content is more probable to appear.
- Other features, or any combination of them.

The personal metadata channel is a communication mechanism to insert extra elements on a video stream, which were not originally present, such as personalized advertisings that may be selected according to user's profiles or geographic locations. The idea of the technique applies for an IPTV system, where many multicast channels (e.g., TV, audio, Electronic Programming Guide—EPG) can be received (although

Figure 2. (a) Virtual channels solution, in which a control channel which solely uses metadata contains information that virtually forms a new channel composed of different pieces of existing channels; (b) personal metadata channel solution, focused on adding extra information to localized regions in the images at certain instants in time. Blocks in the original channel, at time instants t₁...t₆ are video frames. On the personal metadata channel, circles represent extra content that can be text or new images to be inserted.

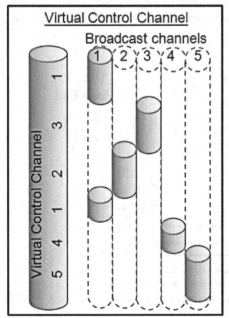
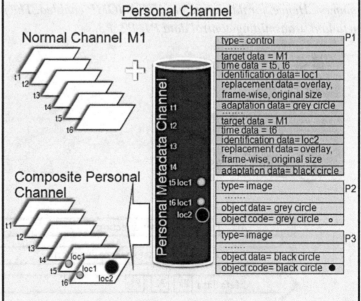

not always simultaneously) on the Set-Top-Box (STB) at viewer/listener premise. Figure 3 shows a very simplified view of an IPTV network, where it can be seen the STB at customer premise and the TV/Audio Head End that collects the channels from the different channel providers and make them available to the viewer/listeners. The video streams might be personalized to clients' needs both at the STBs or at the DSLAM (doing such customizations at the network edge is preferable in order not to increase STB cost and processing requirements), or even at the multimedia content production source.

Typically at start up the STB would tune (e.g., IGMP Join request) to the Announcements multicast channel, and then to the SW multicast channel to download and run the correct/updated STB software. After this step it will be possible for the users to select new channels or the EPG by having the STB tuning/requesting the new channel (e.g., IGMP Join request) probably by releasing (e.g., IGMP leave request) another channel. To better understand the communication approach it is better to understand some basic concepts first:

- **Normal (original) Channel:**
  - A channel coming from TV/audio producers like CNN, VH1, Sky News, TV5, etc.
  - Such channels transmit audio/video contents unlike the Personal Metadata channel.
- **Personal Metadata Channel (PMC):**
  - Personal in the sense of personalized data, i.e. data of the normal channel is modified and transmitted separately, according to "personal" profiles of an end-user, a set of end-users, a community, region or country.
  - Contains extra content and metadata associated to the original video stream.
  - Not a channel *per se*, but a channel that is created by a 3rd party service provider, or by the IPTV service op-

Figure 3. Multicast network overview example. The middle network elements like BRAS, ethernet switches, routers, video servers, etc., have been deliberately removed for simplicity reasons. It is mention the IGMP protocol as the multicast protocol for example purposes only and because it is nowadays the most common. Hence, for this figure, the STB is IGMP enabled. The personal metadata channel includes the metadata, transmitting control data P1, P2, P3.

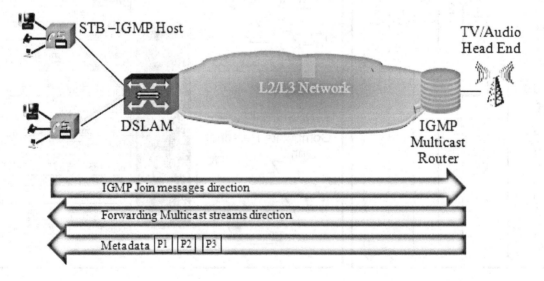

erator, containing either static images of content or short streams of content, or even text, together with associated metadata.

- **Composite Personal Channel:**
  - This is not a content channel *per se*, but a channel that is created automatically from the original normal channel content and from the personal metadata channel content.
  - The available/subscribed broadcasted original normal channel (e.g., CNN, MTV, Discovery, History) is formed by replacing information from the original channel from the one in the Personal Metadata Channel, according to image position and size metadata as defined on the later channel.

Apart from the existing channels like: Video/audio channels (possibly more than one received simultaneously), Announcements channel (controlling channel normally periodically sent), SW version (sent on start up request), and EPG channel (sent on request), it is required for personalization to have one additional type of channel containing extra multimedia content and control metadata: the Personal Metadata Channel.

INTERACTIVE MULTIMEDIA CONTENT DISTRIBUTION

This section describes efficient networked solutions for the distribution of interactive and personalized video streams (Arsenio, 2011). Indeed, one of the challenges concerning interactivity and personalization regards the efficient usage of the multicast network, in order to keep deployment costs at an acceptable level, and to guarantee sufficient levels of Quality of Experience to end-users (Marquet, et al., 2010). Hence, this section discusses new alternative approaches for multimedia content distribution based on virtualization,

namely by employing personal virtual channels to transport metadata for allowing new levels of interactivity and content adaptation. The overall methodology is as follows (Arsenio, 2011), and it will be henceforth detailed:

1. Receiving original multimedia content.
2. Receiving control metadata, and optionally, over the same personal channel, extra multimedia content.
3. Determining replacement specification based on control metadata (and profile).
4. Decompressing and/or decoding original and/or extra multimedia content.
5. Identifying replaceable object in original multimedia content on base of replacement specification.
6. Creating composite multimedia content by adapting replaceable object on base of extra personal multimedia content and replacement specification.
7. Compression/Coding of composite multimedia content according profile.
8. Provision of composite multimedia content according profile.

Communication Protocol

This protocol concerns the information flow over a personal metadata channel providing control metadata and extra multimedia content for adaptation of the original multimedia content. The normal channel multimedia content may be in particular a streamed multimedia content provided over a network e.g. a TV program. A device receiving the multimedia content may be any terminal device connected to a network or any network device being in or part of the network. In particular, the device receiving the original multimedia content may be a network edge device, e.g. a DSLAM, or it may be a terminal device, e.g. a set top box. The first multimedia content may be received as streamed content over a network, but it may as well be downloaded over a network. Control metadata are received as streamed data over a channel (such as a channel of IPTV, mobile TV, satellite TV, terrestrial TV, or cable TV solutions). In the previous Figure 3, the personal metadata channel is provided by a network element (for instance an IGMP multicast, or a satellite or terrestrial antenna providing channel according to DVB-S or DVB-T). The channel is received by a processing element, which may be a terminal device (e.g. a set top box) or a network element (e.g. a DSLAM), for further termination or distribution of channel to terminal devices or other network elements, respectively. As depicted, the personal metadata channel transports metadata P1 together with additional multimedia content P2, P3. The metadata P1 from logical point of view comprise amongst other information a type information (*<type=control>*) indicating that the packet transports control data. The metadata P1 comprises additional control metadata blocks that may contain (Arsenio, 2011):

- *<channel data>* including channel references.
- *<target content data>* a name of or reference to the original multimedia content or an addressable part thereof.
- *<profile data>* which can indicate global/country-wide/regional relevance, user group/family/individual relevance, for a single user or a single terminal device, service provider relevance (e.g. only if a user/terminal device is entitled to use services of a certain service provider). Profile data may be used for deciding upon comparison with locally accessible profile data, if a replacement specification that may be represented by a control metadata block is of relevance for a device receiving the control metadata block.
- *<time data>* such as a start time, an end time, a repetition time, a number of repetitions, a time interval, a trigger condition for at least one event or any equivalent time

or trigger reference information, that is appropriate to limit a replacement specification with respect to time.

- *<identification data>* like coordinate and size information, audio/video detection algorithm reference, start/end event detection parameter data, detection object, detection probability data, detection delay data, detection event duration, detection object class data.
- *<replacement data>* including replacement mode information (e.g. indicating need of total, overlay, constant, frame-wise video replacement and further indicating audio effects as for instance muting, reduced volume, audio signal overlay or substitution). Can further include orientation information, 3D model information, information on replacement according detection per frame/according time or frame based rules.
- *<adaptation content data>* with references to additional multimedia content P2, P3.

The multimedia blocks P2, P3 may comprise amongst other information in particular:

- *<type>* information (in P2, <type=image>, while in P3, <type=audio+video>) indicating that the packet transports image (or audiovisual) data, respectively.
- *<object data>* e.g. inter alia object type, object class, object coding information.
- *<object code>* data.

Image Processing for Content Transformation

A replacement specification on the control metadata is determined on the basis of the aforementioned control metadata and a profile information. The control metadata e.g. may

- Comprise a reference to frames 6 – 60 of the original channel multimedia content.
- Include position coordinate information relative to a frame of the original multimedia content and a size information relative to a frame.
- Include a reference to additional multimedia content (may for instance be a picture, a text, or any combination of a picture and a text, e.g. a logo of a company with a given size that may be used to advertise for products of the company).
- Further comprise a profile information indicating, that the replacement specification is to be executed in all DSLAM nodes in a region. By access to locally available profile information the device executing the method steps is able, to determine if the control data includes a replacement specification with relevance for the device.

With these control metadata, the replacement specification could be for instance the metadata to replace in frames 6-60 the original multimedia content determined by the coordinate information and the size information by a size and position adapted copy of the additional multimedia content. Execution of the replacement specification can for example make company logo appear beginning with frame 6 of the first multimedia content, changing its size and position in the following frames until frame 60 and then disappear. Based on this replacement specification, a replaceable object is identified in the original multimedia content. Given the example above, the identification is already accomplished by the per frame available coordinate and size information together with the frame number information. Considering another example, the replacement specification might define the replaceable object to be a stationary place in the original multimedia content. Such stationary place on the stream of the original multimedia content might for instance be a wall that does not change its position during a scene of

the original multimedia content. The control metadata may also comprise a reference to a detection algorithm capable of detecting a stationary place in the original multimedia content. A composite personal multimedia content is therefore created by adapting the replaceable object in the original multimedia content using the extra multimedia content and the replacement specification. Given the examples described above the composite multimedia content can be created by insertion of company logo according to its coordinates and size information into the original multimedia content, or into a detected stationary place of the original multimedia content, wherein the insertion may be temporarily limited by the control data and the replacement specification respectively. The composite multimedia content is provided according to profile information. In particular, a type of link, connection, or transmission path to a device receiving the composite multimedia content may be a criterion to determine, how the composite multimedia content is to be provided. For example, the composite multimedia content may be provided via A/V or HDMI interface to an output device or via IP based streaming in particular to a terminal or network edge device. Locally available profile information at processing device may also be employed to decide how to provide the composite multimedia content. In order to add the extra content to a sequence of image frames on a channel to create the composite multimedia content, some image processing is required. The overall methodology for this additional processing is shown in Figure 4. This method can be applied both at the STB or at the DSLAM (in the later case, the personalized channel is then sent from the DSLAM to the STB, but in this case extra functionality is needed at the DSLAM/edge server). This component, as shown in Figure 4, is composed by: Configuration Module, Spatial Selection Module, and Object Projection Module.

The *configuration module* is used for building 3D models of objects together with appearance models, as discussed on the previous section. Such

Figure 4. Multimedia content replacement. The service provider, or an algorithm (such as natural places or stationarity estimation) at the network edge equipment determines places to insert extra multimedia content (or whether to transform the original content). Such metadata is thereafter transmitted and distributed through the network, together eventually with extra multimedia content (such as a new object model as constructed by the configuration module). At the network edge (which may occur for instance at the DSLAM for a group of users or at a STB for a single one) the metadata is merged with the original channel, according to profile data and replacement specification, in order to create a new composite personal channel, which is delivered to the (or a group of) end-user(s).

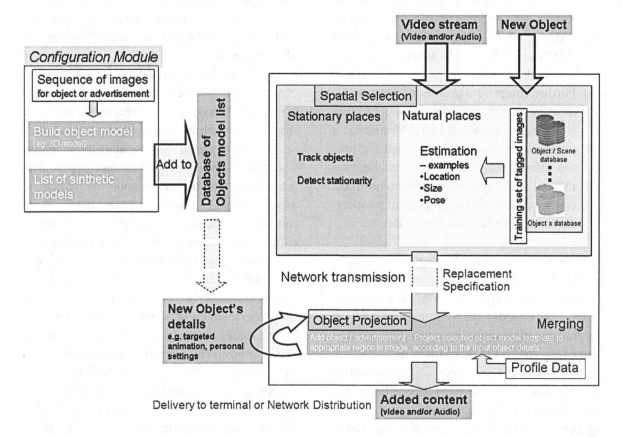

information is necessary whenever the extra content is supposed to contain 3D models of objects which are to be projected into an image at a specific location, in a given size and orientation.

The *spatial selection* module estimates the image regions where the extra content of the personal metadata channel should be placed. Can be done according to different strategies:

- Stationary image region methods: if the metadata information, carried on the personal metadata channel, contains a code flagging this process, then the processing module at the DSLAM or STB will try to put the extra content on stationary regions.

- Natural places of occurrence method: the processing module at the DSLAM or STB will put the extra content on probable image regions where the object should occur, as determined by the corresponding "natural places" algorithm. This algorithm, based for e.g. on Mixture of Gaussians as described previously, outputs the image location where to put the extra content, its size and orientation (pose), using an a-priori model built from a training set of tagged images. A threshold is defined so that, if no image region presents high probability, then the object is not inserted.

- Insertion metadata: information of the personal metadata channel contains a location: then the extra content is directly inserted. This can be the case if any of the two previous methods is applied at the multimedia content producer, so that in such cases the location is already available, and extra image processing is therefore not required at the DSLAM or STBs. Commercially, such strategy can bring many benefits with respect to maintaining terminal equipment costs low, while keeping communication overhead within acceptable levels.
- User specifies target object: through a graphical user interface, the user sets the image object to be replaced on a video stream.

Afterwards, once the regions to place the extra content are determined, if the extra content contains a 3D model of an object then this has to be projected into the frame of the composite personal channel according to the parameters determined on the previous step. Such task is accomplished by the *Object Projection Module*, which outputs the final merged multimedia content, according to different strategies:

- **Content Replacement:** Extra multimedia content on the personal metadata channel is already ready for direct insertion. Image pixels on an image region S_i of the original content at time instant i are replaced by the new content.
- **Content Impersonation:** Replaces objects representing people on a video stream, by rendering 3D models of other objects (such

as 3D models of other people's faces) into the stream.

- **Content Adaptation through Synthetic Characters:** Introduction of artificial personages into a video stream, which replace and behave as the original one.
- **Content Transformation:** Either projects content according to a transformation, or applies a transformation to content already present on the original video stream.

Multichannel Content Composition

Efficient distribution mechanisms are required to take advantage of the methodology just described to enrich content. The personal metadata channel is an efficient approach for implementing the composition of video channels, which may be of several types:

- Broadcasted if content changes are to reach all the end-users.
- Multicasted, if content changes are to reach a subset of users, such as a community or a set of users with same viewing habits or shopping patterns. In this case, only this subset of users will have access to this new, transformed version of the normal channel.
- Unicasted, if extra content is to be personalized to a singles user's profile (such as buying habits, zapping profile, Internet navigation or buying profile, bookmarked information by the user, user's viewing habits, user's service subscription, etc).

Depending on the Personal Metadata Channel transmission type, broadcasted, multicasted, or unicasted, the user will perceive the personal channel as being of the same type. The information transmitted by the Personal Metadata Channel may be produced by a dedicated company, interest group, or individual. It may be an advertiser, a brand company or an arbitrary collection of these or other companies or individuals (the prosumer) providing content. The PMC channel requires less bandwidth to send information when compared with a normal channel, since only the extra content to be inserted, together with associated metadata (on how to insert the extra content) need to be transmitted. Metadata information over the PMC channel can be sent continuously, periodically or at specific time instants. Both channels may be or not synchronized. Different synchronization policies are possible:

- If Personal Metadata Channel is sent continuously, then during buffering of the normal channel, the frame i can be mapped to the metadata from the Personal Metadata Channel.
- If sent periodically, then such pairing only occurs at specific time intervals in time (for instance, from 5 to 5 minutes, it occurs the insertion of adds on stationary regions of an image).
- If sent at specific time instants, then such pairing only occurs at these time instants. An example is the random insertion of adds in stationary regions of the image (or in probable locations, such as a BMW car whenever a road appears on the video stream).

Aggregated Distribution of Personal Multimedia Content

The aim is to allow the composition of content in real-time, according to a given replacement specification (criteria matched for instance to a user or a group of users). On the other hand, service providers, such as advertisers, may contribute to change the context according to some criteria. Of course, aiming at flexible and extensible business models, several different service providers should be allowed to contribute to the final content composition, without imposing unreasonable burdens on the content distribution network. Therefore, we end up this section discussing the aggregation of

multimedia personalized content from different virtual channels. We address the case in which an advertiser is inserting content at different parts and/or different timings than other advertisers. For such case, there can be many Personal Metadata Channels (one or many per composite personal channel) each one transmitting the replacement specification (such as reference location and sizes of advertisements or personalized content with respect to the current channel to be viewed).

In the case of many Personal Metadata Channels per Personal Channel, each Personal Metadata Channel might belong to an advertiser, which is inserting content at different parts and/or different timings than the other Personal Metadata Channel from other advertisers. Of course, for efficiency reasons and optimization of resource usages, all Personal Metadata Channels for a single Personal Channel may be collapsed into a single equivalent one (see Figure 5). Putting together the original channel and the plurality of the metadata channels should work exactly the same way as for the one metadata channel case, after disaggregation of the metadata according to the channel(s) to which they correspond. One Personal Metadata Channel can also aggregate information of many Personal Metadata channels to form several Composite Personal Channels (for various users). For aggregating data to personalize multiple normal channels, one can use the control metadata P_1, P_2, which associate the same extra content and metadata insertion information to a set of time instants (meaning that the same content with the same insertion information is placed in different frames of the composite personal video stream), and to a set of normal channels.

Referring now to Figure 5, the personal channel can be created at the producer and blended to the original channel at the network edge (e.g. DSLAM). For that, the metadata associated to a normal channel k originate n composite personal channels $p_1 \ldots p_n$, according to:

Figure 5. Aggregation and disaggregation of metadata and extra multimedia content. It is shown two personal metadata channels forming one aggregated personal metadata channel with respect to normal channel. Each personal metadata channel might belong to an advertiser, which is inserting content at different parts and/or different timings than the other personal metadata channel from other advertisers. At time t_6 the extra content marked as light grey circle is inserted at loc1 of the composite personal channel ch2 replacing the information of the original video stream. In the figure, this is done at the DSLAM. Simultaneously, the extra content marked as dark grey circle is inserted at loc2 of composite personal ch1. The aggregated channel ch is built by using all the information from the aggregated metadata channel. Notice that the same information in the aggregated channel may be always transmitted separately in two personal metadata channels.

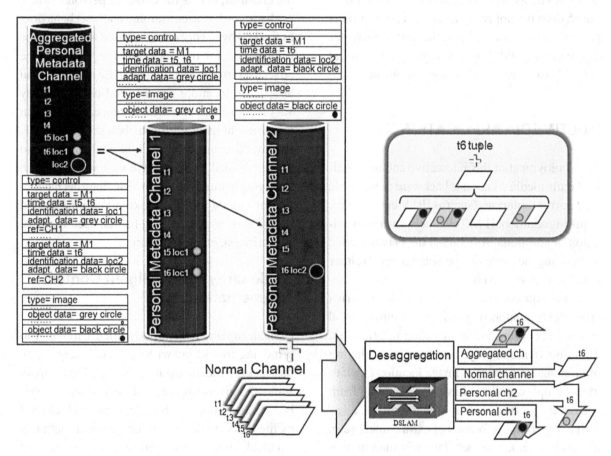

1. Reception of frame *i* of normal channel *k*, and decompression/decoding.
2. Replication of frame *i* of normal channel *k* to all composite personal channels associated to this normal channel.
3. For composite personal channels $p_1...p_n$, inserting into frame *i* the extra content *i* according to metadata information *i*.
4. Compression/coding.
5. Transmission of all channels.

This processing creates a constant delay on all channels (corresponding to the processing times), which is acceptable within certain limits for not to affecting the quality of experience. For processing at the network edge, an aggregated personal metadata channel is received carrying information concerning many personal channels for one normal channel. Hence, the aggregation of this information must occur first for such cases. If this is done at the TV Video Head-end or Video Server, the methodology applies (without step 1), but more data is put into the overall operator network requiring larger capacity.

If composite personal channels are created at the STB (and hence content disaggregation occurs at the STB), then more buffering may be required at the STB, as well as processing capabilities. Step 5 does not apply in this case. Aggregation is possible at this case, either at the content provider or even at the DSLAM, but only for the personal channels to be received by the equipment.

MULTIMEDIA APPLICATIONS

The deployment of this interactive and personalized multimedia solution is backward compatible with current solutions (such as IPTV equipment), requiring additional PMC channels, and the installation of additional software at the network edge for building the composite personal channel (either at the DSLAM or STB).

Let us now consider several possible scenarios on how extra content may be inserted into a normal channel according to the metadata information. This insertion can vary from simple overlaying of an image at a specific image location (x,y), to more complex models of insertion, ranging from considering extra content orientation, size, as well more complex formats of extra content (e.g. text and image). The metadata information may as well not refer explicitly to image coordinates, but instead to stationary image regions or even for placement of extra content in places where this type of content is probable to occur in nature.

Interactive and Customized Advertisements

Consider an actor in a movie (e.g. *Spiderman* movie), who is moving around. The video is being

transmitted on a broadcast channel selected by an IPTV user. On the Personal Metadata Channel, an advertiser defines at specific frames, and during a certain time interval, advertisements that at each instant, and for each frame, are to be located at a given location, with a given size, and in accordance with a specific appearance as provided by the Personal Metadata channel. These definitions are made according to the geographic location of the end-user, or else the end-user personal tastes, or else according to consuming end-user behaviors as tracked by the operator and used to personalize information for the end-user. For instance, the advertiser might select to advertise a given brand at specific locations for that end-user, that may vary for other end-users (for instance, a Spiderman movie might show ads for Reebok shoes in the UK and for Nike in the US). Extra content may also be determined from a single user's profile such as buying habits, zapping profile, Internet navigation or buying profile, bookmarked information by the user, user's viewing habits, user's service subscription, etc. (see Figure 6).

Stationarity for Intelligent Content Augmentation

Consider again a Hollywood actor in a movie who is moving around, but with some static scenery on the background at certain frames, which remain static for a given time interval. An original video is being transmitted on a broadcast channel selected by the IPTV end-user. On the Personal Metadata Channel, an advertiser defines an advertisement that wants to be added to the movie during a time interval T. On the previous application scenario, the location of such advertisement could be predefined on the metadata. But now, either at the DSLAM or at the STB (or even at the content provider premises, by incorporating the location of the stationary region on the personal metadata channel), an algorithm detects stationary portions of the incoming movie images (see Figure 7), with a duration of T. If T is larger than a threshold,

Figure 6. Transmission of a-priori information on a personal metadata channel: extra content information, as well as metadata information on how to insert such content, are provided à-priori by an advertiser or a content provider. The extra content may refer for insertion into a specific normal channel being broadcasted, or else the extra content may be generic and apply to many normal channels. The personal metadata channel may be transmitted, for instance, from the IPTV middleware solution using the information obtained from the content provider. In alternative, additional metadata information (or all of them) may be inserted into PMC at the DSLAM. The composite personal channel is then created at the DSLAM. Alternatively, this last step could be immediately done at the TV head-head, or else postponed for execution at the STB).

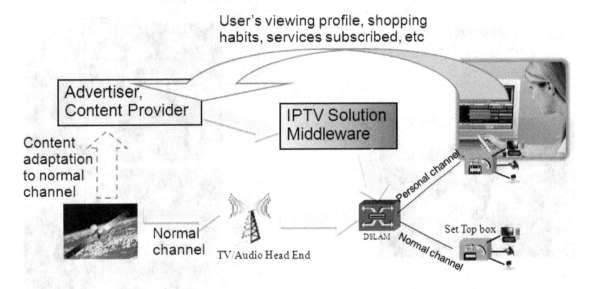

then another algorithm (object projection module) projects the image defining the advertisement into the stationary portion of the image during time interval T. This can also be done for live broadcast, requiring some buffering capability (with an associated delay) at the processing element to allow the stationary detection of image regions and adding the advertisement to such regions.

Interactive Content Placement from Probabilistic Cues

As described before, objects do not appear arbitrarily on nature. There are natural, probable places where to find then. This is exploited to place multimedia information such as advertisements in places not pre-defined. For instance, a car advertiser may want to add an image of a BMW

(or a watch, as shown in Figure 8) on a video stream, without specifying its location or size. An algorithm will then process some features of the image, and based on training data, will learn off-line probable places where to place cars (and their corresponding size on the image) as well as other objects according to the advertiser goals.

DISCUSSION ON EXPERIMENTAL EVALUATION

This chapter described a new methodology for addressing an emergent trend on multimedia content publishing and distribution, namely content interactivity and personalization. Some application scenarios are presented, for which the methodology is shown applied only to a

Figure 7. Using the personal metadata channel to place differentiated advertisements on IPTV channels according to end-user features (such as geographic location). Stationary places on the video stream are selected in order to replace the original background by the advertisement. This is done by the detection of the image regions that do not move above a threshold within a specific period of time. In this figure, the personal metadata channel carries the following information: $\{t_i$ in $[t_k... t_{k+T}]\}$, small advertisement binary image of Nike or Adidas, special code indicating that the way the extra content is to be inserted is through stationary image detection—and optionally the size and/or orientation for placing the advertisement into the stationary region, IDs of normal, and personal channels.

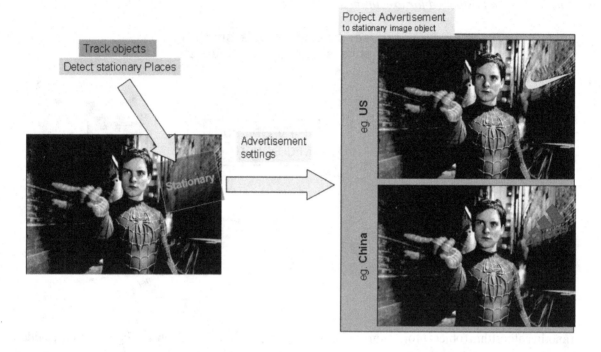

single case. Hence, significant work remains to be done on statistically evaluating the methodology with further scenarios and a large number of experiments. However, several of the individual components of the methodology have already been deeply evaluated:

- The virtual channel methodology is currently a commercial feature for NSN IPTV solution.
- Oliva and Torralba (2001) and Torralba et al. (2003) have presented in-depth experimental results for locating objects in images from probabilistic cues, while Arsenio (2004) addressed the experimental evaluation of object placement from probabilistic cues.
- Stationarity detection has been thoroughly tested in different scenarios. It is widely applied on robotics, as demonstrated by the work of [meu tese] for the humanoid robot Cog. In addition, different methodologies exist for its determination, such as image differencing and optical flow estimation, presenting different properties.

This chapter aims therefore to describe how a different mix of technologies that have been proven to be feasible can be integrated to provide new functionalities, and to discuss the impact of current artificial intelligence and computer vision techniques for the future of multimedia content personalization.

FUTURE RESEARCH DIRECTIONS: EXPLOITING TAGGING MECHANISMS

With respect to tagging objects on a video stream, existing coding formats such as releases 10 or later of MPEG-4, offer significant flexibility on object manipulation and tagging, allowing very efficient implementation approaches for the solutions described on this chapter. With future versions of MPEG-4, the frames on a video stream will come decomposed as tagged objects. Hence, implementation of the mechanisms described will become easier, since objects will not have to be detected and tracked over the video stream. Indeed, these coding mechanisms have very nice

Figure 8. Image features (such as frequency components, as determined using wavelets) are used by a statistical learning algorithm (e.g. mixtures of Gaussians), which, based on training data, will learn (off-line) probable places where to place a watch and a ring (and their corresponding size and projection parameters on the image). One expects based on training data to find rings on fingers and watches on the hand, with an appropriate size. We have already addressed the implementation of these algorithms for determining probable places of occurrence. The personal metadata channel carries information concerning time instants, object model—3D structure and image rendering of the ring, code indicating the method to be employed to insert the extra content (through detection of natural occurring places). The location and size of that content placement is determined locally according to natural places of occurrence of objects on the image.

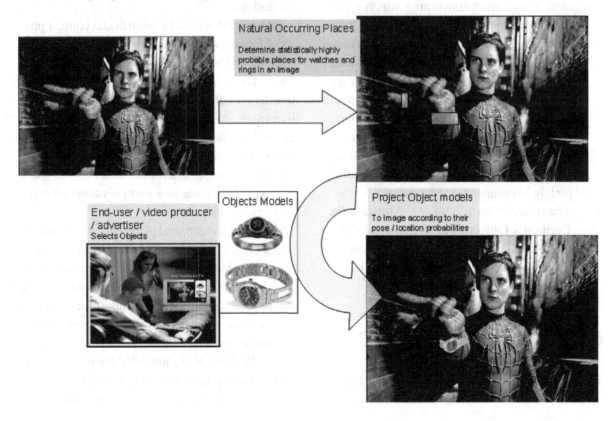

characteristics to be exploited by the framework described in this chapter, namely:

- Video streams' coded regions are objects, instead of a number of pixel elements on squared image regions.
- Objects on a video stream can be tagged.

Therefore, the usage of these tagging mechanisms for multimedia content adaptation and distribution can be exploited in order to distribute the metadata and extra multimedia content, together with the normal video channel, as well as to facilitate the replacement or transformation of the multimedia content:

- **Addition/Removal of Objects to Multimedia Content:** The coding of objects on a video stream, allows a content provider to add new metadata associated to objects into the original video stream. Such information can be efficiently exploited to store information concerning which types of objects can be interactively added to a video stream, as well as how and where they should be inserted. For instance, the user selects the feature to be replaced, which are already isolated on the image (e.g. instead of detecting user-selected face contours, the faces could already come tagged on the MPEG-4 video stream). With new coding mechanisms, the solution might run completely in cheaper home hardware.
- **Transformation of Multimedia Content's Objects:** An alternative for the usage of tagging mechanism is to represent objects as classes of objects that can be replaced by equivalent objects.

CONCLUSION

By developing technologies for adding intelligence to the way multimedia content is created, encoded, and manipulated, a large plethora of networking applications are possible, as well as enriching the user experience while consuming multimedia content. Indeed, multimedia consumers are moving from the traditional TV passive consumers, to the modern active consumers of WebTV, IPTV, social applications, network games, among others. These viewers require increasingly new levels of interactivity and content personalization, so that they also decide on which content they receive, and how this content is delivered. The Internet model is also driving end-users to become prosumers (as they are already on the Internet, using and participating in such applications as social networks, blogs, communication chats, etc).

In addition, service providers such as advertisers need as well new solutions that allow then to have better granularity on the audience targeted by their advertisements. In order to increase revenues, they also need innovative mechanisms to insert advertisements into multimedia content, which make their products more appealing to the end-user.

We have discussed a multimedia content protocol for supporting interactive and personalized content, and a theoretical framework for the creation of such interactive content, and its distribution. Although the description was presented to live video streams, a similar reasoning could be applied in other scenarios (PVR – Personal Video Recording, video on demand, among others). This is especially important for prosumers, since users will typically create new extra content by acting on recorded videos.

In order to create and place additional content into a video stream, we have discussed the usage of machine learning techniques, such as Mixtures of Gaussians. Such approach, although not a technique of multimedia data mining *per se*, aims at mining large corpora of images for estimating location patterns on images associated to a specific object. We have also shown that the placement of extra content requires the application of different techniques commonly used in computer vision.

The combined usage of such image processing and machine learning techniques allows the creation of additional content, which can be personalized according to many different criteria. Interactivity is also achieved, since it is possible to define near real-time information such as users' choices in order to decide on the extra content to be inserted into a multimedia channel (e.g. broadcast) to create a personal channel. Prosumer models, in which the user is simultaneously consumer and producer of multimedia content, become also possible, either by having the user to produce extra content, or by having her to transform existent content.

Of course, the deployment of such solutions is strongly conditioned to efficient network distribution mechanisms, which minimize the overhead of traffic introduced on the network. With this goal in mind, we introduced a multimedia content distribution framework enabling a large range of networking applications targeting various market needs such as multimedia content personalization and interactivity.

REFERENCES

Arsenio, A. (2004). *Cognitive-developmental learning for a humanoid robot: A caregiver's gift.* (PhD Thesis). MIT. Cambridge, MA.

Arsenio, A. (2011). *Method and apparatus for adaptation of a multimedia content.* International Patent Number: EP2341680 A1. Nokia Siemens Networks GMBH & CO.

Aryananda, L. (2002). Recognizing and remembering individuals: Online and unsupervised face recognition for humanoid robot. In *Proceedings of the International IEEE/RSJ Conference on Intelligent Robots and Systems.* IEEE.

Belhumeur, P., Hespanha, J., & Kriegman, D. (1997). Eigenfaces vs. fisherfaces: Recognition using class specific linear projection. *IEEE Transactions on Pattern Analysis and Machine Intelligence, 19*(7), 711–720. doi:10.1109/34.598228

Bobick, A., & Pinhanez, C. (1995). Using approximate models as source of contextual information for vision processing. In *Proceedings of the ICCV 1995 Workshop on Context-Based Vision*, (pp. 13-21). Cambridge, MA: ICCV.

Breazeal, C., & Aryananda, L. (2002). Recognition of affective communicative intent in robot-directed speech. *Autonomous Robots, 12*(1), 83–104. doi:10.1023/A:1013215010749

Chellappa, R., Wilson, C., & Sirohey, S. (1995). Human and machine recognition of faces: A survey. *Proceedings of the IEEE, 83*, 705–741. doi:10.1109/5.381842

Cruz, R., Nunes, M., Patrikakis, C., & Papaoulakis, N. (2010). SARACEN: A platform for adaptive, socially aware multimedia distribution over P2P networks. In *Proceedings of the 4th IEEE Workshop on Enabling the Future Service-Oriented Internet: Towards Socially-Aware Networks, GLOBECOM 2010.* IEEE Press.

Curwen, R., & Blake, A. (1992). Dynamic contours: Real-time active splines. In *Active Vision.* Cambridge, MA: MIT Press.

Damasio, A. (1994). *Descartes error: Emotion, reason, and the brain.* London, UK: Bard.

Dogan, S. (2002). Personalised multimedia services for real-time video over 3G mobile networks. In *Proceedings of the Third International Conference on 3G Mobile Communication Technologies.* IEEE.

Faugeras, O. (1993). *Three - dimensional computer vision: A geometric viewpoint.* Cambridge, MA: MIT Press.

Ferman, A., Errico, J., van Beek, P., & Sezan, M. (2002). Content-based filtering and personalization using structured metadata. In *Proceedings of the 2nd ACM/IEEE-CS Joint Conference on Digital Libraries (JCDL 2002)*, (pp. 393-393). New York, NY: ACM Press.

Ferrell, C. (1998). Emotional robots and learning during social exchanges. In *Proceedings of Agents in Interaction- Acquiring Competence through Imitation: Papers from a Workshop at the Second International Conference on Autonomous Agents (Autonomous Agents 1998)*. Autonomous Agents.

Freeman, W. T., & Adelson, E. H. (1991). The design and use of steerable filters. *IEEE Transactions on Pattern Analysis and Machine Intelligence*, *13*(9), 891–906. doi:10.1109/34.93808

Fu, D., Hammond, K., & Swain, M. (1994). Vision and navigation in man-made environments: Looking for syrup in all the right places. In *Proceedings of CVPR Workshop on Visual Behaviors*, (pp. 20-26). Seattle, WA: IEEE Press.

Fuente, Y., et al. (2011). iDASH: Improved dynamic adaptive streaming over HTTP using scalable video coding. In *Proceedings of the Second Annual ACM Conference on Multimedia Systems (MMSys 2011)*, (pp. 257-264). New York, NY: ACM Press.

Fukunaga, K. (1990). Introduction to statistical pattern recognition. In *Computer Science and Scientific Computing*. New York, NY: Academic Press.

Gershenfeld, N. (1999). *The nature of mathematical modeling*. Cambridge, UK: Cambridge University Press.

Horn, B. K. (1986). *Robot vision*. Cambridge, MA: MIT Press.

Kass, M., Witkin, A., & Terzopoulos, D. (1988). SNAKES: Active contour models. *International Journal of Computer Vision*, *1*(4), 321–331. doi:10.1007/BF00133570

Marquet, A., Monteiro, J., Martins, N., & Nunes, M. (2010). Quality of experience vs. QoS in video transmission. In *Quality of Service Architectures for Wireless Networks* (pp. 352–376). Hershey, PA: IGI Global. doi:10.4018/978-1-61520-680-3.ch016

Murase, H., & Nayar, S. (1995). Visual learning and recognition of 3d objects from appearance. *International Journal of Computer Vision*, *14*(1), 5–24. doi:10.1007/BF01421486

Oliva, A., & Torralba, A. (2001). Modeling the shape of the scene: a holistic representation of the spatial envelope. *International Journal of Computer Vision*, *42*(3), 145–175. doi:10.1023/A:1011139631724

Papageorgiou, C., & Poggio, T. (2000). A trainable system for object detection. *International Journal of Computer Vision*, *38*(1), 15–33. doi:10.1023/A:1008162616689

Pasi, G., & Villa, R. (2005) Personalized news content programming (PENG): A system architecture. In *Proceedings of the Sixteenth International Workshop on Database and Expert Systems Applications*, (pp. 1008-1012). IEEE.

Ramanathan, S., & Venkat Rangan, P. (1994). Architectures for personalized multimedia. *IEEE MultiMedia*, *1*(1), 37–46. doi:10.1109/93.295266

Rao, R., & Ballard, D. (1997). Dynamic model of visual recognition predicts neural response properties in the visual cortex. *Neural Computation*, *9*(4), 721–763. doi:10.1162/neco.1997.9.4.721

Schiele, B., & Crowley, J. (2000). Recognition without correspondence using multidimensional receptive field histograms. *International Journal of Computer Vision*, *36*(1), 31–50. doi:10.1023/A:1008120406972

Sebe, N., & Tian, Q. (2007). Personalized multimedia retrieval: the new trend? In *Proceedings of the International Workshop on Multimedia Information Retrieval (MIR 2007)*, (pp. 299-306). New York, NY: ACM Press.

Shi, J., & Tomasi, C. (1994). Good features to track. In *Proceedings of the IEEE Computer Society Conference on Computer Vision and Pattern Recognition*, (pp. 593-600). IEEE Press.

Sirovich, L., & Kirby, M. (1987). Low dimensional procedure for the characterization of human faces. *Journal of the Optical Society of America, 4*(3), 519–524. doi:10.1364/JOSAA.4.000519

Strang, G., & Nguyen, T. (1996). *Wavelets and filter banks*. Cambridge, MA: Wellesley-Cambridge Press.

Swain, M., & Ballard, D. (1991). Colour indexing. *International Journal of Computer Vision, 7*(1), 11–32. doi:10.1007/BF00130487

Terzopoulos, D., & Szeliski, R. (1992). Tracking with Kalman snakes. In *Active Vision*. Cambridge, MA: MIT Press.

Torralba, A. (2003). Contextual priming for object detection. *International Journal of Computer Vision, 53*(2), 169–191. doi:10.1023/A:1023052124951

Torralba, A., Murphy, K. P., Freeman, W. T., & Rubin, M. A. (2003). Context-based vision system for place and object recognition. In *Proceedings of the IEEE International Conference on Computer Vision (ICCV)*, (Vol. 1, pp. 273-280). Nice, France: IEEE Press.

Turk, M., & Pentland, A. (1991). Eigenfaces for recognition. *Journal of Cognitive Neuroscience, 3*(1). doi:10.1162/jocn.1991.3.1.71

Veiga, J. (2009). *Virtual channels - Automatic TV/audio channel switching upon reception of information from special control channels*. European Patent Application EP2034639. Nokia Siemens Networks GMBH & CO.

Velasquez, J. (1997). Modeling emotions and other motivations in synthetic agents. In *Proceedings of the 1997 National Conference on Artificial Intelligence*, (pp. 10-15). IEEE.

Viola, P., & Jones, M. (2001). *Robust real-time object detection. Technical Report*. Cambridge, MA: COMPAQ Cambridge Research Laboratory.

Wei, Y., Bhandarkar, S., & Li, K. (2009). Client-centered multimedia content adaptation. *ACM Transactions on Multimedia Computing, Communications, and Applications, 5*(3).

Wolfson, H., & Rigoutsos, I. (1997). Geometric hashing: An overview. *IEEE Computational Science & Engineering, 4*(4), 10–21. doi:10.1109/99.641604

Yucel, Z., & Ali Salah, A. (2009). Head pose and gaze direction estimation for joint attention modeling in embodied agents. In *Proceedings of the Annual Meeting of the Cognitive Science Society, COGSCI 2009*. COGSCI.

Zhang, H., & Malik, J. (2003). Learning a discriminative classifier using shape context distance. In *Proceedings of the International Conference on Computer Vision and Pattern Recognition*, (Vol. 1, pp. 242-247). IEEE.

KEY TERMS AND DEFINITIONS

Interactive Multimedia Content Distribution: Methods and Infrastructure for the delivery of multimedia content allowing bidirectional communications between the content source and the viewer, so that the multimedia content consumed by the viewer depends on its own interactive choices while consuming the content.

Machine Learning: Methods and algorithms for mapping the output of a system as a function of its input and/or current state. While widely used in artificial intelligence, plenty of algorithms have been proposed, ranging from supervised or unsupervised learning algorithms or reinforcing learning, among other schemes.

Multimedia Content Adaptation: Modification of multimedia content in order to meet different criteria, such as the viewers' personal tastes, a mobile device's resolution capability, a geographic target or quality of service requirements.

Networking Applications: Applications that are distributed over one or more networks, making use of communication channels and several distributed machines.

Personal Metadata Channel: It contains extra content and metadata associated to an original video stream. Personal in the sense of personalized data, i.e. data of the normal channel is modified and transmitted separately, according to "personal" profiles of an end-user, a set of end-users, a community, region, or country. It is not a channel per se, but a channel that is created by a 3rd party service provider, or by the IPTV service operator, containing either static images of content or short streams of content, or even text, together with associated metadata.

Personalized Content: Customization of content targeting a specific user or community of users.

Virtual Channel: A channel that is created from the aggregation of pieces from other channels, not being a physical transmission channel of multimedia. Hence, this channel contains the identification of a video segment's physical channel, and associated time interval timestamps, used to build the final virtual channel from many such segments from different physical channels.

Chapter 9
A Semantic Generic Profile for Multimedia Document Adaptation

Cédric Dromzée
University of Pau, France

Sébastien Laborie
University of Pau, France

Philippe Roose
University of Pau, France

ABSTRACT

Currently, multimedia documents can be accessed at anytime and anywhere with a wide variety of mobile devices, such as laptops, smartphones, and tablets. Obviously, platform heterogeneity, users' preferences, and context variations require document adaptation according to execution constraints. For example, audio contents may not be played while a user is participating in a meeting. Current context modeling languages do not handle such real life user constraints. These languages generally list multiple information values that are interpreted by adaptation processes in order to deduce implicitly such high-level constraints. In this chapter, the authors overcome this limitation by proposing a novel context modeling approach based on services, where context information is linked according to explicit high-level constraints. In order to validate the proposal, the authors have used semantic Web technologies by specifying RDF profiles and experimenting on their usage on several platforms.

DOI: 10.4018/978-1-4666-2833-5.ch009

INTRODUCTION

A huge amount of multimedia documents can be created and accessed by users. These documents may be composed of different types of contents, such as videos, audios, texts, and images (Jedidi, et al., 2005). For instance, good examples of multimedia documents are Web pages or SMIL presentations (Bulterman, et al., 2008). In those documents, multimedia contents are synchronized and organized according to the graphical layout of the presentations (Roisin, 1998). Moreover, users may be able to interact with presentations by selecting particular elements (e.g., a click on a picture plays a video). Besides, many mobile devices (e.g., laptops, smartphones, and tablets) are able to display multimedia documents. This universal access allows users to consult documents anytime and anywhere (W3C Ubiquitous Web Domain). However, such devices have heterogeneous capabilities and characteristics in terms of hardware (e.g., screen size, battery) and software (e.g., players, codecs) (W3C Device Independence Working Group). Moreover, user's preferences or handicaps may prevent from playing specific multimedia contents (W3C Web Accessibility Initiative). For instance, a user may avoid reading texts written in French and/or avoid playing audio contents, while he is participating at a meeting. All these restrictions introduce constraints that have to be specified with a profile. In a profile, various categories of information have to be managed:

- Device characteristics (hardware and software).
- Context information related to interactions between the user and the device, such as the preferred languages, the bandwidth, and the surrounding devices.
- Document structure like the types of contents that should be played or the preferred presentation organization (e.g., layout, multimedia contents synchronization, hypermedia links).

Consequently, if a multimedia document does not comply with some constraints that are specified inside a target profile, the document may not be correctly executed on the target device. Therefore, in order to display documents on any devices, these ones have to be adapted, i.e., transformed in order to comply with the target profiles.

Since the last decade, a fair amount of research has been conducted on multimedia document adaptation (Adzic, et al., 2011; Laborie, et al., 2011; Lemlouma & Layad'da, 2002; Asadi & Dufourd, 2005; Jannach & Leopold, 2007; Ahmadi & Kong, 2008). Considering some target profiles, these approaches combined multiple operators: transcoding (e.g., AVI to MPEG), transmoding (e.g., text to speech), and transformation (e.g., text summarization). Certainly, each profile expressiveness is exploited by these approaches in order to determine a combination of these operators, such as in Lemlouma and Layaïda (2001), or to optimize their deployments (e.g., for saving battery energy), such as in Laplace et al. (2008) and Girma et al. (2006). However, each proposal exploits specific profile format, which usually contains a list of multiple descriptive information values, such as the screen size, the user languages, and the battery power. Consequently, an adaptation process has to interpret such profile values and to deduce implicitly some constraints. For instance, if a battery power is lower than 10%, an adaptation process may avoid playing hi-quality videos, while another one may provide low-quality videos. Obviously, each adaptation mechanism may deduce different constraints that in many situations might be wrong, thus providing to users incorrect adapted documents. Furthermore, current context modeling languages, e.g., Bolchini et al. (2007), Forough and Reza (2012), do not consider expressing such high-level constraints, while they might be very useful to guide the adaptation process. In this chapter, we overcome this limitation by defining a new profile description model where:

1. Profile information are organized into facets (e.g., device characteristics, context information, and document structure) and composed of services that either provide data or require modifications, and
2. Some profile information are linked by explicit high-level constraints.

Thanks to this proposal, our profiles may migrate from one platform to another, while preserving the specified constraints, thus ensuring profile interoperability between platforms. In order to validate our proposal, we encode profiles in RDF/XML (Klyne & Carroll, 2004) and evaluate several query executions on heterogeneous platforms. Experimental results confirm that adaptation processes can efficiently query our profile descriptions both on a server and on a mobile device.

The chapter is organized as follows. Firstly, the related work section gives an overview of the current existing profiles and context modeling approaches. Then, we detail our service-based profile specification and we illustrate it through real-life examples. Thereafter, we propose a generic description model, named Generic Semantic Profile (SGP), and some use-case instantiations encoded in RDF/XML. Some queries that may be used by adaptation processes are presented in order to show how they may exploit our profile descriptions. Experiment results have been also conducted on several platforms to measure query execution time. Finally, in the last section, we conclude and present some future work.

RELATED WORK

Since the last decade, a fair amount of research (Bolchini, et al., 2007; Forough & Reza, 2012) has been proposed in order to model devices characteristics and users contexts that are further exploited by multimedia document adaptation processes. We have noticed that some of these approaches provide exclusively a descriptive view of context information (e.g., CC/PP, UAProf, WURFL), while others propose enhancements with some constraints expressions (e.g., CSCP, Context-ADDICT). In this section, we present an overview of these approaches.

CC/PP

CC/PP (Composite Capability/Preference Profiles) (Klyne, et al., 2004) is a W3C recommendation for specifying device capabilities and user preferences. This profile language is based on RDF and was maintained by the W3C Ubiquitous Web Applications Working Group (UWAWG) (W3C Ubiquitous Web Applications Working Group). The profile structure is very descriptive since it lists sets of values, which correspond to the screen size, the browser version, the memory capacity, etc. Indulska et al. (2003) have expanded the vocabulary of CC/PP to describe the location, the network characteristics, and application dependencies. However, the CC/PP structure lacks functionality, for instance it limits complex structure description by forcing a strict hierarchy with two levels. Furthermore, it does not consider the description of relationships and constraints between some context information. Finally, it is necessary to extend the vocabulary used in CC/PP to include new elements corresponding to hardware profile (Girma, et al., 2005).

UAProf

UAProf (User Agent Profile) (Forum, 2001) is based on RDF and is a specialization of CC/PP for mobile phones. More precisely, its vocabulary elements use the same basic format as the one used in CC/PP for describing capabilities and preferences for wireless devices. Thus, it describes specific items, such as the screen size, the supported media formats, the input and output capabilities, etc. UAProf is a standard adopted by a wide variety of mobile phones and provides detailed lists of information about the terminal characteristics. However, this standard is limited to the description of wireless telephony equipments. Hence, it does not allow a user to express his/her requirements, such as avoiding playing videos while he/she is participating at a meeting.

CSCP

CSCP (Comprehensive Structured Context Profiles) (Buchholz, et al., 2004) uses RDF and is also based on CC/PP. In contrast to CC/PP, CSCP has a multi-level structure and models alternative values according to predefined situations. Even if CSCP provides a description of the context, which is not limited to two hierarchical levels, this proposal does not allow the specification of complex user constraints (e.g., avoiding playing videos while a user is participating at a meeting) (Indulska, et al., 2003). Moreover, Tarak et al. (2007) stated that this proposal was developed as a proprietary model for specific domains.

Context-ADDICT

Context-aware Data Integration Customization and Tailoring (Context-ADDICT) proposes the Context Dimension Tree (Bolchini, et al., 2006). The context can be represented with a hierarchical structure composed of a root and some level nodes. The authors propose constraints and relationships among values. In Context-ADDICT, the data sources are generally dynamic, transient and heterogeneous in both their data models (e.g., relational, XML, RDF) and schemas (Bolchini, et al., 2011). The Context-ADDICT approach lacks some relevant features for the data-tailoring problem, such as context history, context quality monitoring, context reasoning, and ambiguity and incompleteness management (Bolchini, et al., 2007). Moreover, this model strongly depends on the application used and does not permit portability between platforms.

Generic Profiles for the Personalization of Information Access

Chevalier et al. (2006) propose a generic UML profile for describing the structure and semantics of any type of user profile information. This contribution is used to describe the semantic links between elements and incorporate weighting on elements. The semantic graph is described thanks to a logic-oriented approach (Jouanot, et al., 2003) with RDF, RDFS, and OWL. However, this model does not express actions that an adaptation process may exploit to transform the multimedia document, such as increase audio volume and avoid playing hi-quality videos.

WURFL

WURFL (Wireless Universal Resource File) (Passani, 2007) is an XML description of mobile devices resources. WURFL contains information about the capabilities and the functionalities of mobile devices with more than 500 "capabilities" for each device (divided into 30 groups). This project is intended to adapt Web pages on mobile devices (Veaceslav, 2011). Unfortunately, users cannot specify explicit constraints, such as decrease screen luminosity, if the battery power level is below 10%.

Alternative Approaches

There are alternative approaches to design profiles with markup languages, such as PPDL (Pervasive Profile Description Language) (Ferscha, et al., 2006) or CCML (Centaurus Capability Markup Language) (Kagal, et al., 2001, 1995). However, these frameworks are domain specific and limited to some aspects of the context (e.g., location, environment). Other projects have also been proposed for user and context modeling, such as E^2R (http://www.ist-spice.org) and MAGNET Beyond (https://ict-e3.eu3). However, they do not consider the specific aspects of multimedia document adaptation, and especially the document structure dimension that have to be described inside profile.

In the next section, we propose a new context-modeling framework suitable for adapting multimedia documents. Through real-life examples, we present step by step how we have build such profile containing rich constraints.

FACETS AND CONSTRAINTS

In order to display a multimedia document on multiple devices, several constraints described in a profile have to be satisfied. A profile is usually composed of some characteristics and some constraints that will be used by an adaptation process in order to compute an adapted document complying with all the specified restrictions. In this section, we present our *Semantic Generic Profile* (SGP) that allows rich constraints specifications and permits portability and generality.

Naturally, we illustrate our proposal with some real-life oriented examples.

Facet

A profile should provide information on some devices capabilities, user context and documents characteristics that the target device is able to take into account. Currently, profile descriptions contain descriptive values, such as the device screen size, the user preferred languages, the device model, etc. However, if this type of profile migrates on different platforms, many profiles characteristics have to be reconfigured. For instance, an easy evolving characteristic is a screen resolution or some available codecs. Therefore, to ensure profile portability our profile structure is composed of service descriptions. A profile is most of the time designed as a hierarchical description with some data outputs. Figure 1 illustrates an example of a hierarchy, which is composed of services descriptions concerning the context of a specific user. In Figure 1, each service s_i of

Figure 1. A facet example describing a user context

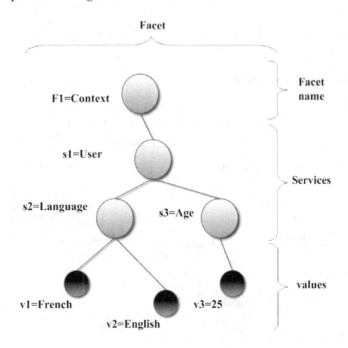

the hierarchy is identified by a resource name, e.g., Language, Age, etc. Each potential value v_j corresponds to a value returned by the related service. For instance, the *Language* service may return two potential values: *French* and *English*. Of course, this means that the user may understand both languages, i.e., French and English. The whole hierarchy of service descriptions and data values is called a *facet*. The root node, i.e., the service name on the top of the hierarchy, will be the name of the facet. In Figure 1, the name of the facet is "Context."

A service may provide some data, i.e., it will be in charge of giving information about a current situation. A service may also require some parameters in order to update its status. In Figure 2, on one hand, the *Battery service* provides some data, here 15, meaning 15% of the remaining battery level. On the other hand, the *Luminosity service* may require a parameter value in order to update the screen luminosity, here 70 means that the luminosity intensity has to be set to 70%.

Obviously, many facets (i.e., multi-facets profile) can be considered, when a user want to play a multimedia document on several devices.

Multi-Facets

A profile has to describe different categories of information:

1. Device characteristics (hardware and software);
2. Context information related to the user and the device; and
3. Document structures that can be executed by the target device.

Consequently, profiles may be composed of several facets. Figure 3 illustrates the description of three facets that corresponds to the three previously mentioned categories.

In this figure, the contextual facet describes some user's information, such as its neighbors, its location and its preferred languages. The document facet specifies the types of multimedia contents that the device is able to execute and particularly video decoding. The hardware facet collects physical and technical characteristics of the device (e.g., RAM availability, battery level, etc.).

Figure 2. A service-based profile example

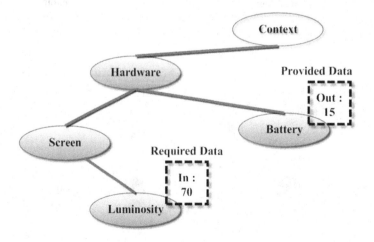

Figure 3. A multi-facets example

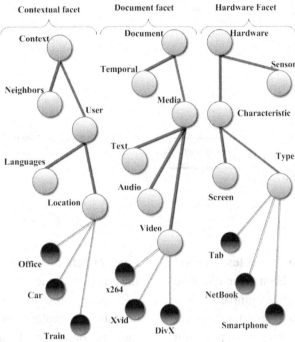

Constraints

At this point, profile information is hierarchically structured and thanks to the services one may migrate the profile from one platform to another. However, it provides only raw data that have to be analyzed further by adaptation processes in order to deduce implicit constraints. For example, let us suppose that a profile specifies a specific smartphone screen size, the adaptation processes have to deduce if they have to transform the document or not. Moreover, based on the user language information, adaptation processes also have to deduce if they have to translate some contents, like texts, sounds, etc. However, deducing implicit constraints from profiles may yield to adapted multimedia documents that do not comply with the real user needs. For example, even if a user has a small screen device, it is possible that the user would like pictures in native resolution display and not screen resolution display. Furthermore, it is usually impossible to deduce some implicit

constraints. For example, adaptation processes cannot deduce what to do if the user is in a specific situation, like in a train. Does the user want to allow the execution of videos or not? To solve such situations, we propose to design explicit constraints. Using facet descriptions, these constraints will associate different categories of profile information.

Explicit Constraints

In order to define explicit constraints between facets, we need to specify conditions and actions. For example, if a user is located in his car (*i.e., the condition*), he may not want to see videos (*i.e., the action*). Hence, we propose to design explicit constraints by associating several facet services to some conditions that have to be satisfied. Obviously, if all conditions are satisfied, an action will be triggered on a facet service. Figure 4 illustrates the specification of an explicit constraint by using the terms that have been defined above.

Figure 4. General scheme of an explicit constraint

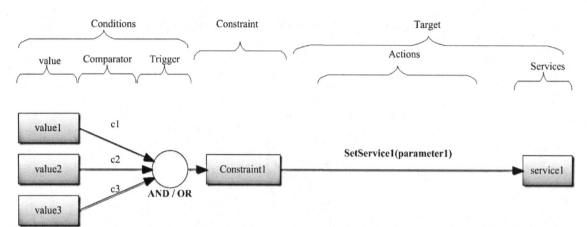

In this figure, an explicit constraint tests several potential values of a service. Each value v_i will be checked with a comparator c_i. A comparator is a binary relation, such as equal, less than, greater than, etc. The comparator compares a potential value with the current situation value provided by the related service. If one (i.e., OR) or all (i.e., AND) conditions are satisfied, it triggers an action which is associated to a service. Of course, several types of actions may be specified depending on the targeted service.

Figure 5 presents a multi-facets profile, which contains an explicit constraint. The contextual facet is composed of information related to the user's location. Let us suppose that a user is located in a car, this information will be provided by the method *getUserLocation* associated to the *UserLocation* service. The hardware facet defines

Figure 5. An explicit constraint example

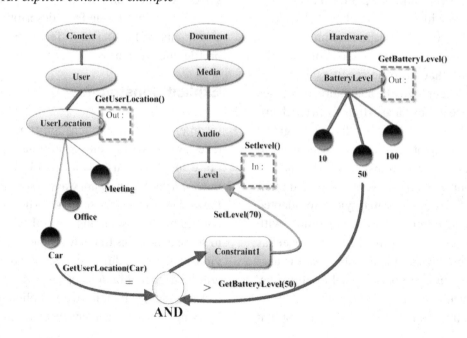

some battery power levels; in the example, the *BatteryLevel* service may provide a remaining battery level of 50%. The document facet specifies that different sound levels may be set through the *setLevel* method associated to the *Audio* service. As shown in Figure 5, an explicit constraint has been defined (see the arrows in the figure). It specifies that if a user is located in a car and if the battery power level is greater than 50%, the audio level of all audio contents of a document has to be set to 70%.

Naturally, a profile may not contain an explicit constraint. Hence, the profile will only describe information about the three facets: device characteristics, context information, and document structure. Each service may be invoked independently in order to collect the current data. If no constraints are specified or if none of the conditions are satisfied, any action will be specified for some adaptation processes. In this case, it is up to each process to make implicitly the adaptation. Formally, an explicit constraint is: $C_e = <S_c, S_a>$ with S_c a set of conditions and S_a a set of actions. The set of conditions $S_c = \{C_1, ..., C_n\}$ is composed of some conditions $C_i = <v_i, c_i>$ with v_i a potential value of a service and c_i a comparator. A condition C_i is satisfied if the value of v_i complies with c_i and the value provided by the related service. An explicit constraint triggers all actions in S_a if all C_i in S_c are satisfied.

Advanced Explicit Constraints

As illustrated in the previous section, a set of conditions may trigger different actions on several services. For instance, Figure 6 shows that actions may be related to several services. Moreover, different actions may be related to the same service (e.g., modify the style of texts and translate texts in French).

Figure 7 illustrates such an example of an explicit constraint triggering several actions. In this example, if the available CPU power is less than 50%, audios and texts are played, while videos are forbidden/removed. As you may see, for a given condition, we may trigger several actions related to different facet services.

Priorities Inside Explicit Constraints

Priorities between actions may be defined in explicit constraints by specifying a weight value. This is useful when a constraint triggers alternative actions. Indeed, it indicates a preference between concurrent actions of explicit constraints (see Figure 8). Each weight is a value between 0 and 1. The more the weight value is close to 1, the more the action of the corresponding explicit constraint is important.

Figure 9 shows an example that illustrates potential weights on actions. Especially, it speci-

Figure 6. An explicit constraint with multiple actions

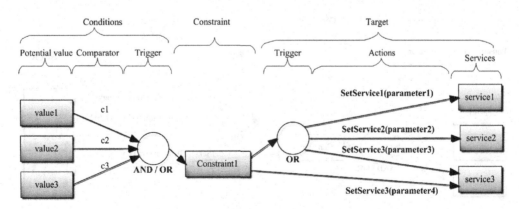

Figure 7. Explicit constraint triggering several actions

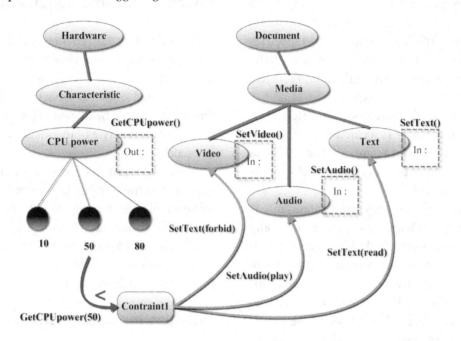

fies the preferred understanding level of a language according to the type of a media. For example, in Figure 9 John prefers reading texts in French than in English. Of course, these weights cannot be set directly by a user. John may specify in advance that he prefers texts written in French than in English, and these weights are then automatically computed during the profile creation.

In this section, we have shown that our profile descriptions may hold rich and complex constraints

that will be further exploited by adaptation processes. In the remainder, profile may hold several facets with a wide variety of services and several explicit constraints.

Profile

A profile is a set of facets that can be enriched by constraints. In this chapter, we promote the use of three facets: device characteristics, context

Figure 8. Priorities between actions of explicit constraints

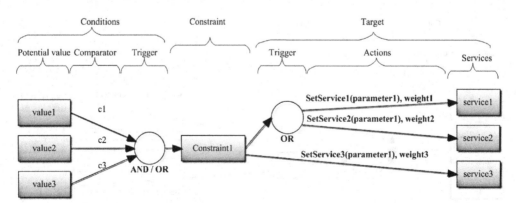

Figure 9. Priorities between actions

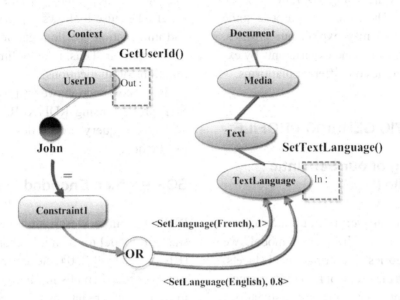

information and document structures. As we have shown previously, complex high-level constraints may be specified. For instance, we may compose several conditions with conjunction and disjunction operators, and we may trigger several actions with priorities. Moreover, actions parameters may be fixed values or may refer to values provided by the profile services. For example, Figure 10 specifies an explicit constraint that set each document image resolution to the same resolution of the terminal screen if its resolution is between 800x600 and 1480x1200.

In the next section, we illustrate a generic model of our proposal in UML and we present an

Figure 10. Multiple conditions constraint example

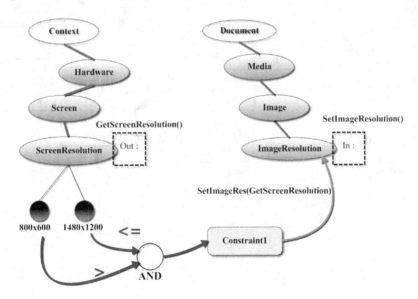

encoding of our profiles using RDF/XML (Klyne & Carroll, 2004). Thereafter, we propose some query examples that may exploit our profiles. Finally, we have made some experiments by executing several queries on different platforms.

THE SEMANTIC GENERIC PROFILE

UML Modeling of our Semantic Generic Profile (SGP)

Figure 11 presents an overview of our Semantic Generic Profile model. In such a model, we identify two categories of information: (1) facets, which consist of a hierarchy of reusable services; and (2) constraints that represent associations between the facets and services. Of course, one

may specify new kind of facets. This model is generic because it allows expressing constraints and information profile regardless of the application using it. Thus, it is not limited to specific functions or applications.

In the next section, we propose to encode our SGP profiles using RDF/XML. Indeed, it will allow us to query profile descriptions with semantic queries.

SGP Profiles Encoded in RDF/XML

RDF (Resource Description Framework) is a standard model for data interchange on the Web (Klyne & Carroll, 2004). Several reasons motivate us to use this formalism: Firstly, RDF allows us to perform aggregations of descriptions, which can be useful if several applications describe a

Figure 11. Our semantic generic profile (SGP) model

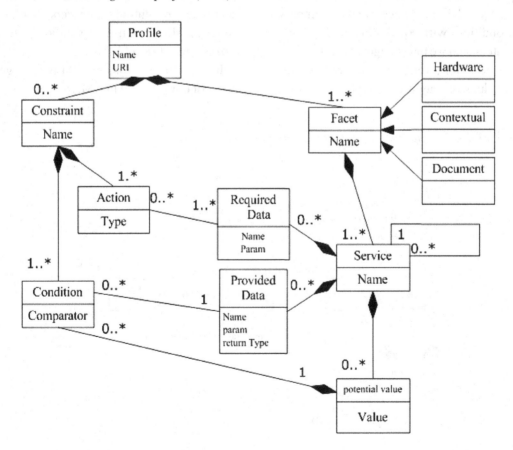

profile. Secondly, RDF can handle semantic concepts described in ontologies, thus enhancing the SGP semantics. For instance, semantics allow us to state that "Sensor" is equivalent to "Captor." Hence, a semantic query on "Captor" will also refer to "Sensor." Thirdly, RDF does not force us to express hierarchies of data as defined in other languages (it has a graph-based structure thanks to triples). Finally, other languages and proposals whose objectives are to describe profiles are based on this formalism (e.g., CC/PP, UAProf and CSCP).

Figure 12 is an example of a SGP profile encoded in RDF/XML. It is composed of a RDF header, some services descriptions (*i.e., Screen, ScreenResolution, ScreenLuminosity, and Battery*) of the hardware facet and an explicit constraint. The *ScreenResolution* service is composed of one input (i.e., *SetScreenResolution*) and one output (*GetScreenResolution*) functions. In this example,

the explicit constraint is composed of the following condition: if the battery power level is less than 0.1 (for 10%). If this condition is satisfied, it will trigger the following action: set the screen luminosity level to 0.3 (for 30%).

In the next section, we briefly introduce some query examples that may be used to retrieve some information contained in our profile SGP structure.

SPARQL Queries

SPARQL is a RDF query oriented language able to retrieve and manipulate data stored in RDF descriptions. It was made a standard by the RDF Data Access Working Group (DAWG) of the W3C, and this query language is considered as one of the key technologies of the Semantic Web. In the following, we present several SPARQL queries that may be specified in order to retrieve important information contained in a SGP profile. Note that

Figure 12. RDF/XML SGP

```xml
<?xml version="1.0"?>
<rdf:RDF xmlns:sgp="http://SGP#" xmlns:rdf="http://www.w3.org/1999/02/22-rdf-syntax-ns#">
    <sgp:Profile rdf:about="http://SGP#Profil_1">
        <sgp:name>John's profile</sgp:name>
        <sgp:describes>
            <sgp:Facet rdf:about="http://SGP#Hardware">
                <sgp:contains>
                    <sgp:Service rdf:about="http://SGP#Screen">
                        <sgp:contains>
                            <sgp:Service rdf:about="http://SGP#ScreenResolution">
                                <sgp:in>
                                    <sgp:InputFunction rdf:about="http://SGP#SetScreenResolution">
                                        <sgp:param rdf:datatype="http://www.w3.org/2001/XMLSchema#string"/>
                                    </sgp:InputFunction>
                                </sgp:in>
                                <sgp:out>
                                    <sgp:OutputFunction rdf:about="http://SGP#GetScreenResolution">
                                        <sgp:return rdf:datatype="http://www.w3.org/2001/XMLSchema#string"/>
                                    </sgp:OutputFunction>
                                </sgp:out>
                            </sgp:Service>
                        </sgp:contains>
                        <sgp:contains>
                            <sgp:Service rdf:about="http://SGP#ScreenLuminosity">
                                <sgp:in>
                                    <sgp:InputFunction rdf:about="http://SGP#SetScreenLuminosity">
                                        <sgp:param rdf:datatype="http://www.w3.org/2001/XMLSchema#string"/>
                                    </sgp:InputFunction>
                                </sgp:in>
                            </sgp:Service>
                        </sgp:contains>
                    </sgp:Service>
                </sgp:contains>
                <sgp:contains>
                    <sgp:Service rdf:about="http://SGP#Battery">
                        <sgp:out>
                            <sgp:OutputFunction rdf:about="http://SGP#GetBatteryLevel">
                                <sgp:return rdf:datatype="http://www.w3.org/2001/XMLSchema#int"/>
                            </sgp:OutputFunction>
                        </sgp:out>
                    </sgp:Service>
                </sgp:contains>
            </sgp:Facet>
        </sgp:describes>
        <sgp:handles>
            <sgp:Constraint rdf:about="http://SGP#C1">
                <sgp:contains>
                    <sgp:Condition>
                        <sgp:on rdf:resource="http://SGP#GetBatteryLevel" />
                        <sgp:comparator>&lt;</sgp:comparator>
                        <sgp:value>0,1</sgp:value>
                    </sgp:Condition>
                </sgp:contains>
                <sgp:trigger>
                    <sgp:Action>
                        <sgp:over rdf:resource="http://SGP#SetScreenLuminosity" />
                        <sgp:param>0,3</sgp:param>
                    </sgp:Action>
                </sgp:trigger>
            </sgp:Constraint>
        </sgp:handles>
    </sgp:Profile>
</rdf:RDF>
```

adaptation processes may use these queries in order to extract explicit constraints. The query in Figure 13 returns a list containing all SGP services.

The query in Figure 14 returns a list containing all provided data.

The query in Figure 15 returns a list of actions, for example *SetImageResolution(400x600)* or *SetScreenLuminosityLevel(0.3)*.

Figure 16 returns the complete hierarchy of services.

We have experiment whether it was possible to execute those queries on mobile devices. In the next section, we present our experimentation based on query execution time.

Experimentation

Using the JENA library (http://jena.apache.org) (available on standard Java and Android), we have tested the query execution performance of several SPARQL queries on some SGP profiles encoded in RDF/XML. We have performed these experiments on two heterogeneous mobile configurations.

Configuration 1: A Laptop running Windows 7 (x64) with 6GB of RAM and i7-2630QM quadruple core processor (2 GHz).

Configuration 2: A Samsung Galaxy Tab running Android 3.2 with 1 GB of RAM and a double core Tegra 2 processor (1 GHz).

Figure 13. SGP services list query

```
PREFIX sgp: <http://SGP#>
PREFIX rdf: <http://www.w3.org/1999/02/22-rdf-syntax-ns#>
SELECT *our SGP.
WHERE {
        ?C rdf:type sgp:Service.
}
```

Figure 14. Provided data query list

```
PREFIX sgp: <http://SGP#>
PREFIX rdf: <http://www.w3.org/1999/02/22-rdf-syntax-ns#>
SELECT *
WHERE {
        ?S rdf:type sgp:Service .
        ?S sgp:out ?F .
        ?F rdf:type sgp:OutputFunction .
}
```

Figure 15. Actions list query

```
PREFIX sgp: <http://SGP#>
PREFIX rdf: <http://www.w3.org/1999/02/22-rdf-syntax-ns#>
SELECT *
WHERE {
        ?S rdf:type sgp:Service .
        ?S sgp:in ?F .
        ?F rdf:type sgp:InputFunction .
}
```

Figure 16. Services hierarchy

```
PREFIX sgp: <http://SGP#>
PREFIX rdf: <http://www.w3.org/1999/02/22-rdf-syntax-ns#>
SELECT *
WHERE {
        ?S1 rdf:type sgp:Service .
        OPTIONAL {
                ?S2 rdf:type sgp:Service .
                ?S1 sgp:contains ?S2 .
        }
}
```

In Figure 17, we compare the execution time of queries on both platforms. We have also performed repeated loops (10, 50, 100, 500, 1000) of query 1 (R1), query 2 (R2) and a five different queries sequence (R1, R2, R3, R4, R5). R1 to R5 is a series of queries such as those previously mentioned. R1 is the most simple query: it returns all triples (i.e., SELECT * WHERE {?x ?y ?z .}).

R2 corresponds to Figure 13, R3 to Figure 14, R4 to Figure 15 and R5 to Figure 16.

We find that query execution on SGP profiles is on average 14 times slower on the first platform to the second. However, query execution time on Android remains below 8ms against under 1.6ms on Windows. The difference between two requests

Figure 17. Query execution time comparison

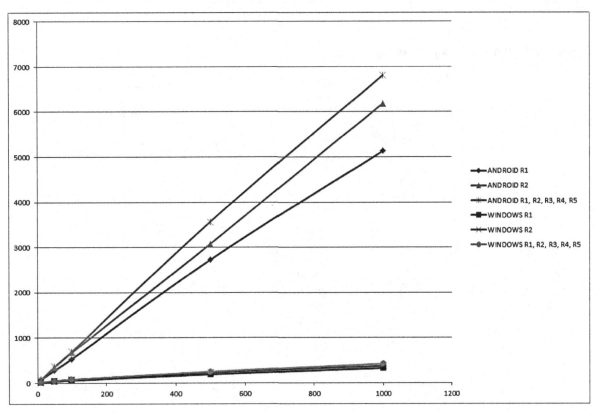

Figure 18. Query execution time based on the size of a profile

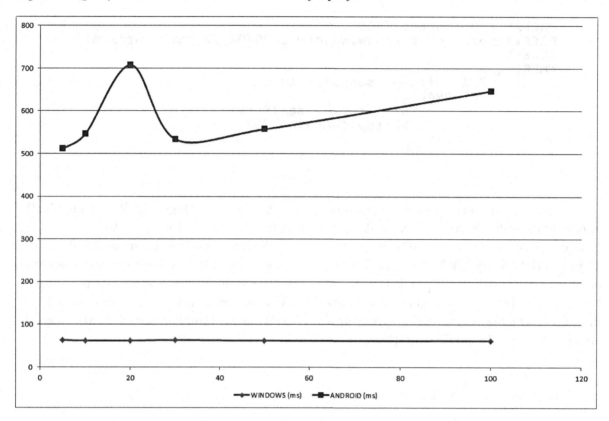

Figure 19. SGP framework positioning

is up to double in time, thus we must create a strategy to optimize querying performance.

In Figure 18, we compare the execution time of 100 R1 queries on profiles composed by 5, 10, 20, 30, 50, and 100 triples. Our objective was to identify performance's differences between profiles with different sizes.

We note that the query execution is nine times faster in configuration 1 than in configuration 2. However, the profile size seems to have a low impact on processing time. Consequently, an adaptation process could query, in a reasonable time, SGP profiles specifying several explicit constraints on many services. It is important to notice that our implementation is based on the JENA framework that executes SPARQL queries through their semantic query engine. Consequently, the query execution times measured on SGP profiles (see Figure 17 and Figure 18) are strongly dependent on the JENA framework. If some optimizations are proposed on this framework, we will also take the benefits of these optimizations. Those experiments comfort us to consider the processing of queries on both configuration 1 and 2. We find that the profile sizes have low impacts on query execution performances. We want to continue the evaluation of queries on a wide variety of profiles, and especially semantic queries, in order to design an optimized strategy.

Figure 20. SGP framework blocks

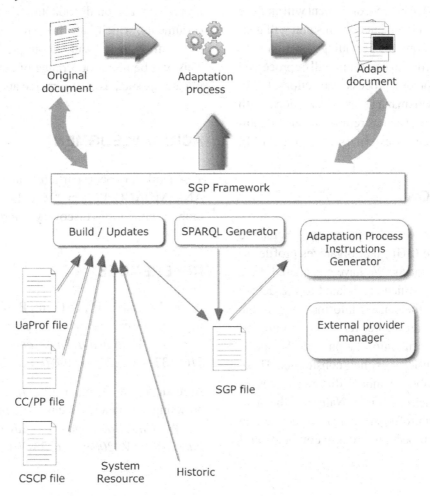

FUTURE RESEARCH DIRECTIONS

In the future, we plan to enhance our model and to design a SGP framework. Indeed, in order to use the SGP, it is necessary to have a software layer able to provide usable adaptation information. Figure 19 shows the SGP framework included into an operational environment. In this workflow, the SGP framework will receive the multimedia document and generate adaptation processes request in order to compute an adapted document that complies with the SGP profile.

In Figure 20, we give more details about the SGP framework. First of all, the framework will build and update the SGP profile from models transformations (e.g., from CC/PP-based profiles) and will access to system resources, user forms, and statistical analysis of the user behaviors (e.g., context history). Another component will analyze the multimedia document contents and will generate SPARQL queries providing constraints to consider. Then, those constraints will be processed to provide adaptation process instructions. Other components will manage the interactions with external services, such as remote distant adaptation process, access to external sensors (e.g., GPS).

CONCLUSION

In this chapter, we have proposed a Semantic Generic Profile (SGP) that organizes profile information into facets. We have considered three kinds of facets, which are related to the device characteristics, the context information, and the document structure. Moreover, we have proposed to link these profile information with the specification of high-level explicit constraints. These constraints enable to model different types of actions under rich conditions. Naturally, the main objective of our profile structure is to better guide the adaptation process in order to compute valid adapted multimedia documents. We have shown that querying the SGP would helps adaptation processes to know which actions have to be done in order to satisfy the profile constraints. We are currently experimenting and making the validation of our proposal in a real application (*implemented with Java/Androïd devices*).

In the future, we plan to develop a global framework that will use our SGP profile structure. This framework will exploit the profile semantics in order to exploit other information, which is contained into other types of profile, such as CC/PP, CSCP, etc. Furthermore, we will develop some efficient methods that will compare some initial multimedia documents with our profile structures. Finally, we will experiment in a real adaptation architecture the benefits of using our profile descriptions. For instance, we will evaluate the user's feedback on the adaptation of multimedia documents, which have been made on different platforms with and without our profile structure. This will be done for a range of networking applications, such as e-Commerce and e-Learning.

ACKNOWLEDGMENT

This work has been partially supported by the ANR MOANO Project "Models and tools for mobile applications discovery of territory" (T2i).

REFERENCES

Adzic, V., Kalva, H., & Furht, B. (2011). A survey of multimedia content adaptation for mobile devices. *Multimedia Tools and Applications*, *51*(1), 379–396. doi:10.1007/s11042-010-0669-x

Ahmadi, H., & Kong, J. (2008). Efficient web browsing on small screens. In *Proceedings of the Working Conference on Advanced Visual Interfaces (AVI 2008),* (pp. 23-30). ACM Press.

Asadi, K. M., & Dufourd, J.-C. (2005). Context-aware semantic adaptation of multimedia presentations. In *Proceedings of the IEEE International Conference on Multimedia and Expo (ICME 2005)*, (pp. 362 – 365). Amsterdam, The Netherlands: IEEE Computer Society.

Bolchini, C., Curino, C., Quintarelli, E., Schreiber, F., & Tanca, L. (2006). *Context-ADDICT. Technical Report*. Milan, Italy: Politecnico di Milano.

Bolchini, C., Curino, C. A., Quintarelli, E., Schreiber, F. A., & Tanca, L. (2007). A data-oriented survey of context models. *SIGMOD Record, 36*(4), 19–26. doi:10.1145/1361348.1361353

Bolchini, C., Orsi, G., Quintarelli, E., Schreiber, A., & Tanca, L. (2011). Context modeling and context awareness: Steps forward in the Context-ADDICT project. *A Quarterly Bulletin of the Computer Society of the IEEE Technical Committee on Data Engineering, 34*(2), 47–54.

Buchholz, S., Hamann, T., & Hübsch, G. (2004). Comprehensive structured context profiles (CSCP): Design and experiences. In I. C. Society (Ed.), *Proceedings of the Second IEEE Annual Conference on Pervasive Computing and Communications Workshops, PERCOMW 2004*, (p. 43). Orlando, FL: IEEE Press.

Bulterman, D., Jack, J., Pablo, C., Sjoerd, M., Eric, H., Marisa, D., et al. W3C. (2008). *Synchronized multimedia integration language (SMIL 3.0), W3C recommendation*. Retrieved from http://www.w3.org/TR/SMIL/

Chevalier, M., Soulé-Dupuy, C., & Tchienehom, P. (2006). Semantics-based profiles modeling and matching for resources access. *Journal des Sciences pour l'Ingénieur, 1*(7), 54–63.

Ferscha, A., Hechinger, M., Riener, A., Schmitzberger, H., Franz, M., Rocha, M., & Zeidler, A. (2006). Context-aware profiles. In *Proceedings of the Autonomic and Autonomous Systems, ICAS 2006*. ICAS.

Forough, S., & Reza, J. (2012). A comparative study of context modeling approches and applying in an infrustructure. *Canadian Journal on Data Information and Knowledge Engineering, 3*(1).

Forum, W. (2001). *WAG UAProf, wireless application protocol*. Retrieved from http://www.wapforum.org

Girma, B., Brunie, L., & Pierson, J.-M. (2006). Planning-based multimedia adaptation services composition for pervasive computing. In *Proceedings of the 2nd International Conference on Signal-Image Technology and Internet-Based Systems (SITIS 2006)*. Hammamet, Tunisia: ACM/IEEE.

Girma, B., Lionel, B., & Jean-Marc, P. (2005). Content adaptation in distributed. *Journal of Digital Information Management, 3*(2), 95–100.

Indulska, J., Robinson, R., Rakotonirainy, A., & Henricksen, K. (2003). Experiences in using CC/PP in context-aware systems. In *Proceedings of Mobile Data Management 4th International Conference (MDM 2003)*, (pp. 247-261). Melbourne, Australia: Springer.

Jannach, D., & Leopold, K. (2007). Knowledge-based multimedia adaptation for ubiquitous multimedia consumption. *Journal of Network and Computer Applications, 30*(3), 958–982. doi:10.1016/j.jnca.2005.12.007

Jedidi, A., Amous, I., & Sèdes, F. (2005). A contribution to multimedia document modeling and querying. *Multimedia Tools and Applications, 25*(3), 391–404. doi:10.1007/s11042-005-6542-7

Jouanot, F., Cullot, N., & Yétongnon, K. (2003). Context comparison for object fusion. *Lecture Notes in Computer Science, 2681*, 1031. doi:10.1007/3-540-45017-3_36

Kagal, L., Korolev, V., Avancha, S., Joshi, A., Finin, T., & Yesha, Y. (1995). Centaurus: An infrastructure for service management in ubiquitous computing environments. *Wireless Networks, 1*.

Kagal, L., Korolev, V., Chen, H., Anupam, J., & Finin, T. (2001). Centaurus: A framework for intelligent services in a mobile environment. In *Proceedings of the 21st International Conference on Distributed Computing System Workshop.* Mesa.

Klyne, G., & Carroll, J. (2004). *Resource description framework (RDF): Concepts and abstract syntax.* Retrieved from http://www.w3.org/TR/2004/REC-rdf-concepts-20040210/

Klyne, G., Reynolds, F., Woodrow, C., Hidetaka, O., Hjelm, J., Butler, M. H., & Tran, L. (2004). *Composite capability/preference profiles (CC/PP): Structure and vocabularies 1.0.* Retrieved from http://www.w3.org/Mobile/CCPP/

Laborie, S., Euzenat, J., & Layaïda, N. (2011). Semantic adaptation of multimedia documents. *Multimedia Tools and Applications, 55*(3), 379–398. doi:10.1007/s11042-010-0552-9

Laplace, S., Dalmau, M., & Roose, P. (2008). Kalinahia: Considering quality of service to design and execute distributed multimedia applications. In *Proceedings of the IEEE/IFIP International Conference on Network Management and Management Symposium.* Salvador de Bahia, Brazil: IEEE Press.

Lemlouma, T., & Layaïda, N. (2001). *NAC: A basic core for the adaptation and negotiation of multimedia services. Opera Project.* INRIA.

Lemlouma, T., & Layaïda, N. (2002). Content adaptation and generation principles for heterogeneous clients. In *Proceedings of the W3C Workshop on Device Independent Authoring Techniques.* St. Leon-Rot, Germany: W3C.

Passani, L. (2007). *Wurfl (wireless universal resource file).* Retrieved from http://wurfl.sourceforge.net

Roisin, C. (1998). Authoring structured multimedia documents. In *Proceedings of the 25th Conference on Current Trends in Theory and Practice of Informatics: Theory and Practice of Informatics,* (pp. 222-239). Springer-Verlag.

T2i. (2012). *Models & tools for pervasive applications focusing on territory discovery (MOANO).* Retrieved from http://moano.liuppa.univ-pau.fr

Tarak, C., Dejene, E., Frédérique, L., & Vasile-Marian, S. (2007). A comprehensive approach to model and use context for adapting applications in pervasive environments. *Journal of Systems and Software, 80*(12), 1973–1992. doi:10.1016/j.jss.2007.03.010

Veaceslav, C. (2011). *Applying next generation web technologies in the configuration of customer designed products.* Stockholm, Sweden: Royal Institute of Technology.

W3C. (2001). *W3C device independence working group.* Retrieved from http://www.w3.org/2001/di/

W3C (2007). *W3C ubiquitous web applications working group.* Retrieved from http://www.w3.org/2007/uwa/

W3C. (2012a). *W3C ubiquitous web domain.* Retrieved from http://www.w3.org/UbiWeb/

W3C. (2012b). *W3C web accessibility initiative.* Retrieved from http://www.w3.org/WAI/

ADDITIONAL READING

Abbar, S., Bouzeghoub, M., & Lopes, S. (2010). Introducing contexts into personalized web applications. In *Proceedings of the International Conference on Information Integration and Web-Based Applications & Services (IIWAS 2010),* (pp. 155-162). Paris, France: IIWAS.

Berners-Lee, T. (2006). *Linked data W3C design issue*. Retrieved from http://www.w3.org/DesignIssues/LinkedData.html

Context-Aware Pervasive Systems. (2012). *Wikipedia*. Retrieved from http://en.wikipedia.org/wiki/Context-aware_pervasive_systems

Gauch, S., Mirco, S., Aravind, C., & Alessandro, M. (2007). User profiles for personalized information access. *The Adaptive Web*, *4321*, 54–89. doi:10.1007/978-3-540-72079-9_2

Grifoni, P. (2009). Multimodal fusion. In *Multimodal Human Computer Interaction and Pervasive Services* (pp. 103–120). Hershey, PA: IGI Global. doi:10.4018/978-1-60566-386-9.ch006

Hancock, J. T., Toma, C., & Ellison, N. (2007). The truth about lying in online dating profiles. In *Proceedings of the ACM Conference on Human Factors in Computing Systems (CHI 2007)*, (pp. 449-452). ACM Press.

Hothi, J., & Hall, W. (1998). An evaluation of adapted hypermedia techniques using static user modelling. In *Proceedings of the 2nd Workshop on Adaptive Hypertext and Hypermedia*. Southampton, UK: IEEE.

INRIA. (2012). *Project WAM (web, adaptation, multimedia)*. Retrieved from http://wam.inrialpes.fr

Jannach, D., & Leopold, K. (2007). Knowledge-based multimedia adaptation for ubiquitous multimedia consumption. *Journal of Network and Computer Applications*, *30*(3), 958–982. doi:10.1016/j.jnca.2005.12.007

Kanellopoulos, D. (2009). Adaptive multimedia systems based on intelligent context management. *Adaptive and Innovative Systems*, *1*(1), 30–43. doi:10.1504/IJAIS.2009.022001

Kanellopoulos, D. (2010). Intelligent multimedia engines for multimedia content adaptation. *Multimedia Intelligence and Security*, *1*(1), 53–75. doi:10.1504/IJMIS.2010.035971

Lingrand, D., & Riveill, M. (2006). Input interactions and context component based modelisations: Differences and similarities. In *Proceedings of Workshop Context in Advanced Interfaces (Context@AVI 2006)*, (pp. 19-22). Venezia, Italy: ACM Press.

Pham, H. Q., Laborie, S., & Roose, P. (2012). On-the-fly multimedia document adaptation architecture. In Proceedings of the 3rd Workshop on Service Discovery and Composition in Ubiquitous and Pervasive Environments, Workshop as Part of MobiWIS. IEEE.

Soylu, A., De Causmaecker, P., & Desmet, P. (2009). Context and adaptivity in pervasive computing environments: Links with software engineering and ontological engineering. *Journal of Software*, *4*(9), 992–1013. doi:10.4304/jsw.4.9.992-1013

Tchuente, D., Jessel, N., Péninou, A., Canut, M. F., & Sèdes, F. (2012). Visualizing the relevance of social ties in user profile modeling. *Web Intelligence and Agent Systems*, *10*(2).

Tong, M.-W., Yang, Z.-K., & Liu, Q.-T. (2010). A novel model of adaptation decision-taking engine in multimedia adaptation. *Journal of Network and Computer Applications*, *33*(1), 43–49. doi:10.1016/j.jnca.2009.06.004

Viviani, M., Bennani, N., & Egyed-Zsigmond, E. (2010). A survey on user modeling in multi-application environments. In *Proceedings of the Third International Conference on Advances in Human Oriented and Personalized Mechanisms, Technologies, and Services CENTRIC 2010*, (pp. 111-116). Nice, France: IEEE Press.

W3C. (2004). *RDF/XML syntax specification (revised)*. W3C Recommendation 10 February 2004. Retrieved from http://www.w3.org/TR/REC-rdf-syntax/

W3C. (2008). *SPARQL query language for RDF*. W3C Recommendation 15 January 2008. Retrieved from http://www.w3.org/TR/rdf-sparql-query/

W3C. (2012). *W3C RDB2RDF working group*. Retrieved http://www.w3.org/2001/sw/rdb2rdf/

Wang, X. H., Zhang, D. Q., Gu, T., & Pung, H. K. (2004). Ontology based context modelling and reasoning using OWL. In *Proceedings of the IEEE International Conference on Pervasive Computing and Communication*, (pp. 18-22). Orlando, FL: IEEE Press.

KEY TERMS AND DEFINITIONS

Constraint: A constraint refers to information contained in a profile. A simple constraint may be a device screen size, while a complex constraint may state that if a user is participating at a meeting, he/she may not want to play video contents.

Context: A context is any information that can be used to characterize the situation of an entity. An entity may include a person, a device, a location, a computing application, etc.

Document Adaptation: A process that transforms a multimedia document in order to comply with a target profile. This process generally combines multiple operators: transcoding (e.g., AVI to MPEG), transmoding (e.g., text to speech), and transformation (e.g., text summarization).

Explicit Constraint: In a profile, one may specify explicitly a constraint. In this situation, there is no ambiguity for adaptation processes to satisfy such a constraint.

Implicit Constraint: From a profile, adaptation processes may deduce implicit constraints. For instance, if a device has a small screen size, an adaptation process may deduce that large images have to be resized to fit the small screen.

Multimedia Document: A multimedia document is an entity that combines pieces of information, which come from various media types named multimedia objects (as known as media items).

Profile: A profile describes a device's delivery context and can be used to guide the adaptation of content presented to that device.

Section 4
Intelligent Multimedia
Applications and Services

Chapter 10
Provisioning Converged Applications and Services via the Cloud

Michael Adeyeye
Cape Peninsula University of Technology, South Africa

ABSTRACT

The cloud is becoming an atmosphere to store huge data and deploy massive applications. Using virtualization technologies, it is economical and feasible to provide testbeds in the cloud. The convergence of Next Generation (NG) networks and Internet-based applications may result in the deployment of future rich Internet applications and services in the cloud. This chapter shows the migration of mobility-enabled services to the cloud. It presents a SIP-based hybrid architecture for Web session mobility that offers content sharing and session handoff between Web browsers. The implemented system has recently evolved to a framework for developing different kinds of converged services over the Internet, which are similar to services offered by Google Wave and existing telephony Application Programming Interfaces (APIs). In addition, the work in this chapter is compared with those similar technologies. Lastly, the authors show efforts to migrate the SIP/HTTP application server to the cloud, which was necessitated by the need to include more functionalities (i.e., QoS and rich media support) as well as to provide large-scale deployment in a multi-domain scenario.

INTRODUCTION

Convergence is taking place across numerous research, industrial, and application areas. Possible notable examples include network-convergence, service-convergence, and device-convergence.

Network-layer convergence is taking place in networks with focus on technology convergence and integration of different heterogeneous solutions. While application-layer convergence is relevantly occurring in the provisioning of telecommunications and Internet services (e.g., for entertainment applications), device-layer convergence is manifestly taking place in hardware and software

DOI: 10.4018/978-1-4666-2833-5.ch010

manufacturing, which has resulted into new hand-sets and computer equipments at relatively low cost (thus enabling a mass market of users). There are three main technology trends that are influencing the future of the converged telecommunications and Internet industries. They are: (1) IP-based networks; (2) the growth of Web 2.0; and (3) the rapid evolution of devices (with increasing local resources) (Sarin, 2007). These three trends have sequentially led to network convergence, service convergence, and device convergence in the last years. The interplay of the trends will determine the kind of services that will be available in the future. These services are anyway envisaged to be converged services, i.e., services offered over the above-converged provisioning scenario. Owing to this technological evolution, *Communications Service Providers* (CSPs) may soon no longer be the primary market drivers of communications services but increasingly drift towards a role of mediator and change-enabler (Saxtoft, 2008).

By delving into slightly finer details, the first major changes taking place in the communications industry can be seen in the variety of network technologies. These changes are faster network speeds or broadband access, introduction of mobility and development of IP-based architectures. The second major changes for CSPs can be noticed in the developments of Web 2.0. These developments have enabled CSPs to expose their services, telco-ICT enablers to bring together numerous applications, and people to produce and consume composite services. The combinations of potential new services are nearly limitless. The third major changes taking place in the communications industry are the advancement in devices, which has allowed customers to fully benefit from converged offerings.

In this scenario, standard solutions for session management in open environments, such as SIP (Rosenberg, et al., 2002; Handley, et al., 1999; Silvana & Schulzrinne, 2008) are crucial. *Extensible Messaging and Presence Protocol* (XMPP), formerly known as *Jabber Protocol*, is

also widely used to achieve interaction between *User Agent Clients* (UACs) (Saint-Andre, 2004). XMPP is commonly used for instant messaging and developing online games between two or more UACs. Although XMPP could be integrated into a Web browser to achieve content sharing and session handoff between Web browsers, using SIP in this work offers the advantage of having an adaptive UAC in which a Web browser could act as a SIP client for voice call and possibly be extended to support other SIP related functionalities. XMPP has however been used in Instant Messaging clients, such as *Pidgin* and *GTALK*.

In addition, the future converged scenario requires innovative concepts for networking and service evolution. Openness, self-organization, self-adaptation, improved flexibility, and hiding management/operation complexities to users and operators are features expected in the future Internet (Manzalini, 2008). Researchers and industry experts have recently started to provide some frameworks and service delivery platforms for converged services provisioning by adopting standard protocols, which are becoming reference points for the above scenario, such as XML, SIP, and XMPP. Only to mention a few notable examples, projects that take advantage of SIP extensibility include the *Akogrimo Project* (Mobile Grids, 2008), which involves embedding Web service data in messages exchanged via the Session Description Protocol (SDP). In addition, in an expired IETF Internet draft (Wu & Schulzrinne, 2001), two approaches are identified for transferring URLs between Web browsers for session mobility purposes: the first proposed approach is by sending the URL via a SIP MESSAGE method.

Another project that exploits SIP extensibility and is similar to the proposal presented in the following was carried out by Munkongpitakkun et al. (2007). Two software packages, related to this work and of some impact in the developers' community, are *Google Browser Sync* (Google Corporation, 2008) and *Mozilla Weave* (Mozilla Corporation, 2008). Both Mozilla Weave and

Google Browser Sync are based on HTTP, which does not provide support for Peer-to-Peer (P2P) interaction, pure asynchronous events (in the case of SIP SUBSCRIBE/NOTIFY, though AJAX does) and multimedia sessions. In addition, XMPP is also widely used to achieve interaction between UACs (Saint-Andre, 2004). However, XMPP could be used in place of SIP to move the Web session data, but it is not capable of providing multimedia services that SIP offers. These solutions are distinctively different from this work, though all of them tend to improve the Web browsing experience.

This chapter presents an original proposal in terms of a SIP-based hybrid architecture that leverages SIP, HTTP, and XML to provide converged services. The project is a novel approach to session mobility in HTTP by using SIP as the carrier of session data. It introduces a new service in the Web browsing context, namely content sharing and session handoff among Web browsers. The service is derived from SIP Mobility mechanisms - Third Party Call Control and Session Handoff mode flow for transfer to a single device (Shacham, et al., 2007). To avoid abuse of this new service, the service is controlled by a SIP server called *CAS*. CAS stands for *Converged Application Server*. CAS has the capabilities of blocking, screening and forwarding a content sharing or session handoff request. In addition, it offers parking and picking up of requests through its session tracking (with history) mechanism.

The project also proposes and implements a new Web browser architecture that integrates SIP to achieve HTTP session mobility and provide SIP functionalities. In the implementation, a new extension, called *TransferHTTP*, is developed for a Web browser, and the extension makes the Web browser act as a SIP client. That is, the extension integrates a SIP stack into the Web browser in order to make voice calls. In addition, the Web browser, with the extension installed, can transfer Web sessions to another Web browser.

The existing schemes used in the academic environment to achieve Web session mobility are client-based (Song, 2002), proxy-based (Canfora, et al., 2005) and server-based (Hsieh, et al., 2006) architectural schemes. However, these schemes have either introduced high complexity during their implementations or violated the Request for Comments (RFCs) 2616 and 2965 specifications from the *Internet Engineering Task Force* (IETF), a body whose goal is to make the Internet work better. This Chapter proposes a hybrid-based architectural scheme, which builds on the advantages of the client-based and proxy-based architectural schemes. The extension of the existing Web client and provision of an application-based proxy make the scheme a hybrid-based architectural scheme. In addition, the work in this chapter uses SIP in its scheme to migrate Web session data.

The remainder of the chapter is organized as follows. The next section presents an overview of the original SIP-based proposal for interoperable session mobility support for converged services. Then, we compare the work with various Web and telephony application development tools and technologies, by highlighting the different targeted objectives and the pros/cons of each solution. After that, the last part of the chapter points out the wide potential area of application of the proposed solution, its easy application to these areas, and the future academic-/industry-oriented research directions that will be followed to extend the work in the next months.

A SIP-BASED HYBRID ARCHITECTURE FOR CONVERGED SERVICES IN WEB SESSION MOBILITY SCENARIOS

Several works have investigated the issue of Web session mobility using client-based, proxy-based and server-based architectural schemes (Canfora, et al., 2005; Song, 2002; Hsieh, et al., 2006), thus demonstrating the research/industrial interest in the topic. Session Mobility enables seamless transfer of a Web session between different devices, based on user preferences and other context data (e.g., availability of a given access network). The reasons for session transfer include cheapest access cost, better user experience, and physical user mobility. As convergence is now moving into the mobile space, there is a strong push from Triple play of voice, video, and data to Quad play of voice, video, data, and mobility services (Mate, et al., 2006).

In our solution, it was decided to adopt SIP as the basic starting point technology because its *User Agent* (UA) could act as both a *User Agent Client* (UAC) and a *User Agent Server* (UAS), thereby making two SIP UAs to interact with each other without a mediator or proxy. SIP is also used because it could offer multimedia services between two or more extended Web browsers in both peer-to-peer and client-server architectures. HTTP UA could only be a client or a server, thereby making it difficult to have two

UAs interact with each other without a server or extending the protocol. In particular, this solution adopts the distributed hybrid-based architecture shown in Figure 1 for content sharing and session handoff between Web browsers.

In the following, the term *'content sharing'* will be used to indicate the ability to simultaneously view the same Web page on two browsers at the same time, by typically transferring only the Web page Universal Resource Locator (URL). In addition, session transfer will be defined as the ability to move the whole Web session, including not only the Web page URL, but also all needed session data (cookies, hidden form elements, and rewritten URLs). An example of content sharing is when "Alice refers Bob" to visit the same news Website that she is browsing: in this case, she would only want the URL to be sent. An example of session handoff, instead, is moving an email session between two devices with the final goal of continuing to check emails, with the same view

Figure 1. The hybrid-based architectural scheme

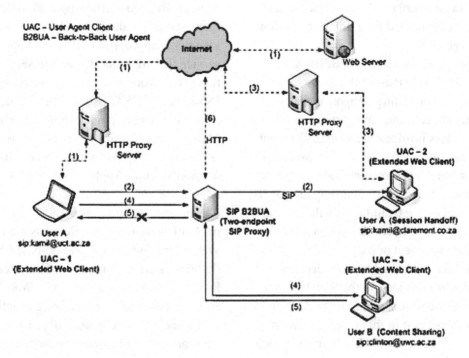

of read/unread/cancelled messages, etc., without having to sign in again.

The Web browser extension in this project (called *TransferHTTP*) was developed for Mozilla Firefox version 2.0-3.0. It required modifying the architecture of a Web browser. In addition, session data are transferred in plain text format between Web browsers, though recommendations are made on the security of session data. The SIP stack in the TransferHTTP extension was wrapped as a shared library and a new *Cross Platform Component Object Model* (XPCOM) was written to interact with it. XPCOM makes it possible to write language-agnostic components thereby separating an implementation from its interface (Boswell et al., 2002). This approach provided a new layer of abstraction to the Web browser in order to integrate the protocol (SIP). JavaScript was used in the implementation to pass user data to the extension's XPCOM, which was written in the C++ programming language. The TransferHTTP extension added a new XPCOM with the contract id "@ngportal.com/SIPStack/SIPStackInit;1" into the existing XPCOMs in the browser. The extension core comprised of an XPCOM and an interface. The interface, named "ImyStack," was defined in an Interface Definition Language (IDL).

CAS (Adeyeye, 2009a) is a SIP B2BUA, which responds to both HTTP and SIP requests. A SIP B2BUA is a call-controlling component that maintains a complete call state and participates in all call requests. It is involved in call establishment, management, and termination. CAS is involved in Web session blocking, screening and forwarding. It also acts as a SIP registrar. Participating Web clients (i.e. users) appear online when they register with CAS. This service could work in a Peer-to-Peer (P2P) environment or a Client-Server (C-S) model. In a C-S model, CAS acts as a server, which offers the control services, to the Web browsers. The control services are available in the converged application developed in this project, as a proof of concept. CAS also acts as a SIP proxy, which

is required when the UACs are behind *Network Address Translation* (NAT) boxes.

In addition, CAS is a logical entity with both UAC and UAS capabilities that has full control over traversing dialogs and SIP messages. The *Mobicents Communications Platform* (Deruelle, 2008; Mobicents, 2009) was used to implement it. It is publicly available for the SIP research community at http://transferhttp.berlios.de. Mobicents is an open Java-based platform that enables creation, deployment, and management of services and applications that integrate voice, video and data across a range of Internet Protocol (IP) and communications network by multiple devices. It implements and delivers both competing and interoperable programming models—JAIN SLEE and SIP SERVLETS—to develop Web and VoIP applications that work together.

CAS implements Web session screening, forwarding and blocking. In addition, it implements Web session mobility parking and pickup. Although applications could be developed using SIP servlets or JSLEE in the Mobicents Platform, the implementation is based on the Mobicents SIP servlets programming model. When the proxy receives a SIP request, the Application Router (AR) in the server is called by a SIP container. The AR selects the appropriate SIP servlet application to service the SIP request. The SIP servlet application in this work responds to these requests—SIP INVITE, REGISTER, and MESSAGE. Figure 2 shows the trigger point, where the SIP Servlets Management Console has been set to execute the application "org.mobicents.servlet.sip.example.SessionBlockingApplication" for the above request methods.

The architecture is a composite system that is made of heterogeneous elements - an extended Web client and an application-based SIP proxy (hybrid-based architecture). The interaction in Figure 1 is between extended Web browsers, enhanced with SIP capabilities, and acting as SIP UACs and a two-endpoint SIP proxy that coordinates browser-to-browser interactions. SIP proxy

Figure 2. The trigger point using the SIP servlets management console

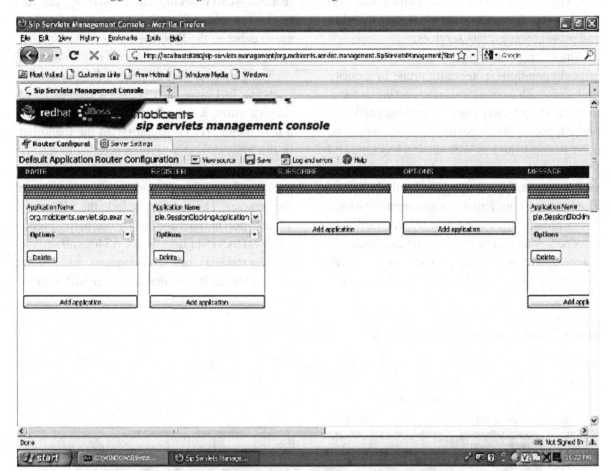

enforces (via SIP, solid arrows in Figure 1) Web session blocking or forwarding by using standard SIP user identities and access policies declared by the users. In the first phases of the project, a novel service, referred to as content sharing and session handoff between Web browsers, was designed and implemented as extensively described in Adeyeye and Ventura (2010). This implementation has provided a fast and efficient way of referring someone else to the same Web page currently viewed by the referrer rather than the slow way of copying, pasting and sending the URL in a chat session or an email. While Canfora et al. (2005) and Hsieh et al. (2006) broke the IETF RFC 2616 and 2965 specifications in their implementations, this work has provided a

way of transferring a stateful Web session from one PC to another that does not break the specifications.

Secondly, a SIP stack was integrated into a Web browser, by offering the advantage of extending a Web browser to act as a SIP client (Adeyeye & Ventura, 2010). In this way, Web browsers can act as SIP clients, thereby setting up multimedia sessions between two or more users. Most notably, the extended Web browsers in this work have unique SIP addresses to interact with one another like PCs. Lastly, a plethora of control services that prevent abuse of the content sharing and session handoff service were introduced (Adeyeye, et al., 2009). The Web browser extension, developed in this work, co-ordinates with the novel SIP-based

Converged Application Server, which was also developed in this work, to enable session mobility and prevent abuse of its services.

Since content sharing and session handoff are critical operations, potentially prone to security problems such as malicious users acting as men-in-the-middle between two interacting parties or possible abuses of services offered by the browsers, some features, such as Web session blocking and forwarding, are introduced at the proxy of the system to control the interaction between the browsers. These features are found in telecommunications, where phone calls can be blocked, forwarded or screened. A feature or service can be defined as a value-added functionality provided by a network to its users (Gurbani & Sun, 2004). Although the

features—call blocking, call forwarding, and the likes—are specific to telecommunications, they are feasible in the Web-browsing context owing to the interactions between two or more browsers.

The graphical user interface of the extension is shown in Figure 3; it is also a screenshot of the extension, when in use. TransferHTTP adds a new menu "HTTP Mobility" to the menu bar in the browser and a "Preferences" submenu to the new menu. The Preferences submenu, when clicked, gives users the opportunity to configure the browser. The settings include the SIP proxy address/port number and the SIP username. Figure 4 shows CAS User Interface. CAS logs all session transfer requests, call setup requests, and actions taken on them. It provides the source SIP address,

Figure 3. The TransferHTTP user interface

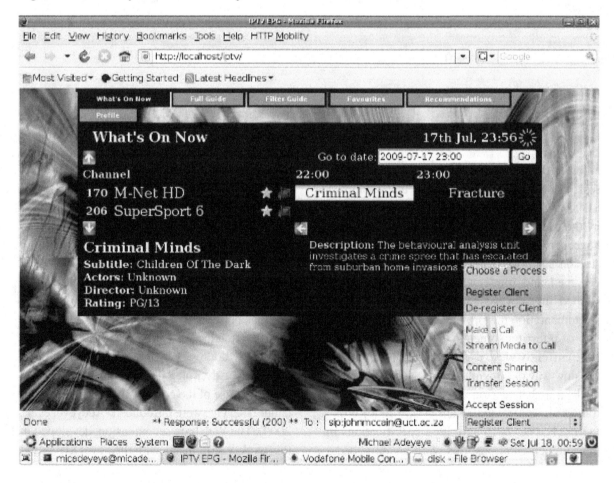

the destination SIP address, the SIP method, date, action taken (also known as status) and the referred URL in the case of a session transfer request.

This information is available under the "Session Tracking and Pickup" page, as shown in Figure 4. In addition, the "My Account" page in Figure 5 enables a user to set their policies. Information available here includes their SIP URIs with their login details and policies. A user could set how every request should be handled. That is, they could set what requests should be blocked and what address should a request go to. They could add/change/delete their identities. On the other hand, the "Buddy List" page contains a list of their contacts. A user could add a new contact

via the page and could also check the Buddy List page to see if their contacts are online or offline.

As a proof of concept, a basic architecture comprising of one SIP proxy, called a Converged Application Server (CAS), and two or more modified Web clients is used. It could however be scaled up to support multiple servers or domains. This is an experimental work that does not ensure the security of data. Although a TLS-supported SIP proxy is recommended in the production environment, the proxy used in this project does not have TLS enabled. The proxy only executes the SIP application developed in this work and responds based on users' policies.

A SIP proxy server (CAS) is also required in a Client-Server (C-S) environment when the UACs

Figure 4. The TransferHTTP proxy user interface (session tracking and pickup page)

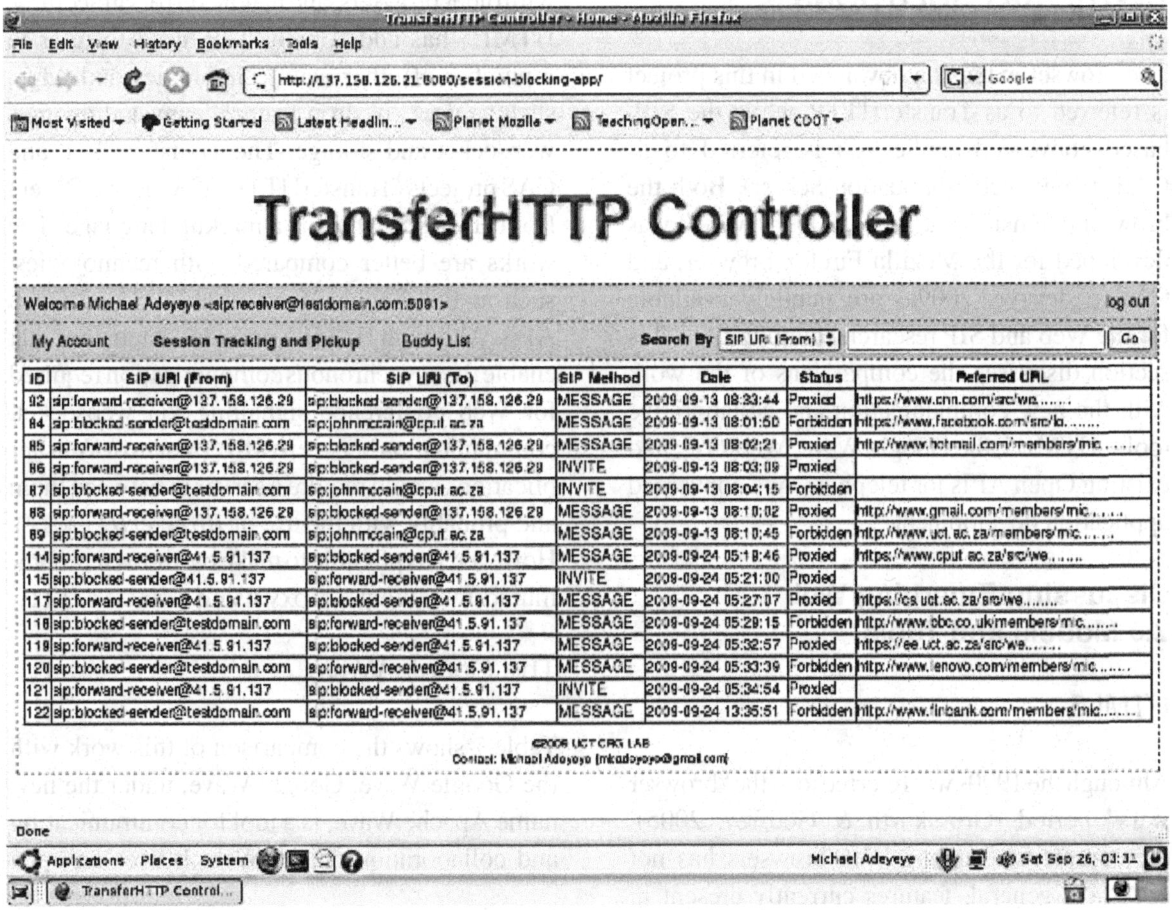

are behind NAT boxes. In this book chapter, emphasis is on the extended Web browser, the Web client services and the control services offered by the proxy. Although this work addresses session mobility or session continuity over heterogeneous network access to the Internet, it does not come close to IEEE 802.21 Media Independent Handover (MIH), Peer-to-Peer solutions via overlay network or terminal mobility using Mobile IP. This is an application layer implementation in which SIP is loosely coupled with HTTP to offer session mobility or session continuity. In addition, session data are transferred in plain text format between Web browsers, though recommendations are made on the security of session data.

EVALUATION AND COMPARISON WITH STATE-OF-THE-ART SOLUTIONS

The browser extension developed in this project is referred to as TransferHTTP, while the SIP-based converged application is referred to as CAS (Converged Application Server). Both the browser extension (Adeyeye, 2009b), which was developed for the Mozilla Firefox browser, and CAS (Adeyeye, 2009a) are publicly available for the Web and SIP research communities. This section discusses the comparisons of the work with the new communication and collaborative tool—HTML5, the Google Wave, WebRTC, and existing Open APIs for telephony and converged application development.

Discussing Emerging Web 2.0 Models and Tools

HTML5

Although the 1990s was referred to as the "browser wars" period (Grosskurth & Godfrey, 2005), the competition among Web browsers has not ended. In general, features currently present in

the Web browsers in order to improve the Web browsing experience include tabbed browsing, pop-up blocking, spell-check, page zooming and bookmarking. The capabilities of Web browsers are also quickly increasing. The World Wide Web Consortium (W3C) has recently standardized new tags and features in Web programming and how Web browsers function. Taking the recently standardized HTML 5 (HTML5, 2009) as an example, Web browsers can play audio files embedded in the <audio> tag without specifying a plug-in in the HTML code.

HTML5 is the latest iteration of the HTML markup language. It includes new features, improvements to existing features and scripting-based APIs. It includes all valid elements from both HTML4 and XHTML 1.0. It has been designed with some primary principles in mind to ensure it works on just about every platform, is compatible with older browsers, and handles errors gracefully. HTML5 has additionally been used to refer to a number of other new technologies and APIs, such as drag 'n' drop feature, canvas drawings, websocket and storage. The TransferHTTP and CAS projects (TransferHTTP+CAS) are different from HTML5, which is a markup language. The works are better compared with technologies, such as the Google Wave, WebRTC and Open APIs. Although HTML5 provides extensions that enable the asynchronous communication required for Web conference signalling via websocket, communication between two browsers or Web applications would require a server in between them and probably with the use of *BrowserID* (2012). However, it is possible to make two browsers communicate without a proxy using TransferHTTP.

The Google Wave

Table 1 shows the comparison of this work with the Google Wave. Google Wave, under the new name Apache Wave, is a tool for communication and collaboration on the Web. It uses an open

Figure 5. The TransferHTTP proxy user interface (my account page)

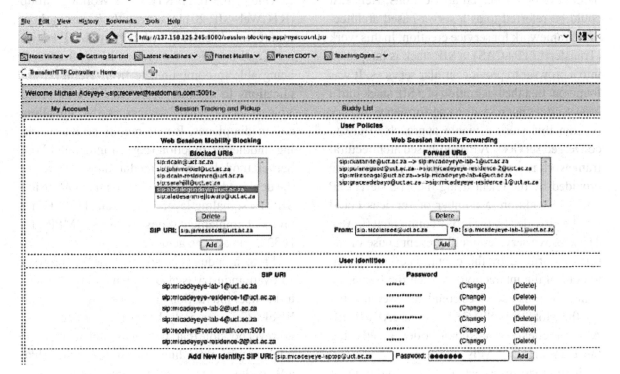

protocol (Google Corporation, 2009a, 2009b), so anyone can build his or her own wave system.

The Google Wave API allows developers to use and enhance Google Wave through two primary types of development, namely Extensions and Embed. The Extensions represent the server-side, while the Embed represents the client-side. The extensions (also called the Robots API) can be developed using the Java Client Library, Python Client Library, or Gadgets API, while the embed, which is embedded into a Web application, is always written in JavaScript.

The Google Wave and this work (TransferHTTP+CAS) provide the same services, though over different architectures. While the Google Wave API is used to develop applications that reside on a Web server, TransferHTTP APIs are used to develop applications that reside at the client end. For example, the Click-to-dial in Google wave (Google Corporation, 2009c) requires the server to set up a call session, while in TransferHTTP, the client sets up the call session. In Google Wave, the robot in the Web server is responsible for the signalling, while in TransferHTTP, the SIP stack in the browser does the signaling.

The Google team has separated the signaling (HTTP and XMPP) in a bid to maintain the current Web architecture. The Extensible Messaging and Presence Protocol (XMPP) stack (Saint-Andre, 2004) in the Google Wave resides at the server, and its APIs are written for third-parties to help

Table 1. Comparison of Google wave and TransferHTTP+CAS

	Google Wave	**TransferHTTP+CAS**
Client technologies	JavaScript in HTML	XUL and JavaScript
Server technologies	Python/Java/ Gadget	Java (HTTP/SIP Servlet)
Architecture	Server-based	Hybrid-based
Protocol	Wave (Extension to XMPP)	SIP

them develop converged applications. Hence, it could be referred to as a server-based architectural framework for service creation. In this work (TransferHTTP+CAS), a SIP stack is integrated into a browser to provide similar services. It was found out that the integration of a SIP stack into a browser does not impede its performance thereby making this work a viable approach to create converged services. A hybrid-based architectural framework is created in which services could be provided by the client using the TransferHTTP APIs. In addition, a number of services could also be provided by the proxy component (CAS). These proxy services could prevent abuse of the services offered by the client. They are meant to control the interaction between the browsers. While the proxy services could be developed using the Mobicents SIP Servlets and JAIN SLEE APIs, the client services could be developed using the TransferHTTP APIs.

In summary, irrespective of the technologies or programming languages used in the Google Wave and this work (TransferHTTP+CAS), the difference between them is that the Google Wave only has a stack (an XMPP stack) in its server, thereby making it a server-based architectural framework for service creation, while TransferHTTP has a stack (a SIP stack) both in its client and proxy, thereby making it a hybrid-based architectural framework for service creation.

WebRTC

WebRTC is an open framework that offers Web application developers the ability to write rich real-time multimedia applications (e.g. video and gaming applications) on the Web without requiring plugins or extensions. Its purpose is to help build a strong *Real Time Communication* (RTC) platform that works across multiple Web browsers and platforms. In an implementation, the WebRTC API will abstract several key components for real-time audio, video, networking, and signal (WebRTC, 2011a, 2011b).

One of the IETF RTCWEB Working Group (RTCWeb-SIP, 2011) is currently discussing how to integrate WebRTC with deployed SIP equipment and domains. An area of its application is being able to communicate from WebRTC applications to existing deployed SIP/RTP-based Voice/Video-over-IP devices at the signalling and media planes. It may require an interworking middlebox function (e.g. an integrated Web Server module) in the media-plane. However, the deployed devices should communicate using SIP at a signaling layer rather than HTTP. Other protocol implementations, such as XMPP and H.323, can also be achieved.

From the industry perspective, the Web browser software industry is also implementing browser-to-browser interaction in various ways. Although WebRTC is currently being standardized, it is however possible that some of its implementations might require extending an existing terminal (like a Web client in this work), a proxy, or a server.

Open APIs

Open standard APIs are desirable for introducing new services because they make the separation between an application and its platform explicit. They allow application portability and allow the functions of the platform to be used by multiple applications easily. APIs are application-centric, while protocols are network-centric. APIs allow programmers to focus on the logical flow of applications using only the necessary functions provided by the platform, rather than concerning themselves with low-level details of messages that must flow across the network. As a result, a well-defined API allows the application programmer to work at a higher level of abstraction than that of the protocol (Jain, et al., 2005).

The SIP API reflects the SIP protocol fairly closely (Bhat & Tait, 2001; Microsystems, 2001a, 2001b). It is useful for situations where the application is rather simple and where the underlying network is known to be an IP network. However, the SIP API is at a lower level of abstraction than the call control APIs, such as JTAPI, JAIN, and Parlay APIs. As a result, it offers the programmer finer grained control, better performance than the call control APIs.

Another messaging and presence protocol that is widely used is Extensible Messaging and Presence Protocol (XMPP). Its APIs have been used to develop XMPP clients, such as the Google Talk and Pidgin. It is gaining wide acceptable in the software industry, where it is being used to develop communication and collaborative tools. Examples of shared applications built with XMPP are shared whiteboard and chessboard (Mirra, 2009). Another work that is currently exploiting XMPP is the Google Wave, which has just been discussed.

The need for Open APIs is greatly increasing (Schonwalder, et al., 2009; Krechmer, 2009; Mulligan, 2009). The APIs are needed for user-generated services. Although there are APIs, such as Google APIs and Parlay-X APIs, for developing Web 2.0 applications and basic Web service APIs for access to Circuit-Switched (CS), Packet-Switched (PS), and IMS networks, they fall short of enabling innovative converged applications or services from users.

The Google Wave project currently looks promising for application developers but could be limited in functionality in the near future. The Parlay-X APIs are already claimed to have very limited functionality (Mulligan, 2009). The reason is that they are not designed to handle the data model for the entire service or signaling in Telecommunications.

This work (TransferHTTP+CAS) presents APIs that expose the signaling in Telecommunications to create innovative applications. The APIs make it possible to create applications that can use the instant messaging and presence features in SIP (Rosenberg, 2004). The TransferHTTP APIs currently expose the SIP REGISTER method so that a browser can register with a SIP network. They also expose the SIP MESSAGE method to

send messages or chat and the SIP INVITE method to make calls between two browsers. In addition, there is a media-to-call function that could be used to create a media broadcast service.

On sample applications that use these APIs, the TransferHTTP APIs have been used to implement HTTP session mobility service. In addition, its media-to-call interface has been used to create a media service. The APIs are available for use when the TransferHTTP extension (Adeyeye, 2009b) is installed in the Mozilla Firefox browser.

The APIs, which were released under the Mozilla Public License (MPL), can be extended to support other SIP methods, notifications, or functionalities, and the technologies used in creating the extension are already shown in Table 1. The underlying component, which was written in C++, is scriptable; that is, a user could develop a JavaScript application that implements the methods and arguments.

To enjoy the full potential of these APIs, developers are advised to develop XUL-based applications. *XML User Interface Language* (XUL) was the language used to develop the Mozilla Firefox user interface. Hence, a XUL-based application developed by any interested user will be able to use the APIs in the Web browser and the extension, unlike Web applications, which are restricted for security reasons.

THE NEED TO MIGRATE CAS TO THE CLOUD

The two factors that necessitated the need to migrate CAS to the Cloud are discussed below.

QoS-Enabled Architectural Scheme

One of the works that explored latency in browser-to-browser interaction is Linner et al. (2010). Latency in browser-based user-to-user interactions in this work was reported in Adeyeye et al. (2012). A QoS-enabled SIP-based architectural scheme would be required in a bi-directional exchange of time-critical data between Web terminals. A QoS-Enabler, as shown in Figure 6, could be integrated into CAS to prioritize bandwidth consumptive or real-time applications on demand. In a multi-domain implementation (such as the IMS), the QoS-Enabler in CAS will play the role of a gateway thereby preprocessing QoS requests of applications and transforming them into requests to the underlying network QoS enforcement.

Two or more CAS in the QoS-Enabled architectural scheme will act as a message overlay and enforce message policies for both text-based data (IM, URLs/Web session transfer) and multimedia data. They can accept configurations for QoS parameters, such as delay, jitter, packet loss, and bandwidth, and send to the QoS-Enabler module. The module can also work with existing QoS entities in the IMS implementation. The configurations can be provided by application providers via the TransferHTTP API from a Web client side

or via the CAS Web interface from a proxy side. The QoS_Enabler in CAS is discussed as follows:

CAS can only serve a limited number of concurrent requests. QoS Module in Figure 6 is a Quality of Service (QoS) module for CAS implementing additional control mechanisms that can provide different priority to different requests. In this case, it is used to ensure that important resources stay available under high server load. While the mobility module is used to accept, redirect or reject session transfer or call requests, the QoS module is used to control access to resources. That is, the QOS module can reject requests to unimportant resources or disable requests to very important resources while grant access to more important services for very important users. The key jobs of the qos module are:

1. **Request Level Control:** It controls the number of concurrent requests to CAS. It is used to define different priorities to different classes of service (messaging, session

Figure 6. CAS with a QoS-enabler

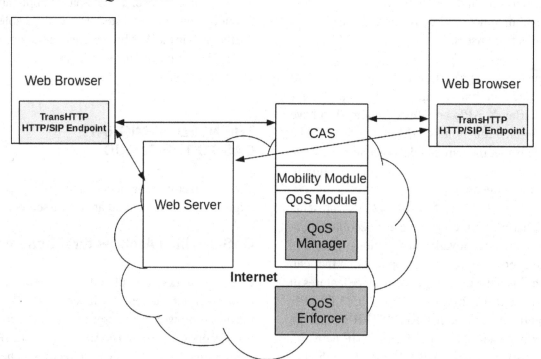

transfer and audio/video call) supported by CAS.

2. **Connection Level Control:** It controls the number of TCP connections to CAS. This helps limit the connections coming from the TransferHTTP Web clients in order to reduce the maximum number of concurrent connections.

3. **Bandwidth Level Control:** It throttles requests/responses to certain SIP addresses of TransferHTTP clients connected to CAS.

4. **Generic request line and header filter:** It drops suspicious requests that can be a threat to CAS.

The QoS manager stores the qos parameters for each user in a domain while the QoS enforcer applies the QoS policies to each request/response from a user. The QoS-related information include available classes of service, bandwidth allocated per class of service, key performance parameters per class (one-way packet delay, packet loss rate), call level quality parameters and control plane parameters. The XACML primary focus is access control and is supported by Mobicents. Hence, it can be used in implementing the qos module. However, using the XACML policy implementation in Mobicents (XACML Policy, 2012) in CAS would increase its complexity. Hence, a new XML format is proposed as shown in Figure 7 for policy setting. A service provider can set the QOS for all users in a domain via the CAS Web interface. Alternatively, the policy file can be downloaded or uploaded to CAS. This approach will make it easier to set similar policies over multiple CAS in a multi-domain implementation. The QoS implementation would require extending the current XML message format from TransferHTTP client (as shown in Adeyeye & Ventura, 2010) to support some additional tags, such as priority level.

Effective Solutions for Large Deployments

Although most of the services in this project mirrored SIP session mobility signaling and GSM supplementary services (e.g. Call blocking and Call forwarding), significant changes were made to the signaling in order to develop a functional system. An example is when the destination SIP address cannot be reach, though it could accept a session transfer request from the source SIP address. In this case, the SIP proxy (CAS) will have to generate and send a 408 Timeout response to the source Web client; but a 408 Timeout is normally generated at a source UAC (User Agent Client) when its request cannot be processed within a specific time.

Since CAS is designed to either block or forward a request, a request could however be picked-up later when the destination Web client registers with the proxy. In addition, the user can control ongoing session movement requests via CAS user interface (session tracking). The signaling however is slightly different from the usual SIP request-response signaling.

Another issue was the implementation of partial blacklisting/whitelisting. In the real world, when a number is blacklisted by a user, it would not be able to send messages or make calls to that user. The situation is different here; CAS could be configured in a way that it could block session transfer requests from a source Web client but allow a call set-up between the source Web client and its intended destination Web client.

The six other possible enhancements that could be looked into to improve this project are multidomain implementation, extending TransferHTTP client to support Instant Messaging and IPTV, extending TransferHTTP and CAS interfaces to support buddy list with presence, smart home services via a SIP-based Web browser, developing a SIP-based Web services model for Internet Telephony services and improving the HTTP

Figure 7. QoS XML message format

```
<?xml version="1.0" encoding="UTF-8"?>
<qos_params>
<sipdomain>cput.ac.za</sipdomain>
<qos_cos_list>
<cos status="accepted">call</cos>
<cos status="accepted">messaging</cos>
<cos status="accepted">session_transfer</cos>
</qos_cos_list>
<qos_request_level>
<request limit="minimum">8000</request>
<request limit="maximum">16000</request>
</qos_request_level>
<qos_connection_level>
<connection limit="minimum">32000</connection>
<connection limit="maximum">64000</connection>
</qos_connection_level>
<qos_bandwidth_level>
<sipuser id="sip:kbaolu001@cput.ac.za">
<bandwidth>100</bandwidth>
</sipuser>
</qos_bandwidth_level>
<qos_priority>
<priority status="accepted">normal</priority>
<priority status="accepted">non-urgent</priority>
<priority status="rejected">urgent</priority>
<priority status="rejected">emergency</priority>
</qos_priority>
</qos_params>
```

Session Mobility service to work on AJAX-based, mashups, frame/iframe-based websites.

However, some of them, in addition to the above two factors compelled us to start migrating the server to the cloud. While some of them are enhancements to the client (TransferHTTP), the other features, such as multi-domain implementation and Web services model for Internet Telephony services require that additional components (e.g. servers) be integrated into the system. In a multi-domain implementation, the deployment of this reference system would involve multiple CASs. To achieve this, the HTTP/SIP application would have to be deployed on a Service Delivery Platform (SDP) that supports multi-domain, such as the IP Multimedia Subsystem (IMS). Figures 8 and 9 show screenshots of the OpenStack dashboard with an instance (Virtual Machine) running on a node and the VMM (Virtual Machine Monitor) showing CAS running at the Command Line Interface (CLI), respectively. The screenshots were taken at the server end, hence the IP address 127.0.0.1 in the address bar in Figure 8. The three prominent cloud service models are Software as a Service (SaaS), Platform as a Service (PaaS) and Infrastructure as a Service (IaaS) (Zhang, et al., 2010). This work presents a *Software as a Service* (SaaS) model, where CAS runs in a cloud to manage the various Web session mobility services offered by TransferHTTP.

OpenStack (2012) can be used to deliver a massively scalable cloud operating system. The

Figure 8. The OpenStack dashboard with a running instance

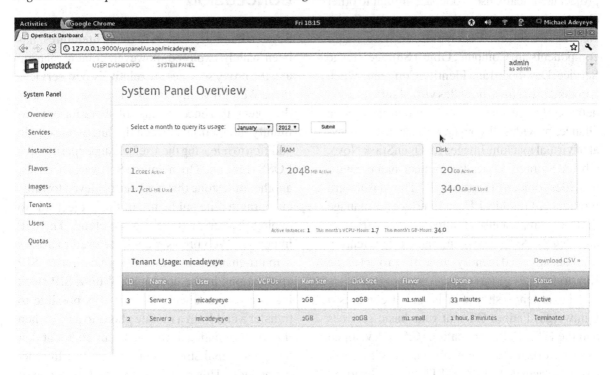

Figure 9. The virtual machine manager and the running instance

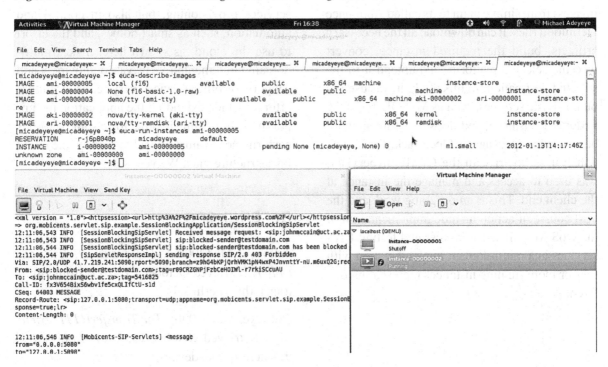

project technically uses codebase similar to other cloud computing platform projects, such as Open-Nebula, DevStack, Eucalyptus. Its five major components are Compute, Object Storage, Image Service, Dashboard and Identity. Compute, which is codenamed Nova, provides virtual servers upon demand. Image service, under the project name Glance, provides discovery, storage, and retrieval of virtual machine images for OpenStack Nova. Object Storage, under the project name Swift, provides object/blob storage. The dashboard provides a Graphical User Interface to manage the three components—*Compute, Object Storage, and Image Service*—using an authentication mechanism called Identity (under the project name Keystone).

The instance shown in Figure 9 contains a lightweight Fedora OS that runs basic services and the HTTP/SIP application (CAS) developed in this work. It has the Mobicent JBoSS server (which executes the CAS HTTP/SIP application) and MySQL (which hosts the database) running on it. The disk image was created with *BoxGrinder* (2011). BoxGrinder creates appliances (also called images) from simple plain texts called Appliance Definition files. It can download all the necessary artifacts, build the required instance, convert it to the selected platform, and upload it to the selected destination. In this work, it was used to bundle basic services, such as MySQL, build the Fedora 16 instance, and deliver the image with basic services/packages to Openstack. The VMM, which uses libvirt with the QEMU hypervisor, was used to access and manage the instance at the client end. This is an on-going work and the next course of action include implementation of the QoS Module and Enforcer and testing CAS services in a multi-domain implementation with four or more running instances representing different domains.

CONCLUSION

Session handoff, as seen in many literatures, has been widely explored, however, content sharing and the proxy services are relatively new services in the Web-browsing context. These services have the potential of encouraging collaboration or community interaction between the Internet users. In addition to releasing the artefacts under permissive FOSS (Free and Open Source Software) licenses, another milestone that has been achieved towards encouraging the public use of this project was to deploy the server (CAS) in the cloud. The SIP integrated Web browser could be used to access a multi-channel/multimodal application or SIP applications in the endpoints. With a SIP stack integrated into a Web browser, it is possible to transfer a call from a mobile phone to a PC, when the caller or callee moves to an environment that has poor signal strength but a very fast Internet connection. This reference architecture is one of the converged applications and services platform that could be delivered via a cloud. SIP is an extensible protocol that is not only used in multimedia services provisioning, but also in control and automation, such as smart homes, and the effort to use the cloud, as shown here, makes multi-domain implementation, private deployment and support for other services very realistic. Future efforts include extending the application to support multi-threading in order to do some rigorous performance evaluation and to manage the cloud infrastructure via a Web console.

REFERENCES

Adeyeye, M. (2009a). *The TransferHTTP controller*. Retrieved October 13, 2011, from http://transferhttp.berlios.de

Adeyeye, M. (2009b). *The TransferHTTP extension*. Retrieved October 13, 2011, from http://transferhttp.mozdev.org

Adeyeye, M., & Ventura, N. (2010). A SIP-based web client for HTTP session mobility and multimedia services. *Computer Communications, 33*(8), 954–964. doi:10.1016/j.comcom.2010.01.015

Adeyeye, M., Ventura, N., & Foschini, L. (2012). Converged multimedia services in emerging web 2.0 session mobility scenarios. *Wireless Networks, 18*(2), 185–197. doi:10.1007/s11276-011-0394-z

Adeyeye, M., Ventura, N., & Humphrey, D. (2009). Control services for the HTTP session mobility service. In *Proceedings of the 3rd IEEE Conference on New Technologies, Mobility and Security (NTMS)*. Cairo, Egypt: IEEE Press.

Akogrimo Project. (2008). *Access to knowledge through the grid in the mobile world*. Retrieved October 13, 2011, from http://www.mobilegrids.org

Bhat, R. R., & Tait, D. (2001). JAVA APIs for integrated networks. In Jespen, T. (Ed.), *Java in Telecommunications: Solutions for Next Generation Networks* (p. 193). New York, NY: Wiley & Sons.

Boswell, D., King, B., Oeschger, I., Collins, P., & Murphy, E. (2002). *Creating applications with Mozilla*. New York, NY: O'Reilly Press.

BoxGrinder. (2011). *The BoxGrinder*. Retrieved October 13, 2011, from http://boxgrinder.org/

Browser, I. D. (2012). *The BrowserID project*. Retrieved October 13, 2011, from https://browserid.org

Canfora, G., Di Santo, G., Venturi, G., Zimeo, E., & Zito, M. V. (2005). Proxy-based handoff of web sessions for user mobility. In *Proceedings of the Second Annual International Conference on Mobile and Ubiquitous Systems: Networking and Services (MobiQuitous 2005)*. IEEE.

Deruelle, J. (2008). JSLEE and SIP-servlets interoperability with mobicents communication platform. In *Proceedings of the 2nd International Conference on Next Generation Mobile Applications, Services and Technologies,* (pp. 634-639). Cardiff, UK: IEEE.

Google Corporation. (2008). *Google browser sync*. Retrieved October 13, 2011, from http://www.google.com/tools/firefox/browsersync/

Google Corporation. (2009a). *The wave protocol*. Retrieved October 13, 2011, from http://www.waveprotocol.org

Google Corporation. (2009b). *The Google wave project*. Retrieved October 13, 2011, from http://wave.google.org

Google Corporation. (2009c). *The Google wave click-to-dial*. Retrieved October 13, 2011, from http://googlewavedev.blogspot.com/2009/06/twiliobot-bringing-phone-conversations.html

Grosskurth, A., & Godfrey, M. (2005). A reference architecture for web browsers. In *Proceedings of the 21st IEEE International Conference on Software Maintenance (ICSM 2005),* (pp. 661-664). IEEE Press.

Gurbani, V. K., & Sun, X.-H. (2004). Terminating telephony services on the Internet. *IEEE/ACM Transactions on Networking, 12*(4), 471–481. doi:10.1109/TNET.2004.833145

Handley, M., Schulzrinne, H., Schooler, E., & Rosenberg, J. (1999). *SIP: Session initiation protocol*. IETF RFC 2543. Retrieved from http://www.ietf.org/rfc/rfc2543.txt

Hsieh, M.-D., Wang, T.-P., Tsai, C.-S., & Tseng, C.-C. (2006). Stateful session handoff for mobile WWW. *Information Sciences, 176*(9), 1241–1265. doi:10.1016/j.ins.2005.02.009

HTML5. (2009). *HTML 5 differences from HTM*. Retrieved October 13, 2011, from http://www.w3.org/TR/html5-diff/

Jain, R., Bakker, J., & Anjum, F. (2005). *Programming converged networks: Call control in Java, XML, and Parlay/OSA*. New York, NY: Wiley Interscience.

Krechmer, K. (2009). Open standards: A call for change. *IEEE Communications Magazine, 47*(5), 88–94. doi:10.1109/MCOM.2009.4939282

Linner, D., Stein, H., Staiger, U., & Steglich, S. (2010). Real-time communication enabler for web 2.0 applications. In *Proceedings of the Sixth International Conference on Networking and Services (ICNS 2010)*, (pp. 42-48). Cancun, Mexico: ICNS.

Manzalini, A. (2008). Tomorrow's open internet for telco and web federation. In *Proceedings of the International Conference on Complex, Intelligent and Software Intensive Systems (CISIS 2008)*, (pp. 567-572). Barcelona, Spain: CISIS.

Mate, S., Chandra, U., & Curcio, I. (2006). Moveble-multimedia: Session mobility in ubiquitous computing ecosystem. In *Proceedings of the 5th International Conference on Mobile and Ubiquitous Multimedia (MUM 2006)*. Stanford, CA: MUM.

Microsystems, S. (2001a). *JAIN SIP release 1.1 specification, 2001*. Retrieved October 13, 2011, from http://www.jcp.org/en/jsr/detail?id=289

Microsystems, S. (2001b). *JAIN SIP release 1.0 specification, 2001*. Retrieved October 13, 2011, from http://www.jcp.org/aboutJava/community-process/final/jsr032

Mirra, M. (2009). *Web sharing and RichDraw*. Retrieved October 13, 2011, from http://hyper-struct.net/2007/2/24/xml-sync-islands-let-the-Web-sharing-begin

Mobicents. (2009). *The Mobicents open source SLEE and SIP server*. Retrieved October 13, 2011, from http://www.mobicents.org/index.html

Mozilla Corporation. (2008). *A prototype of Mozilla weave*. Retrieved October 13, 2011, from http://labs.mozilla.com/2007/12/introducing-weave/

Mulligan, C. E. A. (2009). Open API standardization for the NGN platform. *IEEE Communications Magazine, 47*(5), 108–113. doi:10.1109/MCOM.2009.4939285

Munkongpitakkun, W., Kamolphiwong, S., & Sae-Wong, S. (2007). Enhanced web session mobility based on SIP. In *Proceedings of the 4th International Conference on Mobile Technology, Applications and Systems (Mobility2007)*, (pp. 346-350). Singapore, Singapore: Mobility.

OpenStack. (2012). *The OpenStack project*. Retrieved October 13, 2011, from http://openstack.org/

Policy, X. A. C. M. L. (2012). *Implementation in mobicents*. Retrieved October 13, 2011, from http://www.jboss.org/picketlink/XACML

Rosenberg, J. (2004). *A presence event package for the session initiation protocol* (SIP). IETF RFC 3856. Retrieved from http://www.ietf.org/rfc/rfc3856.txt

Rosenberg, J., Schulzrinne, H., Camarillo, G., Johnston, A., Peterson, J., & Sparks, R. … Schooler, E. (2002). *SIP: Session initiation protocol*. IETF RFC 3261. http://www.ietf.org/rfc/rfc3261.txt

RTCWeb-SIP. I. (2011). *IETF RTCWeb-SIP WG*. Retrieved October 13, 2011, from http://tools.ietf.org/html/draft-kaplan-rtcweb-sip-interworking-requirements-01

Saint-Andre, P. (2004). *Extensible messaging and presence protocol (XMPP): Core*. IETF RFC 3920. Retrieved from http://www.ietf.org/rfc/rfc3920.txt

Sarin, A. (2007). The future of convergence in the communications industry. *IEEE Communications Magazine, 45*(9), 12–14. doi:10.1109/MCOM.2007.4342843

Saxtoft, C. (2008). *Convergence: User expectations, communications enablers and business opportunities*. London, UK: John Wiley & Sons Ltd.

Schonwalder, J., Fouquet, M., Rodosek, G. D., & Hochstatter, I. (2009). Future internet = content + services + management. *IEEE Communications Magazine, 47*(7), 27–33. doi:10.1109/MCOM.2009.5183469

Shacham, R., Schulzrinne, H., Thakolsri, S., & Kellerer, W. (2007). *Session initiation protocol (SIP) Session mobility: Internet-draft: Draft-shacham-sipping-session-mobility-05*. Retrieved October 13, 2011, from ftp://ftp.rfc-editor.org/in-notes/internet-drafts/draft-shacham-sipping-session-mobility-05.txt

Silvana, G. P., & Schulzrinne, H. (2008). SIP and 802.21 for service mobility and pro-active authentication. In *Proceedings of the Communication Networks and Services Research Conference (CNSR 2008)*, (pp. 176-182). Halifax, Canada: CNSR.

Song, H. (2002). Browser session preservation and migration. In *Proceedings of WWW 2002*, (p. 2). Hawaii, HI: IEEE.

WebRTC. (2011a). *WebRTC*. Retrieved October 13, 2011, from http://www.webrtc.org

WebRTC. (2011b). *IETF WebRTC*. Retrieved October 13, 2011, from http://tools.ietf.org/wg/rtcweb

Wu, X., & Schulzrinne, H. (2001). *Use SIP MESSAGE method for shared web browsing*. Retrieved October 13, 2011, from http://www3.tools.ietf.org/id/draft-wu-sipping-Webshare-00.txt

Zhang, Q., Cheng, L., & Boutaba, R. (2010). Cloud computing: State of the art and research challenges. *Journal of Internet Services and Applications, 1*(1), 7–18. doi:10.1007/s13174-010-0007-6

ADDITIONAL READING

Banerjee, N., Acharya, & Das, S. (2006). Seamless SIP-based mobility for multimedia applications. *IEEE Network Magazine, 20*(2), 6-13.

Bond, G., Cheung, E., Fikouras, I., & Levenshteyn, R. (2009). Unified telecom and web services composition. In *Proceedings of the 3rd International Conference on Principles, Systems and Applications of IP Telecommunications - IPTComm 2009* (p. 1). New York, NY: ACM Press.

Cesar, P., Vaishnavi, I., Kernchen, R., Meissner, S., Hesselman, C., Boussard, M., et al. (2008). Multimedia adaptation in ubiquitous environments. In *Proceeding of the Eighth ACM Symposium on Document Engineering - DocEng 2008*, (p. 275). New York, NY: ACM Press.

Chen, C., Chen, G., Jiang, D., Ooi, B. C., Vo, H. T., Wu, S., & Xu, Q. (2010). *Providing scalable database services on the cloud*. Retrieved from http://dl.acm.org/citation.cfm?id=1991336.1991338

Chou, X. S., & Wu, J. J. L. (2002). An architecture of wireless web and dialogue system convergence for multimodal service interaction over converged networks. In *Proceedings of the Eleventh International Conference on Computer Communications and Networks, 2002*, (pp. 69-74). IEEE.

Emeakaroha, V. C., Brandic, I., Maurer, M., & Breskovic, I. (2011). SLA-aware application deployment and resource allocation in clouds. In *Proceedings of the 2011 IEEE 35th Annual Computer Software and Applications Conference Workshops*, (pp. 298-303). IEEE Press.

Fan, P., Wang, J., Zheng, Z., & Lyu, M. R. (2011). Toward optimal deployment of communication-intensive cloud applications. In *Proceedings of the 2011 IEEE 4th International Conference on Cloud Computing*, (pp. 460-467). IEEE Press.

Geneiatakis, D., Dagiuklas, T., & Kambourakis, G. (2006). Survey of security vulnerabilities in session initial protocol. *IEEE Communications Surveys & Tutorials, 8*(3), 68–81. doi:10.1109/COMST.2006.253270

Han, R., Guo, L., Guo, Y., & He, S. (2011). A deployment platform for dynamically scaling applications in the cloud. In *Proceedings of the 2011 IEEE Third International Conference on Cloud Computing Technology and Science,* (pp. 506-510). IEEE Press.

Kataoka, H., Toyama, M., Sueda, Y., Mizuno, O., & Takahashi, K. (2010). *Demonstration of web contents collaborative system for call parties.* Retrieved from http://dl.acm.org/citation.cfm?id=1834217.1834288

Maes, S. H. (2007). A call control driven MVC programming model for mixing web and call or multimedia applications. In *Proceedings of the 4th International Conference on Mobile Technology, Applications, and Systems and the 1st International Symposium on Computer Human Interaction in Mobile Technology - Mobility 2007,* (p. 439). New York, NY: ACM Press.

Mori, T., Nakashima, M., & Ito, T. (2012). A sophisticated ad hoc cloud computing environment built by the migration of a server to facilitate distributed collaboration. In *Proceedings of the 2012 26th International Conference on Advanced Information Networking and Applications Workshops,* (pp. 1196-1202). IEEE Press.

Rosenberg, J., Hiie, M., Audet, F., & Kaufman, M. (2012). *An architectural framework for browser based real-time communications (RTC).* Retrieved from http://tools.ietf.org/html/draft-rosenberg-rtcweb-framework-00

Sbata, K., Khrouf, H., Zander, S., & Becker, M. (2009). Converging web and IMS services. In *Proceedings of the International Conference on Management of Emergent Digital EcoSystems – MEDES 2009,* (p. 315). New York, NY: ACM Press.

Shacham, R., Schulzrinne, H., Thakolsri, S., & Kellerer, W. (2007). Ubiquitous device personalization and use. *ACM Transactions on Multimedia Computing, Communications, and Applications, 3*(2).

Tsai, C.-M. H. (2008). A SIP-based session mobility management framework for ubiquitous multimedia services. *Lecture Notes in Computer Science, 5061,* 636–646. doi:10.1007/978-3-540-69293-5_50

Vijay, K., & Gurbani, X.-H. S. (2005). Inhibitors for ubiquitous deployment of services in the next-generation network. *IEEE Communications Magazine, 43*(9), 116–121. doi:10.1109/MCOM.2005.1509976

KEY TERMS AND DEFINITIONS

Browser Extension: A browser extension is a computer program, which is installed in a browser, to extend its functionality. It is installed by a user to enhance their browsing experience.

Cloud Computing: Cloud computing is the delivering of hosted services over the Internet. Its services are broadly classified into Infrastructure-as-a-Service (IaaS), Platform-as-a-Service (PaaS), and Software-as-a-Service (SaaS).

Google Wave: Google Wave is an online communication and collaboration tool for seamless real-time interactions. Now called Apache Wave, it is a dynamic mix of conversation models, such as text, photos, videos and maps, and highly interactive document creation via a browser.

Open API: Open API applies to collaborative services environments where managed service providers can outsource specific services to other providers via systems integration. It is used to describe sets of technologies that enable websites to interact with each other (and telecommunication platforms) by using Web technologies.

Service Convergence: Service Convergence is an application-layer convergence that involves provisioning of telecommunications and Internet services (e.g., for entertainment applications) over the Internet. The convergence is between Internet telephony services and Web services.

Session Mobility: Session Mobility enables seamless transfer of a Web session between different devices, based on user preferences and other context data (e.g., availability of a given access network). The reasons for session transfer include cheapest access cost, better user experience, and physical user mobility.

SIP: Session Initiation Protocol is an IP telephony signaling protocol used to establish, modify, and terminate VOIP telephone calls. Other SIP applications include video conferencing, streaming multimedia distribution, instant messaging, presence information, file transfer, and online games.

WebRTC: It is an open framework for the Web that enables Real Time Communications (RTC) in a browser. In addition, it is a free and open project that enables Web browsers with RTC capabilities to communicate via simple JavaScript APIs.

Chapter 11
PriorityQoE:
A Tool for Improving the QoE in Video Streaming

Adalberto Melo
Federal University of Pará, Brazil

Antônio Abelém
Federal University of Pará, Brazil

Paulo Bezerra
Federal University of Pará, Brazil

Augusto Neto
Federal University of Ceará, Brazil

Eduardo Cerqueira
Federal University of Pará, Brazil

ABSTRACT

In the next generation of mobile network services, there will be the provision of multimedia services with the desired quality for wireless networks. In the future Internet, an integrated platform of cloud services will be made available within the XaaS (X-as-a-Service) paradigm. In the light of this evidence, the focal point of this study is an area that is very important to analyze, which is how to ensure a satisfactory Quality of Experience (QoE) for applications with video streaming. This chapter shows the PriorityQoE tool, which employs a methodology to establish a hierarchy for video streaming packets that are based on QoE objective metrics. It also outlines an intelligent mechanism for packet discard together with the PriorityQoE. The results of the performance evaluation of the tools showed that the effects of congestion on the network through the QoE of the video streaming were reduced. The QoE mediations were carried out by considering the knowledge of three QoE objective metrics (SSIM, VQM, and PSNR). The evaluation was conducted by means of a simulation of the transmission of multimedia content in IEEE 802.11 networking standards. The tools showed a better buffer handling and discarded the packet that least degrades the QoE of the video streaming.

DOI: 10.4018/978-1-4666-2833-5.ch011

INTRODUCTION

Recent research has shown that the Internet was not devised to support the required functions and performance that are needed by the increasingly wide range of real-time multimedia communication services. The increase in demand for services and multimedia applications together with the popularity of mobile equipment with access to the Internet have transformed the international network into a multimedia mobile network. This situation makes it a real challenge to ensure the efficiency of the multimedia network. The perceptions and satisfaction of the end-users are two factors of great importance. Other factors, like the robustness of the network, platform, and software, are needed to offer handling operations of quality and low cost. Thus, in so far as existing services are constantly evolving, new demands and services are also being introduced since it is necessary for the current networks to allow and provide flexibility, integration and re-usability. Solutions such as cloud computation and service platforms are key terms that are currently attracting the interest of the academic world, industry and government departments throughout the world. Owing to its huge potential for realizing what is desired, cloud computation has the capacity to completely change the models for providing services for the current IT industry. These solutions set out a new criterion for rendering services that start from scratch, with a reduced initial investment, expected performance, high availability, a tolerance capacity for failure and infinite scalability. It opens up a new era where new businesses can create or provide services that are in accordance with this new concept (Zhou, et al., 2010).

There is an important trend in the increase of the cooperativeness and flexibility of digital systems. This trend is the introduction of *X-as-a-Service* (XaaS) *paradigms* such as *Software as a Service* (SaaS), together with the PaaS platform and infra-structure like the IaaS service (Zhou, et al., 2010). These paradigms are already being implemented by employing the virtualization of computational resources as virtual machines. This approach allows innovations like computational grade, which can enable the computational capacity of a range of equipment to be used at times of idleness in an efficient and distributed way. Studies aimed at ensuring the quality of the multimedia network have shown results that have disputed the efficiency of the traditional *Quality of Service* (QoS) metrics, as a means of measuring the quality perceived by the user with regard to the use of multimedia services and applications. For this reason, efficient metrics that can allow the multimedia content perceived by the user to be evaluated is of crucial importance. These metrics can be regarded as *Quality of Experience* (QoE) metrics. It is necessary to consider QoE metrics in solutions for handling the multimedia network to ensure its quality. The Forward Error Correction (FEC) mechanisms and the packet discards are examples of mechanisms for the handling of the network that become more efficient when the QoE is taken into account. These handling tools seek to undertake a more in-depth study of the video streaming by considering specific features, such as the coding parameters of the video. These also include the Group of Pictures (GOP) framework, Codifiers and De-Codifiers (CODECs), the average and maximum bit rate of the video, types of charts and the respective degrees of dependence etc.

In this chapter, we provide a tool called *PriorityQoE*, which employs a methodology to ensure the QoE of video streaming transmitted by the network. PriorityQoE allows an evaluation of the packets transmitted with the video streaming. This is carried out through the hierachization of the video frames on the basis of objective QoE metrics. In the context of the platform services, PriorityQoE can be easily made available to the multimedia service providers. By employing this tool, the providers will economize on bandwidth by conserving and sharing network resources. The performance evaluation of PriorityQoE was carried out through a case study, where the transmission of three video streaming was simulated in a mobile scenario. The three video streaming employed are quite well known among the academic community and are codified in the MPEG-4 standard. Three QoE objective metrics (PSNR, VQM, and SSIM) were employed together with PriorityQoE to hierarchize the packets that were transmitted. The same three objective QoE metrics (PSNR, VQM, and SSIM) were also used to evaluate the QoE of the received video.

In the next section, we show the state of the art used to evaluate the QoE in video streaming. We present the main metrics and the approaches for the optimization of wireless networks. After that, we examine the PriorityQoE solution, the methodology, and the tools that were employed. We present the performance evaluation of PriorityQoE through a case study, the applied scenario, and an analysis of the results. Then, we discuss the challenges and trends in wireless multimedia systems that consider QoE. In the final section, we provide our conclusions and make our suggestions for future work.

BACKGROUND RESEARCH OF QOE

QoE Metrics

Conservative mechanisms that are designed to enhance the quality of the multimedia flows are essential in QoS metrics. Network management mechanisms can be defined to handle the delivery of multimedia flows of high quality in heterogeneous networks. Such mechanisms involve the current measurement indicators based on the packets (Zapater & Bressan, 2007). At the same time, the QoE metrics that were shown were based on advances achieved in multimedia systems. These metrics revealed benefits in the assessment of the quality of multimedia applications since they are an essential part of human perception. The QoE metrics can be divided into (1) subjective and (2) objective categories. These metrics serve to assess the level of quality and identify failures in the multimedia flows such as: blocking, blurring, and errors in color. The subjective methods consist of conducting opinion polls through punctuation schemes, regarding the quality noticed in the processed video compared with the original video. The objective methods consist of mathematical models, which are employed to estimate the performance of the multimedia systems and attempt to approximate to the results of the subjective metrics.

Subjective Metrics

The subjective metrics are satisfaction surveys that are carried out with the users of multimedia desktop applications to find out their "opinions" about the video and audio quality they receive. The surveys must comply with the procedures laid down in International Telecommunication Union, Telecommunication Standardization Sector (2002). The *Mean Opinion Score* (MOS) is the most widely used subjective metric. The MOS adopts a punctuation scale from 1 to 5 and this should reflect the degree of satisfaction of the user with regard to the perceived video and/or audio quality. Lower the punctuation, the worse the quality of the video perceived by the user. Table 1 shows the MOS punctuations.

The *Single Stimulus Continuous Quality Evaluation* (SSCQE) is an example of a subjective metric and allows the televiewers to assess the quality of a long video sequence through sliding equipment combined with a scale of quality. However, the subjective metrics do not possess scalability and their measurements are complex, time-consuming, and expensive, and thus they cannot be used for real-time assessment.

Objective Metrics

The PSNR metric (Nishikawa, et al., 2008) is a complete reference of the metrics employed for the

Table 1. Mean option score

MOS	Quality	Impairment
5	Excellent	Imperceptible
4	Good	Perceptible but not annoying
3	Fair	Slightly annoying
2	Poor	Annoying
1	Bad	Very annoying

assessment of videos. When calculating a ´final´ quality, this metric employs a mean value of the difference between the luminosity value of each pixel of the original frame which is processed so that the calculation of the final quality can be carried out. Apart from the fact that it is widely used because of its low complexity, the PSNR metric provides an indication of the difference between the received frame and a benchmark. However, it fails to take into account aspects of the Human Visual System (HVS) (Wang, et al., 2004), which can have a powerful influence on the quality of the video that is perceived. Their typical values for videos with losses are between 30 dB and 50 dB, since the greater the value of the PSNR, the better the assessment. The PSNR of a video is calculated on the basis of the Mean Squared Error (MSE) only considering the luminance (Y) of the original and processed frames and assuming the pixel width (M) from the height of the pixels (N). The MSE is calculated from Equation 1 (see Table 2).

In Equation 1, the Ys (i,j) represents the pixel position (i,j) in the frame origin, Yd (i,j) represents the pixel position (i,j), that is the equivalent of the frame that was used. On the basis of the MSE 8bits metric/with pixels from a frame MxN, the PSNR is obtained through a logarithmic scale of twenty at base 10. The values obtained through the PSNR objective metric can be related to the punctuation employed by the MOS subjective metric. The mapping of its values is displayed in Table 3.

SSIM is another metric that is often employed in objective assessment. SSIM is based on the PSNR metric but takes into account the HVS features such as the perception of the human eye (Rouse & Hemami, 2008) as well as luminosity, contrast and structural distortions; it is able to combine these three factors which, when calculated, results in a single index. The values for the index vary between 0 and 1, since the higher the level, the better the assessment of the video quality.

The *Quality Video Method* (VQM) is a single measure which combines the effects on perceptions of video failures and takes into account of blurring, spasmodic and artificial movements, global noises, block distortion, and color distortion (Revés, et al., 2006). This metric shows results that are higher than those of PSNR and MSE. They are calculated by means of a linear combination of calculated parameters based on calibration, the

Table 2. MSE and PSNR equation

$$MSE = \frac{1}{M \times N} \sum_{i=0}^{M-1} \sum_{j=0}^{N-1} \left\| Ys(i,j) - Yd(i,j) \right\|^2$$	(1)
$$PSNR = 20 \log_{10} \left(\frac{255}{\sqrt{\frac{1}{M \times N} \sum_{i=0}^{M-1} \sum_{j=0}^{N-1} \left\| Ys(i,j) - Yd(i,j) \right\|^2}} \right)$$	(2)

Table 3. PSNR to MOS conversion

PSNR (db)	MOS
> 37	5 (Excellent)
31 – 37	4 (Good)
25 – 31	3 (Fair)
20 – 25	2 (Poor)
< 20	1 (Bad)

quality of extraction of features and the quality of the calculation of the parameters. This metric possesses a high correlation coefficient with the subjective assessment of the video quality. The results obtained for this metric vary between 0 and 5 where the lower the punctuation obtained, the better the quality of the video.

Another metric that considers features of the HVS model (Lambrecht & Verscheure, 1996) for assessing video quality, is the *Moving Picture Quality Metric* (MPQM). This metric receives as input both from the original video and the distorted video, by first calculating the difference between the original video and the distorted video. Based on the HVS model, the original and codified Video Streaming are separated and then broken down into targeted placements that use uniform areas, and contour and texture classification. Detection calculation is carried out which involves sensitivity to the contrast of the HVS model and masking parameters. To complete the process, the channels produce a single image that shows the levels of perception that are most accentuated. The MPQM requires a very high memory usage because it is a complex metric and carries out a filtered space-time processing. Following the recommendations of Lambrecht and Verscheure (1996), the resulting values can be shown through masking with the PSNR equation (MPSNR) or can be mapped on an MOS scale.

Another metric that considers the HVS model in its assessment is the *Perceptual Evaluation of Video Quality* (PEVQ) (OPTICOM, 2010a). It is necessary to combine the HVS model with the mechanisms for spatial-temporal measurement. The values for assessing the degradation of the video quality are supplied in the MOS values that results from an end-to-end assessment of communication. Information about the level of perception of the luminance-induced distortion and chrominance, as well as the time factors in the assessment of the video, are also shown by the PEVQ metric.

The wireless network can be optimized by being integrated with a content management system that involves deploying sensitive and self-adjustable management mechanisms. Such management mechanisms are suited to Video Streaming and network features and are based on QoE metrics that have been devised and agreed on for the assessment of video quality.

QOE Assessment Approaches

In order to improve wireless systems, both subjective and objective QoE metrics are needed to measure the level of quality of the applications and procedures for optimization in situations where there is congestion or failures of network devices. For this reason, a means of classifying the methods for assessing video quality was drawn up. This classification is based on reference procedures with the video and comprises a Full Reference (FR), a Reduced Reference (RR), and a No-Reference (NR).

- **Full Reference (FR):** The main feature of the FR is that it employs a reference to the original video to obtain the level of quality of the processed video. FR makes a comparison between the differences of each pixel image of the processed video and the corresponding pixel image of the original video. The FR assessment metrics achieve a higher performance although it is difficult to implement them in real-time systems (monitoring agents with other agents), or in other words, for the assessment they depend on the original video or a multimedia sequence as input (which is common in simulation environments). The PSNR, SSIM, and MPQM metrics are examples of FR metrics.

- **Reduced Reference (RR):** The RR only regards selected features of videos like movement information as essential for the assessment of quality. The objective of classified metrics like RR is to match their precision in assessing video quality with the FR metrics and make less use of the network resources and processing. The reference parameters can be transmitted from a secondary channel or assembled with multimedia flows. The VQM metric was developed by National Telecommunications and Information Administration (NTIA) and described by Pinson and Wolf (2004) as being a classified example like RR.
- **No-Reference (NR):** The NR metrics can be employed for network monitoring services and diagnostic operations. In these services, the original sequence of the Video Streaming is not known. Thus, the classified metrics like NR attempt to assess the multimedia service quality without any reference to the original content. Nonetheless, these metrics have certain drawbacks compared with the classified metrics like FR and RR. Some examples of these disadvantages are the low correlation with MOS, the high consumption of CPU and memory usage, and time constraints. The V-Factor (OPTICOM, 2010b) is an example of a classified metric like NR. The classified metrics like RR and NR are those that can be most suitably employed in measuring video quality.

With measurement classifications based at the level of application, time demands, and processing, and the support of feasibility studies based on content, it is possible that the multimedia quality prediction mechanisms can be employed as management tools in the wireless networks as a means of ensuring the quality of the video transmitted. These tools have the capacity to foresee the quality of multimedia content after the transmission, on the basis of the context (codified parameters, inspections of packets and network conditions). Thus, it is not necessary to deal with the original data subsequently and this considerably reduces the consumption of the necessary resources for the assessment because it entails less complex mechanisms. These prediction mechanisms can be regarded as an extension of the current measurement system, which involves the assessment of quality from the perspective of the user, and improves the assessment of required multimedia applications (Mu, et al., 2008).

It is based on approaches to wireless networks where the suitable alternatives for controller assessment in emerging wireless systems are as follows: the assessment of the quality of multimedia content, which basically involves verifying all the transmitted packets that are related to the application and network conditions and where the decodifying process is not necessary. The main question is that these approaches require a thorough analysis to be carried out in the packets.

They have to collect information about the current network conditions, such as the rate of packet loss and delay in the unidirectional packets, so that it can be used in the assessment process. The final decision about assessing the level of quality can be taken on the basis of previous information, together with information about multimedia features such as frame-rate, GOP, the frame type and dependence, and whether it is only available at the level of application. This approach is preferable for a service (real-time) and for multimedia applications where the computational complexity is reduced. The performance is low to average but the feasibility is high. For example, in a simpler scenario, the quality indicators only consist of certain QoS parameters such as the relation of packet losses or the bit error rate, while an inspection of the hybrid content and network measurement approaches have also been recommended (Romaniak, et al., 2008). The main reason for devising this kind of scheme is to allow the network operators to combine the benefits of the previous approaches, while stipulating the degree of complexity, performance and viability as well as addressing the question of the operational costs that might be incurred by the different needs, type of media content, networks, and equipment.

QoE Optimization in Wireless Networks

Real-time multimedia content of quality is one of the stringent conditions for the Multimedia Services of the Future. By requiring a large number of network resources, this matter has become the object of a good deal of research in the area of technology. At the same time, more attention is now being paid to the quality perceived by the end-users, since concern about the influence that multimedia applications in real-time exert on other types of traffic flow in the Internet is becoming increasingly important. However, the use of control algorithms for multimedia traffic envisages the guarantee of QoS and QoE for both fixed and mobile users in the wireless networks. Zhou et al. (2010) investigate some approaches that are adopted in the sphere of cloud platform computing services, which in the context of the Future Internet draw on different concepts directly related to the creation of XaaS services, as well as companies that are already seeking solutions in this area. A study of the impact of the loss of individual packets in four key Video Streaming with H.264 format and employing four metrics (PSNR, SPIC, TPDR, SPRR) is shown in Mu et al. (2009). Kovács et al. (2010) examine an algorithm to calculate the required conditions for the bandwidth for Multimedia Variable Bit Rate services. The calculation is based on the mapping of a QoE descriptor that assesses the quality perceived by the users and a QoS descriptor where the

expected amount of loss is calculated. To archive this, three metrics (PSNR, SSIM, and VQM) were employed to measure the quality perceived by the user. The model for the proposed system is designed to estimate the bandwidth of the flow that is needed for a required level of quality. The bandwidth reserve is divided into two stages:

- In the first stage, the method determines the amount of loss by the end-user, and
- In the second stage, the required bandwidth is calculated on the basis of the maximum loss permitted. This second procedure is modeled by using a Discrete Time Markov Chain (DTMC) queuing system, which determines the arrival process at the video queue based on codec H.264/SVC.

The video model shown by the authors has not been devised to provide an exact description of the original video but to allow an efficient bandwidth reserve. In another study, Serral-Gracià et al. (2010) outline Autonomic Network Management (ANM) techniques that automate resource reservation in traffic engineering through user satisfaction metrics; this means that the assessment of QoE requires definitions of a set of metrics, which are able to provide an objective assessment of the final user satisfaction. However, in practice, the assessment can be complex because it involves a series of subjective factors, which are generally not related to network performance, such as for example, the sense of humor of the user. There is a need to understand the new challenges that QoE assessment is facing with regard to subjective factors. Such factors are related to human perception in the sense of allowing a more precise assessment of the quality experienced by the users. For this reason, Serral-Gracià et al. (2010) assess multimedia traffic through the network by using QoE mechanisms where there is a definition of the kind of metrics that are suitable for the assessment of quality. These metrics have been divided into two distinct categories called:

- **Direct Metrics:** These are obtained on the basis of several types of data in different layers, requiring specific information about the network performance; and
- **Indirect Metrics:** These take into account the properties that affect the experience of multimedia but which are not directly related to the quality of multimedia content.

In another study, Greengrass et al. (2009) conduct tests to assess the impact of different durations of packet losses. In practice, the loss can occur because of the loss of transitory connectivity following link failures or in our own experiences of the network, in the best and worst scenarios of MPEG transmission flow. The tests that were conducted showed that each scenario could verify the type of artifacts displayed and the impact of QoE on the spectator. They were carried out in consecutive periods when there was a loss of IP packets, which varies from 10ms to 500ms for high-resolution video and for Standard-Definition (SD) and high-definition resolutions. It was confirmed by the tests that were conducted, that the same conclusions found for SD also applied to HD; there were no significant differences between SD compared with HD during an equivalent period of packet losses. Costa et al. (2011) examine control mechanisms for packets that are designed to maintain the Video Streaming applications at an acceptable level of quality in WiMax (Worldwide Interoperability for Microwave Access) networks. These mechanisms take account of the degradation of the QoE, depending on the type of frame and its links.

THE PRIORITY QOE SOLUTION

Methodology

This section outlines a methodological proposal and employs the case study referred to in this Chapter. The purpose of this is to hierarchize the Video Streaming packets based on the objective QoE metrics, which can be used together with several types of mechanisms and solutions for the provision of QoE as a service in the context of NaaS. For a better understanding of the methodology outlined in this study, it is divided into stages in the subsections below.

Hierarchization of the Packets

In this stage, the first activity carried out is the drop packet of a single video streaming and the assessment, which employs an objective QoE metric. This stage is repeated for all the other packets and finally, these are arranged in accordance with the degree of importance attached to them by the metrics employed for the video. Subsequently, the results of this assessment of the video with objective QoE metrics reflect the decline in the level of user satisfaction that each discarded packet causes in the video.

Although we obtained the results anticipated, when this procedure was undertaken, the computational cost increased considerably because the drop procedure is repeated for each Video Streaming packet that is transmitted. For this reason, and in view of the block structures for GOP (Group of Pictures) of the MPEG (Moving Picture Experts Group) codification, the videos are generated by a codification involving a closed GOP, where the frames of a GOP are completely independent of the frames of the other video GOPs. In the case of the videos codified with closed GOP, the methodology again discards a single packet from each GOP rather than the whole video and following this, carries out an assessment in the same way by means of an objective QoE metric for each frame of the video received. The sum of the results for the assessment of the frames for GOP is worked out and then an assessment obtained for each GOP. The results of the assessment of GOP with objective QoE metrics reflect the level of degradation of QoE that a single discarded packet can cause the video GOP. The drop procedure will be repeated for the other packets of each video GOP and in the end, the packets are arranged by the assessment results for GOP until they form the hierarchy that was intended. This specialized methodology for codified video with closed GOP allows a considerable reduction in computational costs to be made. The reason for this is that if the methodology is employed without any alteration for a codified video, with closed GOP and 600 packets, (with a maximum of 33 packets for the GOP), it will be necessary to carry out a drop procedure and assessment of the video 600 times. Whereas if the methodology is employed for codified videos with closed GOP, the drop procedure and assessment of the video will only have to be carried out 33 times.

Packet Drop Mechanisms Based on Hierarchization

Control mechanisms for the intelligent drop of packets based on QoE are of vital importance to maintain the quality of multimedia transmissions with the aim of optimizing the bandwidth made available by the network. Within the context of NaaS, where the network is responsible for testing applications that are self-adjusted and offer suitable conditions for the traffic of multimedia content, the measurement of QoE as a service becomes increasingly desirable. In this subsection, we outline a possible control mechanism for selective drop, which is governed by the hierarchization of packets based on the degradation of the QoE Video Streaming. The mechanism entails controlling the packets that will be received by the buffer queue and, as a means of prioritizing the packets, employs the measurement of QoE degradation that the packet causes the video if it is discarded. In this way, instead of drop packets at random during periods of congestion, the mechanism optimizes the drop process by taking account of the importance that the packet that is going to be discarded, has for the video. The mechanism is initiated when the buffer queue is full. It compares the priority of the packet received with the priority of the packets that are queuing up at the buffer. The mechanism discards the packet of least priority. If the received packet has less priority than all the packets in the buffer, it will be discarded. If not, the packet of least priority that is in the queue at the buffer will be discarded and the packet received will enter the queue. The way this mechanism operates is illustrated by the pseudo-code shown in Table 4.

PriorityQoE Tool

In this section, we examine the tools employed for adopting the methodology outlined in this study. The adoption of the methodology was made possible by devising a software called PriorityQoE. Using PriorityQoE together with the ffmpeg (FFmpeg, 2011), mp4box (GPAC, 2011), mp4trace, etmp4 (Klaue, 2007), msu_metric (Video Group of MSU Graphics and Media Lab, 2011) tools, and on the basis of the data described as DBQoE, the following procedures were carried out:

- Import of information from the video in DBQoE.
- Generating codes.

Table 4. Pseudo-code for the selective drop algorithm

Pseudo-code for the selective drop algorithm
01 if queue.is_not_full():
02 queue.enqueue(packet)
03 else:
04 pktToDrop:= NULL;
05 pkt:= queue.firstPacket();
06 priorityQoEpacket = packet.getPriorityQoE();
07 while queue.isNotEof() do
08 if (priorityQoEpacket < pkt.getPriorityQoE())
09 priorityQoEpacket = pkt.getPriorityQoE();
10 pktToDrop = pkt;
11 pkt:= queue.getNextPacket();
12 if pktToDrop:
13 queue.drop(pktToDrop);
14 queue.enqueue(packet);
15 else:
16 drop(packet);

- Employing the dropping procedures.
- Allowing a widespread execution and assessment of videos.
- Importing the results of the assessments for DBQoE and storing them by mean of frames, GOP, and video.

The PriorityQoE Tool, together with the DBQoE database, was developed with the aim of allowing the methodology to be employed for codified videos with closed GOP. Figure 1 shows the PriorityQoE Block Diagram.

The PriorityQoE employs the methodology in four processes: (1) import videos, (2) create test plan, (3) evaluation, and (4) order by metric.

The "import video" process is where the video data, its codification, and transmission are imported to the DBQoE. The input parameters of the PriorityQoE tool are shown in Table 5. Three external commands are executed to import the video, although they are handled and carried out by the PriorityQoE tool. The first is to undertake the conversion of the video (originally in h264 format) to m4v format by using the ffmpeg tool. On the basis of the input parameters, the ffmpeg tool is executed within the parameters shown in Table 6 (line 1).

Later on, the PriorityQoE employs the mp4box tool to convert the video to an m4v format, using as input, the parameters shown in Table 6 (line 2). By following the codified parameters established for the user and using the mp4trace tool, the video transmission is carried out in a mp4 format through the wireless network. The mp4trace tool employs the input parameters shown in Table 6 (line 3). When the "import video" process is finalized on the basis of a generated trace file, the video information, the frames and codification of

Figure 1. PriorityQoE block diagram

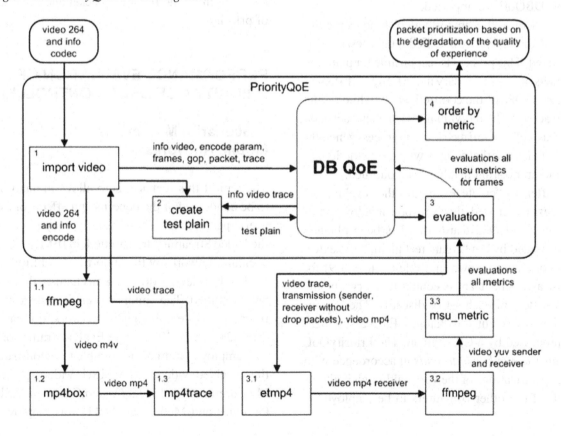

Table 5. Input parameters of the PriorityQoE tool

Parameter	Description
folder	Folder where the file for the H264 video is located
video	Name of the file with the H264 video
GOP	Maximum size of the GOP video
fps	Number of frames per second in the video
cgop	If true, inform that the video will be codified with the closed GOP
sameq	If true, inform that the video will be codified with the same quality as the original video
vcodec	Version of codec that will be used
bf	Maximum number of Type B frames among the Type P frames
size	Video Format Resolutions
hintmtu-size	Maximum size of the packets for bit rate video transmission
yh	Height of the video in pixels
yw	Width of the video in pixels

the packets and the transmission for the Priority-QoE/DBQoE are imported.

As soon as the "import video" process is completed, the next process, called "create test plan" is initiated. This process is responsible for planning the sets of drop necessary in the DBQoE, to assess all the GOPs of the codified video, while taking into account the individual loss of each packet on the basis of the QoE metrics. The process generates sets of discards that comply with the methodology of hierarchization that has been outlined.

After the "create test plain," the "evaluation" process is initiated; this is responsible for assessing each set of discards that has been planned and created by the "create test plain." The external commands shown in Table 6 (lines 5-9) for each assessment are executed to carry out the assessment of each set of discards. These commands are executed in batches. Each assessment is processed by a CPU thread. The PriorityQoE limits the *number of threads* in accordance with the specifications of the CPU, although it allows CPUs from other computers to be employed if

they have been installed with PriorityQoE and are connected to the network.

The external commands execute the etmp4, ffmpeg, and msu_metric tools and are able to reconstruct the received video without the discarded packet. It also converts the video to the YUV format, which is a format that is suitable for assessment, to ensure that the video can be assessed in accordance with the specified objective QoE metrics. Once the processing of each assessment has been completed, the importing of information from the assessment of the videos processed for the DBQoE is carried out.

In the "order by metric" process, the hierarchization of the packet video streaming is undertaken on the basis of assessment measures of the GOP frames, in case the packet is discarded. Since each assessment is based on specific objective QoE metrics, the hierarchization of the packets is effected in accordance with a specified objective QoE metric. This hierarchization is exported to a file containing the ID of the packet and its order of priority.

PERFORMANCE EVALUATION OF PRIORITY QOE DROP CONTROLLER

A Scenario: Multimedia Mobile Networks

The aim of this scenario is to allow an analysis to be conducted of the benefits and effects of the users' perceptions. These benefits may improve the Video Streaming transmitted (in IEEE 802.11 standard networks with support for mobility) as well as the selective dropping mechanism that has been outlined. The simulation environment was represented by employing the Network Simulator 2 (NS2) (NS-2, 2011). The Evalvid framework was employed to make it possible to reconstruct the received videos. The MSU Video quality Measurement Tool 2.7.3 (Video Group of MSU Graphics and Media Lab, 2011) was employed

Table 6. External commands executed by the PriorityQoE

External commands executed by the PriorityQoE			
Stage	**Software**	**External commands**	
		line	**command**
import video	Ffmpeg	1	ffmpeg -s [param.size] -bf [param.bf] -r [param.fps] -flags [param.cgop] [param.sameq] -g [param.gop] -i [param.folder][param.video].264 -vcodec [param.vcodec] [param.video][dbqoe.codi.id].m4v
	MP4Box	2	MP4Box -hint -mtu [param.hintmtusize] -fps [param.fps] -add [param.video][dbqoe.codi.id].m4v [param.video][dbqoe.codi.id].mp4
	mp4trace	3	mp4trace -f -s 192.168.0.2 12346 [param.video][dbqoe.codi.id].mp4 > [param.video][dbqoe.codi.id].trace
evaluation	ffmpeg	4	ffmpeg -i [param.video].264 [param.video]+[dbqoe.codi.id]+.yuv
	etmp4	5	etmp4 -p -0 [param.video]-t[dbqoe.tran.id]-snd.dump [param.video]-t[dbqoe.tran.id]-rcv.dump [param.video]+[dbqoe.codi.id].trace [param.video]+ [dbqoe.codi.id].mp4 [param.video]-t[dbqoe.tran.id]-rcv.mp4
	ffmpeg	6	ffmpeg -i [param.video]-t[dbqoe.tran.id]-rcv.mp4 [param.video]-t[dbqoe.tran.id]-rcv.yuv
	msu_metric	7	msu_metric -f [param.video]+[dbqoe.codi.id].yuv IYUV -yw [param.yw] -yh [param.yh] -f [param.video]-t[dbqoe.tran.id]-rcv.yuv -sc 1 -metr VQM_YYUV -cc YYUV
	msu_metric	8	msu_metric -f [param.video]+[dbqoe.codi.id].yuv IYUV -yw [param.yw] -yh [param.yh] -f [param.video]-t[dbqoe.tran.id]-rcv.yuv -sc 1 -metr PSNR_YYUV -cc YYUV
	msu_metric	9	msu_metric -f [param.video]+[dbqoe.codi.id].yuv IYUV -yw [param.yw] -yh [param.yh] -f [param.video]-t[dbqoe.tran.id]-rcv.yuv -sc 1 -metr SSIM (precise)_YYUV -cc YYUV

to assess the videos with the VQM, SSIM, and PSNR objective QoE metrics.

The scenario is formed by a wireless network containing 4 nodes, two being the Access Points (AP), one the Mobile Equipment (ME), and one the Fixed Equipment (FE). The APs have an operating range of 50m and the distance between each of them is 95m. The FE is connected to the API and will remain fixed. The ME is initially connected to the AP2 at a distance of 56m from the API. The FE was responsible for sending the Video Streaming to the ME in the course of the simulation. The ME moves in a straight line in the direction of the API at a constant speed of 1.2m per second (National Research Council, 2010) and covers 12m in 10 seconds. While the ME is moving, it is disconnected from the AP2 and connected to the AP1. Video Streaming and VoIP flows were used for best-effort and background traffic, since each flow was mapped in an access

category of a different 802.11 standard. The flows were divided into access categories as follows: VoIP for AC_VO, Video Streaming for AC_V1, best-effort traffic for AC_BE and background traffic for AC_BK.

The akiyo video streaming, Foreman and Common Intermediate Format Sequences (CIF) were used. These videos have a 352x288 resolution and a duration of 10 seconds in the MPEG-4 compression standard, codified with the following parameters: 30 frames per second (fps); two B frames between each P frame; keeping the same quality as that of the original video; with a closed GOP; with a variable bit rate; and GOP sizes of 10, 15 and 18. The frames were broken up into blocks of 1024B.

The codifications can be identified as follows: akiyo_cif GOP 10, akiyo_cif GOP 15, akiyo_cif GOP 18, foreman_cif GOP 10, foreman_cif GOP 15, foreman_cif GOP 18, mobile_cif GOP 10, mobile_cif GOP 15 e mobile_cif GOP 18, making a total of 9 codified Video Streaming.

All the identified codifications were processed in advance in the PriorityQoE tool and resulted in the order of priority of each packet that remained in the buffer queue. The order gave priority to the packet that most degraded the video GOP and were based on an objective QoE metric. Three orders of priority were put into effect: one order based on the objective PSNR metric, another order based on the assessment of the SSIM metric and the last based on the VQM metric. The orders of priority were identified as: Order by PSNR, Order by SSIM and Order by VQM, respectively.

The orders of priority were employed together with the discard mechanism that uses the order of priority as the basis for the selection of the packet. Thus, we identify the dropping mechanism (MDoPSNR, MDoSSIM, and MDoVQM) as being when the Order by PSNR, Order by SSIM and Order by VQM are used respectively. As well as the mechanisms already discussed, we also identify the Default mechanism as a dropping mechanism of standard packets. 21 transmissions were carried out for each set of mechanisms and the codified Video Streaming, with congestion rates that varied from 0-50%.

Analysis of Results

Performance evaluation of the drop packet mechanisms based on prioritization was structured to allow analysis of the simulation results of videos akiyo, supervisor and mobile separately. Figures 2, 3, and 4 show the results of the simulations performed by transmitting the video akiyo, supervisor and mobile respectively. In each analysis, we present the means of evaluation results obtained by metrics PSNR, SSIM and VQM, the whole mechanism/video streaming transmitted.

Results for Akiyo Video

Figure 2(a) depicts the results obtained by averaging the ratings extracted with VQM metric akiyo video encoded with GOP size of 10, 15, and 18. The transmissions of akiyo video encoded with GOP size of 10 MDoPSNR, MDoSSIM and MDoVQM mechanisms exceeded Default mechanism in 34.45%, 33.78% and 33.78% respectively. The transmissions of akiyo video encoded with GOP size of 15 the Default mechanism is overcome by MDoPSNR, MDoSSIM, and MDoVQM mechanisms in 40.24%. The transmissions of akiyo video encoded with GOP size of 18 the Default mechanism is overcome by MDoPSNR and MDoSSIM mechanisms in 47.87% and the MDoVQM mechanism in 48.48%.

Figure 2(b) shows the results obtained by averaging the ratings extracted with metric PSNR akiyo video encoded with GOP size of 10, 15, and 18. The MDoPSNR, MDoSSIM, and MDoVQM mechanisms were overcome by Default mechanism the transmissions of akiyo video encoded with GOP size of 10 in 7.40%, 3.70% and 6.61%, respectively. The transmissions of akiyo video encoded with GOP size of 15 the Default mechanism is overcome by MDoPSNR and MDoSSIM mechanisms in 4.55% and 5.89% respectively and surpassed the MDoVQM mechanism in 0.80%. The transmissions of akiyo video encoded with GOP size of 18 the Default mechanism is overcome by MDoPSNR, MDoSSIM and MDoVQM mechanisms in 6.86%, 7.14% and 0.27%, respectively.

Figure 2(c) shows the results obtained by averaging the ratings extracted with SSIM metric akiyo video encoded with GOP size of 10, 15, and 18. The MDoPSNR, MDoSSIM and MDoVQM mechanisms were overcome by Default mechanism the transmissions of akiyo video encoded with GOP size of 10 in 1.02%. The transmissions of akiyo video encoded with GOP size of 15 the mechanisms equaled their evaluations. The transmissions of akiyo video encoded with GOP size of 18 the Default mechanism is overcome by

Figure 2. Average evaluation PSNR, SSIM, and VQM of akiyo video, for each approach and GOP´s

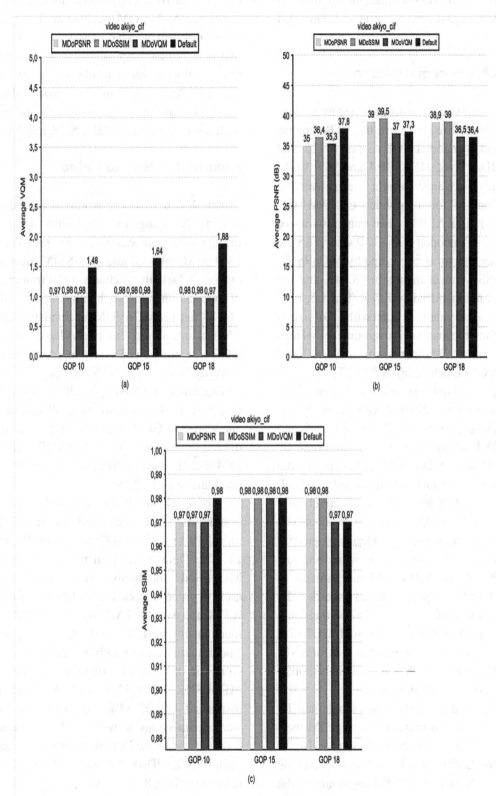

MDoPSNR and MDoSSIM mechanisms in 1.03% and equaled its assessment with the MDoVQM mechanism.

Results for Foreman Video

Figure 3(a) shows the results obtained by averaging the ratings extracted with the VQM metric foreman video encoded GOP size of 10, 15, and 18. The MDoPSNR, MDoSSIM, and MDoVQM mechanisms outweigh Default mechanism the transmissions of foreman video encoded with GOP size of 10 in 51.98%. The transmissions of foreman video encoded with GOP size of 15 the Default mechanism is overcome by MDoPSNR and MDoSSIM mechanisms in 48.94% and the MDoVQM mechanism in 49.47%. The transmissions of foreman video encoded with GOP size of 18 the Default mechanism is overcome by MDoPSNR and MDoSSIM mechanisms in 58.69% and the MDoVQM mechanism in 59.13%.

Figure 3(b) shows the results obtained by averaging the ratings extracted with metric PSNR foreman video encoded GOP size of 10, 15, and 18. The Default mechanism was overcome by MDoPSNR and MDoSSIM mechanisms the transmissions of foreman video encoded with GOP size of 10 in 2.82% and 2.50% respectively and exceeded the MDoVQM mechanism in 0.31%. The transmissions of foreman video encoded with GOP size of 15 the Default mechanism is overcome by MDoPSNR and MDoSSIM mechanisms in 0.93% and 1.56% respectively and exceeded the MDoVQM mechanism in 2.18%. The transmissions of foreman video encoded with GOP size of 18 Default mechanism exceeded MDoPSNR, MDoSSIM and MDoVQM mechanisms in 0.96%, 4.83%, and 6.77%, respectively.

Figure 3(c) depicts the results obtained by averaging the ratings extracted with SSIM metric foreman video encoded GOP size of 10, 15 and 18. The Default mechanism was overcome by MDoPSNR and MDoSSIM mechanisms the transmissions of foreman video encoded with

GOP size of 10 in 1.05% and equal to the obtained evaluation MDoVQM mechanism. The transmissions of foreman video encoded with GOP size of 15 mechanisms equaled their evaluations. The transmissions of foreman video encoded with GOP size of 18 Default mechanism exceeded MDoSSIM and MDoVQM mechanisms in 1.05% and matched with their assessment MDoPSNR mechanism.

Results for Mobile Video

Figure 4(a) depicts the results obtained by averaging the ratings extracted with metric VQM mobile video encoded GOP size of 10, 15, and 18. The MDoPSNR and MDoSSIM mechanisms outweigh Default mechanism the transmissions of mobile video encoded with GOP size of 10 in 66.78% and the MDoVQM mechanism in 69.89%. The transmissions of mobile video encoded with GOP size of 15 the Default mechanism is overcome by MDoPSNR, MDoSSIM and MDoVQM mechanisms in 68.64%, 69.30%, and 71.28%, respectively. The transmissions of mobile video encoded with GOP size of 18 the Default mechanism is overcome by the MDoPSNR and MDoSSIM mechanisms in 70.32% and the MDoVQM mechanism in 71.29%.

Figure 4(b) shows the results obtained by averaging the ratings extracted with metric PSNR mobile video encoded GOP size of 10, 15, and 18. The Default mechanism was overcome by MDoPSNR mechanism the transmissions of mobile video encoded with GOP size of 10 in 0.37% and exceeded MDoPSNR and MDoVQM mechanisms in 0.37% and 18.93% respectively. The transmissions of mobile video encoded with GOP size of 15 the Default mechanism exceeded MDoPSNR, MDoSSIM and MDoVQM mechanisms in 0.76%, 23.37%, and 6.13%, respectively. The transmissions of mobile video encoded with GOP size of 18 the Default mechanism exceeded MDoPSNR, MDoSSIM, and MDoVQM mechanisms in 6.87%, 8.77%, and 17.17%, respectively.

Figure 3. Evaluation PSNR, SSIM, and VQM of foreman video, for each approach and GOP´s

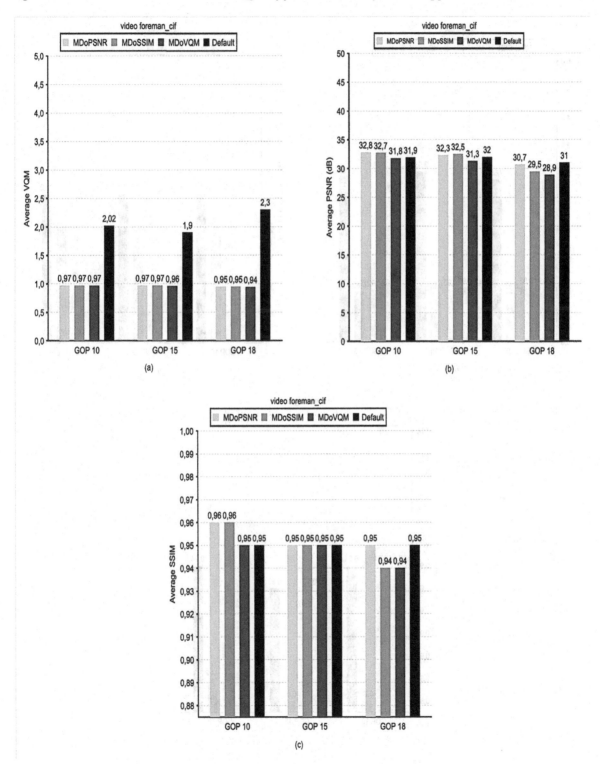

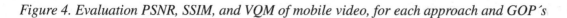

Figure 4. Evaluation PSNR, SSIM, and VQM of mobile video, for each approach and GOP´s

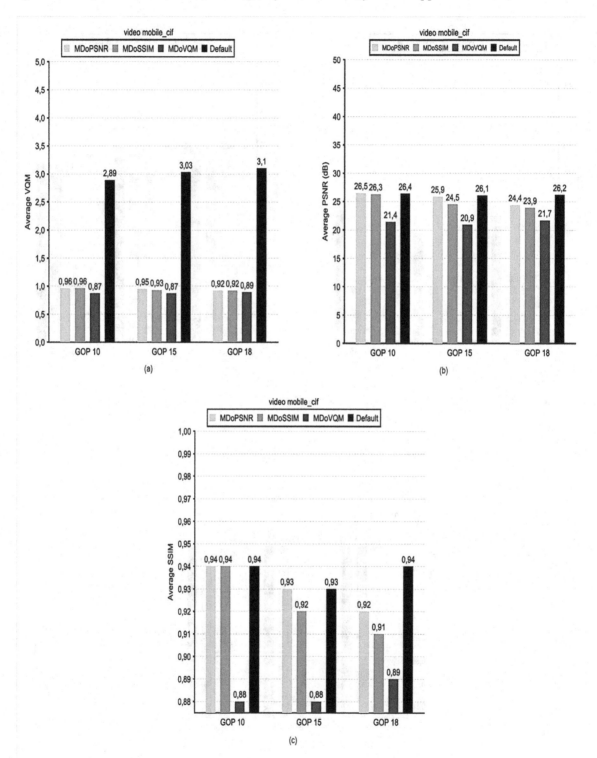

Figure 4(c) shows the results obtained by averaging the ratings extracted with metric SSIM mobile video encoded GOP size of 10, 15, and 18. The Default mechanism exceeded the MDoVQM mechanism the transmissions of mobile video encoded with GOP size of 10 in 6.38% and obtained the same assessment MDoPSNR and MDoSSIM mechanisms. The transmissions of mobile video encoded with GOP size of 15 the Default mechanism exceeded MDoSSIM and MDoVQM mechanisms in 1.07% and 5.37% respectively and equaled its evaluation with the MDoPSNR mechanism. The transmissions of mobile video encoded with GOP size of 18 the Default mechanism exceeded MDoPSNR, MDoSSIM and MDoVQM mechanisms in 2.12%, 3.19%, and 5.31%, respectively.

FUTURE RESEARCH DIRECTIONS

Recent works are being achieved in QoE management solutions for heterogeneous wireless networks. Other studies focus on policies to drop packets and sending error correction, both are based on packet characteristics and various network conditions. Such proposals do not need to access the received video stream. They can be used to predict and measure video quality. In this context, it is important to study novel solutions that make possible to identify the level of importance of each packet, in order to ensure the video quality according to the user's perspective.

There are several mechanisms that can be improved or adapted to provide video quality in multimedia networks. We can cite as an example: QoE-aware applications, transport-level mechanisms, and network optimization, routing, adaption in inter or intra-session, resource reservation, traffic control, full mobility, and selection of multimedia base station/schemes of user experience. These studies consider specific characteristics of each application. Among the most relevant applications, we cite the studies that have been focused on improving the network

transmissions of video streams, 3D videos, audio, games, Web applications, and others. Additionally, there are studies that consider the Regions of Interest (ROI). There is other research that deals with the improvement of video quality evaluation metrics according to the human perception. However, these advances require less processing and memory requirements. To overcome these computational challenges, QoE metrics should apply techniques of cloud and grid computing, in order to make possible the emergence of solutions that will ensure the quality level of real-time multimedia applications.

CONCLUSION

The optimization of control packets in wireless multimedia networks enables the generation of new solutions to maximize the use of network resources. Such a control optimization generates benefits that contribute to ensuring customer satisfaction, where the user's experience is a crucial parameter. Control policies on disposal of buffer queues and forward error correction based packet and network conditions, without the need to access the received video stream, may be used to measure and predict the video quality. In this context, it is of fundamental importance to study new solutions. This chapter has focused on three main areas: (1) the evaluation forms of video streams; (2) policy optimization controls packet; (3) the PriorityQoE technique associated packet drop mechanism and its performance evaluation. The PriorityQoE methodology that prioritizes packets that further degrade the video stream in relation to packages that degrade at least on an QoE objective metric, applied to a drop packet control in wireless network, proved effective in providing quality videos for broadcast, is outstanding in its assessment in VQM metric, which has a high coefficient of correlation with subjective assessment of video quality. In order to address some of the important issues that include intelligence multimedia wireless networks, this chapter has cover multimedia wireless networks along, with some of their challenges. We hope this chapter has enhanced understanding of issues and challenges for the next generation of multimedia networks, serving as a stimulus for researchers to search for innovative solutions to address and resolve these challenges.

REFERENCES

Costa, A., Quadros, C., Melo, A., Cerqueira, E., Abelém, A., Neto, A., et al. (2011). QoE-based packet dropper controllers for multimedia streaming in WiMAX networks. In *Proceedings of the 6th Latin America Networking Conference, LANC 2011,* (pp. 12-19). Quito, Ecuador: ACM Press.

FFmpeg. (2011). *Website.* Retrieved January 20, 2012, from http://ffmpeg.org/

GPAC. (2011). *MP4Box | GPAC.* Retrieved January 10, 2012, from http://gpac.wp.mines-telecom.fr/mp4box/

Greengrass, J., Evans, J., & Begen, A. (2009). Not all packets are equal, part 2: The impact of network packet loss on video quality. *IEEE Internet Computing, 13*(2), 74–82. doi:10.1109/MIC.2009.40

International Telecommunication Union. Telecommunication Standardization Sector. (2002). *Recommendation BT 500-11, methodology for the subjective assessment of the quality of television pictures.* Geneva, Switzerland: ITU-T.

Klaue, J. (2007). *EvalVid - A video quality evaluation tool-set.* Retrieved January 10, 2012, from http://www2.tkn.tu-berlin.de/research/evalvid/

Kovács, Á., Gódor, I., Rácz, S., & Borsos, T. (2010). Cross-layer quality-based resource reservation for scalable multimedia. *Computer Communications, 33*(3), 283–292. doi:10.1016/j.comcom.2009.09.006

Lambrecht, C., & Verscheure, O. (1996). Perceptual quality measure using a spatiotemporal model of the human visual system. In *Proceedings-SPIE The International Society for Optical Engineering* (*Vol. 2668*, pp. 450–461). San Jose, CA: SPIE.

Mu, M., Gostner, R., Mauthe, A., Tyson, G., & Garcia, F. (2009). Visibility of individual packet loss on H.264 encoded video stream--A user study on the impact of packet loss on perceived video quality. In *Proceedings of the Sixteenth Annual Multimedia Computing and Networking (MMCN 2009).* San Jose, CA: SPIE.

Mu, M., Mauthe, A., & Garcia, F. (2008). A utility-based QoS model for emerging multimedia applications. In *Proceedings of the Second International Conference on Next Generation Mobile Applications, Services and Technologies (NGMAST 2008),* (pp. 521-528). Cardiff, UK: IEEE Press.

National Research Council. (2010). Highway capacity manual. In *Transportation Research Board.* Washington, DC: National Research Council.

Nishikawa, K., Munadi, K., & Kiya, H. (2008). No-reference PSNR estimation for quality monitoring of motion JPEG2000 video over lossy packet networks. *IEEE Transactions on Multimedia, 10*(4), 637–645. doi:10.1109/TMM.2008.921849

NS-2. (2011). *Website.* Retrieved January 12, 2012, from http://www.isi.edu/nsnam/ns/

OPTICOM. (2010a). *PEVQ perceptual evaluation of video quality.* Retrieved January 16, 2012, from http://www.pevq.org/

OPTICOM. (2010b). *V-factor quality of experience platform.* Retrieved January 16, 2012, from http://www.pevq.org/

Pinson, M., & Wolf, S. (2004). A new standardized method for objectively measuring video quality. *IEEE Transactions on Broadcasting, 50*(3), 312–322. doi:10.1109/TBC.2004.834028

Revés, X., Nafisi, N., Ferrús, R., & Gelonch, A. (2006). User perceived quality evaluation in a B3G network testbed. In *Proceedings of the IST Mobile Summit*, (pp. 1-5). IST.

Romaniak, P., Mu, M., Leszczuk, M., & Mauthe, A. (2008). Framework for the integrated video quality assessment. In *Proceedings of the 18th ITC Specialist Seminar on Quality of Experience,* (pp. 81-89). Karlskrona, Sweden: Blekinge Institute of Technology.

Rouse, D., & Hemami, S. (2008). Understanding and simplifying the structural similarity metric. In *Proceedings of the 2008 15th IEEE International Conference on Image Processing (ICIP 2008),* (pp. 1188-1191). San Diego, CA: IEEE Computer Society Press.

Serral-Gracià, R., Cerqueira, E., Curado, M., Yannuzzi, M., Monteiro, E., & Masip-Bruin, X. (2010). An overview of quality of experience measurement challenges for video applications in IP networks. In *Wired/Wireless Internet Communications* (pp. 252–263). Luleå, Sweden: Springer. doi:10.1007/978-3-642-13315-2_21

Video Group of MSU Graphics and Media Lab. (2011). *Website*. Retrieved January 10, 2012, from http://compression.ru/video/quality_measure/index_en.html

Wang, Z., Lu, L., & Bovik, A. (2004). Video quality assessment based on structural distortion measurement. *Signal Processing Image Communication*, *19*(2), 121–132. doi:10.1016/S0923-5965(03)00076-6

Zapater, M., & Bressan, G. (2007). A proposed approach for quality of experience assurance of IPTV. In *Proceedings of the 2007 First International Conference on the Digital Society (ICDS 2007),* (p. 25). Guadeloupe, French Caribbean: IEEE Computer Society Press.

Zhou, M., Zhang, R., Zeng, D., & Qian, W. (2010). Services in the cloud computing era: A survey. In *Proceedings of the 2010 4th International Universal Communication Symposium (IUCS),* (pp. 40-46). Beijing, China: IEEE Computer Society Press.

ADDITIONAL READING

Cerqueira, E. C., Veloso, L., Curado, M., Mendes, P., & Monteiro, E. (2008). Quality level control for multi-user sessions in future generation networks. In *Proceedings of IEEE Global Telecommunications Conference (Globecom 2008),* (pp. 1-6). New Orleans, LA: IEEE Press.

Cerqueira, E. C., Zeadally, S., Leszczuk, M., Curado, M., & Mauthe, A. (2011). Recent advances in multimedia networking. *Multimedia Tools and Applications*, *54*(3), 635–647. doi:10.1007/s11042-010-0578-z

Engelke, U., & Zepernick, H. J. (2007). Perceptual-based quality metrics for image and video services: A survey. In *Proceedings of the 3rd Euro NGI Conference on Next Generation Internet Networks,* (pp. 190-197). NGI.

Garcia, M., Canovas, A., Edo, M., & Lloret, J. (2009). A QoE management system for ubiquitous IPTV devices. In *Proceedings of the Third International Conference on Mobile Ubiquitous Computing, Systems, Services and Technologies (UBICOMM 2009),* (pp. 147-152). UBICOMM.

Grega, M., Janowski, L., Leszczuk, M., Romaniak, P., & Papir, Z. (2008). Quality of experience evaluation for multimedia services. *Przegląd Telekomunikacyjnyi Wiadomości Telekomunikacyjne*, 142-153.

Jailton, J. J., Dias, K. L., Cerqueira, E. C., & Cavalcanti, D. (2009). Seamless handover and QoS provisioning for mobile video applications in an integrated WiMAX/MIP/MPLS architecture. *International Journal of Advanced Media and Communication*, *3*(4), 404–420. doi:10.1504/IJAMC.2009.028710

Ke, C., & Chilamkurti, N. (2008). A new framework for MPEG video delivery over heterogeneous networks. *Computer Communications*, *31*(11), 2656–2668. doi:10.1016/j.comcom.2008.02.029

Kishigami, J. (2007). The role of QoE on IPTV services style. In *Proceedings of the Ninth IEEE International Symposium*, (pp. 11-13). IEEE Press.

Monteiro, J. M., & Nunes, M. S. (2007). A subjective quality estimation tool for the evaluation of video communication systems. In *Proceedings of the 12th IEEE Symposium on Computers and Communications (ISCC 2007)*. IEEE Press.

Mu, M., Mauthe, A., & Garcia, F. (2010). A discrete perceptual impact evaluation quality assessment framework for IPTV services. In *Proceedings of the 2010 IEEE International Conference on Multimedia and Expo (ICME)*, (pp. 1505-1510). Singapore, Singapore: IEEE Press.

Parker, A. (2006). Addressing the cost and performance challenges of digital media content delivery. In *Proceedings of the P2P Media Summit*. Santa Monica, CA: P2P.

Schwarz, H., Marpe, D., & Wiegand, T. (2007). Overview of the scalable video coding extension of the H.264/AVC standard. *IEEE Transactions on Circuits and Systems for Video Technology*, *17*(9), 1103–1120. doi:10.1109/TCSVT.2007.905532

Video Traces Research Group. (2008). *YUV 4:2:0 video sequences*. Retrieved from http://trace.eas.asu.edu/yuv/qcif.html

Yamada, H., Fukumoto, N., Isomura, M., Uemura, S., & Hayashi, M. (2007). A QoE based service control scheme for RACF in IP-based FMC networks. In *Proceedings of the E-Commerce Technology and the 4th IEEE International Conference on Enterprise Computing, E-Commerce, and E-Services (CEC/EEE)*, (pp. 611-618). CEC/EEE.

Zhao, H., Ansari, N., & Shi, Y. (2005). Layered MPEG video transmission over IP DiffServ. In *Proceedings of the International Conference on Information Technology: Coding and Computing (ITCC 2005)*, (vol. 1, pp. 63-67). ITCC.

KEY TERMS AND DEFINITIONS

Drop Packet Control Mechanism: A buffer control mechanism that serves to manage the storage of packets in a buffer limited space.

Grid Computing: A collection of computing resources managed by distributed systems used to perform non-interactive workloads.

Mobile Services: These are services designed for mobile devices. Such services consider all the complexity that requires mobility.

Multimedia Streaming: A technology that transmits a multimedia file in packets. It allows the observation of the transmitted multimedia file according to the received packets.

Quality of Experience (QoE): A subjective measure of quality perceived by the user in some service types (game, video transmission, voice transmission, Web browsing).

Quality of Service (QoS): A metric that is usually related to the network parameters necessary to provide quality of services offered by the network. Such services can be file transfer, Web browsing, email, voice over IP, video streaming in real time, etc.

Wireless Systems: These are distributed systems that use wireless networks to communicate.

Chapter 12
The Realisation of Online Music Services through Intelligent Computing

Panagiotis Zervas
Technological Educational Institute of Crete, Greece

Chrisoula Alexandraki
Technological Educational Institute of Crete, Greece

ABSTRACT

This chapter presents an extensive, although non-exhaustive, study on existing Online Music Services (OMSs), which aims at identifying two principal characteristics: (1) the functionalities and interaction capabilities offered to their end-users; and (2) the tools of computational intelligence employed so as to enable these functionalities. The study is predominantly motivated by the ever-growing impact of Music Information Retrieval (MIR) research on the music industry, as new approaches for knowledge acquisition are rapidly integrated in existing online services targeting music consumers, musicians, as well as the music industry. Since MIR is inherently addressing user needs in music aggregation and distribution, the first part of the chapter is dedicated to illustrating user functionalities and accordingly classifying existing OMSs. The second part of the chapter focuses on musical semantics, different methods for harvesting them, and approaches for exploiting them in existing OMSs. Finally, the chapter attempts to foresee functionalities of future OMSs enabled by forthcoming MIR achievements.

INTRODUCTION

Music is possibly the most popular information domain in multimedia technologies and networking applications. Currently, a large number of online music repositories containing tens of millions of music tracks and a notably large number of related applications and services exist. It is interesting to note that these online services have radically altered the way music content is being created, consumed, disseminated, distributed and commercialised. For instance, music downloads are becoming so popular that they are compensating for the decline of CD sales (Casey, et al., 2008).

DOI: 10.4018/978-1-4666-2833-5.ch012

Therefore, a systematic study of the interactions supported in such services is of particular significance, not only to music consumers, but also to the music industry and consequently to music professionals themselves.

The functionalities offered in current Online Music Services (OMSs) range from providing instant access to specific music sources (e.g. Grooveshark.com), or structured information of interest (e.g. allmusic.com), to serving less common needs such as identifying a song played in a music club (e.g. www.soundhound.com) or predicting the market potential of songs to be released (e.g. uplaya.com). The abundance of online music and music related information, in combination with the requirement to efficiently support diverse user requirements, demands for the development of robust computational tools that automatically analyse, structure, cluster and visualise music material. Such tools commonly follow one of two existing approaches, namely 'content-based approaches' and 'context-based approaches.'

Specifically, content-based approaches rely on analysing music sources so as derive music descriptors associated with semantically relevant music properties. In this chapter, the term 'music source,' may refer either to audio files (i.e. signals of digital audio) or to symbolic representations of music (i.e. MIDI or music scores in various data formats). Alternatively, context-based approaches rely on Web-mining techniques for retrieving manually provided information, which under certain conditions, may offer valuable knowledge describing music sources (e.g. artist profile, lyrics album art, etc.).

The rest of this chapter is structured as follows. The next section provides background information on the evolution of OMSs from conventional (static or dynamic) Web 1.0 pages to today's services of the semantic Web. Then, the chapter is split into two parts: The first part, entitled 'Online Music Services and End-User Communities,' presents OMSs from a user oriented perspective,

whereas the second part elaborates on tools of computational intelligence being employed so as to allow for enhanced user functionalities in contemporary music services. Specifically, in the first part different user functionalities are outlined and our classification of OMSs is presented. The section includes our observations of user and community based characteristics from a large number of OMSs. The second part, entitled 'Semantic Knowledge Acquisition,' initially presents the types of metadata used for indexing musical material and then distinguishes between content-based and context-based approaches for automatic metadata acquisition. Finally, the chapter concludes by presenting current research challenges and prominent perspectives in future OMSs that are yet to be realised.

BACKGROUND

Nowadays, there is a vast availability of music services offering the possibility to listen to music, retrieve relevant information, share, annotate and collaboratively interact with music material. In fact, it seems like every month or so an additional Web-based music service appears. It is even more astonishing to observe how quickly new music services are becoming popular and known for widespread usage. Therefore, it is important to examine the sequence of events and technological advancements that has lead to the profound explosion of OMSs. Fundamental reasons contributing to this explosion concern the proliferation of the Internet and the ever increasing availability of high-speed Internet access for consumers, the advancement of audio codecs offering high compression ratios without significantly degrading the perceived sound quality, as well as the availability of affordable storage equipment and the widespread adoption of mobile and ubiquitous devices by consumers.

Figure 1 depicts the advancements in information technology that have contributed to the

Figure 1. The timeline of events that have contributed to the evolution of online music services

widespread development of OMSs in desktop and mobile platforms. The first Web page (World Wide Web Project, 1991) was published in 1991 by Tim Burners-Lee at CERN as a proposal for an information system to facilitate sharing and updating information among researchers. Two years later, in April 1993 CERN announced that the World Wide Web would be offered free to anyone without any fee due (CERN, 2003).

Around the same time in 1993 the first public release of the ISO/IEC 11172-3:1993 (a.k.a. MPEG-1 Layer 3) international standard (Musmann, 2006) was announced. Shortly after the standard was published, it was realised that no provisions had been made for storing data describing the MP3 files. In 1996, Eric Kemp had the idea of adding a small chunk of data to the audio file, thus solving the problem of providing descriptions (O' Neill, 2006). This method of storing content descriptions within MP3 files is known as *ID3 tagging*.

The MP3 encoding grew to widespread use in the late 1990s. In 1998, MP3.com and Audiogalaxy.com music sharing communities were established. In June 1999, Napster was released as a centralised unstructured peer-to-peer system, requiring a central server for indexing and peer discovery. It is generally credited as being the first peer-to-peer file sharing system. Napster provided a service for indexing and storing MP3 file information that Napster users made available on their computers for others to download. MP3 files were transferred directly between the host and client users after authorisation by Napster. Following, Gnutella, eDonkey2000, and Freenet were released in 2000, as MP3.com and Napster were facing litigation. In July 2001, Napster was

sued by several recording companies and lost. Shortly after loss in court, Napster was shut down to comply with court order (Menn, 2003).

At approximately the same time in 2001, the iPod device and iTunes software appeared. Steve Jobs commented during the official launch of iPod by saying, "With iPod, listening to music will never be the same again" (Biersdorfer & Pogue, 2010). iTunes on the other hand is considered to be the music store that created a new music retail experience (Moody, 2011).

Advanced Internet technology emerged with the introduction of Web 2.0. The term was conceived in 1999 (DiNucci, 1999) but became popular in 2004 when the first Web 2.0 conference was launched (O'Reilly, 2005). In the context of Web 2.0, several Web applications that simplify information sharing, interoperability, personalised design, and collaboration capabilities emerged. Virtual settlements offering Web 2.0 technology allow users to interact and co-engage in editing user-generated content. Such capabilities provide great improvement over Web 1.0 technology websites where clients were limited to the passive retrieval of content offered by individual Web servers. Some examples of Web 2.0 services include social networking platforms, blogs, wikis as well as multimedia file sharing and tagging.

In respect with music, functionalities based on semantic information were officially considered at the 1st conference of the International Society of Music Information Retrieval in 2000. In 2002, Last.fm, one of the currently most popular personalised Internet radio stations was launched. Since 2005, Last.fm incorporates Audioscrobbler, a music recommendation system based on usage history of music content and collaborative filtering (i.e. retrieving music recommendations from users of similar taste). Currently, Last.fm offers end-user tagging and labelling of artists, albums and tracks as free unrestricted text. Moreover, it offers a free API allowing researchers as well as application developers to access the information provided by end users. As will be seen in the sections that follow, the possibility of accessing Last.fm metadata has introduced several innovative capabilities in offering music related information.

Another technological development, not yet fully exploited in current OMSs involves the evolution of Networked Music Performance (NMP) communities. These communities started to appear around 2006, with the first software suites (e.g. Jacktrip) allowing musicians to collaboratively participate in geographically distributed music performances through the network. Communities such as eJamming Audiio, Ninjam or the DIAMOUSES research project (Alexandraki & Akoumianakis, 2010) target musicians that are willing to collaborate with like-minded peers. Intelligent tools targeting novel functionalities in NMP communities have not been presented thus far, but are expected to have a great impact on the way music is created and performed online. This perspective is further elaborated in the section entitled 'future directions.'

ONLINE MUSIC SERVICES AND END-USER COMMUNITIES

This part of the chapter concentrates on the available OMSs from a user oriented perspective. Most OMSs aim at a wide range of potential audience, which can be arranged in three groups. Specifically, these user groups are music consumers (MC, i.e. users interested in finding music and music related information), music professionals (MP, i.e. musicians, musicologists, music teachers, and music researchers), and music industries (MI, i.e. record companies, radio broadcasters, music producers, etc.). Various functionalities and inter-

action capabilities are provided to these groups of users. The next section outlines user functionalities and the section that follows attempts to discern different categories of OMSs depending on the target user group and the functionalities aimed at.

Interacting with Music Content

Some of the available music services are tailored to provide a specific user functionality that identifies their purpose. Nevertheless, there are plenty of music services offering multiple functionalities. For instance, Last.fm is an Online Music Service primarily appearing as a personalised radio station, however at the same time offering the possibility to buy music, to tag music pieces, to search for lyrics, etc. This section attempts to outline individual user functionalities, disregarding the fact that they may be offered in combination with functionalities. Table 1 outlines these functionalities roughly ordered from the most conventional to the most innovative ones.

Searching for Music or Music Related Information: This functionality is supported by information systems enabling to search for a specific, artist, album or song. They commonly offer the possibility to play snippets and buy selected songs. The most popular example is allmusic.com, formerly known as All Music Guide (AMG). Today this functionality is offered by most OMSs (e.g. Last.fm, Jango, etc.).

Purchasing Music: Purchasing music is also allowed by most OMSs, however there are dedicated stores and retailers (e.g. Amazon, eMusic) that are appointed by specific OMSs. For instance, purchasing a song from Midomi redirects the user to the iTunes store. Regarding Digital Right Management (DRM) policies (Kasprowski, 2010) sites such as iTunes and Rhapsody encumber tracks with DRM software thus restricting where and how you play your purchased music. On the other hand, music stores such as eMusic, Jamendo and Napster offer DRM-free tracks.

Table 1. User functionalities supported in current OMSs

Index.	User Functionality	Target User Group	Popular Examples
1	*Searching for music or music related information*	MC, MP, MI	www.allmusic.com, www.midomi.com
2	*Purchasing music*	MC	iTunes, Google Music Beta, Amazon, eMusic
3	*Retrieving lyrics*	MC	http://www.lyricsmuse.com/
4	*Retrieving music scores*	MC, MP	Wemix, SoundCloud, indabamusic
5	*Rating music*	MC	eMusic, Youtube, iTunes
6	*Commenting on music (free text)*	MC	Last.fm, Youtube, iTunes
7	*Cloud Music Streaming*	MC, MP	iTunes, Google Music
8	*Uploading music*	MP	SoundCloud, wemix
9	*Promoting independent artists*	MP	Wemix, indabamusic, MySpace Music, Jamendo
10	*Selling music*	MP	Wemix, indabamusic, http://www.tunecore.com/
11	*Tagging music (folksonomy)*	MC	Last.fm
12	*Playing music games*	MC	TagATune, MajorMiner, Herd it (Facebook App)
13	*Listening to personalised radio stations*	MC	www.last.fm, http://www.jango.com
14	*Connecting with people of similar music taste*	MC, MP	wemix
15	*Visualising Music Collections (according to genre, mood, artist, etc.)*	MC	http://www.musicplasma.com/ http://musicartistcloud.appspot.com
16	*Identifying Music Content*	MC	Shazam, SoundHound, Musipedia, Midomi
17	*Receiving location based recommendations*	MC	AOL Radio
18	*Predicting popularity of new songs*	MI, MP	Uplaya.com
19	*Retrieving information using dedicated APIs*	MP, MI	Last.fm, MusicBrainz
20	*Making Music Collaboratively*	MP	eJAMMING AUDiiO, Ninjam, DIAMOUSES, IndabaMusic

Retrieving Lyrics: Some OMSs offer the functionality of retrieving the lyrics of a song, while it is being reproduced. This service is mostly carried out by streaming lyrics from online lyric databases such as Metrolyrics, Gracenote or WikiLyrics. The retrieved lyrics are either embedded in the OMS user interface, such as in Pandora.com, or provided as a link to an external Web page as in Jango Radio.

Retrieving Music Scores: Another functionality encountered in certain OMSs is that of retrieving music scores in various formats (i.e. pdf, ps, tiff and in proprietary notation formats). Such portals provide recordings along with their corresponding sheet music, without copyright restrictions. Examples of OMSs for this type of functionality

are the Mutopia, the Petrucci Music Library, the Choral Public domain library and the Musopen portal. The Musopen besides the aforementioned functionality offers social networking capabilities and streaming of randomly selected tracks from their repository.

Rating Music: Many services offer the possibility to rate songs or albums. For example, YouTube used to allow user rating on scale 1-5, but now it only provides 'Like' and 'Dislike' buttons. Last.fm permits pressing a "Love this track" button, and eMusic allows for rating an album on a 1-5 scale.

Commenting on Music: The possibility of commenting music entities is offered by several OMSs. For instance, MySpace allows for creating threads of comments, whereas in Last.fm comments are referred to as 'shouts.'

Cloud Music Streaming: This functionality provides a solution to the problem of syncing music libraries between users' multiple devices (i.e. laptops, tablets, smart phones). It is offered either as a free or as a paid service for several platforms. Some representative examples of related OMSs are Google Music, Amazon, Spotify and Apple's iCloud. Google Music, freely available and still in beta release, lets users upload their entire music library to the Web and stream it back over Google Web frontend or the Android Music app. Amazon provides 20 GB of space with the user's first mp3 purchase and any item purchased afterwards is saved there and accessed from any device. Spotify streaming service utilises a technology of streaming music from its own central servers as well as from person-to-person. Finally, Apple's iCloud apart from storing and streaming users' music repositories can additionally handle pictures, applications, documents etc.

Uploading Music: Uploading music is offered by the majority of the OMSs for purposes such as social interactions (MySpace, SoundCloud) or for promoting and selling artists' own music (Jango, Wemix).

Promoting Independent Artists: Internet technology allows artists to introduce their music to a potentially enormous audience at a low cost without necessarily affiliating with a major recording label. Services as IndabaMusic, Wemix, MySpace, Jango offer to independent artists functionalities such as message boards, music blogs for communicating to their audience, etc. In particular, Jango offers the Jango Airplay (airplay.jango.com) service which allows artists, with a per play fee, to have access to Jango's recommendations service.

Selling Music: With services such as Wemix, which among other descriptions defines its self as a user-generated record label, or the IndabaMusic artists have the possibility to sell their music and control its copyrights.

Tagging Music (folksomy): This functionality refers to the possibility of users to tag specific music entities such as artists, albums and tracks, in the form of unrestricted text. Last.fm provides inherent support for social tagging, as users may review other users' tags and label music entities as comma separated values. In the relevant literature, these values are referred to as 'cultural features' or 'community meta-data' (Schedl, et al., 2011) and the act of doing so, is called 'folksonomy.' There are no other known OMSs that provide folksonomy as an intrinsic characteristic. However, services such as iTunes and MusicBrainz allow people to enter free text in the 'comment' ID3 field, which is commonly embedded in the sound file.

Playing music games: Music games, also known as annotation games, aim at collecting semantic and emotional information while the user is listening to a certain piece of music. This is an approach to collecting social tags, which is alternative to folksonomy, as they rely on user's motivation to get high scores or give correct replies. Popular examples of such games are Tag-a-Tune (www.tagatune.org), MajorMiner (majorminer. org) and Herd It Facebook Application (Barrington, et al., 2009).

Listening to Personalised Radio Stations: Internet radio stations like Pandora radio, Last.fm and AOL radio have become very popular since they monitor user preferences and automatically create personalised playlists depicting these preferences, in respect with artists, genres, mood etc. Most of them allow for skipping certain songs from the recommended playlists, which is a possibility not available in conventional radio services.

Connecting with people of similar musical taste: This functionality refers to the possibility of users to receive recommendations and socially connect with other users having similar musical taste. Similarity is calculated directly by tracking users' listening habits (i.e. Last.fm, MySpace). Similarity is offered to users as 'People You May Know' information (i.e. MySpace) or as compatibility rating in a scale of 'Very Low' to 'Super' (i.e. Last.fm).

Visualising Music Collections: Novel visualisations and user interfaces for exploring music collections requires clustering artists or music material according to different similarity measures (commonly expressed as genre and mood attributes). As described in following sections, clustering and music similarity detection may follow either content-based approaches (i.e. compute similarity by comparing audio signals) or context-based approaches (i.e. retrieve listener preferences correlations from the Web). An example of an innovative user interface using content-based clustering is the *nepTune* (Knees, et al., 2007a, 2007b) interface, which given a music collection produces a 3D landscape where similar songs are arranged in close regions. Furthermore, nepTune landscapes may be enriched from Web-based retrieval, therefore offering genre information, album art cover, and other complementary information. MusicPlasma.com forms another example of a visualisation, which allows users to enter the name of an artist and then produces a dot. When clicked, this dot becomes a graph node connected with nodes represented by similar artists (see Figure 2).

Identifying Music Content: This functionality closely relates to the MIR research task entitled Query-By-Singing/Humming, and it is concerned with the possibility of retrieving information about a music piece by fetching a sound snippet to the music service. This functionality is most commonly available as mobile phone applications (i.e. Shazam, SoundHound) where the user is presumably in a music club or similar location and wants to know which music piece he or she is listening to. In addition to smart phone applications, Midomi Web portal offers an embedded application for song identification through singing and humming. Musipedia.org, presents a variation of this functionality by allowing users to perform a melodic or rhythmic search, using embedded applications receiving mouse or keyboard events, or the microphone for whistling.

Receiving Location-Based Recommendations: Location information is another feature that can be taken into account for providing recommendations to an OMS user. Recommendations made on the basis of user's location are usually concerned with music venues and gigs, record stores, and other musicians located in proximity. Thus, InbadaMusic.com connects musicians based on their listed skill levels, influences, instruments, location, and previously recorded content. Alternatively, the Soundtracker, combines Internet, radio, social networking and location-awareness therefore allowing to communicate with friends and find music performed in proximity.

Predicting Popularity of New Songs: 'Hit Song Science' concerns the possibility of predicting whether a song will be a hit, prior to its distribution, by using automated means incorporating dedicated machine learning processes. Uplaya.com is such a service that aims at assisting independent artists to investigate the market potential of their musical works.

Retrieving Information Using Dedicated APIs: Many OMSs allow application developers to implement software programs for desktop or mobile devices using their metadata repositories. For this purpose they provide downloadable APIs (Application Programming Interfaces) which allow for posing queries related to usage statistics, user provided data and ID3 tags.

Making Music Collaboratively: This functionality refers to the possibility of musicians to co-engage in online music making. In this respect, the IndabaMusic.com platform offers an online collaboration environment for musicians. Musicians can upload their original sound material and invite other musicians to join and co-engage in basic sound editing and mixing. At the other end of the spectrum, a number of online communities for Networked Music Performance exist. These communities, still in experimental stage due to various technical restrictions, commonly offer

Figure 2. Artist similarity graphs provided by musicplasma.com

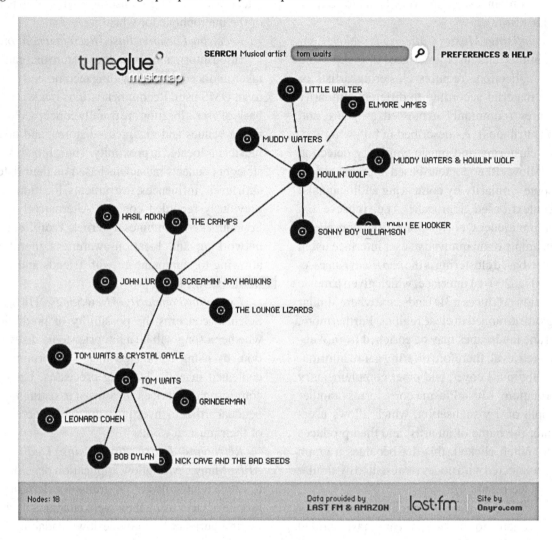

downloadable software suites that allow musicians to collaboratively perform a piece of music over the network as if they were co-located. Examples of such communities are those built around the Ninjam or the eJAMMING AUDiiO platforms.

Classification of OMSs and Representative Examples

Nowadays technology gives artists the opportunity to instantly reach the fans that enjoy our work.— Vicente Fernández

It's amazing how things have changed since I bought my first album and had to catch the bus to Livingston to go to the record shop. Now you just press a button. It's phenomenal the choice people have.—Susan Boyle

We have absolutely done our job as an industry to digitise and make available our repertoire.—Eric Daugan, Warner Music Europe

The previous quotes depict that the digital music revolution is clearly driven by consumer needs.

Consumers worldwide have largely embraced digital media, by adopting new paradigms for accessing entertainment. In response, music labels and service providers are utilising innovative business models and licensing a wide range of novel and sophisticated functionalities. Therefore, it is evident that the process of creation, production and distribution of music content as well as ensuring intellectual property rights is built around consumer requirements and trends. The plethora of OMSs, more than four hundred licensed worldwide (IFPI Digital Music Report, 2011), has as its "raison d'être" to promote and distribute the works of music professionals. In other words it represents the contemporary music market. Hence the 'Creation – Distribution – Access' dynamics depicted in Figure 3, focus on supporting and facilitating consumer access to music industry products. Details of various structural elements

of the aforementioned dynamics were presented in the previous section and listed in Table 1.

Acknowledging that the creation of music concerns music professionals, distribution concerns the music industry as well as music professionals (as there are several independent artists directly promoting their works) and access concerns primarily music consumers, we discern the categories of current OMSs portrayed in Figure 4.

Table 2 presents examples of OMSs for each category, while depicting the functionalities they support. This classification of services was based either on the service providers' formal descriptions or in cases where descriptions were not available, determined on the basis of the most prominent functionality they support (e.g. InbadaMusic allow musicians to collaborate online). The categories depicted in Table 2 are Personalised Radio Sta-

Figure 3. Creation, distribution, and access driving the consumption of music material

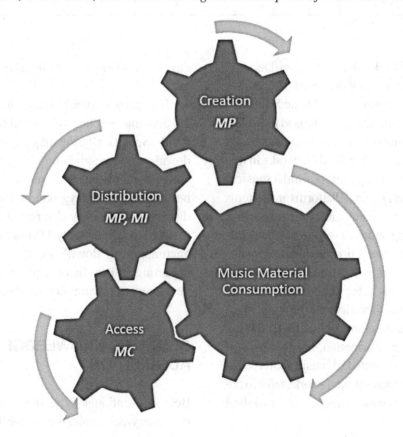

Table 2. Types of music services and the functionalities they support

Type of Service	Service Name	Users	Subscription	Buy music	Social Recom/tion	Cloud Player Capabilities	Streaming Audio	Lyrics Retrieval	Music Score Retrieval	Location Based Services	Mobile Apps	API/Developer Platform	Extra functionalities
Personalised Radio Stations	Last.fm	MC	✓	✓	✓		✓	✓		✓	✓	✓	Statistics, Social Network
	Jango	MC MP	✓	✓	✓		✓	✓			✓		Social Network, Jango Airplay
	Pandora	MC	✓	✓	✓		✓	✓			✓		Expert annotated data
	AOL Radio	MC	✓	✓	✓		✓				✓	✓	Recommendations by taste-making DJs
	Slacker Radio	MC	✓				✓				✓		Powers AOL radio
Social Networks	MySpace	MC MP		✓	✓		✓				✓	✓	-
	Indieland	MC MP			✓								-
	Wemix	MP	✓		✓		✓	✓	✓					Music creation community User-powered record label
Music Streaming on Demand	Spotify	MC MP	✓	✓	✓	✓	✓	✓		✓	✓	✓	DRM-based music streaming service
	Sound-cloud	MC MP	✓	✓	✓	✓	✓			✓	✓	✓	Allows collaboration, promotion & distribution recordings.

continued on following page

tions, Social Networks, Streaming on Demand Platforms, Music Information Systems, Music Collaboration Frameworks and Music Stores.

As already mentioned, personalised radio stations aim at offering personalised playlists based on user preferences while Social Networks in this context aim at connecting users of similar musical taste. Streaming on Demand Platforms are systems that provide a direct search and retrieval mechanism for streaming audio as well as the possibility to synchronise among private music repositories. (i.e. cloud music streaming). Music information systems refer to systems that are organised as digital libraries and provide functionalities such as searching, uploading, indexing based on semantic information, music scores or textual information related to music sources. Finally, music collaboration frameworks refer to the systems that offer functionalities related to online

engagement in music making and music performing.

The data presented in Table 2 provides evidence for drawing conclusions related to current OMS trends. Specifically, the majority of services offer downloadable applications for mobile devices, which demonstrates that mobile computing has become the new playground for music creation, distribution and access. Moreover, it is worthwhile to see that increasingly OMSs are interested in encompassing downloadable APIs, therefore expecting to benefit by widespread application development relying on their data repositories.

SEMANTIC KNOWLEDGE ACQUISITION

Both conventional as well as more innovative music services require the availability of semantic

Table 2. Continued

Type of Service	Service Name	Users	Subscription	Buy music	Social Recom/tion	Cloud Player Capabilities	Streaming Audio	Lyrics Retrieval	Music Score Retrieval	Location Based Services	Mobile Apps	API/ Developer Platform	Extra functionalities
Music Information Systems	Music-Brainz	MC	required for editing data permission								✓	✓	Online music metadata library
	Musipe-dia	MC MP						✓			✓		Identifying pieces of music (audio, whistle, pitch contour, virtual keyboard, Parson code)
	Allmusic	MC MP		✓									Online music guide service
	Shazam	MC	✓	✓			✓				✓	✓	Commercial mobile phone based music identification
	Midomi	MC	✓	✓			✓			✓	✓	✓	Music identi-fication
	Mutopia	MC MP							✓				-
	Petrucci Music Library	MC MP							✓				-
	Choral Public Domain Library	MC MP							✓				-
	Musopen	MC MP					✓		✓				Teaching sessions and textbooks
Music Collaboration Frameworks	Ninjam	MP											-
	eJam-mingAu-diio	MP	✓										Collaborative musical jam-ming
	LiveMu-sicPortal	MP											-
	Indaba-music	MP	✓		✓	✓			✓	✓		✓	Music col-laboration environment, personal store
Music Stores	iTunes	MI	✓	✓	✓	✓	✓				✓		Ping & Genius services
	eMusic	MI		✓		✓	✓				✓		-
	Amazon	MI		✓		✓	✓	✓			✓		-
	Napster	MI		✓			✓				✓		-
	Google Music Beta	MI		✓	✓	✓	✓				✓	✓	Connection to Google interface
	Uplaya	MP	✓	✓			✓						Song potential for commer-cial success

knowledge associated with specific music sources. Such human meaningful knowledge is mainly used by machine processes to allow for indexing and clustering music collections not only for direct search and retrieval but also for the purposes of playlist generation, music recommendations and so on.

This second part of the chapter concentrates on the semantic information required for offering the various interaction capabilities outlined in

Figure 4. Classification of OMSs based on their purpose

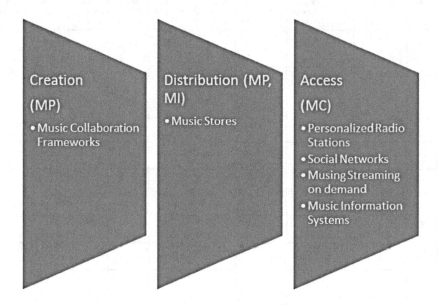

previous sections and gives an overview on how such information may be harvested, organised and checked for validity.

Metadata and Complementary Information

The term metadata is used to refer to data describing the contents of data instances (i.e. data files). Metadata is usually created by humans and used by machine processes for the purposes of automatic file indexing and clustering. File indexing is a prerequisite for efficient search and retrieval in large data repositories. The term originates from the <meta> tag of the HTML language, where it represents a page indexing methodology (Debaecker & El Hadi, 2011). This methodology was later extended to further textual resources with the aim of supporting technical and semantic concepts of information. The standardisation of metadata is a complicated task, especially due to the variety of data sources and the diversity of information content to be described.

As previously mentioned, musical metadata appeared shortly after the first official release of

the MP3 codec as the ID3 tagging scheme. This scheme allows for storing descriptive information along with the data stream and is currently supported by most audio formats, including uncompressed formats (e.g. wav, aiff). ID3 tagging comes in two unrelated versions, ID3v1 and ID3v2. ID3v1 provides tags for simple editorial information of a music piece such as track, title, artist, album, year, comment. ID3v2 provides less apparent information such as ratings, chapters, recording locations, cover images etc. Despite the popularity of the ID3 tagging scheme, there are several music descriptors that instead of being included in the audio file are stored in dedicated OMS repositories. This may be due to various reasons including: more efficient processing of external descriptors by dedicated APIs (e.g. Last. fm social tags), the use of descriptors that are not yet supported by ID3 scheme (e.g. scores), copyright restrictions (e.g. iTunes user rating does not use the ID3v2 rating tag), or simply due to the ambiguity of information stored in descriptors and the need for validity checking.

The proliferation of the semantic Web introduced new types of collaborative applications.

Among other things user participation in music tagging was expected to significantly contribute to several MIR tasks such as genre classification and mood detection, therefore increasing ease of access to music sources. A straightforward example of the contribution of social tagging to accessing music, concerns the possibility of developing intuitive visualisations of clusters residing in large music collections, such as tag clouds (e.g. as represented at musicartistcloud.appspot.com). Social tagging has lead to the conceptualisation of the term 'folksonomy' referring to the outcome of the act of tagging by the person consuming the information (Van Der Wal, 2007).

Social tags comprise of unstructured labels assigned to a resource without any restriction, as opposed to traditional keyword assignment where terms are often selected from a controlled vocabulary (Lamere, 2008). Although it is highly believed that folksonomy provides a powerful means for gaining information on how to connect source material (i.e. cluster or classify), there is strong evidence that it may be responsible for spreading noisy, spurious and irrelevant information concerning music sources. This is in fact the main criticism of Web 2.0 systems. As an indication, it is interesting to note that MusicBrainz, a popular and open source music encyclopaedia based on user contributions of music descriptions, does not support genre information as it is considered highly subjective and error prone (MusicBrainz – FAQ).

Web 2.0 technologies such as the Ontology Web Language (OWL) and the Resource Description Framework (RDF), are aimed at structuring the 'Internet information repository' in a more predetermined, less chaotic manner. In the music domain, adoption of this paradigm has lead to a number of research initiatives well depicted in the omras2 (www.omras2.org) and the easaier (www.elec.qmul.ac.uk/easaier/) research projects. Omras2 proposes an RDF compliant Music Ontology (musicontology.com) aiming at providing a vocabulary for linking music sources.

In this respect, the Music Ontology Language defines three levels of expressiveness - from the simplest one to the most complex one:

Level 1: Aims at providing a vocabulary for simple editorial information (tracks/artists/releases, etc.)

Level 2: Aims at providing a vocabulary for expressing the music creation workflow (composition, arrangement, performance, recording, etc.)

Level 3: Aims at providing a vocabulary for complex event decomposition, to express, for example, what happened during a particular performance, what is the melody line of a particular work, etc.

Clearly, the specification of the Music Ontology considers various music properties including both explicit as well as highly subjective information, while at the same time it makes provisions for maintaining information to be temporally aligned with a music piece, such as score, lyrics or other musicological properties.

Content-Based vs. Context-Based Approaches

Ideally, semantic knowledge describing music sources should be provided by expert users such as musicologists and other music professionals. However, expert tagging is extremely time consuming and requires heavy human labour. It has been estimated that it takes 20-30 minutes per track of one expert's time to enter the metadata (http://en.wikipedia.org/wiki/Music_Genome#cite_note-2). Given the enormous availability of music sources, the cost of expert tagging is unaffordable. Moreover, this approach does scale to future music collections that are expected to contain tens-of millions of tracks (Casey, et al., 2008). A straightforward solution to this problem would be social tagging, which however requires to sacrifice the quality of the provided information.

The alternative to manual human (expert or user) tagging is to harvest metadata using tools of computational intelligence. In respect with intelligent information retrieval, two approaches have been devised, namely 'Content-based MIR' and 'Context-Based MIR.'

Content-Based Music Information Retrieval

Content-based approaches attempt to derive semantic information related to music sources by analysing the sources themselves, which as previously mentioned may be either audio signals, or symbolic representations such as MIDI, MusicXML, etc. The common characteristic of Content-based MIR approaches is that music features are computed from short audio segments. In majority, content-based methods are inspired from the text information retrieval area of research.

Generally, content-based approaches to semantic knowledge acquisition follow a two phase procedure. The first phase includes the extraction of features from the music source. In the case of audio signals, such features would be either temporal or spectral qualities of the signal that are computed on every audio frame (a signal segment of 5-50ms). The second phase of content-based approaches concerns the use of statistic models and machine learning techniques so as to identify patterns of feature vectors (the set of attributes per audio frame) as classes of mid-level (i.e. melodic or rhythmic patterns) or higher-level information (e.g. artist, genre, mood). These models (also known as classifiers), are 'trained' using supervised learning techniques. Supervised learning requires a large pool of correctly annotated and well-balanced data, the dataset. A key issue in content-based approaches is to obtain datasets that are appropriately annotated for specific information retrieval tasks.

Figure 5 illustrates the typical workflow in a content-based MIR system. The left part depicted as 'offline procedure' illustrates the process of obtaining a dataset from a set of annotated music sources and further obtaining a trained model, appearing as 'classification function.' The classification function is obtained using supervised learning techniques, which roughly means the computation of probability distributions of features for the various labels available in the dataset. The right part of the diagram, 'online procedure,' shows the retrieval of relevant results from a posed query, commonly a given music source. The process initially involves feature extraction from the music source. Then these features are fed to the pre-trained classification function which produces the label that yields the maximum probability for the provided feature vectors. The produced label is then used for querying a music library for relevant results. Depending on the type of query (i.e. identification of content, similarity matching etc.) the results may be either textual information or lists of pointers to additional music sources carrying the same label. For an explicit example, consider the task of query-by-humming. The online procedure takes as input an audio waveform, produces a vector of features which are then fed to the classification function (the result of offline training as depicted in the figure). The classification function produces a label which is then used to query the music library and retrieve the requested information about the specific piece such as title, band, discography, etc.

In respect with feature vectors to represent audio signals, a variety of attributes have been proposed and further organised in related taxonomies for grouping them. Weihs et al. (2007) have classified the audio features into four subcategories which are short-term, long-term, semantic and compositional features. Scaringella et al. (2006) presented an alternative taxonomy for audio features, which were used for the task of genre classification and comprised of three groups related to timbre, rhythm, and pitch. Finally, Fu

Figure 5. The typical workflow of a content-based MIR system

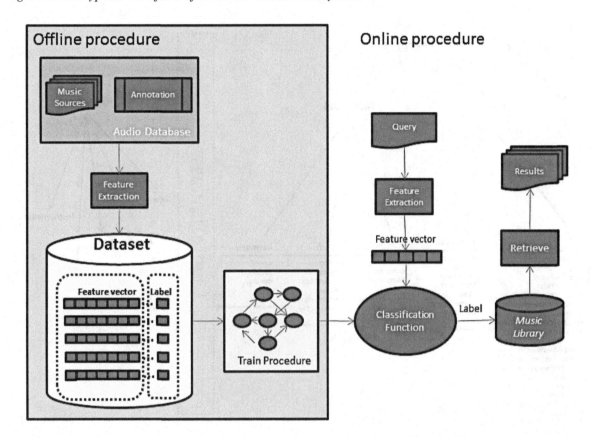

et al. (2011) proposed the division of features into low-level and mid-level features, therefore reserving high-level for semantically relevant information. As shown in Figure 6, the semantic gap between features and labels requires for statistical inference, i.e. classification learning which accounts to providing the most probable label given the features. In other words, the statistical analysis of low-level and mid-level information allows for crossing the semantic gap, and thus providing semantically meaningful, higher level information.

In this respect, low-level features are further subdivided to timbral and temporal features. Timbral features serve as indicators for the timbral qualities, namely the instrumentation, of a given audio frame, while the second depict the temporal evolution of such qualities. On the other hand, mid-level features account for non-timbral music

properties, which are rhythm, pitch and harmony of a piece. Rhythm, i.e. the sequence of strong and soft beats within a piece can largely characterise the music genre. This becomes obvious if we consider that certain rhythmic patterns are exclusively used in certain music styles (e.g. blues, traditional music). Pitch accounts for the melodic structure, namely the horizontal evolution of a piece, while harmony is informative about chords and harmonic progressions, namely simultaneous notes or otherwise viewed as the vertical content, which are also informative for the style of music. A definition of the features presented in Figure 6 is beyond the scope of the present chapter. The interested reader may refer to Jensen (2009). The following paragraphs attempt to outline some research initiatives that facilitate

Figure 6. Classification of audio features

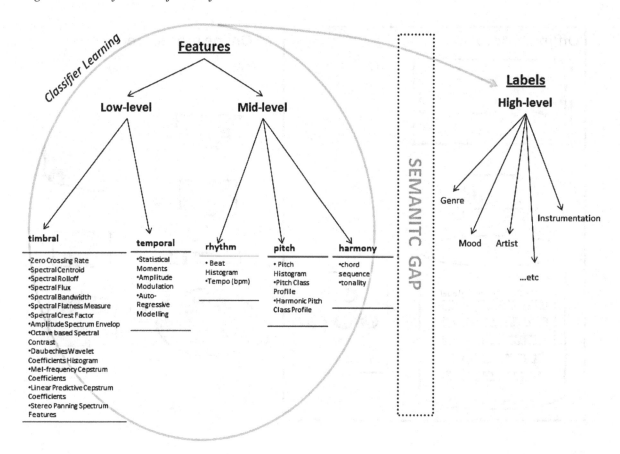

different types of features in order to perform content-based classification.

Spectrum characteristics and amplitude envelope are the physical characteristics of sound that determine the perception of timbre (Iverson & Krumhansi, 1993). Focusing on timbral attributes of music several research efforts have been conducted. Logan and Salomon (2001) proposed an approach for Content-based MIR that facilitates Mel Frequency Cepstrum Coefficients (MFCCs) (Fang, et al., 2001). MFCCs are a representation of the short-term power spectrum of a sound, based on a linear cosine transform of a log power spectrum on a nonlinear MEL frequency scale. In the work of Aucouturier and Pachet (2002), that followed MFCC feature datasets were utilised in order to estimate the parameters of a probability

density function represented as the weighted sum of Gaussian component densities known as Gaussian Mixture Model (GMM) (Reynolds, 2009). In another attempt (Mandel & Ellis, 2005), music similarity was estimated by using Support Vector Machines (SVM) which finds the maximum margin hyperplane separating two classes of data. This approach resulted in increased classification performance as well as optimised computation time. Thus, it is considered one of the standard methods in computing music similarity.

Datasets depending on rhythm properties have been used in Foote and Uchihashi (2001). In specific, this work employed 'beat-spectrum' for modelling rhythmic patterns. The beat-spectrum is defined as a measure of acoustic self-similarity according to a time lag. Thus, highly structured or repetitive music is described by strong beat spectrum peaks at the repetition times. This reveals tempo and the relative strength of particular beats, and therefore analysing the beat-spectrum allows for distinguishing between different rhythms performed at the same tempo. Rhythm features for the calculation of music similarity were also facilitated by Pampalk et al. (2002) where fluctuation patterns (i.e. two dimensional representations of periodicities in the audio signal per frequency band) were exploited. In Pohle et al. (2009), certain improvements of the fluctuation patterns were achieved most notably by facilitating onset patterns.

Besides algorithms explicitly modelling rhythmic and timbral qualities, a number of music similarity approaches try to combine multiple features to improve their performance. In the pioneering work of Tzanetakis and Cook (2002), a similarity signature combining various aspects of audio representation was presented. Each music signal was described by a 30-dimensional vector comprising of of timbre, rhythm and pitch features. Timbre and rhythm features were also combined for measuring similarity in the work of Neumayer et al. (2005). In Pampalk (2006), fluctuation pattern similarity was combined with the single Gaussian timbre features of Mandel and Ellis (2005). Finally, a number of different music features describing musical timbre, rhythm, onsets or tonality was proposed by Seyerlehner et al. (2010). The aforementioned acoustic features were extracted on a block basis instead of very short frames, as it has been done in most methods before.

Context-Based Music Information Retrieval

Context-based approaches also referred to as 'Web-based' or 'community-based' approaches, do not use information extracted from the audio file itself, but rather employ external resources. They generally employ Natural Language Process-ing (NLP) techniques to derive information from Web sources, such as lyrics or manually annotated data (user or expert tagging) or music collections owned by users (Chakrabarti, et al., 1999).

The general approach is concerned with analysing a number of textual resources containing a given artist name or a song name. The extracted information is checked for validity by computing various probabilities of tag occurrence related to Document Frequency (DF) and Term Frequency (TF) (Schedl, et al., 2011). The output of this process involves information about music artists such as biographies, song lyrics, members of a band or instrumentation for specific songs (Schedl, 2011). The user functionalities aimed at, include enriching information in music players, providing novel user interfaces for browsing, automatically creating playlists and so on. In music recommendation systems, context-based approaches are often combined with Collaborative Filtering (CF) techniques (Cohen & Fan, 2000). CF-based systems attempt to predict the taste of a user by comparing his or her preferences with the preferences of similar, like-minded users (Sarwar, et al., 2001).

Although context-based approaches have been used successfully in several cases, however they suffer from a few shortcomings that should be essentially considered prior to algorithm implementation (Celma, 2006). These shortcomings are summarised in the following: (1) *"the cold start problem"* (Eck, et al., 2007), which concerns the fact that inferences cannot be drawn from insufficient data relating to certain entities such as songs, artists/bands, etc., (2) *"the long tail problem or popularity bias"* (Anderson, 2006), referring to the problem of certain entities having little related data in comparison to others being over-informed due to their popularity or long term use. The following paragraphs provide an overview of the main research approaches in context-based music information retrieval, namely Web-based queries, co-occurrence analysis of textual data, collaborative tagging and tags retrieved from P2P networks.

Research in context-based MIR was initially focused on Web data extracted from search engines or provided by specialised Web crawlers (Chakrabarti, et al., 1999). Whitman and Lawrence (2002) attempted to infer artist similarity based exclusively on textual Web resources.

The frequency of occurrence of a pair of terms from a text corpus alongside each other in a certain order has also been utilized for music information retrieval tasks. This approach is known as co-occurrence analysis and in a linguistic sense can be interpreted as an indicator of semantic proximity. Examples of research based on this approach can be found in the work of Schedl et al. (2006) for genre classification of artists, and in Miotto et al. (2010) for the purpose of automatic annotation and retrieval of music. Alternatively, the works of Logan et al. (2004), Mahedero et al. (2005), and Laurier et al. (2008) propose song lyrics as a further source of deriving music similarities. Specifically, Logan et al. (2004) tried to use lyrics for artist similarity estimation, Mahedero et al. for song similarity retrieval and Laurier et al. (2008) utilized song lyrics for mood detection.

Methods that utilize data from collaborative tagging systems (Last.fm API, Pandora Genome Project, etc.) are yet another category of context based information retrieval. Research paradigms utilizing this approach is the work of Geleijnse et al. (2007) and Green et al. (2009). Geleijnse in his work facilitated the API provided by Last.fm for the construction of a "tag ground truth" set for artists by filtering redundant and noisy tags. Green et al. (2009) proposed a collaborative-filtering method utilizing Last.fm data for music recommendation. Next to online repositories of tagged data the idea of gathering tags through online games was proposed. One of the first games was Tagatune by Law et al. (2007).

Last but not least, data from user collections that reside in P2P networks (Shen, et al., 2009) has been considered as an invaluable resource for various music information retrieval tasks, including music similarity (Shavitt & Weinsberg, 2009), recommendation and trend prediction (Goto, 2012). Some examples of P2P network topologies that have been employed in MIR research are Napster (http://www.napster.com), Kazaa (http://www.kazaa.com), and Chord (Stoica, et al., 2001).

Comparison and Discussion

The main argument promoting the use of context-based over content based approaches is that the later do not take into account aspects such as the cultural context of an artist, the semantics of the lyrics, and the emotional impact of a song on the listeners, and therefore content-based approaches reach an upper bound on the performance of recommendations that can be made (Schedl, et al., 2011). Based on this argument, a number of hybrid (content and context) systems have been proposed (Boletsis, et al., 2011; Knees, et al., 2007a).

There are several examples of OMSs adopting one, the other or hybrid approaches to semantic knowledge retrieval. For instance, OMSs offering the functionality of music content identification use purely content-based approaches. In these systems a technique known as 'audio fingerprinting' is facilitated in order to quickly identify perceptual similarity in audio sources. An audio fingerprint can be thought of as a compact digital summary of the contents of an audio signal comprising of several perceptually relevant characteristics such as the zero-crossing rate, the estimated tempo, the average spectrum etc. Examples of such content-based processing solutions are the MusicDNS service incorporated in MusicBraiz, the Midomi song identification functionality as well as its mobile equivalents SoundHound, Gracenote'sMusicID and Shazam. In respect with content-based music recommendations the Mufin music player (www.mufin.com), also available as an iTunes plugin, estimates content-based similarities based on feature vectors from private music libraries.

A representative example of context-based music recommendations is the one presented by iTunes Genius (Mims, 2010). Genius processes information regarding iTunes usage such as song sharing, play counts and star ratings and performs recommendations using Collaborative Filtering techniques. Another example is the MusicPlasma visualisation tool, which as described in previous section, produces graphs connecting similar music artists based on metadata mined from Amazon and Last.fm.

In respect with hybrid approaches, the Echonest Platform adopted by several popular OMSs, combines large-scale data mining, natural language processing, acoustic analysis and machine learning. For example, in Spotify, the Echonest platform allows to present a slideshow of pictures while users are listening to music, while in eMusic, the Echonest playlist and taste profiling engine is used to build song lists based on song attributes.

FUTURE RESEARCH DIRECTIONS

Despite the vast availability of OMSs claiming to offer advanced functionalities that are beyond conventional search and retrieval and despite the rapid growth of related information technologies, the effectiveness of the offered functionalities is debatable. For example, automatic playlist generation (or personalised radio services) do not sufficiently reflect users' personal and contextual preferences. This phenomenon is either due to technologies being immature or due to service providers' not being up-to-date with contemporary technological advancements.

Personalisation is a major issue when interacting with music services. Therefore it is important to comprehend aspects of user profiles with regard to the perceptual and emotional impact of music to individual users. Relevant research in user modelling should more closely address cognitive, cultural, social and psychological attributes to user profiling. Moreover, contextual attributes

such as user's local time and place as well as the connecting device (portable or desktop), should be more explicitly reflected in the decision making processes of intelligent OMSs.

The First International ACM Workshop on Music Information Retrieval with User-Centered and Multimodal Strategies (MIRUM), held in November 2011 (http://mirum11.tudelft.nl/), seems to explicitly address these requirements and to offer a valuable perspective in this respect. MIRUM is expected to disseminate research achievements and raise awareness to OMS providers about areas such as user (context) models and personalisation, interactive music systems and retrieval, adaptive user interaction and interfaces and so on (Liem, et al., 2011).

At the other end of the spectrum, the ever growing availability of music sources in current OMSs requires addressing issues of scalability and robustness of the techniques facilitated for acquiring data descriptions. In this respect, a major requirement is to devise benchmarking measures for evaluating knowledge acquisition relevant to offering effective user functionalities. For example, the MIREX framework provides a successful approach for evaluating research oriented MIR tasks (Downie, 2008). We believe that such evaluation procedures should be extended to more explicitly address the needs of the three target user groups of current and future OMSs.

Another promising perspective in future OMSs is expected to emerge from the proliferation of Networked Music Performance communities. Currently, several NMP communities exist (e.g. eJAMMING AuDiiO, Ninjam, LiveMusicPortal, etc.). However, their growth is severely restricted by technical hurdles related to network bandwidth consumption, communication delays due to long geographical distances, synchronisation problems, etc. (Alexandraki & Akoumianakis, 2010). Focusing too much on the problem of alleviating technical deficiencies in audio communication has constrained the development of this area towards other directions. For instance, the integration of tools of computational intelligence and information retrieval has hardly been considered. In this respect, we believe that integrating semantic knowledge in NMP systems and communities would allow for expanding NMP applications in multiple directions. Relevant user functionalities may involve the enhancement of musicians' virtual co-presence by providing real-time contextual information about music performances (e.g. audio-to-score alignment, synchronisation by revealing instant rhythmic progressions), the development of social networking among music performers (i.e. automatic similarity matching of musicians' performances, may allow for connecting musicians that would otherwise be impossible to collaborate), the enrichment of information sharing by Web-mining music repositories, etc. Ultimately, the enhancement of NMP communities by offering advanced user functionalities may lead to surprisingly interesting results in music making therefore influencing not only musicians but also music consumption and consequently the music industry.

CONCLUSION

This chapter has presented an exploratory study on existing Online Music Services focusing on user functionalities and interaction capabilities offered to different groups of users. Aspects such as the possibilities offered by Web 2.0 frameworks as well as the sequence of consecutive advancements in information technology and digital music that have lead to the present evolution of OMSs were discussed. Initially our investigation focused on presenting OMSs from a user oriented perspective. Thus, user functionalities were elaborated along with representative examples of popular online services. Subsequently, OMS users were classified in three groups depending on their purposes in the "creation, distribution, access" chain. Based on

the three groups of users, a tentative classification of the OMSs was presented. Then the chapter attempted to elucidate the technical approaches that are facilitated in order to enable convetional and advanced user functionalities.

The chapter concludes with exposing the need of improvement in current OMSs and attempts to foresee novel paradigms of making, distributing and accessing music based on the emergence of online music making communities.

REFERENCES

Alexandraki, C., & Akoumianakis, D. (2010). Exploring new perspectives in network music performance: The DIAMOUSES framework. *Computer Music Journal*, *34*(2), 66–83. doi:10.1162/comj.2010.34.2.66

Anderson, C. (2006). *The long tail*. New York, NY: Hyperion Press.

Aucouturier, J. J., & Pachet, F. (2002). Music similarity measures: What's the use. In *Proceedings of the 3rd International Conference on Music Information Retrieval (ISMIR 2002)*, (pp. 157-163). Paris, France: ISMIR.

Barrington, L., O'Malley, D., Turnbull, D., & Lanckriet, G. (2009). User-centered design of a social game to tag music. In *Proceedings of the ACM SIGKDD Workshop on Human Computation*. New York, NY: ACM.

Biersdorfer, J. D., & Pogue, D. (2010). *IPod: The missing manual*. Sebastopol, CA: O'Reilly Media.

Boletsis, C., Gratsani, A., Chasanidou, D., Karydis, I., & Kermanidis, K. (2011). Comparative analysis of content-based and context-based similarity on musical data. In *Proceedings of the EANN/AIAI (2) 2011*, (pp.179-189). Corfu, Greece: EANN/AIAI.

Casey, M. A., Veltkamp, R., Goto, M., Leman, M., Rhodes, C., & Slaney, M. (2008). Content-based music information retrieval: Current directions and future challenges. *Proceedings of the IEEE, 96*(4), 668–696. doi:10.1109/JPROC.2008.916370

Celma, O. (2006). Foafing the music: Bridging the semantic gap in music recommendation. *International Journal of Web Semantics: Science. Services and Agents on the World Wide Web, 6*(4), 250–256. doi:10.1016/j.websem.2008.09.004

CERN. (2003). *10 years public domain.* Retrieved February 7, 2012, from http://tenyears-www.web. cern.ch/tenyears-www/Welcome.html

Chakrabarti, S., Van den Berg, M., & Dom, B. (1999). Focused crawling: A new approach to topic-specific web resource discovery. *Computer Networks, 31*(11-16), 1623-1640.

Cohen, W. W., & Fan, W. (2000). Web-collaborative filtering: Recommending music by crawling the web. *WWW9/Computer Networks, 33*(1–6), 685–698.

Debaecker, J., & El Hadi, W. M. (2011). Music indexing and retrieval: Evaluating the social production of music metadata and its use. In *International Society for Knowledge Organisation*. Dublin, Ireland: Emerald Group Publishing Ltd.

DiNucci, D. (1999). Fragmented future. *Design & New Media, 53*(4).

Downie, J. S. (2008). The music information retrieval evaluation exchange (2005–2007): A window into music information retrieval research. *Acoustical Science and Technology, 29*(4), 247–255. doi:10.1250/ast.29.247

Eck, D., Lamere, P., Bertin-Mahieux, T., & Green, S. (2007). Automatic generation of social tags for music recommendation. In *Proceedings of the 21st Annual Conference on Neural Information Processing Systems*. Vancouver, Canada: IEEE.

Fang, Z., Guoliang, Z., & Zhanjiang, S. (2001). Comparison of different implementations of MFCC. *Journal of Computer Science and Technology, 16*(6), 582–589. doi:10.1007/BF02943243

Foote, J., & Uchihashi, S. (2001). The beat spectrum: A new approach to rhythm analysis. In *Proceedings of the IEEE International Conference on Multimedia and Expo*. Tokyo, Japan: IEEE Press.

Fu, Z., Lu, G., Ting, K., & Zhang, D. (2011). A survey of audio-based music classification and annotation. *IEEE Transactions on Multimedia, 2*(13), 303–319. doi:10.1109/TMM.2010.2098858

Geleijnse, G., Schedl, M., & Knees, P. (2007). The quest for ground truth in musical artist tagging in the social web era. In *Proceedings of the Eighth International Conference on Music Information Retrieval (ISMIR 2007)*, (pp. 525-530). Vienna, Austria: ISMIR.

Goto, M. (2012). Grand challenges in music information research. In Mueller, M., Goto, M., & Schedl, M. (Eds.), *Multimodal Music Processing* (pp. 217–226). Schloss Dagstuhl-Leibniz-Zentrum fuer Informatik.

Green, S. J., Lamere, P., Alexander, J., Maillet, F., Kirk, S., & Holt, J. … Mak, X. W. (2009). Generating transparent, steerable recommendations from textual descriptions of items. In *Proceedings of the Third ACM Conference on Recommender Systems, (RecSys 2009)*, (pp. 281–284). New York, NY: ACM Press.

IFPI. (2011). Digital music report 2011. *International Federation of the Phonographic Industry*. Retrieved February 7, 2012, from http://www.ifpi. org/content/library/DMR2011.pdf

Iverson, P., & Krumhansi, C. (1993). Isolating the dynamic attributes of musical timbre. *The Journal of the Acoustical Society of America, 94*(5), 2595–2603. doi:10.1121/1.407371

Jensen, J. (2009). *Feature extraction for music information retrieval*. Aalborg, Denmark: Aalborg University.

Kasprowski, R. (2010). Perspectives on DRM: Between digital rights management and digital restrictions management. *Bulletin of the American Society for Information Science and Technology*, *36*(3), 49–54. doi:10.1002/bult.2010.1720360313

Knees, P., Pohle, T., Schedl, M., & Widmer, G. (2007a). A music search engine built upon audio-based and web-based similarity measures. In *Proceedings of the 30th Annual International ACM SIGIR Conference on Research and Development in Information Retrieval*, (pp. 447-454). New York, NY: ACM Press.

Knees, P., Schedl, M., Pohle, T., & Widmer, G. (2007b). Exploring music collections in virtual landscapes. *IEEE MultiMedia*, *14*(3), 46–54. doi:10.1109/MMUL.2007.48

Lamere, P. (2008). Social tagging and music information retrieval. *Journal of New Music Research*, *37*(2), 101–114. doi:10.1080/09298210802479284

Laurier, C., Grivolla, J., & Herrera, P. (2008). Multimodal music mood classification using audio and lyrics. In *Proceedings of 7th International Conference on Machine Learning and Applications (ICMLA 2008)*, (pp. 688 –693). ICMLA.

Law, E. L. M., Von Ahn, L., Dannenberg, R. B., & Crawford, M. (2007). Tagatune: A game for music and sound annotation. In *Proceedings of the 8th International Conferenceon Music Information Retrieval (ISMIR 2007)*, (pp. 361–364). Vienna, Austria: ISMIR.

Liem, C. S., Müller, M., Eck, D., & Tzanetakis, G. (2011). 1st International ACM workshop on music information retrieval with user-centered and multimodal strategies. In *Proceedings of the 2011 ACM Multimedia Conference and Co-Located Workshops*, (pp. 603-604). Scottsdale, AZ: ACM Press.

Logan, B., & Salomon, A. (2001). A music similarity function based on signal analysis. In *Proceedings of the IEEE International Conference on Multimedia and Expo*. Tokyo, Japan: IEEE Press.

Mahedero, J. P. G., Martinez, A., Cano, P., Koppenberger, M., & Gouyon, F. (2005). Natural language processing of lyrics. In *Proceedings of the 13th Annual ACM International Conference on Multimedia, MULTIMEDIA 2005*, (pp. 475–478). New York, NY: ACM Press.

Mandel, M., & Ellis, D. (2005). Song-level features and support vector machines for music classification. In *Proceedings of the 6th International Conference on Music Information Retrieval (ISMIR 2005)*, (pp. 594-599). Barcelona, Spain: ISMIR.

Menn, J. (2003). *All the rave: The rise and fall of Shawn Fanning's Napster*. New York, NY: Crown Business.

Mims, C. (2010). *How itunes genius really works*. Retrieved February 7, 2012, from http://www.technologyreview.com/blog/mimssbits/25267/

Miotto, R., Barrington, L., & Lanckriet, G. (2010). Improving auto-tagging by modelling semantic co-occurrences. In *Proceedings of the International Society of Music Information Retrieval Conference (ISMIR 2010)*, (pp. 297-302) Utrecht, The Netherlands: ISMIR.

Moody, N. M. (2011). iTunes great for Apple but was it for music biz? *Associated Press*. Retrieved February 7, 2012, from http://technology.inquirer.net/5209/itunes-great-for-apple-but-was-it-for-music-biz/

MusicBrainz. (2012). *General FAQ*. Retrieved February 7, 2012, from http://musicbrainz.org/doc/General_FAQ#Why_does_MusicBrainz_not_support_genre_information.3F

Musmann, H. G. (2006). Genesis of the MP3 audio coding standard. *IEEE Transactions on Consumer Electronics*, *52*(3), 1043–1049. doi:10.1109/TCE.2006.1706505

Neumayer, R., Lidy, T., & Rauber, A. (2005). Content-based organization of digital audio collections. In *Proceedings of the 5th Open Workshop of MUSICNETWORK*. Vienna, Austria: MUSICNETWORK.

O'Neill, D. (2006). *ID3v1 – ID3.org*. Retrieved February 7, 2012, from http://www.id3.org/ID3v1

O'Reilly, T. (2005). *What is web 2.0?* Retrieved February 7, 2012, from http://oreilly.com/web2/archive/what-is-web-20.html

Pampalk, E. (2006). Audio-based music similarity and retrieval: Combining a spectral similarity model with information extracted from fluctuation patterns. In *Proceedings of the 3rd Annual Music Information Retrieval eXchange (MIREX 2006)*. Victoria, Canada: MIREX.

Pampalk, E., Rauber, A., & Merkl, D. (2002). Content-based organization and visualization of music archives. In *Proceedings of the 10th ACM International Conference on Multimedia*, (pp. 570–579). New York, NY: ACM Press.

Pohle, T., Schnitzer, D., Schedl, M., Knees, P., & Widmer, G. (2009). On rhythm and general music similarity. In *Proceedings of the 10th International Conference on Music Information Retrieval (IS-MIR 2009)*. Kobe, Japan: ISMIR.

Reynolds, A. D. (2009). Gaussian mixture models. In Li, S. Z., & Jain, A. K. (Eds.), *Encyclopedia of Biometrics* (pp. 659–663). New York, NY: Springer.

Sarwar, B. M., Karypis, G., Konstan, J. A., & Reidl, J. (2001). Item-based collaborative filtering recommendation algorithms. In *Proceedings of the World Wide Web*, (pp. 285–295). ACM Press.

Scaringella, N., Zoia, G., & Mlynek, D. (2006). Automatic genre classification of music content- A survey. *IEEE Signal Processing Magazine*, *23*(2), 133–141. doi:10.1109/MSP.2006.1598089

Schedl, M. (2011). Web-based and community-based music information extraction. In Li, T., Ogihara, M., & Tzanetakis, G. (Eds.), *Music Data Mining* (pp. 219–249). Boca Raton, FL: CRC Press. doi:10.1201/b11041-11

Schedl, M., Pohle, T., Knees, P., & Widmer, G. (2006). Assigning and visualizing music genres by web-based co-occurrence analysis. In *Proceedings of the 7th International Conference on Music Information Retrieval.ISMIR 2006*, (pp. 260-265). Victoria, Canada: ISMIR.

Schedl, M., Widmer, G., Knees, P., & Pohle, T. (2011). A music information system automatically generated via web content mining techniques. *Information Processing & Management*, *47*(3), 426–439. doi:10.1016/j.ipm.2010.09.002

Seyerlehner, K., Widmer, G., & Pohle, T. (2010). Fusing block-level features for music similarity estimation. In *Proceedings of the 13th International Conference on Digital Audio Effects (DAFx 2010)*. Graz, Austria: DAFx.

Shavitt, Y., & Weinsberg, U. (2009). Song clustering using peer-to-peer co-occurrences. In *Proceedings of the 11th IEEE International Symposium on Multimedia*, (pp. 471–476). San Diego, CA: IEEE Press.

Shen, X., Yu, H., Buford, J., & Akon, M. (Eds.). (2009). *Handbook of peer-to-peer networking*. New York, NY: Springer.

Stoica, I., Morris, R., Karger, D., Kaashoek, M. F., & Balakrishnan, H. (2001). Chord: A scalable peer-to-peer lookup service for internet applications. In *Proceedings of the 2001 Conference on Applications, Technologies, Architectures, and Protocols for Computer Communications*, (pp. 149-160). New York, NY: ACM Press.

Tzanetakis, G., & Cook, P. (2002). Musical genre classification of audio signals. *IEEE Transactions on Speech and Audio Processing, 10*(5), 293–302. doi:10.1109/TSA.2002.800560

Van Der Wal, T. (2007). *Folksonomy coinage and definition*. Retrieved February 7, 2012, from http://www.vanderwal.net/folksonomy.html

Weihs, C., Ligges, U., Morchen, F., & Mullen-siefen, D. (2007). Classification in music research. *Advances in Data Analysis and Classification, 1*(3), 255–291. doi:10.1007/s11634-007-0016-x

Whitman, B., & Lawrence, S. (2002). Inferring descriptions and similarity for music from community metadata. In *Proceedings of the 2002 International Computer Music Conference, ICMC 2002*, (pp. 591–598). Goteborg, Sweden: ICMC.

World Wide Web Project. (1991). *Website*. Retrieved February 7, 2012, from http://www.w3.org/History/19921103-hypertext/hypertext/WWW/TheProject.html

ADDITIONAL READING

Alexandraki, C., & Valsamakis, N. (2008). Enabling virtual music performance communities. In Akoumianakis, D. (Ed.), *Virtual Community Practices and Social Interactive Media: Technology Lifecycle and Workflow Analysis*. Hershey, PA: IGI Global.

Anderies, J. (2005). The promise of online music. *Library Journal*. Retrieved February 7, 2012, from http://www.libraryjournal.com/article/CA602662.html

Bainbridge, D. (1998). MELDEX: A web-based melodic index service. *Melodic Similarity: Concepts, Procedures, and Applications. Computing in Musicology, 11*(12), 223–230.

Byrd, D. (1994). Music notation software and intelligence. *Computer Music Journal, 18*(1), 17–20. doi:10.2307/3680518

Chen, A. L. P., & Chen, J. C. C. (1998). Query by rhythm: An approach for song retrieval in music databases. In *Proceedings of the IEEE Eighth International Workshop on Research Issues in Data Engineering: Continuous-Media Databases and Applications (RIDE)*, (pp. 139–146). IEEE Press.

Cooper, M., Foote, J., Pampalk, E., & Tzanetakis, G. (2006). Visualization in audio-based music information retrieval. *Computer Music Journal, 30*(2), 42–62. doi:10.1162/comj.2006.30.2.42

Dannenberg, R. (1993). Music representation issues, techniques, and systems. *Computer Music Journal, 17*(3), 20–30. doi:10.2307/3680940

Dannenberg, R. (2001). Music information retrieval as music understanding. In *Proceedings of the Second International Symposium on Music Information Retrieval*, (pp. 139-142). IEEE.

Dannenberg, R., Birmingham, W., Tzanetakis, G., Meek, C., Hu, N., & Pardo, B. (2004). The MUSART testbed for query-by-humming evaluation. *Computer Music Journal, 28*(2), 34–48. doi:10.1162/014892604323112239

Downie, J. S., West, K., Ehmann, A., & Vincent, E. (2005). The 2005 music information retrieval evaluation exchange (MIREX 2005): Preliminary overview. In *Proceedings of the 6th International Conference on Music Information Retrieval (ISMIR 2005)*, (pp. 320-323). London, UK: ISMIR.

Ellis, D. P. W. (2006). Extracting information from music audio. *Communications of the ACM, 49*(8), 32–37. doi:10.1145/1145287.1145310

Ghias, A., Logan, J., Chamberlin, D., & Smith, B. C. (1995). Query by humming: Musical information retrieval in an audio database. In *Proceedings of ACM International Conference on Multimedia*, (pp. 231-236). ACM Press.

Good, M. (2001). MusicXML for notation and analysis. In Hewlett, W. B., & Selfridge-Field, E. (Eds.), *The Virtual Score: Representation, Retrieval, Restoration* (pp. 113–124). Cambridge, MA: MIT Press.

Krasic, C., Li, K., & Walpole, J. (2001). The case for streaming multimedia with TCP. In *Proceedings of the 8th International Workshop on Interactive Distributed Multimedia Systems*, (pp. 213-218). London, UK: Springer-Verlag.

Li, T., Ogihara, M., & Tzanetakis, G. (Eds.). (2011). *Music data mining book*. Boca Raton, FL: CRC Press.

Masataka, G., & Takayuki, G. (2005). Musicream: New music playback interface for streaming, sticking, sorting, and recalling musical pieces. In *Proceedings of the 6th International Conference on Music Information Retrieval (ISMIR 2005)*, (pp. 404-411). London, UK: ISMIR.

Moore, B. C. J. (2004). *An introduction to the psychology of hearing* (5th ed.). London, UK: Elsevier.

Müller, M. (2007). *Information retrieval for music and motion*. Berlin, Germany: Springer Verlag. doi:10.1007/978-3-540-74048-3

Nekesa, M. M. (2011). iTunes great for Apple but was it for music biz? *Associated Press*. Retrieved February 7, 2012, from http://technology.inquirer.net/5209/itunes-great-for-apple-but-was-it-for-music-biz/

OMRAS. (2002). *Online music recognition and searching*. Retrieved February 7, 2012, from http://www.omras.org

Orio, N. (2006). Music retrieval: A tutorial and review. *Foundations and Trends in Information Retrieval*, *1*(1), 1–90. doi:10.1561/1500000002

Pachet, F. (2002). Playing with virtual musicians: The continuator in practice. *IEEE MultiMedia*, *9*(3), 77–82. doi:10.1109/MMUL.2002.1022861

Patrikakis, C. Z., Papaoulakis, N., Stefanoudaki, C., & Nunes, M. S. (2009). Streaming content wars: Download and play strikes back. In *Proceedings of the Personalization in Media Delivery Platforms Workshop*, (pp. 218-226). Venice, Italy: IEEE.

Raimond, Y., & Sandler, M. (2008). A web of musical information. In *Proceedings of the 9th International Conference on Music Information Retrieval (ISMIR 2008)*, (pp. 263-268). Philadelphia, PA: ISMIR.

Sawyer, S., & Tapia, A. (2005). The sociotechnical nature of mobile computing work: Evidence from a study of policing in the United States. *International Journal of Technology and Human Interaction*, *1*(3), 1–14. doi:10.4018/jthi.2005070101

Schwarz, D. (2004). *Data-driven concatenative sound synthesis*. (Unpublished Doctoral Dissertation). Universite Paris 6. Paris, France.

Seyerlehner, K., & Schedl, M. (2009). Block-level audio feature for music genre classification. In *Proceedings of the 5th Annual Music Information Retrieval eXchange (MIREX 2009)*. Kobe, Japan: MIREX.

Turnbull, D., Barrington, L., & Lanckriet, G. (2008). Five approaches to collecting tags for music. In *Proceedings of the 6th International Conference on Music Information Retrieval (ISMIR 2008)*, (pp. 225-230). Philadelphia, PA: ISMIR.

Tzanetakis, G., Essl, G., & Cook, P. R. (2001). Audio analysis using the discrete wavelet transform. In *Proceedings of the WSES International Conference of Acoustics and Music: Theory and Applications (AMTA 2001)*, (pp. 318-323). Skiathos, Greece: WSES.

Zbigniew, R., & Wieczorkowska, A. (Eds.). (2010). *Advances in music information retrieval*. Berlin, Germany: Springer-Verlag.

KEY TERMS AND DEFINITIONS

Content-Based Retrieval: Retrieving metadata directly from music sources using tools of computational intelligence.

Context-Based Retrieval: Retrieving metadata from the World Wide Web using Web-ming techniques.

Music Source: The files containing music content, which may be either audio files (i.e. signals of digital audio) or to symbolic representations of music (i.e. MIDI, Music XML, etc.).

Online Service: A service that is provided from a Web application.

Pattern Recognition: Machine learning task aiming at assigning a label to a given input value.

Streaming Audio: Audio that is reproduced on the Web client while being downloaded from a Web server.

Web 2.0: Web technology that facilitates information sharing, interoperability, user-centred design, and collaboration on the World Wide Web.

APPENDIX

URLs of the Presented OMSs

This appendix provides an alphabetic list of all the OMSs mentioned in this chapter along with their URLs:

- Allmusic: http://www.allmusic.com/
- Amazon: http://www.amazon.com/
- AOL Radio: http://music.aol.com/radioguide/bb
- AudioScrobbler: http://www.audioscrobbler.net/
- Choral Public Domain Library: http://www.cpdl.org/
- eJammingAudiio: http://ejamming.com/
- eMusic: http://www.emusic.com/
- Google Music Beta: http://music.google.com
- IndabaMusic: http://www.indabamusic.com/
- Indieland: http://indieland.com/
- iTunes: http://www.apple.com/itunes/
- Jamendo: http://www.jamendo.com/
- Jango Radio: http://www.jango.com/
- Last.fm: http://www.last.fm/
- Live Music Portal: http://www.livemusicportal.eu/
- Midomi: http://www.midomi.com/
- MusicBrainz: http://musicbrainz.org/
- Musipedia: http://www.musipedia.org/
- Musopen: http://musopen.org/
- Mutopia: http://www.mutopiaproject.org/
- MySpace: http://www.myspace.com/
- Napster: http://www.napster.com/
- NINJAM: http://www.cockos.com/ninjam/
- Pandora Radio: http://www.pandora.com
- Petrucci Music Library: http://imslp.org/wiki
- Shazam: http://www.shazam.com/
- Slacker Radio: http://www.slacker.com/
- SoundCloud: http://soundcloud.com/
- Spotify: http://www.spotify.com
- The Echonest Platform: http://the.echonest.com/
- Uplaya: http://uplaya.com/
- Wemix: http://www.wemix.com/

Chapter 13
Intelligent IPTV Distribution for Smart Phones

Miguel García
Universidad Politécnica de Valencia, Spain

Jaime Lloret
Universidad Politécnica de Valencia, Spain

Irene Bellver
Universidad Politécnica de Valencia, Spain

Jesús Tomás
Universidad Politécnica de Valencia, Spain

ABSTRACT

Current advances in embedded hardware for mobile devices, jointly with new type of batteries that let smart phones have more power during longer periods of time, allow them to offer new services to customers. Internet Protocol Television (IPTV) is the Internet Protocol (IP) service that is experiencing highest demand. Many mobile telephone companies are adding this service to their supply. In this chapter, the authors show all steps to transmit the IPTV service from the provider to the smart phone. First, they introduce the current hardware and operating systems for smart phones. Then, they describe the main parts of the IPTV architecture and the main protocols used for IPTV transmission in order to show how it works. Next, the authors show where intelligent systems can be deployed in the IPTV network in order to provide better QoE (Quality of Experience) on the end-user side. Then, they discuss the limitations and requirements for IPTV reception on smart phones. Finally, the authors explain the IPTV implementation in smart phones and describe an IPTV player on Android.

INTRODUCTION

The Internet Protocol (IP) is currently the starting point for all integrated services, also known as *Triple play* (Hellberg, et al., 2007) and *Quad play* (inCode, 2006). The Triple play concept in telecommunications' world is a marketing concept that could be defined as the audiovisual content and services integration (voice, data, and television). Unlike traditional phone services, broadband and television are available to users through an independent communications network. Triple-play

DOI: 10.4018/978-1-4666-2833-5.ch013

service is based on providing all services using IP technology by a single connection (e.g., coaxial cable, fiber optic, twisted pair cable, power line communication, or radio). Besides, recent quad play services include mobile facilities to triple-play services.

Users regularly use triple play services in their IP SetTopBox or personal computer. This is a normal procedure to request reception of such data in their handsets. We will see soon a movie in our last generation mobile, although many people think that we will never want to see a movie on the screen of our mobile phone. A recent study has shown that the terminal screen has the same aspect that TV, when we have it at 15-30 centimeters of our face.

Mobile IPTV (Park, et al., 2008) was born going after this business opportunity (see a movie in your mobile phone at any place). It is a technology that enables users to transmit and receive multimedia traffic including television signal, video, audio, text, and graphic services through IP-based with support for Quality of Service (QoS)/Quality of Experience (QoE), security, mobility, and interactive functions. Mobile IPTV users can enjoy IPTV services anywhere even while they are on the move.

Several works of the related literature focus on the study of quality of service in mobile telephony. For example, Florido et al. (2009) study QoS in the HSDPA networks used in 3G telephony. In particular, they propose an admission control algorithm that takes into account that this guard power varies according to the channel conditions of each user. Díaz et al. (2010) analyze video streaming QoS using smart phones on live mobile networks. The goal of their study is to validate current deployments and simulations.

According to the International Telecommunication Union (ITU) (IPTV, 2012), IPTV is the set of multimedia services (television, video, audio, text, graphics and data) which are distributed through an IP network. These services must possess an adequate level of quality service, security,

interactivity, and reliability. When it comes to the provider, IPTV includes video acquisition, video processed, and video secure distribution on the IP network infrastructure. The main IPTV characteristics are:

- **Interactive Television (TV) support:** IPTV systems have two channels. These channels let the service provider to distribute interactive TV applications. The live television, High Definition Television (HDTV), interactive games, quick searches on the Internet, etc. are some indicative examples.
- **Time Shifting:** This service can be used to record TV, so the user is able to see these contents later.
- **Personalized Content:** IPTV has two-way communications. This feature allows the user to indicate, what and when does (s)he want to see.
- **Requires low bandwidth:** IPTV technology does not broadcast all channels to each end-user, it allows only sending the channel requested by the user. It only multicasts each channel when the user requests it, otherwise this bandwidth is saved. This is an important feature as it helps to save bandwidth in their networks.
- **Accessibility using several devices:** The IPTV contents can be viewed with several types of devices such as computers or mobile devices, not only with televisions.

In the beginning of 2009, more than 55% of American adults had a broadband connection (Pew Internet, 2008). USA had 50 million broadband lines approximately, while China had 46 million broadband lines. Japan had about 25 million lines, followed by South Korea with about 13 million. China has been the country with greater growth of broadband lines in the last years (Lloret, et al., 2009a). Taking into account all broadband connections (approximately 53.7) million subscribers have contracted the IPTV service (Tadayoni & Sigurdsson, 2006). This is a low value according to the IPTV advantages.

The main purpose of this chapter is to describe the IPTV architecture by focusing on the aspects of IPTV. Although the IPTV in mobile environment is the same as generic IPTV, there are several differences among them. For example, the channel is very variable, the encoded video is different according the devices, and the transmission can use several protocols. Finally, we explain how to implement easily an IPTV client in a smart phone using the Android platform.

The remaining part of the chapter is structured as follows. In the next section, we introduce the IPTV architecture and explain the network protocols used in IPTV. Then, we show where an intelligent system can be deployed in order to provide better QoE at the end-user side. The limitations and requirements for IPTV reception on smart phones are discussed in the next section. Moreover, we compare the mobile operative systems that support multimedia flows. The IPTV implementation of IPTV players in smart phones, specifically in the Android Operating system, is detailed after the previous section. Finally, in the last section we conclude the chapter and draw some future research directions.

IPTV Architecture

Telephone companies were the first ones who offered IPTV service. Later, this technology was expanded to television providers. IPTV uses IP networks and sends less information than analog or digital TV. IPTV also allows lower costs for providers and lower prices for consumers. An IPTV network is formed by several broadband accesses that are capable of supporting the required bandwidth for video delivery. IPTV network topology can be split into five parts (Lloret, 2009b):

- Network header.
- Backbone network.
- Distribution network.
- Access network.
- Customer network.

A typical IPTV network is shown in Figure 1.

- The *network header* is a service provider's network that sends the video content to the IPTV network. It is essentially the core components of the IPTV infrastructure layer. This network is formed by devices that receive, transform, and distribute the content to the customers. The network header is the main point of the infrastructure. It receives all the subscriber requests and provides content to the set-top boxes. The network header is the most critical point of the service provider network. In addition, it may need several actions to ensure that it has a controlled access. Only authorized users should exchange information with it.

- The *backbone network* is the part of the network that distributes the video flows from the header to the distribution network. It interconnects the service provider's network header to the distribution networks. The backbone network often uses Gigabit Ethernet (Cunningham, et al., 1999), SONET/SDH (Cavendish, et al., 2002), xWDM technologies (Iwatsuki, et al., 2004). In addition, this part of the network may have different architectures: peer to peer, ring, double ring, etc. In the IPTV backbone network, routing and switching between the aggregation routers and end routers is the most important part of the network infrastructure. The IPTV routers should be scalable and high-performance devices. They should also be able to mix interfaces.

- The *distribution network* goes from the end of the backbone network to the aggregation router, where the access network starts.

Figure 1. IPTV network

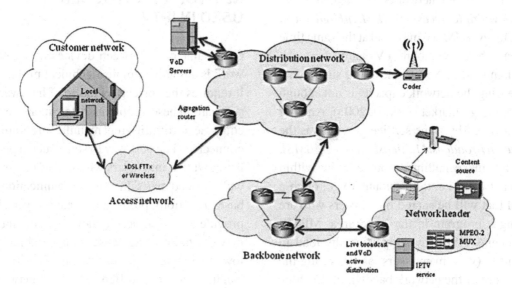

The distribution network performs data transmission and switching tasks. Its main function is to multiplex the information provided by different service providers and to adapt the transport system to the specific characteristics of the subscriber loop.

- The *access network* consists of all elements that transport the multimedia content to the user and manages the user demands by a return channel. The first requirement of an access network is to have enough bandwidth to support multiple IPTV channels for each subscriber, while it allows other services (telephony and data). Currently, the most used technologies are xDSL and FTTx (Keiser, 2006). The IPTV channels transmission is sent through multicast groups to the distribution network and to the access network.

- The *customer's network* enables communication and information exchange between connected computers, as well as access to available resources in the network. The sharing means at the customer's network may be wired or wireless. Each device connected to this network may enjoy the services through the residential gateway. This gateway connects the customer's network with the service provider's network. The technologies used in this network tend to be Fast Ethernet and Wi-Fi (IEEE 802.11b/g).

Using IPTV on Mobile Networks

Wireless mobile networks are evolving to support higher transfer rates. Although they are the most used type of network, it does not mean that they are the most adequate ones to support the IPTV service. Mobile TV is the television service provided to end-users of mobile networks. Such system combines mobile telephony services with TV content. An important feature available on mobile TV is to have a personal TV on the end-users, but it is always adapted to mobile devices, in addition to the advantage of mobility. There are currently two methods for providing television service on mobiles:

- The first method uses two channels of the mobile network.
- The second method uses a dedicated channel in broadcast or multicast mode.

The use of 3G networks (*Wideband Code Division Multiple Access (WCDMA)/High Speed Packet Access (HSPA)*) may lead at the same time to increase the use of mobile TV mainly due to the higher bandwidth. Nevertheless, the simple fact of increasing the network capacity is not enough to serve a large market (Uskela, 2003). Another service offered by some service providers is the *Multimedia Broadcast Multicast Service (MBMS)*. Its goal is to make multimedia broadcasting within the 3G network, i.e. to distribute TV through a channel that will be shared by all users who are watching a program in the same area. MBMS complements HSPA by supporting the load in dense areas (with many users) and ensuring the efficient use of the network bandwidth (Correia, et al., 2007).

Although there are several multicast solutions, the 90% of the 120 mobile TV providers are using the method of the two channels based on a unicast transmission (Lloret, 2009a). Therefore, each server sends the information to each user needing more resources on the server and increasing the bandwidth costs of the network.

In most wireless technologies, the network administrators are integrating an Integrated Multimedia Subsystem (IMS), which simplifies the interface between IPTV applications and other IP services such as high-speed data connection and VoIP. The IMS is an emerging part of the architecture that enables network operators to accelerate and simplify the deployment of IP-based services (Camarillo & García-Martín, 2011). These wireless mobile networks must be able to support advanced mechanisms for quality of service, since it is a basic feature used in IPTV and improves real-time in other applications. Another fundamental issue to bear in mind in this type of network is the user's mobility. It is a difficult feature when it comes to achieve high TV and Voice quality. One of the biggest problems is to maintain the reception of a video stream, when a user is moving between different base stations. Today, many companies try to offer this service.

NETWORK PROTOCOLS USED IN IPTV

Communication between devices in a data network is a highly complex technical problem and it requires the correct operation of hardware and software. These problems often increase when equipment from different manufacturers are interconnected. The best way to solve such a problem is to divide it into layers with the aim of creating well-defined parts in the communication. The basic functions produced in each layer should provide a basis to design, develop, and diagnose network failures. Moreover, a layered model allows making changes easily. The TCP/IP model can interoperate in different types of networks. It consists of four layers:

- The first layer is called the *network access layer*. It undertakes all issues to send an IP packet to the physical link. This layer includes the details of Local Area Network (LAN) and Wide Area Network (WAN) technology and all the details of the physical layer and data link of the OSI model. On the one hand, it is responsible of transmitting the data and giving the mechanical, electrical and functional transmission specifications, and to maintain, activate and deactivate the physical link (Cali, et al., 1998).
- The second layer is called *network layer*. It provides connectivity and path selection between two distant devices. Best route selection and packet switching are performed in this layer.
- The third layer is called *transport layer*. It refers to service quality issues with respect to reliability, flow control and error correction. It segments the data sent by the sender device, encapsulates the data in transport PDUs (also called transport package), and provides a transport service.

- The fourth layer is called *application layer*. It handles high-level protocols, issues of representation, encoding and dialog control. In this layer is placed the specific protocols for IPTV transmission. This layer also defines the interface with users.

The protocol stack used for IPTV is defined in ITU-T Rec J.281 (ITU-T, 2005). This protocol stack is divided into two parts:

- The set of layers below the Real-time Transport Protocol (RTP). These are responsible for the transmission of any type of information (they do not have to be for IPTV purposes (Schulzrinne, 1992)).
- The set of layers above the *Moving Picture Experts Group Layer 2 (MPEG-2) Transport Stream (TS)* those are responsible for IPTV services transmission for standard definition, but not in case of high definition (Solucionesytecnología.net, 2007).

These two parts are high-related layers, so the network operations and service operations are highly dependent on each other. We can assume the following regarding the structure of the IPTV transmission protocols: Services operate over the MPEG-2 TS and RTP, where many programs (hundreds of programs) can be sent and IP multicast is used to distribute video content.

MPEG-2 is designed for the generic coding of moving pictures and the associated audio to create a video stream using three types of data frames (intra, predictable and bi-directional frames) arranged in a specific order called the *Group of Pictures (GOP)* structure. Currently, MPEG-2 and MPEG-4 are the standards used for *Digital Video Broadcasting (DVB)*. The objective of DVB-IP technology is to specify the interface between an IP based network and the DVB-IP Set-Top-Box, which uses a protocol stack for DVB IP services. Once DVB services are encoded, video content

is packaged and encapsulated. This involves the integration and organization of video data in individual packets. The encapsulation of the video content is performed using MPEG-2 TS, where all the MPEG-2 TS is encapsulated in RTP as the transport layer protocol according to the RFC 1889 (Schulzrinne, et al., 1996) joined with RFC 2250 (Hoffman, et al., 1998) and RFC 768 (Postel, 1980). The message fields used in the transport stream MPEG-2 based on the DVB content over IP are the following: a standard IP header, a User Datagram Protocol (UDP) header, RTP header and a whole number of 188 bytes MPEG-2 TS packets (see Figure 2). The maximum size of an IP datagram is limited and depends on the network access technology.

UDP

UDP (Postel, 1980) is used to provide non connection-oriented and non-reliable service. It allows sending datagrams through the network without having any connection previously established. The datagram includes enough addressing information in its header. UDP does not have error correction or flow control as Transmission Control Protocol (TCP). It is mainly used when connections are not necessary or the information is valuable. IPTV traffic, supported by the UDP protocol, has highly variable data rates and stringent QoS requirements in terms of delay and loss. This protocol is suitable for real time audio and video transmission applications, where it is impossible to relay the messages because of the stringent delay requirements.

TCP

TCP (Wan, et al., 2008) has its own congestion control loop to adjust the sending rate, so the traffic load is also highly dynamic. If IPTV and TCP traffic is simply multiplexed, their performance is difficult to predict and the competition between them will jeopardize their QoS. To efficiently

Figure 2. RTP/RTCP over transport layer protocols

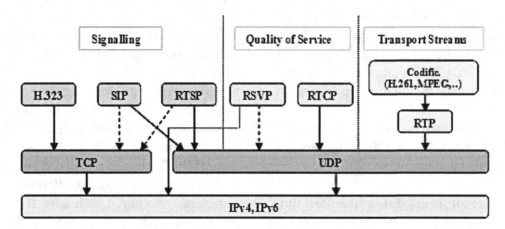

utilize network resources and provide satisfactory QoS for both traffic types, we propose multiplexing IPTV and TCP traffic. By multiplexing IPTV and TCP traffic appropriately, network resources can be more efficiently utilized, the QoS of IPTV can be maintained, and TCP flows can obtain higher throughputs.

RTP/RTCP

RTP (Schulzrinne, et al., 1996) is the transport protocol for streaming media on the Internet. It was designed to work with IP Multicast, but can be used in a unicast, to provide temporal information and synchronization of multimedia streams. RTP is a lightweight protocol without any type of error control mechanisms and flow control. Moreover, it does not provide resource reservation or quality service control. However, it is independent of the network technology on which it is used. RTP/RTCP provides all the features for transmitting multimedia data flows in real time and provides the necessary mechanisms for running applications because allows local synchronization of incoming flows. A RTP session is the association of a group of participants that are exchanged by the same RTP stream in a multimedia session. The data flow transmitted in the RTP session is defined by a particular pair of

port destination addresses (one for RTP and one for RTCP). In order to identify the source of an RTP packet flow, some fields of *Synchronization Source (SSRC)* are used; they are unique in each session and are included in the RTP packet header. All packages from the same source will have the same timing space and sequence numbers. The functions provided by RTP include:

- Identify the type of information transmitted.
- Add temporal markers (timestamps) and sequence numbers to the information transmitted.

The RTCP is based on the periodic transmission of control packets to all participants within a session, using the same distribution mechanism than that for RTP data packets. This protocol performs four functions:

- Its main function is to provide feedback on the distribution of data.
- Transport a unique and global identifier at the transport level for an RTP source called the *canonical name* or CNAME.
- Check the bitrate of control information to allow scaling to a larger number of participants.

- Transport the minimum information of the session control (for example the identification of participants, who are usually displayed in the GUI user applications.

In multicast transmissions, the control information can consume considerable bandwidth, thus limiting the amount of RTCP traffic to a small percentage of RTP traffic.

RTSP

The Real Time Streaming Protocol (RTSP) is defined in the RFC 2326 (Schulzrinne, et al., 1998), and it was developed by the IETF in 1998. RTSP is not a connection-oriented protocol for streaming systems. It allows establishing and controlling one or multiple synchronized data streams, e.g. audio or video. The streaming server maintains a session associated with an identifier and, in most cases, RTSP uses TCP for player control data and UDP for multimedia data flows. RTSP allows a client to perform remote operations on a streaming server. It is not a protocol for sending information (RTP is used for this purpose). RTSP has some similarities with HTTP (Hypertext Transfer Protocol) such as the format of the requests and responses, the status codes, security mechanisms, the format of the Universal Resource Locator (URL), and so on. Otherwise, while HTTP is a stateless protocol, RSTP has states (it needs them to maintain state of the connections). The RTSP servers and clients can make requests. Typically, RTSP requests are sent from clients to servers, but there are exceptional situations where the server sends them to clients. Other features of RTSP are:

- **It is expandable:** You can add new methods and parameters easily.
- **It is secure:** It re-uses Web security mechanisms. All forms of HTTP authentication are directly applicable.
- **Multi-server capability:** Each media stream within a session can reside on a different server. The client will establish several concurrent control sessions with different servers and will carry out the synchronization transport layer.
- **Control of recording devices:** Recording and playback devices can be controlled (e.g. RTSP IP cameras).
- **It is suitable for professional applications:** It supports frame level resolution using timestamps to enable digital edition.

The main RTSP requests are the following ones:

- **DESCRIBE:** Its objective is to obtain a description of a multimedia presentation or object linked by a RTSP URL placed in a streaming server.
- **SETUP:** Allows the server to know that the client wants to view content, where can it be viewed and how it can be reached. It contains the multimedia stream URL and a transport specification, with the used port (normally by RTP).
- **GET COMMAND:** Obtains parameters from the server. It works in a similar way than the GET and POST requests of the HTTP.
- **PLAY:** Asks the server to initiate the specified streams transmission, using the ports previously configured during its initiation.
- **PAUSE:** Pauses data transmission, but without releasing the associated resources. The transmission can be resumed using the PLAY request.
- **TEARDOWN:** It ends the transmission and releases all the resources associated to the stream.

MPEG TS (Transport Stream)

In MPEG-2 Transport Stream, control data packets and user data packets are multiplexed into a single bit stream. This is standardized in ISO 13818-1, ISO 13818-2, and ISO 13818-3. The multiplexed stream can be transmitted in any of the following mediums and technologies:

- Radio Frequency Links (Ultra High Frequency (UHF), Very High Frequency (VHF)).
- Digital Satellite Links.
- Wired networks.
- Wireless networks.
- *Plesiochronous Digital Hierarchy (PDH)* and *Synchronous Digital Hierarchy (SDH)*.
- *Digital Subscriber Lines (ADSL)*.
- Packet or cell switched networks (*Asynchronous Transfer Mode (ATM)*, IP, IPv6...).

The main components of the MPEG bit stream are:

1. **Elementary Stream (ES):** It is formed by the data obtained from a video or audio encoder. There are several forms of ES in a program or a track (e.g. DVD):
   a. Digital Control Data
   b. Digital Audio (sampled and compressed)
   c. Digital Video (sampled and compressed)
   d. Digital data (synchronous or asynchronous)

2. **Packetized Elementary Stream (PES):** The ES is packetized by encapsulating sequential data bytes from the ES inside PES packet headers. A PES is a whole number of ES. Then, PES packets are encapsulated inside *Transport Stream (TS)* packets or program stream. The TS packets can then be multiplexed and transmitted. A PES packet may have a fixed block size (or variable), but with a maximum of 65536 bytes per block. There are two multiplexing methods:
   a. **MPEG Program Stream:** It is a combination of PES packets referenced to the same time stamp. These plots are created for error-free environments and enable the processing of the data received via software.
   b. **MPEG Transport Stream:** Each PES packet is divided into fixed size

transport packets, consisting of one or more frames. This type of streams is used for transmission, where packets may be lost or may have packets corrupted by noise.

The Digital Video Broadcasting (DVB) uses MPEG-2 transport streams. There is interoperability between both formats at the PES level (Reimers, 1998). Although MPEG-2 is usually used to transmit information on several types of media, such as DVB, it can also be used in an IP communications network. The streams are short and use a strict correction mechanism. To correct these errors, MPEG-2 uses higher layer protocols that are responsible for the reliability of the data transported. The header of the transport stream contains the PID (Packet Identification), so the receiver can accept or reject the package at a higher level without overloading the receiver with too many processes. A transport stream may correspond to a single television program (a video PES and audio PES), this is called a Single Program Transport Stream (SPTS) and it contains all the information needed to reproduce the encoded TV channel. Each stream must be synchronized (required for TV or radio) or unsynchronized (games or software). In the transport scheme must send time stamps to help synchronization. The time stamps used for this purpose are the following ones:

c. **Reference Time Stamp:** It indicates the current time. It is found in the Packet *Elementary Stream (PES)* syntax, in the program syntax or in the adjustment of the transport package (particularly in the Program Clock Reference).

d. **Decoding Time Stamp and Presentation Time Stamp:** They are close to the material, which they refer to (normally in the PES header). They indicate the exact moment when the stream must decode audio and video.

LIMITATIONS AND REQUIREMENTS FOR IPTV RECEPTION ON SMART PHONES

Limitations of Mobile IPTV

Mobile IPTV needs at least one wireless link for each device. A minimum of 2 -3 Mbit/s of bandwidth for a standard definition stream must be provided due to the characteristics of the IPTV service. Wireless technologies for short-range use or control purposes are not fully considered yet (Park, et al., 2008). There are many obstacles on the path to the successfully launch and widespread use of Mobile IPTV businesses. Some of them are technical issues, but we should not forget the business issues. Since Mobile IPTV assumes as a minimum one wireless link between the source (e.g., stream server) and the destination (e.g. mobile terminal), most of the technical obstacles are related to the usage of the wireless link. From the business perspective, customer would like to watch TV programs on the go, but the actual technology does not offer it with enough QoE. Moreover, TV content might be a major player. Let us see in Table 1 the major limitations of mobile IPTV.

QoS Requirements of IPTV Networks

IPTV is distributed through a closed network infrastructure. Such a closed network can be controlled by the IPTV service provider according to a required bandwidth and quality of service. TV services cannot be provided using a standard broadband connection of 2 Mbit/s. Moreover, the high-definition programming and the tendency to watch TV on large flat screens, if there is a reduction of bandwidth, cause important effects on image quality.

Quality of Service (QoS) is the ability to give good service, i.e. when service is guaranteed by one or more network QoS parameters. *Service Level Agreement (SLA)* is the contract that specifies the QoS parameters agreed between the service provider and the customer. The QoS has a major impact on operational costs. Poor service quality can lead to increased complaints and calls from users. Each call from the customer involves an economic cost to the provider. A large number of calls mean a lot of discontent users. The nonconformity of users is increased if failures occur when there is a live sport event (Park & Jeong, 2009).

The main issue to take into account in the IPTV service is to provide QoS. The quality of service can be offered in two main parts of the network: The backbone network and the access network. The number of TVs per household grows continuously and the demand for high definition IPTV services creates the need to provide quality service and high bandwidths. The most common parameters of IPTV service quality are:

1. **Bandwidth:** Minimum bandwidth that the operator guarantees to the final users of the network.
2. **Packet loss:** Maximum lost packets (always if the final user does not excess guaranteed flow).
3. **Jitter:** Fluctuation in the round-trip delay.
4. **Round-Trip Delay:** Average round-trip delay in the packets.
5. **Availability:** Minimum time that the operator ensures that the network will be operational.

The best-known standards used to provide QoS are:

1. IEEE 802.1p QoS
2. IEEE 802.1q VLAN
3. VLAN Mapping CEA2007

Telecommunication Standardization Sector (ITU-T) Recommendations for QoS are:

1. ITU-T Rec Y.1541
2. ITU-T Rec J.241

Table 1. Limitation on mobile IPTV

Limitation	Explanation
Capability limitation	Most of the smart phones have limited capabilities compared to the fixed devices. This mainly ascribes to portability considerations, which lead to small displays, low power processors, and limited storage. Moreover, only a few set of technologies can be considered proper for Mobile IPTV.
Bandwidth limitation	Even though the effective bandwidth of wireless links is growing rapidly, the fully deployment of the 4G wireless network is required to have enough bandwidth in the wireless link in order to accommodate High Definition quality video services. The wireless link will always have less bandwidth than the wired link. Mobile IPTV requires a minimum bandwidth of 2-3 Mbps.
Vulnerable wireless link	The wireless link is very vulnerable. Even if mobile terminals are stationary, temporal reflectors and obstacles around the mobile terminals can affect the received signal and cause burst packet losses. Taking into account the movement of mobile terminals, packets delivered through the wireless link are exposed to a great variety of signal degradation such as shadowing, fast/slow fading, etc. Packet losses are intrinsic and inevitable in the wireless link.
Coverage implication	The main reason for carrying mobile terminals is to get access to any service at anytime and anywhere. Services are restricted to some areas because it is practically impossible to deploy wireless networks to cover all geographical areas with no dead spots. However, by adopting vertical handovers (handovers between different networks), the coverage issue can be considerably mitigated.
Dynamic environment	The wireless link is highly dynamic compared to the wired link. The characteristics of the wireless can vary due to many causes and rate of change could be very abrupt. For example, vertical handover can change almost everything in a blink. Therefore many solutions deployed for the relatively static wired computer network environment may not work properly.
Middleware concern	Middleware is one of the key functions of the IPTV service. A service provider can control the usage of IPTV service remotely by deploying middleware. The middleware acts as a transparent solution for adopting IPTV services in different platforms. There are several well-known middleware solutions on a set-top box, but they are too heavy to be implemented on a smart phone, so existing middleware solutions should be slimmed down.
Business issue	The most significant problem of the Mobile IPTV business is the quality of service provided while the consumer is watching the TV program on the go. There is normally limited time for enjoying visual services on the go because of signal disruption and roaming failures. This requires mobile IPTV services to be far more attractive than the consumers' expectation. Due to the small size of the smart phones, it is not possible to adopt fancy User Interface methods. So that, a highly creative and new way of interfacing is required. The lack of mobile contents frustrates the early adopters of Mobile TV. Currently customers have many channels to enjoy their favorite TV programs whenever they want. Moreover, events such as sports are demanded. Contents must be tailored for mobile environments. For example, small display sizes, random and short watching time should be considered.

The video stream does not need SLAs per subscriber, only network SLAs are needed. If the delay is not critical, many applications are not affected. The jitter is not critical because the set-top box can store up to 200 msec. However, packet loss is critical. There are different levels of QoS:

1. *Best effort service* is the basic connectivity without adding any additional service. It is the simplest. An application sends data whenever it wants, in any amount, without requiring any permission and without informing the network. The flow, the delay and the reliability are no guarantied. The queuing model *First-In First-Out (FIFO)* is used.

2. *Differentiated service* is a model of multiple classes of services that can satisfy different QoS requirements. It does not use signals to specify in advance the required services by the network. In this model, the network attempts to make a deal based on a set of QoS classes. The user marks packets with a certain level of priority. The routers add the

users' demands and they are spread along the route.

3. *Guaranteed service* offers two types of QoS, Integrated services (IntServ) and Resource Reservation Protocol (RSVP). It ensures full available resources as long as the network has sufficient resources. In this model, an application makes a request for a specific service class to the network before starting to send information.

Service providers and network equipment vendors must verify that IPTV services will meet the quality expectations of the users before deploying triple play services. Users will not tolerate service interruptions, picture degradation, or long waiting periods when changing IPTV channels. The four main systems responsible for distributing IPTV services are:

1. The video header, where applications and the program content are stored.
2. The network, which is the transport mechanism that distributes television content and interactive services from the header to the customer device.
3. The "middleware," which is the software responsible for controlling television content and interactive services from the header to the customer's device.
4. The client device, which is the receiver of the IPTV service (in a house, it is the set-top box and is connected to the television).

Each of these four systems can influence on the IPTV QoS. Each one has its own considerations of test and measurement. Before testing the system from the beginning to the end, and the integration of the entire system, and administrator should test each part of the system (and the individual elements within each system) independently. This will verify whether the performance meets the expectations and clarify where the problem may exist.

There are several key areas where the quality of the IPTV flow could be tested:

1. **Zapping Measurements:** This type of test shows how fast customers can change channels and verify that they are receiving the correct channel. A delay of 1 second in total is considered an acceptable channel zapping. When there is between 100 ms and 200 ms, it is considered instantaneous. In order to keep the delay below 1 second, the delay to join and leave a multicast group should be between 10 ms and 200 ms.
2. **Audio and video quality metrics:** There are many factors that can compromise the perceived audio and video quality.
3. The *extent and behavior of IPTV customers* and the convergence of the other triple play traffic, competing for the limited resources of the network, have a significant impact on the timely and accurate forwarding of the IPTV package.
4. *The deterioration of the network* (packet loss and sequence errors, latency, and jitter) can have several detrimental effects on the visible video quality, such as blurred, distorted, visual noise, etc.

The distributed digital video via IP multicast does not ensure a consistent video quality to all users while viewing the same channel. Therefore, it is quite difficult to ensure the properly TV reception by the customer. The processing resources and bandwidth of the IPTV distribution network are finite, for this reason when there is a lot of customer IPTV service could be worst. In a real environment of triple play users, many customers can simultaneously change from one channel to another or request for Internet connections while there are multiple VoIP conversations. When this is scaled to the customer, this may increase the demand on the control layer of the IPTV network. It can risk the *QoE (Quality of Experience)* of the IPTV receivers. In order to perform a test of IPTV service quality, we can choose several systems:

1. Create a network testbench with data, VoIP, and IPTV.

2. Create a real environment
3. Simulate a triple play environment

The audio and video distribution rate *Media Delivery Index (MDI)* is gaining acceptance from the industry for testing the quality of the audio and video through the network devices in a video distribution infrastructure. MDI is a standard defined in RFC 4445 (Welch & Clark, 2006) and supported by the *Video Quality Alliance IP (IPVQA)*.

MOBILE OS COMPARISON

Nowadays, the number of smart phones available in our society is quite large. This number has increased from the mid-80s. It is still growing a lot more from the second half of the 90s until today. This market has been diversified and there are today different types of devices. In this subsection are presented several comparative tables with the main features of the mobile operating systems that support the transmission of the multimedia over IP.

Table 2 shows the basic features (Kernel Type, Adaptability, Platform Age, First-party Enterprise Support and Wireless Technologies) for Android, BlackBerry OS 4.7, iPhone OS 4, S60 5th Edition, Palm WebOS, Windows Mobile 6.5. Table 2 shows user interface features (Screen Gestures, Screen Technology, Multitouch, *User Interface (UI)* Skinning, and Input Methods) for Android, BlackBerry OS 4.7, iPhone OS 4, S60 5th Edition, Palm WebOS, Windows Mobile 6.5. Table 2 shows the main core functionality features (Notification Style, Contact Integration/Management, Multitasking, Copy/paste, Media Support/ Ecosystem, Global Search, Firmware Updates, Browser Engine, Tethering and Stereo Bluetooth) for Android, BlackBerry OS 4.7, iPhone OS 4, S60 5th Edition, Palm WebOS, Windows Mobile 6.5. Table 2 shows the main third-party development features (SDK Availability/Support, Official App Store, App Availability, Native Applications

and On-Device App Management) for Android, BlackBerry OS 4.7, iPhone OS 4, S60 5th Edition, Palm WebOS, Windows Mobile 6.5.

INTELLIGENT SYSTEMS FOR IPTV NETWORKS

In this section, we explain where an intelligent system can be added to the network in order to improve the distribution of IPTV and enhance the QoE of the end-user. We can distinguish the following places:

Zone 1: Network header.
Zone 2: Distribution network.
Zone 3: Access network .
Zone 4: Smart phone network.

These places can be seen in Figure 3.

Network Header

An adequate place to add an intelligent system for IPTV streaming is the *network header*. The system should be based on an adaptive scheme that takes into account the QoS parameters measured from the devices of the network, when the videos are transmitted through IPTV channels. We propose an implementation based on a video controller that monitors the network in order to estimate: the bandwidth consumption, packet delay, jitter, lost packets, and more parameters given by the IPTV channels. This information is gathered by the SNMP messages sent from the core and the aggregation routers of the IPTV network. For example, when bandwidth allocation is needed, the video controller runs an intelligent algorithm in order to advise the network header servers, or the surrogate video servers (placed in the distribution network) which would take an action. An example that uses an intelligent algorithm in the network header is shown in Lloret et al. (2011a).

Table 2. Mobile OS features

	Android 4.0	BlackBerry OS 7	iPhone OS 5.1	Symbian 9.5	PalmHP WebOS 5.5	Windows Phone 7
Company	Open Handset Alliance	RIM	Apple	Symbian foundation	Hewlett Packard	Microsoft
License	Open source	Proprietary	Proprietary	Open source	Proprietary	Proprietary
Kernel Type	Linux	Mobile OS	OS X	Mobile OS	Linux	Windows CE
Adaptability	Excellent	Good	Bad	Excellent	Excellent	Excellent
Initial realise	2008	2003	2007	1997	2009	2010
CPU Family	ARM, MIPS, x86	ARM	ARM	ARM	ARM	ARM
Programming Lenguaje	Java, C++	Java	Objective-C, C++	C++	C++	C#, ...
Screen Gestures	Yes	Yes	Yes	Limited	Yes	Limited
Screen Technology	Capacitive	Capacitive	Capacitive	Resistive/Capacitive	Capacitive	Resistive
Multitouch	Yes	Yes	Yes	No	Yes	No
UI Skinning	Yes	Yes	No	Yes	No	Yes
Flash	Yes	Yes	No	Yes	Yes	No
Notification Style	Tray	Pop-up, background	Pop-up	Pop-up	Tray	Tray, pop-up
External estorage	Yes	Yes	No	Yes	Yes	No
Multitasking	Yes	Yes	Yes	Yes	Yes	Yes
Copy/paste	Yes	Yes	Yes	Yes	Yes	Yes
Media Support/ Ecosystem	Amazon	Non-DRM iTunes	iTunes	Ovi	Amazon	Windows Media Player
Global Search	Yes	No	Yes	Yes	Yes	No
Firmware Updates	OTA	Tethered, OTA	Tethered	Tethered, OTA	Unknown	Tethered, OTA
Browser Engine	WebKit	WebKit	WebKit	WebKit	WebKit	Internet Explorer
Tethering	Yes	Yes	Yes	Yes	Yes	Yes
Stereo Bluetooth	Yes	Yes	Yes	Yes	Yes	Yes
SDK Availability/ Support	Yes	Yes	Yes	Yes	Yes	Yes

continued on following page

Table 2. Continued

Official App Store	Google Play	BlackBerry App World	App Store	Ovi Store	App Catalog	Windows Marketplace
App Availability	High	Medium	High	Medium	Low	Medium
Native Applications	Yes	No	Yes	Yes	No	No
On-Device App Management	Excellent	Good	Excellent	Good	Excellent	Good

Distribution Network

We can place an intelligent IPTV management system to improve the QoE of the end-user between the ISP core network and the access network, precisely on the distribution network. The fact of having the intelligent system between two networks let it gather information from both. Moreover, we can place a server in every network of the IPTV service provider in order to collect data from each IPTV channel (multicast group). This system could user the QoS parameters of the network and the QoE of the end user as input parameters of the intelligent system. These parameters (jitter, delay, zapping time, etc.) could come from different parts of the network for each channel. A system developed is shown in Lloret et al. (2011). In this case, Lloret et al. (2011) propose to place a server with a QoE process that compares the parameters measured from the network. If the parameters measured are within a range, the QoE process does not perform any task, but simply stores the data. If the QoE data are not in the range, then the QoE process must perform some changes in the network by sending SNMP messages to the appropriate devices in order to improve the IPTV service. Another intelligent system to improve the IPTV service is developed by the same authors of this chapter in Lloret et al. (2011b). In this case, we explain the protocol needed by intermediate devices (routers, switches, servers, etc.) in order to implement a correct IPTV streaming. The intermediate devices perform several actions according to the control messages and their available resources.

Access Network

Generally, there are several wireless technologies coexisting in the access network. Moreover, all these wireless networks are connected to a common IPTV network infrastructure and they are able to offer IPTV service. Usually, the customers have multiband smart phones and several wireless interfaces. Thus, we can place an intelligent system in the access network to let them know the best wireless network to connect with. In those situations, where the smart phone can join two or more wireless technologies, the device may be able to measure the RSSI (Radio Signal Strength Indicator) of each available wireless network and send the measured values to the server in order to decide the best.

We proposed to place an intelligent system in the access network in Lloret (2011a). Every time a device joins a wireless network, it sends the SSID and MAC address, of the detected APs, and the QoS parameters to the QoE test server. The QoE test server has a database with all wireless networks of access network, and intelligently chooses the best wireless network for each smart phone. It is a dynamic system, in which decisions are based on the values of the parameters. This system has the purpose of providing the highest QoE to the end-user.

Smart Phone Network

The intelligent system can also be placed in the smart phone. For example, it can decide to handover to the best available wireless network,

Figure 3. Zones to place intelligent systems in an IPTV network

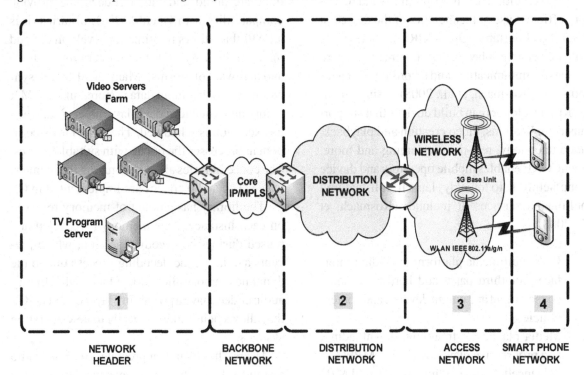

when the QoS parameters measured by the end-user reaches a threshold or because it has higher RSSI. The purpose of this system is to guarantee anytime the highest QoS and QoE to the user to provide high IPTV quality. The smart phone can decide intelligently the best wireless network based on its experience and previous analysis.

In Canovas et al. (2011), we developed an intelligent system to perform vertical handover between different WLANs to receive the best IPTV quality. The measurements taken from a real deployment show that mobile devices roam between different wireless networks without noticing the end-users.

IPTV IMPLEMENTATION IN SMART PHONES

Throughout this section, we delve into the programming in a smart phone. We describe the implementation of an IPTV-capable multimedia player. We selected the Android Operating System as the development platform. The justification for this choice is discussed in the following section. After a list of the features of the Android platform and the IPTV capabilities of the Android, we then describe the code of a sample application.

IPTV Capabilities of the Android Operating System

Android is now becoming a serious alternative to other platforms like iPhone, Symbian or Windows Mobile. Actually, today Android is the second platform after the iPhone, with respect to traffic on the Internet. In addition, even it surpasses the iPhone in USA (Admob, 2010). Android offers an easy way to deploy powerful and innovative mobile applications (Android, 2010). The OpenCORE Framework provides an interface that allows modular use and easy configuration. It was designed only for audio and video codecs to support multimedia. It is optimized and makes very efficient use of the memory of the device; it is solid and portable. OpenCORE includes software that enables playing and streaming standard formats, communication, and recording of images and video (Kosmach, et al., 2008). Using Open-CORE, developers can build devices that support music applications, video creation and playback, video telephony, real-time streaming and more. OpenCORE enables mobile operators and device manufacturers to quickly launch full-featured multimedia services. It includes (Kosmach, et al., 2008):

1. PV's multimedia platform, including interfaces for third party and hardware media codecs, input and output devices, and content policies.
2. Media playback, streaming, downloading, and progressive playback: 3GPP, MPEG-4, Advanced Audio Coding (AAC), and MP3 containers.
3. Media streaming, downloading, and progressive playback: 3GPP, HTTP, and RTSP/RTP.
4. Video and image encoders and decoders: MPEG-4, H.263 and AVC (H.264), and JPEG.
5. Speech codecs: AMR-NB and AMR-WB.
6. Audio codecs: MP3, AAC, AAC+.
7. Media recording: 3GPP, MPEG-4, and JPEG.
8. Video telephony based on 324-M standard.
9. PV test framework to ensure robustness and stability; profiling tools for memory and CPU usage.
10. Support for Khronos OpenMAX™ specification.

OpenCORE is modular and expandable and allows the combination of components such as format file, codecs, streaming protocol and other elements that permit to implement a large variety of applications. The audio codecs for Android are: MP3 1/2/2.5, AAC LC/LTP, HE-AACv1, HE-AACv2, AMR-NB and AMR-WB. OpenCORE provides OMX components, which are used to integrate audio and video codecs and provide compatibility with OpenMAX IL libraries (it is an API that allows multimedia development and integrate it easily in different operating systems and hardware platforms). Many of APIs are asynchronous, so that input data is stored in the OMX components queue and it is later processed, when its execution is scheduled. Then, the OMX component processes the queue with suitable codecs. The codec libraries allow optimizing the memory usage. The audio codecs are implemented in C.

The library has amount of memory reserved on each instance. This memory is recycled and reused during the decoding session, which prevents a delay in the decoding process due to the dynamic memory allocation. OpenCORE library audio codecs have a portability layer called OSCL that allows being adapted easily to new operating systems.

Users should not appreciate possible audio and video loss. There are some standard specifications that provide mechanisms to compensate these losses, for example, AMR-NB (repeats the last valid frame). There are other solutions such as store multiple frames to mask the possible data loss. These solutions can lead to increase the code size, memory use and computational complexity. For speech, signal is not as continuous as the audio. OpenCORE performs silent insertion of pre-established frames. Packet loss is hidden using less memory and less processing cycles than the standard approach. OpenCORE provides download support and progressive playback of different multimedia formats. Progressive download involves download multimedia files from the Web server by using HTTP. For progressive download, playback starts while the file is still being downloaded in the background. Progressive playback is similar to progressive download. Their difference is that during the progressive playback, the entire file is not stored on the client; only part of file is stored in the cache memory of the device (PacketVideo, 2003). You can see an overview in Figure 4.

Implementation of an IPTV Player in Android

In this section, we present a video player application software that lets the user watch IPTV in smart phones using the Android operating system. The video player application is based on the MediaPlayer object (MediaPlayer, 2012). It is responsible for media playback in the Android operating system. Let us review the most important features of the MediaPlayer class. A MediaPlayer object is able to go through a variety of states. Table 3 shows the methods used for state transitions. There are two types of methods: the underlined and the not underlined. Not underlined represent methods invoked synchronously, while the underlined represent methods called asynchronously.

In order to develop the IPTV receiver application, to install the Android SDK and the Eclipse development environment with the plug-in for Android is recommended (Android, 2012). A new project can be created in this environment with the following information:

- **Project Name:** VideoPlayer
- **Build Target:** Google APIs 1.5
- **Application Name:** VideoPlayer
- **Package Name:** org.example.videoplayer
- **Create Activity:** VideoPlayer

In the folder *res/drawable,* we should put four image files that will be used for the icons of the main actions of the media player. These four files must have the following names: *pause.jpeg, play. jpeg, reset.jpeg,* and *stop.jpeg.*

The visual aspect of an Android software application is designed via XML files. In this IPTV application we use the text shown in Box 2, which must be stored in *res/layout/main.xm.*

Figure 5 shows the IPTV software application running in the Android operating system.

The program code used to run the IPTV receiver application is shown in Box 3. It is in Java programming language Java. We can observe that the application extends the class Activity. An *Activity* is the Android element used to encapsulate the different windows of an application. In addition, we implemented four interfaces that

Figure 4. Android OpenCORE

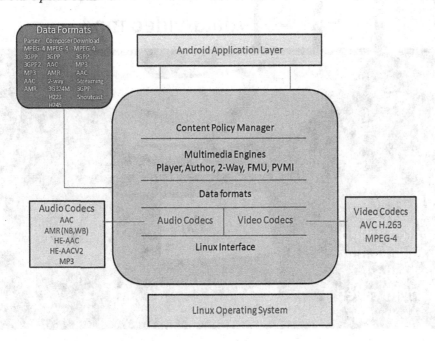

Table 3. Methods for the state transitions of the media player object

		Input state							
		Idle	Initialized	Preparing	Prepared	Started	Paused	Stoped	Playback Completed
Ouput state	**Initialized**	setData-Source							
	Preparing		prepareAs-inc					prepareAs-inc	
	Prepared		prepare	onPrepared	seekTo			prepare	
	Started				start	seekTo start	start		start
	Paused					pause	seekTo pause		
	Stoped				stop	stop	stop	stop	stop
	Playback Completed					onCompletion			seekTo
	End	release	release	release	release	release	release	release	release
	Error	onError	onError	onError	onError	onError	onError	onError	onError

Figure 5. IPTV software application running in the Android operating system

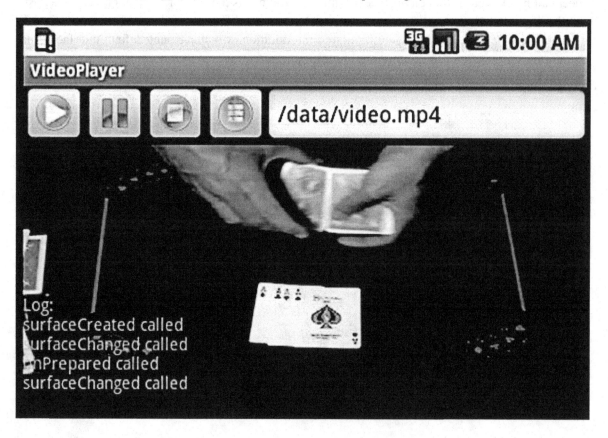

Box 2. File for the visual aspect of the media player

```xml
<?xml version="1.0" encoding="utf-8"?>
<RelativeLayout
     xmlns:android="http://schemas.android.com/apk/res/android"
     android:layout_height="fill_parent"
     android:layout_width="fill_parent"
     android:orientation="horizontal">
    <LinearLayout
          android:id="@+id/ButonsLayout"
          android:layout_height="wrap_content"
          android:layout_width="fill_parent"
          android:orientation="horizontal"
          android:layout_alignParentTop="true">
       <ImageButton android:id="@+id/play"
                  android:layout_height="wrap_content"
                  android:layout_width="wrap_content"
                  android:src="@drawable/play"/>
       <ImageButton android:id="@+id/pause"
                  android:layout_height="wrap_content"
                  android:layout_width="wrap_content"
                  android:src="@drawable/pause"/>
       <ImageButton android:id="@+id/stop"
                  android:layout_height="wrap_content"
                  android:layout_width="wrap_content"
                  android:src="@drawable/stop"/>
       <ImageButton android:id="@+id/logButton"
                  android:layout_height="wrap_content"
                  android:layout_width="wrap_content"
                  android:src="@drawable/log"/>
       <EditText    android:id="@+id/path"
                  android:layout_height="fill_parent"
                  android:layout_width="fill_parent"
                  android:text="/data/video.3gp"/>
    </LinearLayout>
    <VideoView android:id="@+id/surfaceView"
                  android:layout_height="fill_parent"
                  android:layout_width="fill_parent"
                  android:layout_below="@+id/ButonsLayout"/>
    <ScrollView android:id="@+id/ScrollView"
                  android:layout_height="100px"
                  android:layout_width="fill_parent"
                  android:layout_alignParentBottom="true">
       <TextView    android:id="@+id/Log"
                  android:layout_height="wrap_content"
                  android:layout_width="fill_parent"
                  android:text="Log:"/>
    </ScrollView>
</RelativeLayout>
```

Box 3. Program code used to run the IPTV receiver application

```
public class VideoPlayer extends Activity implements
OnBufferingUpdateListener, OnCompletionListener, MediaPlayer.OnPreparedListener,
SurfaceHolder.Callback {
        private MediaPlayer mediaPlayer;
        private SurfaceView surfaceView;
        private SurfaceHolder surfaceHolder;
        private EditText editText;
        private ImageButton bPlay, bPause, bStop, bLog;
        private TextView logTextView;
        private boolean pause;
        private String path;
        private int savePos = 0;
        public void onCreate(Bundle bundle) {
                super.onCreate(bundle);
                setContentView(R.layout.main);
                surfaceView = (SurfaceView) findViewById(R.id.surfaceView);
                surfaceHolder = surfaceView.getHolder();
                surfaceHolder.addCallback(this);
                surfaceHolder.setType(SurfaceHolder.SURFACE_TYPE_PUSH_BUFFERS);
                editText = (EditText) findViewById(R.id.path);
                editText.setText(
                        "http://personales.gan.upv.es/~jtomas/video.3gp");
                logTextView = (TextView) findViewById(R.id.Log);
                bPlay = (ImageButton) findViewById(R.id.play);
                bPlay.setOnClickListener(new OnClickListener() {
                        public void onClick(View view) {
                                if (mediaPlayer != null) {
                                        if (pause) {
                                                mediaPlayer.start();
                                        } else {
                                                playVideo();
                                        }
                                }
                        }
                });
                bPause = (ImageButton) findViewById(R.id.pause);
                bPause.setOnClickListener(new OnClickListener() {
                        public void onClick(View view) {
                                if (mediaPlayer != null) {
                                        pause = true;
                                        mediaPlayer.pause();
                                }
                        }
                });
                bStop = (ImageButton) findViewById(R.id.stop);
                bStop.setOnClickListener(new OnClickListener() {
                        public void onClick(View view) {
                                if (mediaPlayer != null) {
                                        pause = false;
```

continued on following page

Box 3. Continued

```
                                    mediaPlayer.stop();
                        }
                }
        });
        bLog = (ImageButton) findViewById(R.id.logButton);
        bLog.setOnClickListener(new OnClickListener() {
                public void onClick(View view) {
                        if (logTextView.getVisibility()==TextView.VISIBLE)
{
                                logTextView.setVisibility(TextView.INVISIBLE);
                        } else {
                                logTextView.setVisibility(TextView.VISIBLE)
                        }
                }
        });
        log("");
}
```

correspond to various event listeners. After the declaration of variables, we defined the *pause* variable, which allows the user to pause the video, the *path* variable, which allows the path to the video and *savePos* variable, which stores the play position.

Box 4 shows the code for the *playVideo* method. This method takes the reproduction path,

Box 4. PlayVideo method

```
private void playVideo() {
                try {
                        pause = false;
                        path = editText.getText().toString();
                        mediaPlayer = new MediaPlayer();
                        mediaPlayer.setDataSource(path);
                        mediaPlayer.setDisplay(surfaceHolder);
                        mediaPlayer.prepare();
                        // mMediaPlayer.prepareAsync(); Used for streaming
                        mediaPlayer.setOnBufferingUpdateListener(this);
                        mediaPlayer.setOnCompletionListener(this);
                        mediaPlayer.setOnPreparedListener(this);
                        mediaPlayer.setAudioStreamType(AudioManager.STREAM_
MUSIC);
                        mediaPlayer.seekTo(savePos);
                } catch (Exception e) {
                        log("ERROR: " + e.getMessage());
                }
        }
```

Box 5. Playback buffer method

```java
public void onBufferingUpdate(MediaPlayer arg0, int percent) {
            log("onBufferingUpdate percent:" + percent);
    }
        public void onCompletion(MediaPlayer arg0) {
            log("onCompletion called");
    }
```

Box 6. Program code used for running the IPTV receiver application

```java
public void onPrepared(MediaPlayer mediaplayer) {
            log("onPrepared called");
            int mVideoWidth = mediaPlayer.getVideoWidth();
            int mVideoHeight = mediaPlayer.getVideoHeight();
            if (mVideoWidth != 0 && mVideoHeight != 0) {
                    surfaceHolder.setFixedSize(mVideoWidth, mVideoHeight);
                    mediaPlayer.start();
            }
    }
```

Box 7. Code to implement the viewing area

```java
public void surfaceCreated(SurfaceHolder holder) {
            log("surfaceCreated called");
            playVideo();
    }
        public void surfaceChanged(SurfaceHolder surfaceholder, int i, int j,
int k) {
            log("surfaceChanged called");
    }
        public void surfaceDestroyed(SurfaceHolder surfaceholder) {
            log("surfaceDestroyed called");
    }
```

Box 8. Code to destroy the activity

```java
@Override
        protected void onDestroy() {
            super.onDestroy();
            if (mediaPlayer != null) {
                    mediaPlayer.release();
                    mediaPlayer = null;
            }
    }
```

Box 9. Methods to stop playing the video and continue it

```
@Override
        public void onPause() {
                super.onPause();
                if (mediaPlayer != null & !pause) {
                        mediaPlayer.pause();
                }
        }
        @Override
        public void onResume() {
                super.onResume();
                if (mediaPlayer != null & !pause) {
                        mediaPlayer.start();
                }
        }
```

Box 10. Code to restore the activity

```
@Override
        protected void onSaveInstanceState(Bundle guardarEstado) {
                super.onSaveInstanceState(guardarEstado);
                if (mediaPlayer != null) {
                        int pos = mediaPlayer.getCurrentPosition();
                                guardarEstado.putString("ruta", path);
                                guardarEstado.putInt("posicion", pos);
                }
        }
        @Override
        protected void onRestoreInstanceState(Bundle recEstado) {
                super.onRestoreInstanceState(recEstado);
                if (recEstado != null) {
                        path = recEstado.getString("ruta");
                        savePos = recEstado.getInt("posicion");
                }
        }
```

Box 11. Code to display the log information

```
private void log(String s) {
                logTextView.append(s + "\n");
        }
}
```

creates a new object called MediaPlayer. Then, it assigns the path to the viewing area, and prepares the video playback. To play a streaming from the network, this function may take some time, in which case you should use instead the method *prepare ()* the method *prepareAsync()*, which allows to continue execution of the program even without waiting for the video. Next, the code assigns to the object, multiple event listeners to be described later. After preparing the audio type, it is positioned to play the position indicated in *savePos* variable. This position is indicated in milliseconds. When there is a new play this variable will be zero.

The methods shown in Box 5. It implements the interfaces *OnBufferingUpdateListener* and *OnCompletionListener*. The first will show the percentage of the playback buffer, while the second will be invoked when the video playback reaches the end.

Box 6 shows the code for the interface *OnPreperedListener*. It is called once the video is ready for playback. At this point, we know the height and width of the video.

Box 7 shows the code to implement the *SurfaceHolder.Callback* interface. This method is invoked when the viewing area is created, changed, or destroyed. It corresponds to the actions of the life cycle of an activity.

Box 8 shows the code when the activity will be destroyed. Since an object of class *Media-Player* consumes many resources, it is needed to release it as soon as possible.

Box 9 shows two methods that are invoked when the activity goes into the background and when it returns to the surface. Since we need to stop playing the video and continue it, *pause()* and *start()* methods should be invoked respectively. Do not confuse this action with the variable *pause* to what it indicates is that the user has pressed this button.

When the system needs memory, it may decide to remove the activity. In order to restore it, we developed the code shown in Box 10. First, the system calls the method *onSaveInstanceState()* to give us the opportunity to store some important information. If the user later returns to the application, this information will be reloaded, by invoking the method *onRestoreInstanceState()*. In our case, the information to be saved is the *path* and *savePos* variables, representing the video source and the position where the video is playing. The same process occurs, when the user changes the position of the phone. That is, when the phone flips, activities are destroyed and recreated, so these methods have to be also invoked.

We also developed a method to display log information that can be used by multiple event listeners. It is shown in Box 11. This information can be displayed or not, by clicking in the appropriate button.

CONCLUSION

In this chapter, we have introduced the current hardware and operating systems for smart phones in order to know the available devices in the market for mobile IPTV. After the description of the main parts of the IPTV architecture and the main protocols used for IPTV transmission, we have discussed the limitations and requirements for IPTV reception on smart phones. Next, we have explained where several intelligence methods can be incorporated for video transmission over IP fulfilling good video quality to the end user. Finally, we have explained the IPTV layer implementation in an Android operating system. Moreover, we provided an overview of how to carry out transmission in mobile networks. The limitations of mobile IPTV have been analyzed in this work. We focus our attention on mobile IPTV architectural aspects, transmission, and implementation for mobile IPTV. Finally, we have explained the main steps to make an IPTV player. This player is able to take several multimedia flows using different network protocols.

REFERENCES

Admob. (2010). *Mobile metrics report*. Retrieved from http://metrics.admob.com/wp-content/uploads/2010/06/May-2010-AdMob-Mobile-Metrics-Highlights.pdf

Android Developers. (2010). *What is Android?* Retrieved from http://developer.android.com/guide/basics/what-is-android.html

Android Developers. (2012). *Download the Android SDK*. Retrieved from http://developer.android.com/sdk/index.html

Cali, F., Conti, M., & Gregori, E. (1998). IEEE 802.11 wireless LAN: Capacity analysis and protocol enhancement. In *Proceedings of the 1998 IEEE Seventeenth Annual Joint Conference of the IEEE Computer and Communications Societies (INFOCOM 1998)*, (vol. 1, pp. 142-149). IEEE Press.

Camarillo, G., & García-Martín, M. A. (2011). *The 3G IP multimedia subsystem (IMS): Merging the internet and the cellular worlds* (3rd ed.). New York, NY: Wiley.

Canovas, A., Bri, D., Sendra, S., & Lloret, J. (2011). Vertical WLAN handover algorithm and protocol to improve the IPTV QoS of the end user. In *Proceedings of the IEEE International Conference on Communications (ICC 2011)*, (pp. 10-15). Ottawa, Canada: IEEE Press.

Cavendish, D., Murakami, K., Yun, S.-H., Matsuda, O., & Nishihara, M. (2002). New transport services for next-generation SONET/SDH systems. *IEEE Communications Magazine, 40*(5), 80–87. doi:10.1109/35.1000217

Correia, A., Silva, J., Souto, N., Silva, L., Boal, A., & Soares, A. (2007). Multi-resolution broadcast/multicast systems for MBMS. *IEEE Transactions on Broadcasting, 53*(1), 224–234. doi:10.1109/TBC.2007.891705

Cunningham, D., Lane, B., & Lane, W. (1999). *Gigabit ethernet networking*. Indianapolis, IN: Macmillan Publishing Co., Inc.

Díaz, A., Merino, P., & Rivas, F. J. (2010). QoS analysis of video streaming service in live cellular networks. *Computer Communications, 33*(3), 322–335. doi:10.1016/j.comcom.2009.09.007

Florido, G., Liberal, F., & Fajardo, J. O. (2009). QoS-oriented admission control in HSDPA networks. *Network Protocols and Algorithms, 1*(1), 52–61. doi:10.5296/npa.v1i1.178

Hellberg, C., Greene, D., & Boyes, T. (2007). *Broadband network architectures: Designing and deploying triple-play services*. Upper Saddle River, NJ: Prentice Hall.

Hoffman, D., Fernando, G., Goyal, V., & Civanlar, M. (1998) *RTP payload format for MPEG1/MPEG2 video*. RFC 2250. Retrieved from http://tools.ietf.org/rfc/rfc2250.txt

inCode Telecom Group Inc. (2006). *The quadplay – The first wave of the converged services evolution*. White paper. Retrieved from http://www.techrepublic.com/whitepapers/the-quadplay-the-first-wave-of-the-converged-services-evolution/288116

IPTV Focus Group. (2012). *Website*. Retrieved from http://www.itu.int/ITUT/IPTV/

ITU-T Rec J.281. (2005). *Website*. Retrieved from http://www.itu.int/rec/T-REC-J.281-200503-I

Iwatsuki, K., Kani, J., Suzuki, H., & Fujiwara, M. (2004). Access and metro networks based on WDM technologies. *Journal of Lightwave Technology, 22*(11), 2623. doi:10.1109/JLT.2004.834492

Keiser, G. (2006). *FTTX concepts and application*. New York, NY: John Wiley & Sons, Inc. doi:10.1002/047176910X

Kosmach, J., Neff, R., Sherwood, G., Tapia, J., & Veselinovic, D. (2008). Introduction to the opencore video components used in the Android platform. In *Proceedings of the 34th International Conference: New Trends in Audio for Mobile and Handheld Devices*. IEEE.

Lloret, J., Atenas, M., Canovas, A., & Garcia, M. (2011b). A network management algorithm based on 3D coding techniques for stereoscopic IPTV delivery. In *Proceedings of the IEEE Global Communications Conference (IEEE Globecomm 2011)*. Houston, TX: IEEE Press.

Lloret, J., Canovas, A., Rodrigues, J. J. P. C., & Lin, K. (2011c). A network algorithm for 3D/2D IPTV distribution using WiMAX and WLAN technologies. *Journal of Multimedia Tools and Applications*. Retrieved from http://rd.springer.com/article/10.1007/s11042-011-0929-4

Lloret, J., Garcia, M., Atenas, M., & Canovas, A. (2011). A QoE management system to improve the IPTV network. *International Journal of Communication Systems, 24*(1), 118–138. doi:10.1002/dac.1145

Lloret, J., Garcia, M., Atenas, M., & Canovas, A. (2011a). A stereoscopic video transmission algorithm for an IPTV network based on empirical data. *International Journal of Communication Systems, 24*(10), 1298–1329. doi:10.1002/dac.1196

Lloret, J., García, M., & Boronat, F. (2009b). *IPTV: La televisión por Internet*. Malaga, Spain: Publicaciones Vértice.

Lloret, J., Garcia, M., Edo, M., & Lacuesta, R. (2009a). *IPTV distribution network access system using WiMAX and WLAN technologies*. New York, NY: ACM Press.

MediaPlayer Android. (2012). *Website*. Retrieved from http://developer.android.com/reference/android/media/MediaPlayer.html

PacketVideo Corporation. (2009). *OpenCORE multimedia framework capabilities*. Retrieved from http://opencore.net/files/opencore_framework_capabilities.pdf

Park, S., & Jeong, S. (2009). Mobile IPTV: Approaches, challenges, standards, and QoS support. *IEEE Internet Computing, 13*(3), 23–31. doi:10.1109/MIC.2009.65

Park, S., Jeong, S.-H., & Hwang, C. (2008). Mobile IPTV expanding the value of IPTV. In *Proceedings of the 2008 Seventh International Conference on Networking (ICN 2008)*, (pp. 296-301). Cancun, Mexico: ICN.

Pew Internet. (2008). *55% of adult Americans have home broadband connections*. Retrieved from http://www.pewinternet.org/Press-Releases/2008/55-of-adult-Americans-have-home-broadband-connections.aspx

Postel, J. (1980). *User datagram protocol*. RFC 768. Retrieved from http://tools.ietf.org/html/rfc768

Reimers, U. (1998). Digital video broadcasting. *IEEE Communications Magazine, 36*(6), 104–110. doi:10.1109/35.685371

Schulzrinne, H. (1992). *The real time transport protocol (RTP)*. In *Proceedings of the MCNC 2nd Packet Video Workshop*. MCNC.

Schulzrinne, H., Casner, S., Frederick, R., & Jacobson, V. (1996). *RTP: A transport protocol for real-time applications*. RFC 1889. Retrieved from http://www.ietf.org/rfc/rfc1889.txt

Schulzrinne, H., Rao, A., & Lanphier, R. (1998). *Real time streaming protocol (RTSP)*. RFC 2326. Retrieved from http://www.ietf.org/rfc/rfc2326.txt

Soluciones & Tecnologia. (2007). *Tendencia del mercado IPTV mundial*. Retrieved from http://www.tecnologiahechapalabra.com/datos/soluciones/enlaces/articulo.asp?i=827

Tadayoni, R., & Sigurdsson, H. M. (2006). IPTV market development and regulatory aspects. In *Proceedings for the International Telecommunications Society Conference (ITS 2006)*. Beijing, China: ITS.

Uskela, S. (2003). Key concepts for evolution toward beyond 3G networks. *IEEE Wireless Communications*, *10*(1), 43–48. doi:10.1109/MWC.2003.1182110

Wan, F., Cai, L., & Gulliver, A. (2008). Can we multiplex IPTV and TCP? In *Proceedings of the 2008 IEEE International Conference on Communications (ICC 2008)*, (pp. 5804-5808). IEEE Press.

Welch, J., & Clark, J. (2006). *A proposed media delivery index (MDI)*. RFC 4445. Retrieved from http://tools.ietf.org/html/rfc4445

ADDITIONAL READING

Asghar, J., Le Faucheur, F., & Hood, I. (2009). Preserving video quality in IPTV networks. *IEEE Transactions on Broadcasting*, *55*(2), 386–395. doi:10.1109/TBC.2009.2019419

Carlsson, C., & Walden, P. (2007). Mobile TV — To live or die by content. In *Proceedings of the 40th Annual Hawaii International Conference on System Sciences (HICSS 2007)*, (pp. 51-60). IEEE Press.

Hartung, F. (2007). Delivery of broadcast service in 3G networks. *IEEE Transactions on Broadcasting*, *53*(1), 188–196. doi:10.1109/TBC.2007.891711

Issa, O., Li, W., Liu, H., Speranza, F., & Renaud, R. (2009). Quality assessment of high definition TV distribution over IP networks. In *Proceedings of the 2009 IEEE International Symposium on Broadband Multimedia Systems and Broadcasting (BMSB 2009)*, (pp. 1-6). IEEE Press.

Kanumuri, S., Cosman, P., Reibman, A., & Vaishampayan, V. (2006). Modeling packet-loss visibility in MPEG-2 video. *IEEE Transactions on Multimedia*, *8*(2), 341–355. doi:10.1109/TMM.2005.864343

Kerpez, K., Waring, D., Lapiotis, G., Lyles, J., & Vaidyanathan, R. (2006). IPTV service assurance. *IEEE Communications Magazine*, *44*(9), 166–172. doi:10.1109/MCOM.2006.1705994

Maisonneuve, J., Deschanel, M., Heiles, J., Wei, L., Hong, L., Sharpe, R., & Yiyan, W. (2009). An overview of IPTV standards development. *IEEE Transactions on Broadcasting*, *55*(2), 315–328. doi:10.1109/TBC.2009.2020451

Mikóczy, E., Vidal, I., & Kanellopoulos, D. (2012). IPTV evolution towards NGN and hybrid scenarios. *Informatica*, *36*(1), 3–12.

Montpetit, M.-J., Ganesan, S., & Joyce, J. (2007). IPTV: Any device, anytime, anywhere. In *Proceedings of SCTE 2007 Emerging Technologies*. SCTE.

Montpetit, M.-J., Klym, N., & Mirlacher, T. (2011). The future of IPTV. *Multimedia Tools and Applications*, *53*(3), 519–532. doi:10.1007/s11042-010-0504-4

Mu, M., Cerqueira, E., Boavida, F., & Mauthe, A. (2009). Quality of experience management framework for real-time multimedia applications. *International Journal of Internet Protocol Technology*, *4*(1), 54–64. doi:10.1504/IJIPT.2009.024170

Mu, M., Mauthe, A., Haley, R., & Garcia, D. (2011). Discrete quality assessment in IPTV content distribution networks. *Signal Processing Image Communication*, 26(7), 339–357. doi:10.1016/j.image.2011.03.002

Müller, C., & Timmerer, C. (2011). *A test-bed for the dynamic adaptive streaming over HTTP featuring session mobility*. Retrieved from www-itec. uni-klu.ac.at/bib/files/mueller_A_Test_Bed_for_DASH_featuring_Session_Mobility.pdf

Park, J., & Lim, Y. (2011). An association management scheme for seamless IPTV services in a WLAN. In *Proceedings of the 2011 IEEE International Conference on Consumer Electronics (ICCE 2011)*, (pp. 899–900). IEEE Press.

Radha, H. M., van der Schaar, M., & Chen, Y. (2001). The MPEG-4 fine-grained scalable video coding method for multimedia streaming over IP. *IEEE Transactions on Multimedia*, 3(1), 53–68. doi:10.1109/6046.909594

Reibman, A. R., Kanumuri, S., Vaishampayan, V., & Cosman, P. (2004). Visibility of individual packet losses in MPEG-2 video. In *Proceedings of the International Conference on Image Processing*, (Vol. 1, pp. 171-174). IEEE.

Schulzrinne, H. (1998). *Real time streaming protocol*. Retrieved from http://www.ietf.org/rfc/rfc2326.txt

Shihab, E., & Lin, C. (2007). IPTV distribution technologies in broadband home networks. In *Proceedings of the 2007 Canadian Conference on Electrical and Computer Engineering (CCECE 2007)*, (pp. 765-768). CCECE.

Sloup, P. (2011). *WebGL earth*. Retrieved from http://is.muni.cz/th/325196/fi_b/thesis.pdf

Soomro, T. R., Zheng, K., & Pan, Y. (1999). HTML and multimedia web GIS. In *Proceedings of the 1999 Third International Conference on Computational Intelligence and Multimedia Applications, ICCIMA 1999*, (pp. 371-382). ICCIMA.

Staelens, N., Moens, S., Van den Broeck, W., Marien, I., Vermeulen, B., & Lambert, P. (2010). Assessing quality of experience of IPTV and video on demand services in real-life environments. *IEEE Transactions on Broadcasting*, 56(4), 458–466. doi:10.1109/TBC.2010.2067710

Vedantham, S., Kim, S., & Kataria, D. (2006). Carrier-grade ethernet challenges for IPTV deployment. *IEEE Communications Magazine*, 44(4), 24–31. doi:10.1109/MCOM.2006.1632644

Wiegand, T., Noblet, L., & Rovati, F. (2009). Scalable video coding for IPTV services. *IEEE Transactions on Broadcasting*, 55(2), 527–538. doi:10.1109/TBC.2009.2020954

Yi-zhe, S. (2008). Power-up performance research of the mobile smart phone. *Computer Knowledge and Technology*. Retrieved from http://en.cnki.com.cn/Article_en/CJFDTOTAL-DNZS200825091.htm

Yuste, L. B., & Melvin, H. (2012). A protocol review for IPTV and webtv multimedia delivery systems. *Komunikacie*, 14(2), 33–41.

KEY TERMS AND DEFINITIONS

HDTV (High Definition Television): It provides a resolution that is until five times greater than the standard-definition television. It provides higher quality and sharpness image.

HSPA (High Speed Packet Access): It is the combination of complementary and further down technologies to third generation mobile. It supports data rates up to 14.4 Mb/s downstream and 2 Mb/s upstream.

IMS (Integrated Multimedia Subsystem): It is a standardized architecture for telecommunication operators that want to provide mobile and fixed multimedia services. It defines the functional architecture that uses a Voice-over-IP (VoIP) implementation based on a 3GPP standardized implementation of Session Initiation Protocol (SIP), and runs over the Standard Internet Protocol (IP).

INTSERV (Integrated Services): It is an architecture that specifies the elements to guarantee Quality of Service (QoS) on networks. It was defined in IETF RFC 1633, which proposed the resource reservation protocol (RSVP).

IPTV (Internet Protocol Television): It is a system through which television services are delivered using the Internet protocol suite over a packet-switched network such as the Internet, instead of being delivered through traditional terrestrial, satellite signal, and cable television formats.

MBMS (Broadcast Multicast Service): It is a user service, which is a combination of both a broadcast service and a multicast service, developed by the third generation partnership project (3GPP). The data is transmitted once through a common channel to all cell subscribers.

QOE (Quality of Experience): It evaluates the acceptability of a service from the client side. Moreover it assesses the customer satisfaction objectively and subjectively.

QOS (Quality of Service): It is the network capacity to provide better service to selected traffic. it guarantees the transmission of certain amount of information at a given time. It is very important for some applications such as streaming video or voice.

WCDMA (Wideband Code Division Multiple Access): It is the radio access technology that supports all multimedia services that are available through 3rd generation terminals. It is used to increase the capacity and coverage of wireless communication networks.

Chapter 14
An XML–Based Customizable Model for Multimedia Applications for Museums and Exhibitions

Mersini Paschou
University of Patras, Greece

Athanasios Tsakalidis
University of Patras, Greece

Evangelos Sakkopoulos
University of Patras, Greece

Giannis Tzimas
Technological Educational Institute of Messolonghi, Greece

Emmanouil Viennas
University of Patras, Greece

ABSTRACT

Inclusion of Information and Communication Technologies (ICTs) and multimedia in a museum can result in a functional upgrade of the visitor experience in an exhibition. The added value of ICTs includes promoting the enhancement of educational, research, and entertainment purposes towards which a museum has already been designed. To better understand the potential role of ICTs in museums, the authors introduce an XML-based customizable system. In this chapter, they present an XML-based customizable multimedia solution for museums and exhibitions. The proposed approach serves multimedia content solutions for the assistance of visitors and researchers. A modular approach is adopted in order to provide a User Interface abstraction and operation-business logic isolation from the data. The key advantage of the proposed solution is the separation of concerns for User Interface, business logic, and data retrieval. The proposed solution allows the dynamic XML-based customization of museum multimedia applications to support additional data from new seasonal or one-time exhibitions at the same museum, re-arrangement of the exhibits in the museum halls, addition of new digitized halls with the respective multimedia data and any additional documentation or multimedia extras for existing exhibits. The authors present a case study at the digital exhibition for the history of the ancient Olympic Games at the Older Olympia Museum. Several hundreds of exhibits have been included and the dynamic management was successful after a careful digitization procedure. The results have been encouraging, the users and administrators' feedback was positive, and the full-scale deployment was successful.

DOI: 10.4018/978-1-4666-2833-5.ch014

INTRODUCTION

Multimedia tools used in museums facilitate communication of large amounts of information in a user-friendly and interesting manner. In parallel, multimedia tools allow visitors to access the information they require at their own pace (Allison & Gwaltney, 1991). However, multimedia tools are applications that do need careful structuring in a normalized way through a model for the application subsystem layer (Zafirovic-Vukotic & Niemegeers, 1992). Without this, often many multimedia tools offer services using many times either the same or similar elementary capabilities and functionalities at the application layer, when usually new data arrive or data have been re-arranged positions.

Nowadays, electronic information destined to be delivered to applications takes the form of multimedia, which integrates different types of media including text, images, video, and audio. This is the reason why many techniques are being exploited for effective manipulation of this great amount of information by multimedia applications in the cultural heritage sector (Kanellopoulos, 2011). In this chapter, we propose an effective way of handling multimedia data in order to facilitate organization, which allows for future expansion and effective organising of multimedia content. For this purpose, XML (eXtensible Mark-Up Language) is used, because it has become a standard mark-up language. To achieve this goal, a multi-tier architecture has been devised. The key aim is to achieve re-usability of the business logic layer in order to be able to deliver new exhibits or digital exhibitions as a whole using the same already deployed multimedia application. The proposed concept targets minimization or elimination of application logic tuning, implicit user interface re-construction and extension of fresh and updated multimedia data delivered. In parallel, the robustness of the application is guaranteed, as the business logic stays intact following the initial abstraction and the design of open source code.

There are several cases where Web-based multimedia solutions have been employed in order to give re-usability of the application and logic layers (Styliaras, 2007). However, the robustness of the delivered multimedia tools and applications cannot be claimed as soundproof of issues and malfunctions. This is caused by lots of factors such as: (1) the inherent features of HTML that cannot support full multimedia, (2) the unguaranteed stability in implementations for collaboration between Web and external application such Flash technology (Adobe Flash Platform, 2012), and (3) the non inherent support of kiosk mode isolation and UI security in Web browsers. All the previous factors make it clear that Web technologies may fall short of expectations, when it comes to deliver multimedia application for the public (Kanellopoulos, 2011), as in the case of museums. Furthermore, in our proposal, we deal with another aspect that allows data and code re-usability, delivers open data manipulation with no need for updated of the deployed application, and gives an open framework to present any number of exhibits, exhibitions at a thematic topic.

The proposed solution is presented analytically, while it is implemented at a large-scale end-user case study delivered at the Digital Exhibition of the History of Ancient Olympic Games. This exhibition is hosted at the respective Museum in Olympia, Greece (Information Society, 2012; Culture.gr, 2012). In the deployed case, a large number of exhibits have been digitized and transformed into multimedia. The multimedia content is effectively organised and delivered to avoid maintenance cost and time using the proposed solution, based on XML structures and separation of concerns in the multi tier architecture given. The real life deployment shows the generality of the proposed solution and potentials that it has to be delivered and implemented for additional real life digital exhibitions and multimedia museum applications.

MOTIVATION AND RELATED WORK

There is much hidden behind the use of multimedia content. Typical examples of multimedia manipulation are met amongst others in museums, art galleries and so on. A multimedia paradigm is the presentation of online or offline electronic information in form of text, pictures and videos in a way that they become well organized and expandable. Clearly, the most important factor for the end-user has to do with the time it takes for the application to perform the operations on datasets and to deliver its results that is also known as *"user experience."* However, when it comes to administrators and those who are responsible for maintaining the content of the applications, much can be done to ensure sustainability, scalability and Quality of Service (QoS) (Kanellopoulos, 2011). In many cases, there are external records which store data or metadata that require maintenance. This information has to be well organised in order to be expandable, scalable and compatible for information retrieval.

Moreover, there is a need for dynamic discovery structures that will be always up-to-date providing efficient and available results independent of the dataset involved (Makris, et al., 2004). The applications' design should offer a number of capabilities in order to facilitate better data processing performance, inclusion of new capabilities, obvious at both development and execution time. Issues to be addressed include:

- Scalability of applications,
- Effective maintenance,
- Multilingualism,
- Content formation according to user types.

The main obstacle affecting such mechanisms is heterogeneity between services and datasets (Smith, et al., 1999). A high-level approach has to be considered and each examined solution has to include ways to overcome different aspects of this heterogeneity in order to match the best solution available. The identification of different kinds of heterogeneity gives a notion on what has to be taken into consideration in order to avoid problems or mitigate them. In the last few years, several research activities have been undertaken in the field of introducing new technologies to museums and exhibitions. These activities include the implementation of multimedia information systems that offer background information about exhibition objects (Martin & Trummer, 2005).

Wang et al. (2009) detected a need of the users for interface customization in the Web and presented an XML-based interface customization model, used in the construction of digital museum. Their model aims at the separation of interface and function. The model employs XML as the main technology for the system construction, at an attempt to make the Web interface customized and dynamic. Implementations of Web interfaces for the Archaeological Digital Museum of Shandong University and the Ancient Digital Technologies Museum are presented. However, Web technologies inherent features do not provide the same level of robustness and kiosk mode isolation/security for public use within a museum.

Ardito et al. (2010) based on the observation that the effort spent to create multimedia resources is considerable, pointed out that it is worth reusing them to produce applications suited to other types of visitors. Thus, they present an on-going work to provide tailored applications that support different types of visitors. Such applications are developed according to a model that describes how multimedia resources can be combined, also depending on the type of users and devices. However, in this chapter we provide a solid schema and multi tier architecture to deliver in organized global manner reusability results. Schreiber et al. (2008) describe a semantic Web application, used for semantic annotation and search in large virtual collections of cultural heritage objects, indexed with multiple vocabularies. This application was developed in the *Multimedia Netherlands E-Culture project* and its architecture is based on open Web standards, including XML. A set of e-culture demonstrators was developed for providing multimedia access to distributed collections of cultural heritage objects. These demonstrators show various levels of syntactic and semantic interoperability between collections.

Meyer et al. (2007) focus on the exploitation of intra-site cultural heritage data, and thus they developed a Virtual Research Environment dedicated to it. Their information system is based on open-source software modules, and enables the user to do exploratory spatial and temporal analyses of the data. The system is compliant to every kind of cultural heritage Website and allows the management of diverse data types. Therefore, its users can register and exploit data located at different computers. Liew (2006) focuses on development and management of online exhibition projects on the Web as well as strategic management and technical issues related to it. In particular, technical issues such as interface design and functionality, format and technology standards were considered. For the emerging Web 2.0 technologies, Barak et al. (2009) describe a system, a toolbox based on Web 2.0, dedicated to the preservation and presentation of cultural heritage. The system is called *MOSAICA* and provides a generic framework for users, regardless of culture and religion, to engage actively in preserving and nurturing their heritage via different kinds of activities.

Chang and Kim (2006) presented a digital information retrieval system, which provides services in the Web. This system is used for a digital museum and can support unified retrieval on XML documents, based on both document structure and image content. Retrieval based on document structure is achieved by indexing XML documents that describe items for the digital museum, based on the basic unit of elements of the documents. Color and shape features of the images are used for indexing based on image content.

DESIGN PRINCIPLES OF MULTIMEDIA APPLICATIONS FOR MUSEUMS

In this section, the rules and general principles that were followed are set out by three different angles:

1. As Museological Principles,
2. As Objectives for the Report to be achieved according to the Museological Principles, and
3. As Concepts in Learning.

For each principle or rule, where feasible, an example of the proposed solution is presented in order to achieve full understanding. Museological principles; an attempt is made to formulate the principles prevailing in the preparation of the proposals on this matter.

1. **Principle of Interventionist Action:** Because some things are promoted due to the objectives, without a dissent and right reason and desires and possibilities sometimes drift each other, ensuring a good result requires sufficient theoretical proposals and plans from the perspective of human values, which ultimately leads to the following general principle.

2. **General Museological Principle for Digital Technology:** In a historical museum, information is the main purpose and digital technology may simply be a means to achieve it. As a consequence of this principle, all digital media is discreetly separated from archaeological artefacts and the two are not presented in parallel.

3. **General Museological Principle for Information:** Because the natural recipient of the information, the individual, exists both as an individual guest and as a member of groups of visitors, the information must meet on the one hand various levels of knowledge or interest of people and on the other hand in

visibility conditions dependent on the size of objects and the number of local visitors. All content applications provide information on at least three depth levels, while each topic is presented in a manner adjusted to the visitor (e.g. children, adults, different educational levels), varying from simple narrative for children to scientific presentation for archaeologists or researchers. Moreover, the visitor is given the ability to dynamically adjust the desired pace of information he/she wishes to follow.

4. **General Museological Principle for the Language of the Information:** When the audience comes from various countries, it is highly desirable that the information is available to the largest possible number of languages (a measure of the importance of this principle is the usual bilingual and sometimes trilingual stable document annotation, and a multilingual tour guide, printed or recorded, in numerous archaeological and other museums). To achieve this principle, all digital instruments of this system operate in four languages (English, Greek, French, and German). The Architectural Design of the Applications' Content allows for future expansion in more languages, without putting any restriction to the number of languages, with the XML infrastructure.

5. **Principle for the Individuality and Sociality of Visitors of Museums:** Every museum (as house of common goods and public institution) must develop sociability in the way of information intake, which predominantly is shared. Moreover, the history of the Games is inextricably tied to the need for social action of individuals. The digital exhibition in the Museum follows this principle to promote sociability of people, by including as main exhibits two digital theatres, and other exhibits (Digital Imaging Banks, Groups of Info kiosks, Timeline Imaging), target smaller groups of visitors.

CASE STUDY OBJECTIVES

Museums and Collections' Exhibitions around the world have adopted ICTs at an ever increasing pace for several years (Koshizuka & Sakamura, 2000; Lonsdale, et al., 2004). The goal of technology usage in a museum is offering multiple new dimensions to the experience of the traveler/visitor. For the case of "Digital Exhibition of History of Ancient Olympic Games," technologies were exploited to achieve a threefold objective:

- Participation of visitors in collaborative activities with the implementation of digital media that encourage and emphasize collective action.
- Emergence of the Games as a social good and an essential component of ancient Greek life, the perpetual struggle of opposites, of personal freedom, of play.
- Distinctive separation of all digital media from the exposed archaeological exhibits of the ground and create a digital space in the basement of the museum.

Hereafter, we present the architectural design, implementation aspects, and functional details that were taken into consideration in the implementation of a real world, large-scale system for the Digital Exhibition of the History of Ancient Olympic Games. The subsystems that were installed and are operating in the museum are summarized below.

On the ground floor of the museum, where you can find all the archaeological exhibits, the digital exhibition features a number of advanced features such as:

- 'System of Introductory Tour' whose main objective is to present an introduction to the nature of the museum as well as its points of differentiation.

- The 'Portable Information System' for informing visitors about the archaeological exhibits of this archaeological exhibition.
- The 'Tactile Interaction System' specially designed for the visually impaired. Through tactile interaction, these people experience the exhibits through touch.
- 'Visitor Information Station' whose aim is to give visitors digital information, through the opportunity of navigating through the audiovisual application on all Topics presented in the exhibition of the museum.

The visitor has an array of digital media available, which is the key to a smooth transition from the Archaeological Exhibition itself to the Digital Gallery and aims at attracting visitors to the main Digital Gallery. Inside the main Digital Gallery three separate areas are formed. Two digital theaters, the first of which is an Interactive Virtual Reality Theater for organized digital projections while the other is a Panoramic Theatre, for free screenings. The subject in both theaters is the Greek Culture, as revealed in the Games.

"Antiquity" is the place where guests are invited to learn about the Games and the Greek culture through interactive media. These interactive modules-platforms include Interactive Discussion Tables, delivering interactive application that visitors are able to interact with, by touching the monitor. Moreover, a Timeline System is available, which consists of a vertically mounted display, and can be moved along the horizontal axis by the visitor. On the wall behind the screen, various dates related to the fact the celebration of the ancient Olympic Games are carved. For the date corresponding to a specific screen position, dynamic multimedia information is presented, depending on theme chosen. Finally, a Researcher Information Station is provided, and its purpose is to enable specialized visitors (e.g. archaeologists, researchers and students) the opportunity to explore in depth the history of the area and the exhibits and be able to print items of interest.

The main objective is to provide high quality services and to render a visit to the Museum an experience that visitors will remember with pleasure. Therefore, a set of special services and facilities are provided, which are outlined below and are divided chronologically into three categories as "Before the visit," "during the visit" and "after the visit." All these services combined both facilitate visitor management and also improve the overall experience of the visitor. Before the visit, reservation of the theater for a special presentation, e.g. for school classes or groups of elderly is provided. Bookings by travel agents handling groups are recommended in order to avoid overcrowding and delays on during tours. Moreover, configuration of a special program for students is available. The escort of the class (teacher) can review the program in collaboration with the Museum in order to personalize the sections of the program for the visit.

During the visit, recording anything interesting pointed by the visitor is possible. As the visitor wanders the halls, he/she can point at any exhibit or program of interest, in order to learn more. The content will be e-mailed to the address specified by the visitor. Recording the path of the visitor and he/she saw, in order to get this material in commemorative CD is available, before leaving the museum or sent by electronic mail. After the visit, the visitor is provided with the opportunity to see his recorded tour of the Museum, at the comfort of his/her home, over the Internet, thus creating the illusion of a revisit of the Museum.

A great number of services and facilities are provided. To begin with, all applications run on four languages (Greek, English, French, and German) and the design of the systems allows the integration of new languages with no restrictions (see Figure 1). The following main objectives are served; each visitor can hear and see the contents in one of the four languages provided and multilingualism is available both in visual and in audio content. The visitor can select the language of presentation for the visual and audio content

Figure 1. Language selection screenshot

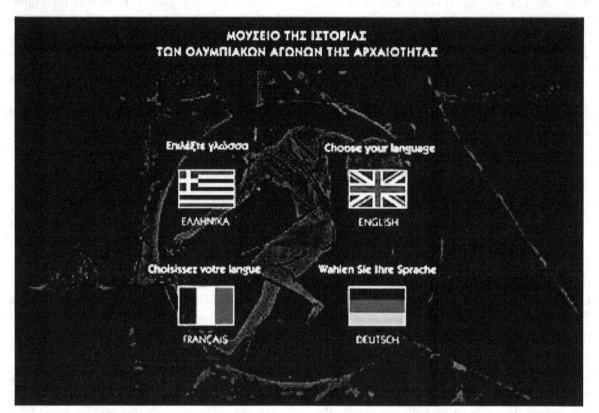

upon arrival, but also can change whenever he/she wishes. Moreover, there is no noise disturbance by other visitors on the same site, due to appropriate technical characteristics that will be described in the section of the technical description of the system.

In locations of information (info-kiosks) as well as above the interactive tables, sound cones are placed that prevent the diffusion of sound. One can hear the sound only when situated in front of the pavilions. In the Panoramic Theatre, in terms of sound, the portable device is tuned to the theater program and the spectator-visitor hears the audio content in the language of his/her choice. Last but not least, in the Interactive Theater, the audio unit is embedded in goggles, supplied upon the entrance in the theater and the language is chosen by the controller located in the seat. The visual program presented on the screen is accom-

panied by a coordinated program of sound emitted simultaneously in four languages.

TECHNICAL DESCRIPTION OF THE SYSTEM

The system is specially designed for indoor and its composition varies depending on the needs of usage. The computing unit and the screen are embedded in a single module (see Figure 2). The presented system is freestanding without keyboard or printer. It has the following features:

- It occupies a small surface area.
- The level is suitable for visitors of a wide range in height, which includes the disabled.
- Built-in flat screen 15.

Figure 2. Free standing kiosk and sound emission unit

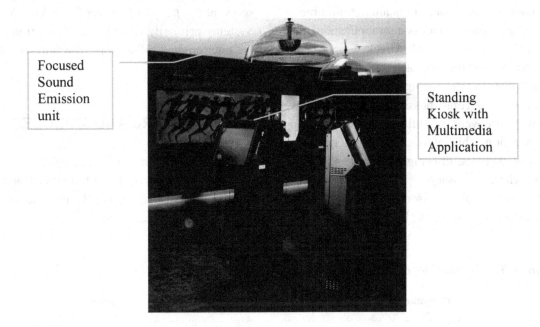

Focused Sound Emission unit

Standing Kiosk with Multimedia Application

- 1.2 GHz CPU.
- 128 MB memory.
- Network Card.
- Operating system Windows XP.

Macromedia Flash MX and Flash MX Professional are professional standard authoring tools for producing high-impact Web experiences. ActionScript is the language that is used when developing an application within Flash. A developer doesn't have to use ActionScript to use Flash, but if one wants to provide user interactivity, work with objects other than those built into Flash (such as buttons and movie clips), or otherwise turn a SWF file into a more robust user experience, the best way to approach it is using ActionScript. This is the reason why, in the presented system, ActionScript was used for the business logic scripts.

To meet the need for focused sound emission units (see Figure 2) are offered with the following specifications:

- Max Power: 100 W,
- The unit has a parabolic shape,
- The unit weight is around 5 Kg.

The system consists of info kiosks (interactive computing unit and display device) located in a museum. The main purpose of the system is to give visitors (who have not purchased portable

Figure 3. Multi layer architecture

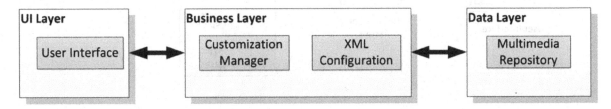

UI Layer

User Interface

Business Layer

Customization Manager

XML Configuration

Data Layer

Multimedia Repository

digital information device) the ability to navigate through the audio-visual application in all subject areas. The audiovisual subsystem application runs in four languages (English, Greek, French, and German) a short tour for each of the twelve (12) Topics, interpretative commentary of the major objects of each Topic, possibly evidence of how they were discovered as well as an anthropological interpretation. This information is enriched by a series of pictures, drawings and representation video, where necessary.

To avoid auditory discomfort, the sound in these locations is broadcast by three guided audio systems, which are placed over the information spots, at the height of the roof. Directed Sound Systems primarily aim at sound isolation from the environment, of visitors who will use these info kiosks.

ARCHITECTURE AND MODULES

The proposed architecture (see Figure 3) is composed of the following three layers: (1) the User Interface (UI) layer, (2) the Business layer, and (3) the Data Layer.

Figure 4. The interface layout template

Navigation Icons	Title of the Exhibition		Navigation Icons
	Exhibition Room ID		
	Exhibition Room Description		
	Multimedia Presentation(image, Video, etc)		
	Multimedia Item Description		
Application Navigation Icons	Multimedia Navigation Items		Application Navigation Icons

Figure 5. Architecture modules

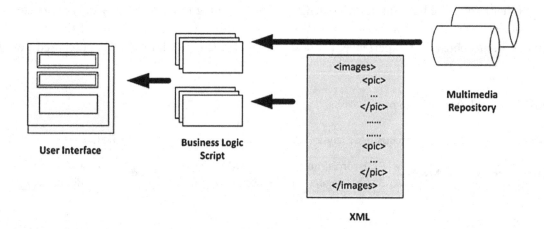

Figure 6. Captions of the screens that utilize the xml files for template of Figure 4

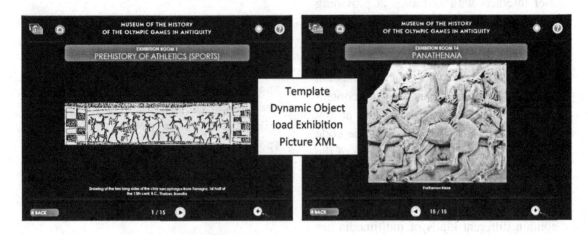

Figure 7. Video dynamic template and dynamic exhibition menu and thematic screens in English and German language through the proposed XML framework

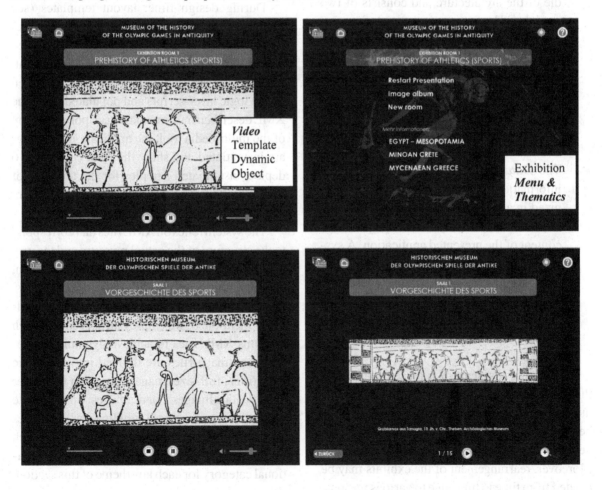

- The *User Interface layer* is the one that the user interacts with and aims at providing clear and well-structured guidance for the user-visitor of the museum. The framework of the UI is standard and its content changes according to the structure and data of the XML files. In this way, the content of the application can be updated at any time, without any intervention to the code of the system. Anyone who is aware of the multimedia files schema is capable of making alterations or additions to the presented content of the museum. UI modules that contain different kinds of multimedia are independent, and thus can be handled separately.

- The *Business layer* is located in the middle of the architecture and consists of two modules. The first one is the customization manager, which is the part that retrieves the appropriate XML files according to the user requests. The other part (XML configuration) is responsible for organizing the XML in the appropriate folders and subfolders and for any modifications to the relative files.

- The *Data layer* includes the multimedia information presented in the system, organized in a way that can be exploited by the business module to dynamically update the content of the presented application. A system administrator must be aware of the fact that, any alterations that are made in the structure and nomenclature of the folders containing the multimedia need to be represented in the XML documents as well.

Organizing a system of a museum in a way that its modules are separated can offer a great number of advantages. Exhibits in museums may be altered and new ones may be added permanently or for a short period of time exhibitions. Moreover, rearrangement of the exhibits may be needed from time to time, due to various reasons.

Algorithm 1. Fraction of code for easy adaptation to new UI elements

```
title5=HELP
&flagsUsage=Language selectio
&helpUsage=Help
&homeUsage=Home
&mapUsage=Interactive map
&playUsage=Start presentation
&pauseUsage=Pause
&stopUsage=Stop
&soundControl=Volume adjustment
&zoomUsage=Zoom
```

In any case, having separated modules in a system offers flexibility and enhances efficiency during maintenance and evolution.

During design time, layout templates (see Figure 4) are set in order to facilitate the User Interface Layer. These templates are delivered within the application core during deployment. They are the abstraction of the UI that is driven by the logic layer-scripts and populated using the XML described multimedia data. Upon execution of the multimedia tool, the application templates are instantiated and show the final UI (Figure 6 depicts instantiated samples for the template of Figure 4). The steps followed for this procedure are described below (depicted in Figure 5).

The system relies on XML files that correspond to the structure of the actual museum and the way it is divided in different exhibition rooms. Then, each exhibition room is divided into subsections, which are represented by a reference key. As already mentioned, we employ an approach, which reclaims XML files to achieve dynamic characteristics. Image files are divided into two basic categories, full-size images, placed in the folder "ApplicationImages" and images on the state zoomed categorized as "LargeImages." This goes on as a recursive regular organization for each section, which contains its photos and an additional category for each key theme of this section.

The recovery of this material by the application is performed in a different way than in other similar applications. In a folder, we organize the xml files corresponding to the topics or to one of the key themes.

The multimedia elements that are used in the user interface of the application are also represented by variables. This approach allows for easy adaptation to new UI elements. This is presented in Algorithm 1.

The layout and the relationships among the multimedia elements used in the application result from the XML files. These files can be easily created, using an authoring tool. The formation of names of folders and files follows a specific policy, corresponding to the one used in the video files. More specifically, for the topics we have files of form images<Section_Number>.

xml, for example images01.xml, while for the thematic keys the file name format becomes images<Section_Number>key<Key_Number>. xml, for example images01Key1.xml. According to the choice that will be made by the user, the appropriate .xml will be retrieved. These files are structured as shown in Algorithm 2.

As shown above, XML files contain the relative position of the images in the folder of the application as well as the caption that accompanies each of them. In this case, updating the material can be performed in two ways; (1) either by direct replacement of the images in their files with files of the same type and name or (2) with alteration of information contained in the XML files. In this way, it can be easily adjusted for other similar application in any other museum, allowing for

Algorithm 2. Structure of authoring tool

```
<images>
        <pic>
                <image>ApplicationImages\Section01\01_Sx158b. jpg</image>
                <legend> Drawing of the two long sides of the clay sarcophagus
from            Tanagra, 1st half of the 13th cent. B. C., Thebes, Boeotia
</legend>
        </pic>
        <pic>
                <image>ApplicationImages\Section01\02_Sx104. jpg</image>
                <legend> Drawing of an inscribed scarab of Amenofis III, Egypt
</legend>
        </pic>
        . . .
        . . .
        . . .

        <pic>
                <image>ApplicationImages\Section01\15_F42. jpg</image>
                <legend>Fragment of an Attic black-figured dinos of Sophilos,
ca. 580-570 B. C., National Archaeological Museum, Athens </legend>
        </pic>
</images>
```

interface updates as well as multimedia content alteration.

Using the proposed multi-layered architecture, the multimedia application delivers a robust and solid UI with a single deployment. In the case of Figure 7, additional templates are displayed instantiated for multimedia data and exhibition presentation menus and thematic rooms. The number of exhibition objects, menus, rooms, and languages is practically limitless as XML representation is modular. Therefore, as long as storage exists, the multimedia application can be updated and extended only by delivering the respective XML described data.

CONCLUSION AND FUTURE STEPS

In this chapter, we presented an XML-based customization model that facilitates usage and enables flexibility towards reusability and total management in multimedia systems. Our goal is to separate interface and functionality based on configuration and content management using XML. When employing our approach, the multimedia content may be easily updated at any time, with minimum management effort. The proposed model is bound to dominate as XML becomes increasingly important and dynamic solutions are preferred. As a formalized language, XML is not easy to use and manipulate for a general user. Moreover, due to its formal nature it is likely to cause errors during the process of directly altering these files. This is the reason why a graphical interface is of great importance for a convenient entry process for the user as well as for reduction of errors. The key advantage of the proposed solution is the separation of concerns for User Interface, business logic and data retrieval.

The proposed system allows the dynamic XML based customization of museum multimedia applications to support additional data from new seasonal or one-time exhibitions at the same museum. We presented a case study of this system at the digital exhibition for the history of the ancient Olympic Games at the Older Olympia Museum. The Digital Exhibition of History of Ancient Olympic Games in Ancient Olympia is equipped with high quality innovative technology applications and efficient operational framework that can be a model for the integration of digital services for other museums and exhibitions. All subsystems are provided as a platform, which provides the ability to extend the Museum Exhibition in the future and to become connected to other museums around the world.

Future steps include the research on compressed XML solutions for over the air delivery of XML data for the case of kiosk and multimedia applications that are not locally and directly connected to a LAN. Finally, another promising aspect is the study for the integration of the proposed framework towards an XML ready content management system.

REFERENCES

Adobe Flash Platform. (2012). *Website*. Retrieved March 2012 from www.adobe.com/flashplatform

Allison, D., & Gwaltney, T. (1991). How people use electronic interactives: Information age - People, information & technology. [Pittsburgh, PA: ICHIM.]. *Proceedings of ICHIM, 1991*, 62–73.

Ardito, C., Costabile, M. F., Lanzilotti, R., & Simeone, A. L. (2010). Combining multimedia resources for an engaging experience of cultural heritage. In *Proceedings of SAPMIA 2010*. Firenze, Italy: SAPMIA.

Barak, M., Hersoviz, O., Kaberman, Z., & Dori, Y. (2009). MOSAICA: A web-2.0 based system for the preservation and presentation of cultural heritage. *Computers & Education, 53*(3), 841–852. doi:10.1016/j.compedu.2009.05.004

Chang, J.-W., & Kim, Y.-J. (2006). XML document retrieval system based on document structure and image content for digital museum. *Lecture Notes in Computer Science, 3842*, 107–111. doi:10.1007/11610496_12

Culture.gr. (2012). *Museum of the history of the Olympic games of antiquity*. Retrieved February 01, 2012 from http://odysseus. culture. gr/h/1/eh155. jsp?obj_id=3488

Information Society SA. (2012). *Digital "wreath" for ancient Olympia*. Retrieved February 01, 2012 from http://www. infosoc. gr/infosoc/en-UK/culture/specials/default. htm

Kanellopoulos, D. (2011). Quality of service in networks supporting cultural multimedia applications. *Program: Electronic Library and Information Systems, 45*(1), 50–66. doi:10.1108/00330331111107394

Koshizuka, N., & Sakamura, K. (2000). Tokyo university digital museum. In *Proceedings of the 2000 Kyoto International Conference on Digital Libraries: Research and Practice*, (pp. 85-92). IEEE.

Liew, C. (2006). Online cultural heritage exhibitions: A survey of strategic issues. *Program: Electronic Library and Information Systems, 40*(4), 372–388. doi:10.1108/00330330610707944

Lonsdale, P., Baber, C., Sharples, M., Byrne, W., Arvanitis, T. N., Brundell, P., & Beale, R. (2004). Context awareness for MOBIlearn: Creating an engaging learning experience in an art museum. In *Proceedings of the 3rd MLEARN Conference*, (pp. 115-118). Rome, Italy: MLEARN.

Makris, C., Panagis, Y., Sakkopoulos, E., & Tsakalidis, A. (2004). *Efficient search algorithms for large scale web service data*. Technical Report No. CTI-TR-2004/09/01. Patras, Greece: Research Academic Computer Technology Institute.

Martin, J., & Trummer, C. (2005). SCALEX – A personalized multimedia information system for museums and exhibitions. In *Proceedings of the First International Conference on Automated Production of Cross Media Content for Multi-Channel Distribution (AXMEDIS 2005)*. AXMEDIS.

Meyer, E., Grussenmeyer, P., Perrin, J.-P., Durand, A., & Drap, P. (2007). A web information system for the management and the dissemination of cultural heritage data. *Journal of Cultural Heritage, 8*(4), 396–411. doi:10.1016/j.culher.2007.07.003

Schreiber, G., Amin, A., van Assem, M., de Boer, V., Hardman, L., & Hildebrand, M. (2008). Semantic annotation and search of cultural-heritage collections: The MultimediaN e-culture demonstrator. *Web Semantics: Science, Services, and Agents on the World Wide Web, 6*(4), 243–249. doi:10.1016/j.websem.2008.08.001

Smith, J. R., Mohan, R., & Li, C.-S. (1999). Scalable multimedia delivery for pervasive computing. In *Proceedings of the Seventh ACM International Conference on Multimedia (Part 1) (MULTIMEDIA 1999)*, (pp. 131-140). New York, NY: ACM Press.

Styliaras, G. (2007). A web-based presentation framework for museums. In *Proceedings of the 2007 Euro American Conference on Telematics and Information Systems (EATIS 2007)*. New York, NY: ACM Press.

Wang, R., Yang, C., Xu, J., Yang, C., & Meng, X. (2009). An XML-based interface customization model. *Lecture Notes in Computer Science, 5940*, 190–202. doi:10.1007/978-3-642-11245-4_17

Zafirovic-Vukotic, M., & Niemegeers, I. G. (1992). Multimedia communication system: Upper layers in the OSI reference model. *IEEE Journal on Selected Areas in Communications, 10*(9), 1397–1402. doi:10.1109/49.184869

ADDITIONAL READING

Adistambha, K., Davis, S. J., Ritz, C. H., & Burnett, I. S. (2010). Efficient multimedia query-by-content from mobile devices. *Computers & Electrical Engineering*, *36*, 626–642. doi:10.1016/j.compeleceng.2008.11.016

Barry, C., & Lang, M. (2001). A survey of multimedia and web development techniques and methodology usage. *IEEE MultiMedia*, *8*(2), 52–60. doi:10.1109/93.917971

Brut, M., Laborie, S., Manzat, A. M., & Sedes, F. (2011). Generic information system architecture for distributed multimedia indexation and management. *Lecture Notes in Computer Science*, *6909*, 347–360. doi:10.1007/978-3-642-23737-9_25

Gavalas, D., & Kenteris, M. (2011). A web-based pervasive recommendation system for mobile tourist guides. *Personal and Ubiquitous Computing*, *15*, 759–770. doi:10.1007/s00779-011-0389-x

Kuflik, T., Stock, O., Zancanaro, M., Gorfinkel, A., Jbara, S., Kats, S., Sheidin, J., & Kashtan, N. (2011). A visitor's guide in an active museum: Presentations, communications, and reflection. *ACM Journal on Computing and Cultural Heritage, 3*(3).

Lian, S., Kanellopoulos, D., & Ruffo, G. (2009). Recent advances in multimedia information system security. *Informatica*, *33*(1), 3–24.

Micha, K., & Economou, D. (2005). Using personal digital assistants (PDAs) to enhance the museum visit experience. *Lecture Notes in Computer Science*, *3746*, 188–198. doi:10.1007/11573036_18

Moen, W. E. (1998). Accessing distributed cultural heritage information. *Communications of the ACM*, *41*(4), 44–48. doi:10.1145/273035.273046

Shang, J., Yu, S., Gu, F., Xu, Z., & Zhu, L. (2011). A mobile guide system framework for museums based on local location-aware approach. In *Proceedings of the International Conference on Computer Science and Service System (CSSS)*, (pp. 1935-1940). CSSS.

Sylaiou, S., Liarokapis, F., Kotsakis, K., & Patias, P. (2009). Virtual museums: A survey and some issues for consideration. *Journal of Cultural Heritage*, *10*, 520–528. doi:10.1016/j.culher.2009.03.003

Tonta, Y. (2008). Libraries and museums in the flat world: Are they becoming virtual destinations? *Library Collections, Acquisitions & Technical Services*, *32*(1), 1–9. doi:10.1016/j.lcats.2008.05.002

Wang, Y., Yang, C., Liu, S., Wang, R., & Meng, X. (2007). A RFID & handheld device-based museum guide system. In *Proceedings of the 2nd International Conference on Pervasive Computing and Applications, ICPCA*, (pp. 308-313). ICPCA.

KEY TERMS AND DEFINITIONS

Data Organization: Arrangement of physical records of data that enables easy data access and management through pre-defined procedures.

Extensible Markup Language (XML): It is a markup language that defines a set of rules for encoding documents in a common way that allows sharing both format and data.

Multi Layer Architecture: Architecture model in which the presentation, application and data are logically separated. This architecture provides a model for developers to create flexible and reusable applications.

Multimedia: Electronic information destined to be delivered to applications and takes a form, which integrates different types of objects including text, images, video, and audio.

Multimedia Applications: Applications that incorporate and handle multimedia data such as video, audio and voice, in various areas including art, education, entertainment, engineering, medicine, business.

Museum Visit Experience Model: Procedures of enhancing the quality of the museum visit, leading to happier and more satisfied visitors.

Scalable Multimedia Model: Organizational structure of multimedia that allows for progressive presentation of data and extensibility.

Section 5
Multimedia Social Networks and Geo-Social Systems

Chapter 15
Multimedia Social Networks and E-Learning

Andrew Laghos
Cyprus University of Technology, Cyprus

ABSTRACT

The purpose of this chapter is to investigate Multimedia Social Networks and e-Learning, and the relevant research in these areas. Multimedia Social Networks in e-Learning is an important and evolving study area, since an understanding of the technologies involved as well as an understanding of how the students communicate in online social networks are necessary in order to accurately analyze them. The chapter begins by introducing Multimedia Social Networks and Online Communities. Following this, the key players of e-Learning in Multimedia Social Networks are presented, including a discussion of the different roles that the students take. Furthermore, Social Interaction research is presented concentrating on such important areas as factors that influence social interaction, peer support, student-centered learning, collaboration, and the effect of interaction on learning. The last section of the chapter deals with the various methods and frameworks for analyzing multimedia social networks in e-Learning communities.

INTRODUCTION

The importance of multimedia social networks in online learning communities is reflected by its subsequent increase in support between peers, higher retention rates, and greater learning outcomes. The Internet and Multimedia play a vital role in Social Networks. Multimedia applications enhance the communication and learning activities by enabling a mix of media including graphics, text, animation,

audio, and video. The Internet is currently a widely used medium for distance education since it is a 24/7 globally accessed medium. Email, bulletin boards, chatting, dialogues, newsgroups, research, and interactive conferencing are all easily available with the Internet. Many courses are now offered online through a website. WWW is very important because it enables the teacher to include content like course information, assignments, tests, and lecture notes, and allows communication with the students either by email or live conferencing. Online materials also include journals, articles,

DOI: 10.4018/978-1-4666-2833-5.ch015

databases, software libraries, past examination papers, FAQs, and notice boards.

The objective of this chapter is to provide a thorough insight into multimedia social networks in e-learning, their benefits and limitations, factors that influence them and the different methods available to analyze them. It is important to understand how students communicate in online social networks before attempting to analyze them.

BACKGROUND

Communication is the Internet's most important asset (Metcalfe, 1992). Through communication services like the Internet, written communication has for many people supplanted the postal service, telephone, and fax machine (Jones, 1995). All these applications where the computer is used to mediate communication are called *Computer-Mediated Communication* (CMC). December (1997) defines CMC as "a process of human communication via computers, involving people, situated in particular contexts, engaging in processes to shape media for a variety of purposes" (p. 1). "Studies of CMC can view this process from a variety of interdisciplinary theoretical perspectives by focusing on some combination of people, technology, processes, or effects. Some of these perspectives include the social, cognitive/psychological, linguistic, cultural, technical, or political aspects; and/or draw on fields such as human communication" (December, 2004, p. 1).

Through the use of CMC applications, online communities emerge. By the end of 2011, just on the social network website Facebook, there were over 800 million registered members (Pingdom, 2012). As Korzenny pointed out even as early as 1978, the new social communities that are built from CMC, are formed around interests and not physical proximity (Korzenny, 1978). CMC gives people around the world the opportunity to communicate with others who share their interests, as unpopular as these interests may be, which does not happen in the 'real' world where the smaller the interest in a particular scene is, the less likely it will exist. This is due mainly to the Internet's connectivity and plethora of information available posted by anyone anywhere in the world.

An Online Learning Community can be defined as: "A group of people who communicate with each other across the Internet (or sometimes by intranet) to share information, learn more about a topic, or work on a project of mutual interest" (Porter, 2004, p. 193). The relevance of certain attributes in the descriptions of online communities, like the need to respect the feelings and property of others, is debated (Preece, 2000). Online communities are also referred to as cyber societies, cyber communities, Web groups, virtual communities, Web communities, virtual social networks, and e-communities among several others.

Social networking on multimedia e-learning websites has its benefits as well as its limitations. For instance, a benefit is that the discussions are potentially richer than in face-to-face classrooms (Scotcit, 2003), but on the other hand, users with poor writing skills may be at a disadvantage when using text-based CMC (Scotcit, 2003). Furthermore, asynchronous discussions allow for "reflective study followed by complex exchanges and genuine collaboration in the application of theory" (Sumner & Dewar, 2002, p. 1).

When it comes to website designers, choosing which CMC to employ (for instance, forum or chat-room) is not a matter of luck or randomness. The determining factor when selecting a CMC is whether the communication should be synchronous or asynchronous. In the case of e-Learning, the choice of the appropriate mode of CMC will be made by asking and answering questions such as (Bates, 1995; CAP, 2004; Heeren, 1996):

- Are the users spread across time zones?
- Can all participants meet at the same time?
- Do the users have access to the necessary equipment?
- What is the role of CMC in the course?

- Are the users good readers/writers?
- Are the activities time independent?
- How much control is allowed to the students?

MULTIMEDIA SOCIAL NETWORKS IN E-LEARNING

This section provides a discussion of important aspects of multimedia social networks in e-learning. More specifically, the key players of these environments are identified along with their characteristics. Furthermore, relevant social interaction research is presented and various analysis methods are discussed.

Key Players

Preece (2000) identified the following as the key players in online communities:

- **Moderators and Mediators:** Who guide discussions/serve as arbiters.
- **Professional Commentators:** Who give opinions/guide discussions.
- **Provocateurs:** Who provoke.
- **General Participants:** Who contribute to discussions.
- **Lurkers:** Who silently observe.

In a study by Nielsen (2006), it has been shown that there is a participation inequality in online communities. More specifically, in most online communities, 90% of users are lurkers who never contribute, 9% of users contribute a little, and 1% of users account for almost all the action (Nielsen, 2006).

With regards to e-Learning social networks, there are four key players:

- **Instructor:** Whose role is to provide the content while at the same time take

into consideration the diverse audiences (DLBOIS, 2002).

- **Facilitator:** Whose role is to link the instructor with the students (University of Idaho, 2002).
- **Support Staff and Administrators:** Whose role is to handle tasks such as student registration and grading, distribution of the material and copyright issues, while the administrators make sure that the technology is used efficiently to meet the students' and academic institutions' needs (DLBOIS, 2002).
- **Students:** Whose role is to learn (Behnke, 2003). Since the students will not have face-to-face interactions with the instructors and other students, they must learn to use the technologies which try to bridge this gap.

Students can be broken down further into 3 student roles types (Laghos & Zaphiris, 2007; O' Murchu, 2005):

- **Self-Learner:** These students need to see their own goals, organize their own work and manage their own time.
- **Team Member:** These students work collaboratively, their social interaction is in teams, and they are actively involved in their projects.
- **Knowledge Manager:** The focus of the knowledge manager role is on the development of knowledge products and these can be in the form of reports and newspapers, while their activities include searching for information, collecting and analyzing data, and designing reports.

Social Interaction

The section deals with social interaction research in e-learning concentrating on such important areas as factors that influence social interaction, peer support, student centered learning, collaboration and the effect of interaction on learning. In most online communities, time, distance, and availability are no longer disseminating factors. Given that the same individual may be part of several different and numerous online communities, it is obvious why more and more online communities keep emerging and increasing in size. There are many reasons that bring people together in online groups. These include hobbies, ethnicity, education, beliefs and just about any other topic or area of interest. Wallace (1999) points out that meeting in online communities eliminates prejudging based on someone's appearance, and thus people with similar attitudes and ideas are attracted to each other.

There have been several studies, which investigate the factors that influence social interaction in online courses. Vrasidas and McIsaac (1999) examined a university graduate online course in the use of telecommunications for instruction. Their analysis showed that four major factors influenced interaction: course structure, class size, feedback, and prior CMC experience (Vrasidas & McIsaac, 1999). For course structure, the authors mention that required activities and collaboration on peer editing of students' papers led to more interactions. Regarding class size, the authors concluded that there would have been more interactions in the course if there were more students enrolled in it. Furthermore, the students felt that the feedback they got in the course was not adequate, and this kept interaction levels low. Finally, the authors provide some insights into the relevance of prior CMC experience with interactivity. They found that students with no previous experience felt more comfortable using asynchronous communication where they had time to reflect on their ideas, as opposed to synchronous communication, which they found hard to keep up with. Students with prior CMC experience enjoyed both modes of communication and used emoticons more frequently. The authors state that online interaction is solely constructed through language and

suggest that educators should structure online courses for dialogue and interaction, that timely feedback should be provided to the students, and that students should be trained early in the course to use the conferencing systems and emoticons (Vrasidas & McIsaac, 1999). In comparison to face-to-face interaction, "online interaction may be slower and 'lacking' in continuity, richness and immediacy... however in some ways online interaction may be as good or even superior to face-to-face interaction" (Vrasidas & Zembylas, 2003, p. 1). Group activities planned in advance can increase the feeling of social presence and learner-learner interaction (Vrasidas & McIsaac, 2000), while gaining access and status in a setting and socializing are examples of intentions that drive interaction (Vrasidas, 2002).

"Peer support is a system of giving and receiving help founded on key principles of respect, shared responsibility, and mutual agreement of what is helpful" (Mead, Hilton, & Curtis, 2001, p. 140). When people find others that they feel are like them, they feel a connection and a deep understanding based on mutual experience (Mead, et al., 2001). "Student centered learning is supported theoretically by various overlapping pedagogical concepts such as self-directed learning (Candy, 1991), student-centered instruction or learning (Felder & Brent, 2001), active learning (Ramsden, 1992), vicarious learning (Lee & McKendree, 1999), and cooperative learning (Felder & Brent, 2001)" (Kurhila, et al., 2004, p. 1). Examples of contemporary ways to support collaborative learning include: awareness of others; joint building of knowledge; and matching unknown actors or resources, and these can lead to positive interdependence within the learning community as well as engagement, autonomy and independence. It is important to have tools that allow easy and straightforward ways for community members to interact with and support each other in a peer-to-peer fashion (Kurhila, et al., 2004). Online peer support occurs through the use of Computer-Mediated-Communication.

The importance of students learning from their study peers is increasingly being recognized by the eLearning community. "In some instances eLearning can foster a greater degree of communication and closeness among students and tutors than face-to-face learning" (Sumner & Dewar, 2002, p. 1). Furthermore, studies show that students would prefer to contact their peer students (rather than their tutor) when they have difficulty with coursework, difficulty understanding lectures and difficulty assessing facilities (Lockley, Pritchard, & Foster, 2004). Thus, the peer-support is an important aspect of e-Learning in both the findings of researchers and the opinions of students.

Collaboration produces a 70% retention rate (Chi, et al., 1989). Online courses are cited as having an average completion rate between 25% and 70% and it was found that a key driver for completion is codependency (Chi, et al., 1989). It is suggested that learners should be informed of who is present in the online sessions and how the group is composed in order to recognize each other and to develop a sense of direction (Hamburg, et al., 2003). The authors also emphasize the importance of e-moderators helping the students by initiating and supporting chats and online socializing, since this makes the students feel at home and more willing to contribute. Finally, they state that compared to individual and competitive learning, collaborative learning raises the students' achievement level and problem-solving activities and enhances the development of personal traits (Hamburg, et al., 2003). Interaction benefits and motivates the learners and facilitates higher order learning (McLoughlin, 2004). Some authors argue that e-Learning systems do not sufficiently acknowledge the importance of the social process and rely on passive material limiting interactivity. They suggest that e-Learning should be socially situated thus providing active interaction with the users (Angehrn, Nabeth, & Roda, 2001).

Interaction is a fundamental process for learning (Vygotsky, 1978; Vrasidas, 2001) and knowledge is constructed in communities of practice through social interaction (De Angeli & Sue, 2005). Social problems affecting online communities include social loafing which leads to low participation rates, disinhibited behaviour (like flaming and abuse), and diffusion of responsibility. A solution to this is the presence of a moderator who can reduce the antisocial behaviours that are triggered by anonymity (De Angeli & Sue, 2005). In addition, the authors believe that the sense of community is greatest for the student when there is a sense of connectedness with the course, and it is engendered by both social and learning dimensions (De Angeli & Sue, 2005). They also note that people who interact more in an online course tend to achieve higher marks at exams, as opposed to lurking, which is not as successful (De Angeli & Sue, 2005). Furthermore, learners perceive the content of communication as an information source (Aviv, 2000). The Social Interdependence Theory of Cooperative Learning (Johnson & Johnson, 1999) suggests that the way social interdependence is structured determines how individuals interact, which in turn determines their learning outcomes. Cooperative experiences promote greater social support than competitive or individualistic efforts (Johnson & Johnson, 1999), while stronger effects exist for peer support than for superior (teacher) support (Aviv, 2000).

Analysis

This section deals with the various methods and frameworks for analyzing multimedia social networks in e-Learning communities. This was done in order to find out what the existing methods are, identify their characteristics, benefits and limitations, and investigate whether or not they are applicable to e-Learning environments and especially for the evolution of social networks in e-Learning environments. The virtual communities that emerge have complex structures, social dynamics, and patterns of interaction that must be better understood. There are various aspects and attributes of e-Learning CMC that can be studied.

Content analysis is a social science methodology where recorded human communications are studied (Babbie, 2004). It is a technique for compressing many words of text into fewer content categories (Stemler, 2001; Weber, 1990). There have been several frameworks created for studying the content of messages exchanged in e-Learning CMC. Examples include work from Archer et al. (2001) and McCreary's (1990) behavioural model, which identifies different roles and uses these roles as the units of analysis. Furthermore, in Gunawardena, Lowe, and Anderson's (1997) model for examining the social construction of knowledge in computer conferencing, five phases of interaction analysis are identified and they are: (1) Sharing/Comparing of Information; (2) The Discovery and Exploration of Dissonance or Inconsistency among Ideas, Concepts, and Statements; (3) Negotiation of Meaning/Co-Construction of Knowledge; (4) Testing and Modification of Proposed Synthesis or Co-Construction; (5) Agreement Statement(s)/Applications of Newly Constructed Meaning. Henri (1992) has also developed a content analysis model for cognitive skills and is used to analyze the process of learning within the student's messages. Furthermore, Mason's (1991) work provides descriptive methodologies using both quantitative and qualitative analysis. In the case of e- Learning for example, a useful framework is the Transcript Analysis Tool (Fahy, 2003) as it offers:

- A student-centered approach.
- It works with Gunawardena's model.
- It was built on weaknesses of other models.
- It uses the sentence as the unit of analysis.

Over the years, there have been several techniques by different researchers for analyzing interaction. It is important to note that the type of interaction studied in this case is interpersonal interaction, more specifically the human-human interaction that takes place through the use of e-Learning CMC. Examples of Interaction Analysis models include Bale's Interaction Process analysis (Bales & Strodbeck, 1951), the SIDE model (Spears & Lea, 1992), a four-part model of cyber-interactivity (McMillan, 2002), Vrasidas's (2001) framework for studying human-human interaction in Computer-Mediated Online Environments and a technique called Social Network Analysis (SNA).

A working definition of Human-Computer Interaction (HCI) as provided by ACM SIGCHI (2002, p. 8) is: "Human-computer interaction is a discipline concerned with the design, evaluation and implementation of interactive computing systems for human use and with the study of major phenomena surrounding them." The focus is on the interaction between one or more humans and one or more computational machines (ACM SIGCHI, 2002). HCI is a multidisciplinary subject, which draws on areas such as computer science, sociology, cognitive psychology and so on (Schneiderman, 1998). The concept of HCI consists of many tools and techniques that are used for information gathering and evaluation. Important HCI techniques include Questionnaires, Interviews, Personas, and Log Analysis.

Analyzing the evolution of e-Learning communities is important since these people networks continuously evolve and change over time, and therefore keeping track of the network changes will enable educators to predict how certain actions will affect their network, and to incorporate various methodologies to alter their state. One such evolution analysis method is FESNeL (Framework for assessing the Evolution of Social Networks in e-Learning). FESNeL (Laghos, 2005) allows e-educators and online course instructors/maintainers to perform in-depth analyses of the communication patterns of the students of their e-Learning courses and to follow their course CMC progression. FESNeL assesses the social network of the students over the duration of the course thus mapping out the changes and evolution of these social structures over time. It is useful for monitoring the networks and keeping track of their changes, while investigating how specific course amendments, participation in computer-mediated communication, and/or conversation topics positively or negatively influence the dynamics of the online community. When using this framework to assess their e-Learning community, e-educators are able to predict how certain actions will affect their network, and can incorporate various methodologies to alter the state of their network. FESNeL is a unified framework compiled of both qualitative and quantitative methods where the unit of analysis is the Social Networks. The four components of FESNeL are: the Attitudes Towards Thinking and Learning Survey (ATTLS), Social Network Analysis (SNA), Topic Relation Analysis (TRA), and the Constructivist On-Line Learning Environment Survey (COLLES). An explanation of the FESNeL components follows.

Social Network Analysis (SNA)

"Social Network Analysis (SNA) is the mapping and measuring of relationships and flows between people, groups, organizations, computers or other information/knowledge processing entities. The nodes in the network are the people and groups while the links show relationships or flows between the nodes. SNA provides both a visual and a mathematical analysis of human relationships" (Krebs, 2004, p. 1). Preece (2000) adds that it provides a philosophy and set of techniques for understanding how people and groups relate to each other. It is concerned about dyadic attributes between pairs of actors (like kinship, roles, and actions), and has been used extensively by sociologists (Wellman, 1982, 1992), communication researchers (Rice, 1994; Rice, et al., 1990) and others. Analysts use SNA to determine if a network is tightly bounded, diversified, or constricted, to find its density and clustering, and to study how the behaviour of network members is affected by their positions and connections (Garton, et al., 1997; Wellman, 1997; Scott, 2000; Knoke & Kuklinski, 1982). Apart from analyzing learning environments analysts have used SNA in a number

of other areas as well. For instance, Holme et al. (2004) used SNA to study the structure of the users' activities in an Internet dating community, Takhteyev et al. (2012) studied the geography of Twitter networks, while Bohn et al. (2011) carried out a content-based social network analysis of mailing lists to find people's interests. Other examples of the diverse use of SNA include a study by Wey et al. (2008) on animal behavior, and a study by Niazi and Hussain (2011) on the trends in the consumer electronics domain.

Topic Relation Analysis (TRA)

The TRA model (Laghos, 2005; Laghos & Zaphiris, 2006) is a content analysis tool. Content analysis is a technique used in qualitative analysis to study written material by breaking it into meaningful units (Babbie, 2004). The data is collected directly from the discussion boards of an e-Learning class and then sorted into the TRA categories. The TRA is a newly developed tool where the units of analysis are the threads and messages of each of the discussions of the forum. The data collected includes the messages per thread, the participants per thread, the discussion topic, and its relevance to the course. The tool assists us in understanding the messages and communication between the learners, and how important the discussed topics are for the learners to remain and complete the online course. The TRA was developed to group the messages that the students post into categories, enabling us to determine which type of messages (peer-support, off-topic conversations, etc) engage the student more in the course and contribute to course retention. TRA is compromised of 3 main categories some of which have sub-categories. These categories were deduced by observations of e-Learning discussion boards and the different types of conversations that take place. The TRA categories are:

- **A: Course Material Related:** Category A deals with conversations in the discussion boards of the e-Learning courses that are related to the course material, and is broken down into two further categories, A1 and A2.

- **A1: Related to Current Lesson:** Threads that belong in A1 are conversations that have to do with the course material of the current Lesson. Examples of such topics include questions and answers and correcting peers mistakes.

- **A2: Related to Course (but not current lesson):** Threads that belong in A2 are conversations that have to do with the course, but their subject is not in the current lesson's syllabus. For example, a conversation about an exercise of Lesson 3, posted in the discussion forum of Lesson 1, would go in this category. Also, a general question about mathematics (in an area that is not included in the Mathematics lesson's syllabus) would also go in A2.

- **B: Course Website/Technical Related:** Category B is specific to conversations regarding the course website, and technical issues. Problems listening to audio files, accessing specific parts of the site, or usage issues are all in this category.

- **C: Not Related to Course:** Finally, posts that are categorized in C are those that have nothing to do with the course in hand or its usage. Category C has two sub-categories:

  - **C1: Peer Socializing:** C1 is a broad category that covers conversation types where peers socialize with each other. Examples include students introducing themselves, discussions about football games and concerts, making new friends and so on.

  - **C2: Other:** Category C2 basically includes all the other off-topic conversations that are not about peers so-

cializing with each other. Examples of posts that belong in this category are spam and advertisements.

Constructivist On-Line Learning Environment Survey (COLLES)

The Constructivist On-Line Learning Environment Survey (COLLES) measures students' perceptions and preferences and was designed to help teachers assess, from a social constructivist perspective, the quality of their online learning environment (Taylor & Maor, 2000). The COLLES electronic questionnaire was designed to support the use of the Web for teaching programs for which social constructionism is a key pedagogical referent and can be used to monitor the quality of innovative online teaching and learning (Taylor & Maor, 2000). It consists of 24 questions arranged into 6 scales (Dougiamas & Taylor, 2003):

- **Relevance:** How relevant is online learning to students' professional practices?
- **Reflection:** Does online learning stimulate students' critical reflective thinking?
- **Interactivity:** To what extent do students engage online in rich educative dialogue?
- **Tutor Support:** How well do tutors enable students to participate in online learning?
- **Peer Support:** Do fellow students provide sensitive and encouraging support?
- **Interpretation:** Do students and tutors make good sense of each other's communications?

Attitudes Towards Thinking and Learning Survey (ATTLS)

The Attitudes Towards Thinking and Learning survey (ATTLS) is used to measure the quality of discourse within a course. It measures the extent to which a person is a 'Connected Knower' (CK) or a 'Separate Knower' (SK). People with higher CK scores tend to find learning more enjoyable, and are often more cooperative, congenial and more willing to build on the ideas of others, while those with higher SK scores tend to take a more critical and argumentative stance to learning (Galotti, et al., 1999). The two different types of procedural knowledge (separate and connected knowing) were identified by Belenky et al. (1986). Separate knowing involves objective, analytical, detached evaluation of an argument or piece of work and takes on an adversarial tone, which involves argument, debate, or critical thinking (Galotti, et al., 1999). "Separate knowers attempt to 'rigorously exclude' their own feelings and beliefs when evaluating a proposal or idea" (Belenky, et al., 1986, p. 111; Galotti, et al., 1999). Separate knowers look for what is wrong with other people's ideas, whereas connected knowers look for why other people's ideas make sense or how they might be right, since they try to look at things from the other person's point of view and try to understand it rather than evaluate it (Clinchy, 1989; Galotti, et al., 1999). These two learning modes are not mutually exclusive, and may "coexist within the same individual" (Clinchy, 1996, p. 207). Differences in SK and CK scores "produce different behaviors during an actual episode of learning, and do result in different descriptions of, and reactions to, that session" (Galotti, et al., 1999, p. 435).

FUTURE RESEARCH DIRECTIONS

Suggestions to Researchers

This study explored multimedia social networks in e-learning. Characteristics of the key players and their online communication patterns were presented. As this is a continuously evolving field, which relies on technological advances, constant research on the participants and their relationship with the multimedia technologies are needed. By utilizing the Internet and the communication technologies of the World-Wide-Web, e-Learning has evolved into a state where students from around the globe become members of online communities where they can work collaboratively and cooperatively with their peers to learn and share knowledge. As new teaching methods and different learning activities emerge, new types of interaction take place and therefore new types of evaluation are necessary. It is recommended that specialist software be written to collect the students' interactions automatically from any kind of discussion board thus making the data collection stage faster and more efficient. Future search directions include applying FESNeL to e-learning courses of various subjects like for example history, math, economics, music and so on, in order to investigate the relation of the course topic with student communication in the online learning environment.

Another future direction is the investigation of CMC usage and student retention in online courses. Does increased use of CMC actually motivate students to remain in and finish an online course? or is it irrelevant? Finally, as a future research direction, it would be beneficial to analyze and compare different types of online courses (for example courses with a teacher and courses without a teacher) to realize the possible differences and similarities in the communication patterns the students undertake in different online learning settings.

Suggestions to Practitioners

It has been shown that students rely on online communication with their peers when part of an e-Learning environment. Designers of such courses should ensure that their environment will provide the functionality to allow the students to interact with each other as this promotes their sense of presence in the e-Learning communities and positively influences their learning outcomes.

CONCLUSION

It is important to understand the characteristics and technologies surrounding multimedia social networks and e-learning before attempting to analyze the way students use them. The formation and characteristics of online communities were discussed. Key players and student roles were identified, while research about social interaction and communication in e-learning was presented, including the factors that drive interaction and a number of other interaction case studies. Research has shown that CMC is important for knowledge building. Studies have shown that interaction is influenced by four major factors: course structure, class size, feedback and prior CMC experience. In addition, students prefer to contact their peer students rather than their tutor when they have difficulties with the online lessons. Student collaboration is seen to produce high retention rates, and online interaction with fellow peers has been linked with greater learning outcomes. These findings stress the importance of students interacting with each other when taking part in online learning courses. Furthermore, the analysis methods of such environments were discussed. As more and more people are engaging daily in social networking activities, new opportunities arise to enhance e-Learning using multimedia social networks.

REFERENCES

ACM SIGCHI. (2002). *Curricula for human-computer interaction.* New York, NY: The Association for Computing Machinery.

Angehrn, A., Nabeth, T., & Roda, C. (2001). *Towards personalised, socially aware and active e-learning systems.* CALT White Paper. Washington, DC: CALT.

Archer, W., Garrison, R. D., Anderson, T., & Rourke, L. (2001). *A framework for analysing critical thinking in computer conferences.* Paper presented at the European Conference on Computer-Supported Collaborative Learning. Maastricht, The Netherlands.

Aviv, R. (2000). Education performance of ALN via content analysis. *Journal of Asynchronous Learning Networks, 14*(2).

Babbie, E. (2004). *The practice of social research* (10th ed.). Belmont, CA: Thomson/Wadsworth Learning.

Bales, R. F., & Strodbeck, F. L. (1951). Phases in group problem-solving. *Journal of Abnormal and Social Psychology, 46,* 485–495. doi:10.1037/h0059886

Bates, A. W. (1995). *Technology, open learning and distance education.* London, UK: Routledge.

Behnke, W. (2003). *Online/eLearning - An overview.* Vancouver, Canada: Vancouver Community College.

Belenky, M. F., Clinchy, B. M., Goldberger, N. R., & Tarule, J. M. (1986). *Women's ways of knowing: The development of self, voice, and mind* (2nd ed.). New York, NY: Basic Books.

Bohn, A., Feinerer, I., Hornik, K., & Mair, P. (2011). Content-based social network analysis of mailing lists. *Remediation Journal, 3*(1), 11–18.

Candy, P. (1991). *Self-direction for lifelong learning.* San Francisco, CA: Jossey-Bass.

CAP. (2004). *E-guide: Using computer mediated communication in learning and teaching.* Retrieved November 8, 2004 from http://www2.warwick.ac.uk/services/cap/resources/eguides/cmc/cmclearning/

Chi, M., Bassok, M., Lewis, M., Reimann, P., & Glaser, R. (1989). Self-explanations: How students study and use examples in learning to solve problems. *Cognitive Science, 13*, 145–182. doi:10.1207/s15516709cog1302_1

Clinchy, B. M. (1989). The development of thoughtfulness in college women: Integrating reason and care. *The American Behavioral Scientist, 32*, 647–657. doi:10.1177/0002764289032006005

Clinchy, B. M. (1996). Connected and separate knowing: Toward a marriage of two minds. In Goldberger, N., Tarule, J., Clinchy, B., & Belenky, M. (Eds.), *Knowledge, Difference, and Power: Essays Inspired by Women's Ways of Knowing*. New York, NY: Basic Books.

De Angeli, A., & Sue, K. (2005). Learning conversations: A case study into e-learning communities. In *Proceedings of the Interact 2005 eLearning and Human-Computer Interaction Workshop*. IEEE.

December, J. (1997). Notes on defining of computer-mediated communication. *Computer-Mediated Communication Magazine, 3*(1).

December, J. (2004). *What is computer-mediated communication?* Retrieved October 19, 2004, from http://www.december.com/john/study/cmc/what.html

DLBOIS. (2002). *Distance learning benefits organizations, individuals and society*. Retrieved November 4, 2002, from http://www.ciscoworld-magazine.com/monthly/2001/04/distance.shtml

Dougiamas, M., & Taylor, P. (2003). Moodle: Using learning communities to create an open source course management system. In *Proceedings of EDMEDIA 2003*. EDMEDIA.

Fahy, P. J. (2003). Indicators of support in online interaction. *International Review of Research in Open and Distance Learning, 4*(1).

Felder, R., & Brent, R. (2001). Effective strategies for cooperative learning. *Journal of Cooperation &Collaboration in College Teaching, 10*(2), 69–75.

Galotti, K. M., Clinchy, B. M., Ainsworth, K., Lavin, B., & Mansfield, A. F. (1999). A new way of assessing ways of knowing: The attitudes towards thinking and learning survey (ATTLS). *Sex Roles, 40*(9/10), 745–766. doi:10.1023/A:1018860702422

Garton, L., Haythorthwaite, C., & Wellman, B. (1997). Studying on-line social networks. In Jones, S. (Ed.), *Doing Internet Research*. Thousand Oaks, CA: Sage.

Gunawardena, C., Lowe, C., & Anderson, T. (1997). Analysis of a global online debate and the development of an interaction analysis model for examining social construction of knowledge in computer conferencing. *Journal of Educational Computing Research, 17*(4), 397–431. doi:10.2190/7MQV-X9UJ-C7Q3-NRAG

Hamburg, I., Lindecke, C., & Thij, H. (2003). Social aspects of e-learning and blending learning methods. In *Proceedings of the 4th European Conference E-COMM-LINE*. Bucharest, Romania: E-COMM-LINE.

Heeren, E. (1996). *Technology support for collaborative distance learning*. (Doctoral Dissertation). University of Twente. Enschede, The Netherlands.

Henri, F. (1992). Computer conferencing and content analysis. In Kaye, A. R. (Ed.), *Collaborative Learning through Computer Conferencing: The Najaden Papers* (pp. 117–136). Berlin, Germany: Springer-Verlag. doi:10.1007/978-3-642-77684-7_8

Holme, P., Edling, C. R., & Liljeros, F. (2004). Structure and time evolution of an internet dating community. *Social Networks, 26*, 155–174. doi:10.1016/j.socnet.2004.01.007

Johnson, D. W., & Johnson, R. T. (1999). *Learning together and alone, cooperative, competitive and individualistic learning*. Needham Heights, MA: Allyn and Bacon.

Jones, S. (1995). Computer-mediated communication and community: Introduction. *Computer-Mediated Communication Magazine, 2*(3), 38.

Knoke, D., & Kuklinski, J. H. (1982). *Network analysis*. Beverly Hills, CA: Sage Publications.

Korzenny, F. (1978). A theory of electronic propinquity: Mediated communication in organizations. *Communication Research, 5*, 3–23. doi:10.1177/009365027800500101

Krebs, V. (2004). *An introduction to social network analysis*. Retrieved November 9, 2004 from http://www.orgnet.com/sna.html

Kurhila, J., Miettinen, M., Nokelainen, P., & Tirri, H. (2004). The role of the learning platform in student-centered e-learning. In *Proceedings of the 4th IEEE International Conference on Advanced Learning Technologies*. IEEE Press.

Laghos, A. (2005). FESNeL: A methodological framework for assessing the evolutionary structure of social networks in e-learning. In *Proceedings of the Junior Researchers of the European Association for Research on Learning and Instruction 11th Biennial JURE/EARLI Conference*. Nicosia, Cyprus: JURE.

Laghos, A., & Zaphiris, P. (2006). *Sociology of student-centred e-learning communities: A network analysis*. Paper presented at the e-Society 2006 Conference. Dublin, Ireland.

Laghos, A., & Zaphiris, P. (2007). Investigating student roles in online student-centered learning. *International Council for Educational Media Conference (ICEM 2007)*. Nicosia, Cyprus: ICEM.

Lee, J., & McKendree, J. (1999). Learning vicariously in a distributed environment. *Journal of Active Learning, 10*, 4–10.

Lockley, E., Pritchard, C., & Foster, E. (2004). Professional evaluation: Students supporting students – Lessons learnt from an environmental health peer support scheme. *Journal of Environmental Health Research, 3*(2).

Mason, R. (1991). Analyzing computer conferencing interactions. *Computers in Adult Education and Training, 2*(3), 161–173.

McCreary, E. (1990). Three behavioural models for computer-mediated communication. In Harasim, L. (Ed.), *Online Education: Perspectives on a New Environment*. New York, NY: Praeger.

McLoughlin, C. (2004). *A learning conversations: Dynamics, collaboration and learning in computer mediated communication*. Washington, DC: TAFE Media Network.

McMillan, S. J. (2002). A four-part model of cyber-interactivity: Some cyber-places are more interactive than others. *New Media & Society, 4*(2), 271–291.

Mead, S., Hilton, D., & Curtis, L. (2001). Peer support: A theoretical perspective. *Psychiatric Rehabilitation Journal, 25*, 134–141.

Metcalfe, B. (1992). *Internet fogies to reminisce and argue at interop conference*. New York, NY: InfoWorld.

Niazi, M. A., & Hussain, A. (2011). Social network analysis of trends in the consumer electronics domain. In *Proceedings of the IEEE International Conference on Consumer Electronics (ICCE)*. Las Vegas, NV: IEEE Press.

Nielsen, J. (2006). *Participation inequality: Encouraging more users to contribute*. Retrieved Oct 16 2011 from http://www.useit.com/alertbox/participation_inequality.html

O'Murchu, D. (2005). New teacher and student roles in the technology-supported language classroom. *International Journal of Instructional Technology & Distance Learning, 2*(2).

Pingdom. (2012). *Internet 2011 in numbers*. Retrieved February 3, 2012, from http://royal.pingdom.com/2012/01/17/internet-2011-in-numbers/

Porter, L. R. (2004). *Developing an online curriculum: Technologies and techniques*. Hershey, PA: IGI Global.

Preece, J. (2000). *Online communities: Designing usability, supporting sociability*. Chichester, UK: John Wiley and Sons.

Ramsden, P. (1992). *Learning to teach in higher education*. London, UK: Routledge. doi:10.4324/9780203413937

Rice, R. (1994). Network analysis and computer mediated communication systems. In Galaskiewkz, S. W. J. (Ed.), *Advances in Social Network Analysis*. Newbury Park, CA: Sage. doi:10.4135/9781452243528.n7

Rice, R. E., Grant, A. E., Schmitz, J., & Torobin, J. (1990). Individual and network influences on the adoption and perceived outcomes of electronic messaging. *Social Networks*, *12*, 17–55. doi:10.1016/0378-8733(90)90021-Z

Schneiderman, B. (1998). *Designing the user interface: Strategies for effective human-computer interaction*. Reading, MA: Addison Wesley Longman. doi:10.1145/25065.950626

Scotcit. (2003). Enabling large-scale institutional implementation of communications and information technology (ELICIT). *Using Computer Mediated Conferencing*. Retrieved November 2, 2004 from http://www.elicit.scotcit.ac.uk/modules/cmc1/welcome.htm

Scott, J. (2002). *Social network analysis: A handbook* (2nd ed.). London, UK: Sage.

Spears, R., & Lea, M. (1992). Social influence and the influence of the social in computer-mediated communication. In Lea, M. (Ed.), *Contexts of Computer Mediated Communication* (pp. 30–65). London, UK: Harvester Wheatsheaf.

Stemler, S. (2001). An overview of content analysis. *Practical Assessment, Research & Evaluation*, *7*(17).

Sumner, J., & Dewar, K. (2002). Peer-to-peer elearning and the team effect on course completion. In *Proceedings of ICCE*, (pp. 369-370). ICCE.

Takhteyev, Y., Gruzd, A., & Wellman, B. (2012). Geography of Twitter networks. *Social Networks*, *34*(1), 73–81. doi:10.1016/j.socnet.2011.05.006

Taylor, P., & Maor, D. (2000). Assessing the efficacy of online teaching with the constructivist on-line learning environment survey. In A. Herrmann & M. M. Kulski (Eds.), *Flexible Futures in Tertiary Teaching: Proceedings of the 9th Annual Teaching Learning Forum*. Perth, Australia: Curtin University of Technology.

University of Idaho. (2002). *Distance education: An overview*. Retrieved November 13, 2002, from http://www.uidaho.edu/evo/dist1.html

Vrasidas, C. (2001). Studying human-human interaction in computer-mediated online environments. In Y. Manolopoulos & S. Evripidou (Eds.), *Proceedings of the 8th Panhellenic Conference on Informatics*, (pp. 118-127). Nicosia, Cyprus: IEEE.

Vrasidas, C. (2002). A working typology of intentions driving face-to-face and online interaction in a graduate teacher education course. *Journal of Technology and Teacher Education*, *10*(2), 273–296.

Vrasidas, C., & McIsaac, M. (2000). Principles of pedagogy and evaluation of web-based learning. *Educational Media International*, *37*(2), 105–111. doi:10.1080/095239800410405

Vrasidas, C., & McIsaac, S. M. (1999). Factors influencing interaction in an online course. *American Journal of Distance Education*, *13*(3), 22–36. doi:10.1080/08923649909527033

Vrasidas, C., & Zembylas, M. (2003). The nature of technology-mediated interaction in globalized distance education. *International Journal of Training and Development, 7*(4), 1–16. doi:10.1046/j.1360-3736.2003.00186.x

Vygotsky, L. S. (1978). *Mind and society: The development of higher mental processes.* Cambridge, MA: Harvard University Press.

Wallace, P. (1999). *The psychology of the internet.* Cambridge, UK: Cambridge University Press. doi:10.1017/CBO9780511581670

Weber, R. P. (1990). *Basic content analysis* (2nd ed.). Newbury Park, CA: Sage.

Wellman, B. (1982). Studying personal communities. In Lin, P. M. N. (Ed.), *Social Structure and Network Analysis.* Beverly Hills, CA: Sage.

Wellman, B. (1992). Which types of ties and networks give what kinds of social support? *Advances in Group Processes, 9,* 207–235.

Wellman, B. (1997). An electronic group is virtually a social network. In Kiesler, S. (Ed.), *Culture of the Internet* (pp. 179–205). Hillside, NJ: Lawrence Erlbaum.

Wey, T., Blumstein, D. T., Shen, W., & Jordán, F. (2008). Social network analysis of animal behaviour: A promising tool for the study of sociality. *Animal Behaviour, 75,* 333–344. doi:10.1016/j.anbehav.2007.06.020

ADDITIONAL READING

Choi, J. H., & Danowski, J. (2002). Cultural communities on the net - Global village or global metropolis? A network analysis of usenet newsgroups. *Journal of Computer-Mediated Communication, 7*(3).

Egan, M., Sebastian, J., & Welch, M. (1991). *Effective television teaching: Perceptions of those who count most...distance learners.* Nashville, TN: Academic Press.

Fakas, G. J., Nguyen, A. V., & Gillet, D. (2005). The electronic journal: A collaborative and co-operative learning environment for web-based experimentation. *Computer Supported Cooperative Work (CSCW). The Journal of Collaborative Computing, 14*(3), 189–216. doi:10.1007/s10606-005-3272-3

Ferris, P. (1997). What is CMC? An overview of scholarly definitions. *Computer-Mediated Communication Magazine, 4*(1).

Fortner, R. S. (1993). *International communication: History, conflict, and control of the global metropolis.* Belmont, CA: Wadsworth.

Laghos, A., & Zaphiris, P. (2005). Computer assisted/aided language learning. In Howard, C., Boettcher, J. V., Justice, L., Schenk, K., Rogers, P., & Berg, G. A. (Eds.), *Encyclopedia of Distance Learning* (Vol. 1, pp. 331–336). Hershey, PA: IGI Global. doi:10.4018/978-1-59140-555-9.ch050

Maier, P., Barnett, L., Warren, A., & Brunner, D. (1998). *Using technology in teaching and learning.* London, UK: Kogan Page.

McLuhan, M. (1964). *Understanding media: The extension of man.* New York, NY: McGraw Hill.

Metz, J. M. (1994). Computer-mediated communication: Literature review of a new context. *Interpersonal Computing and Technology, 2*(2), 31–49.

Oliver, E. L. (1994). Video tools for distance education. In Willis, B. (Ed.), *Distance Education: Strategies and Tools.* Englewood Cliffs, NJ: Educational Technology Publications.

Preece, J., Rogers, Y., & Sharp, H. (2002). *Interaction design: Beyond human-computer interaction.* New York, NY: John Wiley & Sons.

Rheingold, H. (1993). *The virtual community: Homesteading on the electronic frontier.* Reading, MA: Addison-Wesley.

Souder, W. (1993). The effectiveness of traditional vs. satellite delivery in three management of technology master's degree program. *American Journal of Distance Education, 7*(1). doi:10.1080/08923649309526809

SRIC-BIG. (2002). *Technology evolution in e-learning.* Palo Alto, CA: Stanford Research Institute Consulting Business Intelligence Group.

Suler, J. (2004). *The final showdown between in-person and cyberspace relationships.* Retrieved November 3, 2004 from http://www1.rider.edu/~suler/psycyber/showdown.html

Taylor, R. (2002). *An introduction to e-learning.* New York, NY: Creative Learning Media.

Ward, C. (2003). eLearning training: Catching up with the future. In *Proceedings of the EDUCAUSE 2003 Conference.* Adelaide, Australia: EDUCAUSE.

Wasserman, S., & Faust, K. (1994). *Social network analysis: Methods and applications.* Cambridge, UK: Cambridge University Press. doi:10.1017/CBO9780511815478

Wenger, E. (1998). *Communities of practice; learning, meaning and identity.* Cambridge, UK: Cambridge University Press.

KEY TERMS AND DEFINITIONS

Computer-Mediated Communication: Human communication using computers.

E-Learning: Learning using electronic technologies (usually delivered online).

Knowledge Manager: The students whose activities focus on the development of knowledge products.

Self Learner: The students, who need to see their own goals, organize their own work, and manage their own time.

Social Network: A network of people.

Social Network Analysis: The analysis of people networks.

Team Member: The students who work collaboratively, interact in teams, and are actively involved in their projects.

Chapter 16
GWAP as a Tool to Analyze, Design, and Test Geo-Social Systems

Martina Deplano
University of Turin, Italy

Giancarlo Ruffo
University of Turin, Italy

ABSTRACT

In this chapter, the authors discuss the state-of-the-art of Geo-Social systems and Recommender systems, which are becoming extremely popular for users accessing social media trough mobile devices. Moreover, they introduce a general framework based on the interaction among those systems and the "Game With A Purpose" (GWAP) paradigm. The proposed framework/platform can help researchers to understand geo-social dynamics in order to design and test new services, such as recommenders of places of interest for tourists, real-time traffic information systems, personalized suggestions of social events, and so forth. To target the governance of such complexity, relevant data must be collected by the investigators, shared with the community, and analyzed to find dynamical patterns that correlate spatial-temporal information with the user's preferences and objectives. The authors argue that the GWAP approach can be exploited to successfully satisfy many of these tasks.

INTRODUCTION

For the scientific community, geo-social services are becoming increasingly important because they represent a new tool to shed light on the basic rules that govern human mobility, social behaviors, and context-aware attitudes. For example, understanding and predicting the position and the itinerary of (a group of) persons can help us to find ways to stop a virus contagion on a global scale, or to avoid traffic jams in metropolitan areas. Moreover, the optimization of the mobility of thousands or millions of people can influence urban design as well as ecological policies, because we can understand new ways to reduce superfluous air pollution and common resources consumption. In a nutshell, geo-social services can help us to build smarter cities, healthier citizens, and better quality of life.

DOI: 10.4018/978-1-4666-2833-5.ch016

Geo-Social services are usually implemented as Web applications as well as native apps for mobile environments, like iOS and Android. We can see these applications as platforms that make available different information which converge spontaneously into complex aggregated data (such as locations, social links, user's interests and tagging activity over multimedia resources). To govern this complexity, researchers must collect and analyze users' geo-located data, in order to understand geo-social dynamics.

In the procedure of collecting data, one of the problems that can teach us something about human dynamics, is that we usually deal with a kind of information that is sensible to be protected and we cannot violate user's privacy in order to gather as much data as needed. Therefore, researchers must find ways to convince the user to communicate such data on *a volunteering* basis. If the user downloads an application that collects relevant information and sends it to a central storage system, we must be sure that the volunteer is willing to use such service. This means that the application should implement an *appealing* service, such as a game. In this scenario, a game is not just a side-product, but part of the process of acquiring information about users' interests, attitudes, behaviors, and geographical position. Dynamical processes are really relevant, too: temporal data can be important as well as spatial observations, and real-time information is sometimes essential together with instruments for localization and social awareness. Another important feature of a geo-social application that addresses the collection of relevant data is the *personalization* of the service itself: the user must be aware that the service is returning information that is useful and enjoyable for him/her, in that location, and in the specific moment. Therefore, to give an example, if a car navigation system proposes a modification in the planned route, a valid reason must be communicated to the user. For example, a close friend is in the proximity, the temporary exhibition of her favorite painter is just few meters ahead,

or—furthermore—a car incident has been detected two blocks ahead.

In this chapter, we propose an integrated framework, which can be used to analyze, design and test geo-social services that can meet the above requirements. In fact, we believe that the GWAP (game with a purpose) paradigm in the geo-social scenario can attract many volunteers by a way of appealing applications. Moreover, Recommender Systems (RS) are mature enough to provide personalization techniques that can be easily integrated with the proposed framework in order to filter out not relevant notifications.

The chapter is organized as follows. In the next section, we present the background and state of the art of the given research field. We present an overview of Geo-Social networks and describe Recommender Systems (focusing on the Collaborative Filtering approach) and their use to suggest useful information to the user. In addition, we present the GWAP approach as one of the implementation of the Crowdsourcing methodology for the gathering of big amount of users' data. Merging together these three paradigms, afterwards we propose a framework that is helpful to analyze, design and test geo-social systems in their complexity. In particular, we describe the issues, the controversies, and the problems that arise, when we study such systems. Finally, we describe the building blocks of the integrated framework.

BACKGROUND

Geo-Social Networks

Because of the huge spreading of mobile devices in the last period, users can always be connected with online social networking applications. In addition, due to the GPS technology and other localization technologies, users are able to share their geographic position with friends. This new type of service is often referred as *Geo-Social Networking*. In this new context, the location that users can share with friends through "check-in" activity became a very important feature that allows researchers to gather more information about users' spatial and temporal behavior. As we can see in Scellato (2011), *"the trend is progressively going from specialized providers offering location-as-a-service to a widespread new concept of location-as-a-feature, where every online social platform integrates geographic information into their services."*

Two new metrics have been introduced to investigate how social and geographic structure of the social network merge together. With these new measures, Scellato et al. (2010) have shown how geographic distance affects social structure. For example, different social networks exhibit different geo-social properties according to the nature of the service. That means that services based mainly on location-advertising largely foster local ties and clusters, while services used mainly for news and content sharing present more connections and clusters on longer distances.

Some of the existing geo-applications, in which the geo-location is the primary feature, are well known:

- **Facebook Places[1]:** It is a location-based service of *Facebook* that enables users to share their geographic position (usually points of interest) with friends in the social network by using a mobile device.
- **Foursquare[2]:** It is a location-based social networking application for mobile devices that offers to its users the possibility to make the so-called check-in in a specific venue (e.g. restaurants, airports, shops, buildings), chosen from a list of places that the system finds near the user. For every check-in, the user gains points that are useful to earn virtual awards like badges and to have the opportunity to become the mayor of that venue. As a result, the check-in activity allows users not only to share location information with friends that use

the application, but also to compete for the achievement of the best score among its friends.

- **Gowalla[3]:** It is a location-based social network similar to *Foursquare*, in which users are able to check-in venues in the neighborhood and gain rewards for this activity such as virtual items.
- **Geocaching[4]:** As the authors have written in the description of the game in the website, it is a *"real-world, outdoor treasure hunting game using GPS-enabled devices. Participants navigate to a specific set of GPS coordinates and then attempt to find the geocache (container) hidden at that location. (…) Geocaches can be found all over the world. It is common for geocachers to hide caches in locations that are important to them, reflecting a special interest or skill of the cache owner. These locations can be quite diverse. They may be at your local park, at the end of a long hike, underwater or on the side of a city street."* (www.geocaching.com)

Other applications collect geographic information that is used as an additional feature:

- **Twitter[5]:** It is a social networking and micro-blogging service that allows users to share short messages of 140 characters called *tweets*, within which it is possible to insert location information.
- **LiveJournal[6]:** It is a social media platform where users can keep a blog as a journal and share posts with friends. Furthermore, users can communicate them geographic information related to their new articles.
- **BrightKite[7]:** *"The app allows users to syndicate their current location to their friends, meet nearby Brightkite users, and lifestream with the equivalent of geo-encoded Tweets. The application is tied to Yahoo's Fire Eagle[8], which allows users*

to manage their location from a number of other services. The site also uses databases to automatically associate POI's and cross streets with GPS locations, so user positions aren't simply displayed as coordinates" (Kincaid, 2008).

An example of important research on the field of location-based applications is shown by Chang and Sun (2011) on *Facebook places*. Chang and Sun (2011) analyzed the dataset of check-ins and points of interest in the application to discover how users share and respond to location-based data. The result is a model that allows reaching three important achievements:

- *Predicting future check-ins* based on previous check-ins of the user, friends' check-ins, demographic data (in particular age and gender of people that checked-in the specific venue), time of the day, distance of place to user's usual location, distance of place to user's previous check-ins and miscellaneous characteristics of places (e.g. how many people have liked it).
- *Predicting how user's friends will respond to his/her check-in* by writing a comment or clicking or the *Like* button: they have shown that the distance is the best predictive feature for friends' feedback to a new check-in, especially for *Like* actions.
- *Predicting friendship starting from check-in data* based on the strong homophily that they have found in the data. In fact, two users that frequently check-in the same points of interest are much more likely to be friends of each other.

Recommender Systems

Want to know what movie you should see this weekend? Ask Facebook. What should you eat for dinner tonight? Yelp knows! Should you wear the *pink sweater or the blue one? You don't have to decide – you can just ask Twitter! (Murphy, 2011).*

People usually rely on recommendations from other people about books, movies, music, and so forth. *Recommender Systems* (RSs) (also known as Recommendation Systems) use a type of *Information Filtering* technique that assists and augments this natural social process to help people finding the most interesting and valuable information for them by generating meaningful recommendations (Su & Khoshgoftaar, 2009). As shown in Resnick and Varian (1997), RSs use the opinions of a community of users to help individuals in that community to identify content of interest from a potentially overwhelming set of choices. To achieve this goal, designers of RS can adopt one of the main approaches for designing a RS architecture:

- The *Collaborative Filtering*
- The *Content-based Filtering*
- The *Knowledge-based*, and
- The *Hybrid approaches* that merge together some of the previous ones.

All these techniques integrate the three fundamental parts of a RS described by Burke (2002):

- The background data (viz. the information that the system has before the recommendation process begins).
- The input data (viz. the information that user sends to the system such as rates and preferences), and
- The algorithm to merge background and input data to create a recommendation.

Collaborative Filtering (CF) Approach

The term "Collaborative Filtering" (CF) was created by the developers of *Tapestry* (Goldberg, et al., 1992). Tapestry was the first commercial RS, designed to recommend only selected documents to the users by leveraging social collaboration to prevent them from getting flooded by a large volume of streaming documents. This approach is based on the idea to collect user's previous ratings for items (such as like/dislike or a specific vote from a scale) and look for similarities in rating behavior among other users, in order to provide the most likely interesting recommendation for the user (Melville & Sindhwani, 2010). Consequently, the recommendations for the active user (the user to whom we want to make a recommendation) are based on the past rating activity of all users collectively. The fundamental assumption of CF is that if users X and Y give a similar rate for n items, or have similar behaviors (e.g., buying a product, watching a movie, listening to a song), it is likely that they will rate or act on other items in a similar way (Golberg, et al., 1992). To predict new rates or behaviors of the active user, CF techniques use a database of preferences consisting of a list of m users $\{u_1, u_2, \ldots, u_m\}$ and a list of n items $\{i_1, i_2, \ldots, i_n\}$. Each user, u_i, has a list of items, Iu_i, which the user has rated and we can see these user's lists as a *rating matrix*. Some examples of CF algorithms are employed to well-known recommender systems like the ones which are present in *Amazon* (based on the recommendation of additional products that other buyers have bought along with the currently-selected item of the active user) (Linden, et al., 2003), *Last.fm* (based on the suggestion of songs often played by other users with similar interests), *Facebook,* and *LinkedIn* (both are based on the analysis of social connection among the users of the network in order to recommend new friendship links).

There are two main categories of techniques to implement a CF system:

- **Memory-Based CF:** It is based on the behavior of the whole community, and it generates recommendations based on statistical notion of similarity between users or between items. In the first case (called *User-based CF*), a subset of users is chosen based on their similarity to the active user and a weighted combination of their ratings is used to produce predictions or suggestions for this user. In the second case (called *Item-based CF*), user's rated items are matched to similar items. The system generates recommendations using item similarity scores for pairs of items, which are based on the likelihood of the pair being purchased by the same customer. This technique is very useful, when we have a huge database of users and items, and the system would take too much time to analyze every single user to find similarity.

- **Model-Based CF:** Instead of relying on statistical notions of similarity, this technique assumes that *"the similarity between users and items is induced by some hidden lower-dimensional structure in the data"* (Melville & Sindhwani, 2010). For example, we can assume that the rating given to a movie by the user depends on implicit factors, such as the user's preferences on movie genres. Therefore, a model-based collaborative filtering algorithm provides *"recommendations by estimating parameters of statistical models for user ratings"* (Melville & Sindhwani, 2010).

The collaborative filtering is a very useful paradigm for researchers that want to take advantage of the power of community behaviors. The widespread diffusion of smartphone, sensor, wireless communication, and information infrastructure (such as GPS, RFID, and Wi-Fi) among users, has made possible an increasing availability of large-scale information about users' location-based activity. Thanks to the use of a CF approach, the analysis of the similarity among users will contribute to the spreading of useful recommendations and suggestions to a specific group of users, also by avoiding to flood them with useless and unwanted information to which they are not interested. One of the implementations of the CF paradigm for mobile applications is *MobHinter*, an epidemic collaborative filtering and self-organization recommender system developed for mobile ad-hoc networks (Schifanella, et al., 2008). The important contribution of this system refers to the demonstration of the possibility to spread ratings and recommendations also over self-organized community of users, based not only on geographic proximity, but also on the ratings similarity among users that are in the neighborhood.

Content-Based Filtering Approach (CBF)

Collaborative filtering systems base their recommendations only on the user ratings matrix without taking into account the specifics of individual users or items. Through the Content-Based Filtering (CBF) technique, one can make a better-personalized recommendation by knowing more details and information about a user (e.g. preferences and tastes or user profile information) or about an item (e.g. description, director, and genre of the movie) (Melville, et al., 2002). These details are extracted by the system from the analysis of textual information by finding regularities in the content (Pazzani, 1999). Then, the recommendations or predictions are made by using heuristic methods or classification algorithms (Pazzani & Billsus, 1997). For example, if a user liked *Minority Report* and *Paycheck*, we can infer that he/she likes science fiction movies, and so we can suggest him/her to watch *I Robot*. A common problem of this approach is the *cold-start problem* because, to be able to classify in the right way all the information, the system needs a large amount of feedback data to analyze. Obviously, it is not always possible to have enough statistically significant data in the early stage of the use of the system. For a detailed discussion of this problem, see the section about "*Limitations and problems of Recommender Systems.*"

Knowledge-Based Recommendations

This technique is based on the creation of user profiles with needs and preferences of all of them. The system knows how a particular item meets a specific user's need and thanks to this basic knowledge, it can infer the relationship between a user's need and a possible recommendation (Burke, 2002).

Hybrid Approach

The hybrid approach combines the strengths of both previous approaches, proving to be (in some cases) more accurate and effective than pure Collaborative Filtering or Content-based Filtering approaches. Hybrid recommender systems can be designed in some different ways: (1) by adding CF characteristics to content-based models; (2) by adding content-based characteristics to CF models; (3) by combining CF with content-based systems or other systems. The simplest method is to produce separate ranked lists of recommendations by allowing both content-based and collaborative filtering methods, and then merge their results to produce a new list (Cotter & Smyth, 2000) or combining the two predictions using an adaptive weighted average, where the weight of the collaborative component increases as the number of users accessing an item increases (Claypool, et al., 1999). Melville et al. (2002) used the hybrid approach to propose a general framework for *content-boosted collaborative filtering* by applying content-based predictions to fill a sparse user ratings matrix making it a full ratings matrix, and then a collaborative filtering method is used to provide recommendations. In particular, "*they use a Naïve Bayes classifier trained on documents describing the rated items of each user, and replace the unrated items by predictions from this classifier. They use the resulting pseudo ratings matrix to find neighbors similar to the active user, and produce predictions using Pearson correlation, appropriately weighted to account for the overlap of actually rated items, and for the active user's content predictions. This approach has been shown to perform better than pure collaborative filtering, pure content-based systems, and a linear combination of the two*" (Melville & Sindhwani, 2010).

A well-known implementation of the hybrid paradigm of recommendation is *Netflix*[9] (NDH, 2012). Netflix is an American provider of on-demand streaming media that predicts movies that a user might like to watch, according to his/her previous behaviors (e.g. ratings) compared to other similar users and the intrinsic characteristics of the movies (e.g. genre, director, actors). The utility of these combined approaches is also related to several problems that are common limitations of recommender systems, which are designed with a pure approach. Such problems are the "cold-start" and the sparsity problems. For a detailed discussion of these problems, see the following section.

Limitations and Problems of Recommender Systems

When a researcher (or a practitioner) decides to implement a recommender system, (s)he must take into account the various challenges posed by its architecture. Especially, when (s)he has to do with systems that deploy pure approaches (*collaborative filtering* or *content-based filtering*), (s)he must consider that these systems usually suffer from several problems which are just related to the chosen framework. Hereafter, we present the main limitations of recommender systems with the relative proposed solutions.

- **Sparsity of Data:** In a collaborative filtering system, the user ratings matrix is typically sparse, especially when the number of users' ratings is very small, compared with the large number of items in the system (and most users do not rate most of them) or when the system is in an early stage of use. Data sparsity decreases the probabil-

ity of finding users with similar ratings. Therefore, the recommender system may be unable to generate recommendations for users. To mitigate the data sparsity problem, many solutions have been proposed. Hybrid CF algorithms, such as the content-boosted CF algorithm (Melville, et al., 2002), are found helpful to address this problem, in which additional content information is used to produce predictions for new users or new items. Another way to alleviate the data sparsity problem is made possible by the *dimensionality reduction techniques*, such as *Singular Value Decomposition* (*SVD*) (Billsus & Pazzani, 1998) that removes unrepresentative or insignificant users/items to reduce the dimensionalities of the user-item matrix. However, it is noteworthy that, when certain users or items are discarded, useful information for recommendations related to them may get lost and recommendation's quality may be degraded (Law, et al., 2007; Sarwar, et al., 2000).

- **Cold-Start Problem:** In the early stage, a new user or item join the system, and it is difficult to find similarity with other users or items due to the lack of information about them. Regarding new items, until some user rates an item, it cannot be recommended to other people, while, for new users, the lack of a rating history makes unlikely that they have already given a big amount of recommendations. To alleviate the cold-start problem, Schein et al. (2002) proposed the *aspect model latent variable method* for cold start recommendation. Their method combines information in model fitting taken from both collaborative and content-based approaches. Another approach, followed to minimize the problem, has been suggested by Kim and Li (2004), who proposed a "*probabilistic model in which items are classified into groups and* *predictions are made for users considering the Gaussian distribution of user ratings.*"

- **Scalability:** Another issue of recommender systems (that use pure collaborative filtering algorithms) occurs when the number of users and items drastically grow and, hence, the system suffers of scalability problems related to the computational resources needed to manage the huge amount of data. One of the well-known methods to manage this problem is the *dimensionality reduction*, of which the *SVD* (see the previous "*Sparsity of data*" section for a more detailed description) is the most used technique: it can quickly produce good recommendations thanks to the reduction of the user-item matrix, but it requires a very expensive offline computation for the matrix factorization step (Su & Khoshgoftaar, 2009; Guy, 2011). Two other techniques, the first one based on *Memory-based CF algorithms* and the second one based on *Model-based CF algorithms*, are proposed by several authors. In particular, for Memory-based CF algorithm, Sarwar et al. (2001) and Law et al. (2007) have used the *item-based Pearson correlation CF algorithm* to calculate the similarity between pair of co-rated items by a user, instead of calculating the similarity between all pairs of items. Thanks to this approach, they can achieve a satisfactory scalability of the system. O'Connor and Herlocker (1999), Chee et al. (2001), Sarwar et al. (2002), and Xue et al. (2005) have shown a partial solution for the problem by using Model-based CF algorithms, called *clustering CF algorithms*. Instead of using the entire database to search users for recommendations, they employed smaller and highly similar clusters of users: the initial problem is solved, but it is important to note that there are tradeoffs between scal-

ability and prediction performance (Su & Khoshgoftaar, 2009).

- **Synonymy:** This problem is related to the propensity of some items to be identified with different names, even though they are all references to the same (or very similar) item. For example, a user could talk about comedy *film* and another one could talk about comedy *movie*. The item is obviously the same, but, for a recommender system based on pure collaborative filtering algorithms; this equivalence is far from being evident. Hence, in cases with lot of synonyms, the performance of the recommender system could dramatically decrease. One solution has been proposed by Deerwester et al. (1990) through the use of the *Latent Semantic Indexing*, a method of the *SVD* techniques that acts by constructing a semantic space, where terms and documents are placed close to each other when they are closely associated, starting from a matrix of term-document associated data (Su & Khoshgoftaar, 2009).

- **Gray Sheep:** This problem refers to those users with completely different opinions from every other user of the system (Su & Khoshgoftaar, 2009). In a practical demonstration of this limitation, the recommender system has no way to take advantage of the collaboration of the community to make useful suggestions for the active user, precisely due to the deep difference between him and the others. Claypool et al. (1999) proposed a hybrid approach with weighted predictions that determines the optimal mix of content-based and collaborative filtering recommendation for each user.

- **Privacy Attacks:** Some problems of RSs are related to the users' privacy protection. For its correct functioning, the system needs a lot of information about users' behaviors, and users could feel this as a violation of their privacy. Furthermore, there are different targeted attacks that may directly or indirectly affect the users of a RS: active attacks and passive attacks (Calandrino, et al., 2011). With an *active attack*, the attacker creates a fake profile of the target user and manipulates his/her activities such as adding, modifying or deleting his/her rated and preferences (or even purchases). Instead, through a *passive attack*, the attacker has access to the public outputs of the recommender system, such as item similarity lists, item-to-item covariance, and/or relative popularity of items.

Other limitations and challenges related to recommender systems, e.g. providing recommendation in a short time period and managing the increasing noise due to the growth of users' and items' diversity are considered in Su and Khoshgoftaar (2009).

Geo-Social and Mobile Recommender Systems

Recommendations connected to places and points of interest can largely benefit from location features. RSs can dramatically improve the usefulness of the suggestions given to the users, based not only on their activity and social network's structure, but also on their geographic position. Because of the huge spreading of mobile devices, the necessity of specific RSs able to manage in a mobile scenario is irrefutable. Actually, a *geo-social RS* must have also capabilities to overcome typical problems related to mobile environments, such as the sparsity of mobile devices, the limitations of wireless networks and so forth. As shown in Ricci (2011), the characteristics of the *mobile context* have a fundamental relevance. In particular, it is important to take into account the evolution of mobile devices, the ubiquitous availability of wireless communication services and the development of position detection techniques (e.g. GPS – *Global Positioning System* and RFID – *Radio*

Frequency Identification). According to Shiller (2003), the three major dimensions that characterize a mobile RS are:

- **User Mobility:** It refers to the fact that the user can access the system everywhere with different devices. Thus, the mobile RS must be able to adapt its recommendations to the specific context of the user.
- **Device Portability:** It refers to the fact that the used devices are mobile, such as smart phone, tablet, or laptop. Previous researches have focused on this dimension and in particular on the physical limitations of mobile devices (e.g. screen size and computational power).
- **Wireless Connectivity:** It refers to the fact that mobile devices use different wireless technologies to connect with other systems, such as UMTS, Wi-Fi, and Bluetooth. For instance, to avoid the limitation of (temporary) unavailability of wireless broadband connectivity, Schifanella et al. (2008) proposed *MobHinter*, an epidemic collaborative filtering system designed for mobile devices that can communicate via ad-hoc networking (see previous "*Collaborative Filtering (CF) Approach*" section).

An online prototype of a geo-social recommender system is described in Papadimitriou et al. (2011). This system is able to provide recommendations on locations, activities and friends that users may be interested in, based on users' check-in history and friendship network. Quercia et al. (2010) show different ways of recommending events by identifying patterns in people's movements from their phone and they found that "*recommending nearby events is ill-suited to effectively recommend places of interest in a city. By contrast, recommending events that are popular among residents of the area is more beneficial.*" Braunhofer et al. (2011) describe another type of recommender system related to geo-localization, a context-aware recommender system that selects music suited for a touristic place of interest, which the user is currently visiting. The recommendations are based on users similarity created by the selection of the favorite music for the specific place. Hooper and Rettberg (2011) give a different point of view about geo-social networking. Instead of studying how people use check-in to share location information, Hooper and Rettberg have analyzed *"why checking into places gives people pleasure or is useful to them."* They have shown that this activity has competitive aspects and mechanisms to encourage exploration (such as defining sequences of locations, gaining rewards and using visualization tools such as *Google Maps*).

Evaluation Metrics

There are several metrics in literature to evaluate the goodness of a RS, especially related to a specific domain of use (e.g. e-commerce RS). Hereafter, we describe those metrics that are more generally applicable in various fields of research.

- **Accuracy**: To measure how relevant are recommendations provided by a RS (how close are them to the ideal choices of the user), two metrics are usually adopted, known as recall and precision. The *recall* measures the ability of the system to present *all relevant items* (See Formula 1). The *precision* measures the ability of the system to present *only relevant items* (See Formula 2). A related measure to (and complementary of) the accuracy is the *Mean Absolute Error* (MAE): it refers to the average of the absolute errors between the predicted value/item and the actual value/item chosen by the user.

Formula 1: (*recall*):

$$\frac{\text{\# of relevant items retrieved}}{\text{\# of relevant items in collection}}$$

Formula 2: (*precision*):

$$\frac{\text{\# of relevant items retrieved}}{\text{total \# of items retrieved}}$$

- **Usability:** It refers to the feature of the system to explain to the user the reasons for which it is recommending that particular item to her/him. A well-known example is the Amazon's explanation[10] *"Customers Who Bought This Item Also Bought book x, book y and book z"* after the view of a specific book. This is a very important metric of evaluation related to the user-experience. The RS could function perfectly by giving the best recommendation for that user but, if he/she is not aware of the explicit motivation of the system's suggestion, (s)he could feel it only as an invasive action or something that (s)he cannot understand. Therefore, giving an explanation of why the RS is recommending an item to a user is a very useful characteristic of the system's goodness.

- **Serendipity**: A recommender system that makes suggestions based on user preferences is very easy to develop and implement. It only needs a user profile with his/her tastes and the labels that describe the items. By matching the two features, finding an interesting item to propose is a really elementary functioning and it is quite obvious that the user will like it. However, more useful RSs try to provide recommendations of items that the user does not expect: because of serendipity (or novelty), an unexpected suggestion that reveals to be interesting for the user, is usually seen much more rewarding than an expected one. Hence, for this reason, a RS that can create unexpected happiness for the user can be considered a good one.

Crowdsourcing and GWAPs

When researchers want to empirically evaluate their approach, one of the most relevant issues is the possibility to access significant data. In the context we want to explore (the development of a geo-social network analyzed through users' behaviors as reactions of a recommender system's suggestions), we need to gather dynamic information about user locations, interests, friendships and so on. A new and promising way to accumulate large amount of data is the adoption of the *Crowdsourcing* paradigm, in particular through the implementation of appealing systems that are widely spread throughout the users' social network. The crowdsourcing approach is related to the concept of *human computation*, coined and made famous by von Ahn (2005, 2006): it refers to a type of computation that outsources certain steps of the process to humans, in order to overcome the difficulties that come from the computerization of some tasks.

Instead, the concept of *crowdsourcing* was coined by Howe (2006) and refers to a distributed problem-solving model: as he explains in his website, it is "*the act of taking a job traditionally performed by a designed agent (usually an employee) and outsourcing it to an undefined, generally large group of people in the form of an open call*" (Howe, 2012).

These two approaches might seem similar to each other for the use of human capabilities in place of computer ones. However, the primary difference between them is that human computation can also refers to a single user's activity, instead crowdsourcing is always related on the activity of a *group* (the "crowd") of users (Quinn & Bederson, 2011). We focused our attention on the crowdsourcing paradigm. After a study of the state of the art on this crucial topic, we ascertained that there is not yet a shared uniform classification of its concrete applications. Due to the fact that this topic is still in a growth phase, many authors classify implementations of the crowdsourcing paradigm using not uniformly shared categories. For example, Yuen et al. (2011) distinguish among "voting systems," "information sharing systems" (including *Wikipedia*[11], *Yahoo! Answer*[12], and *del.icio.us*[13]), "creative systems" (e.g. *Foldit*[14], that take advantage of the challenge to use human intuition to help with protein folding algorithms) and "games" (e.g. *ESP Game*[15], *TagATune*[16], and *Verbosity*[17]). On the other hand, Cooper et al. (2010) refer to *Foldit* as an online game (and not as a creative system), officially described in its website[18] as a "*revolutionary new computer game enabling you to contribute to important scientific research.*" However, regardless of the necessity of an objective classification (that for the moment is not possible to achieve due to the early stage of the field), we have chosen the *game* approach to propose an integrated framework to analyze, design and test geo-social networks. Starting from the point that so many people in the world like playing computer games to be entertained, a game application can attract people to play voluntarily and take advantage from their work that can produce useful metadata as a by-product (Chen, et al., 2010). Actually, thanks to the game paradigm, a huge amount of data can be easily available to researchers interested in extremely various fields such as networks analysis, mobility, Web search, life and data science and so forth. There are two main typologies of games able to help researchers to solve open problems thanks to the power of the "crowd" or that can teach us a lot of things about human behaviors: (1) *serious games* and (2) *Games With A Purpose* (the so-called GWAPs). In the following sections, we present them in details, but we mainly focus on the GWAP approach.

Serious Games

The aim of a *serious game* is the achievement of a serious goal in a training context, by taking advantage of the voluntary of large amount of people to spend time with a computer game.

Such computer game is not always funny. There are many applications of this approach and they are of interests for several fields of research, such as language learning, military and public health simulations (Zyda, 2005). In particular, serious games are used to stimulate the learning process due to the attractive components of a game.

Serious games offer a powerful, effective approach to learning and skills development. (...) They are designed with the intention of improving some specific aspect of learning, and players come to serious games with that expectation. They are used in emergency services training, in military training, in corporate education, in health care, and in many other sectors of society. (...) They also have other names, including immersive learning simulations, digital game-based learning, gaming simulations (Derryberry, 2012).

Game With a Purpose (GWAP)

A GWAP is primarily designed to solve an open problem and gather useful metadata for future analysis from the users' activities, but the other important characteristic is the creation of enjoyment for the users. Actually, people play because they wish to be entertained, not to explicitly solve a computational problem. The resulting solutions given by the users are only a side effect of the gaming. This approach has shown promises in solving a variety of problems which computers are not yet able to resolve, thanks to the collaboration that emerges from the use of the application by a large amount of players. Several implementations are well known and some of them go back to the coining of the GWAP term by von Ahn (von Ahn, 2006). For instance, *ESP game*, *Verbosity* and *TagATune* have been inserted in the www.gwap.com website, which collects several different games with a purpose. In the following subsections, we describe the characteristics of the most famous ones, preceded by a screenshot of the website that explicitly explains the aim of the GWAP approach (see Figure 1).

- **The ESP Game:** This game was created to ask people to label random images, as they see them from the Web. The aim was to improve the accessibility and the Web search of the images by taking advantage of the appropriate textual description that users gave of what objects are present in them. Obviously, a computer is not able to manage this type of job, so human computation is necessary and also very useful to understand how people think, when they see an image. The goal for the user is to give to a set of images the same labels given from a different and anonymous user that is playing simultaneously with him/her. Users gain points for every matching label and, at the end, the global score of either is updated in the system and they can play again with a different new user. The start point of this game is that, if two unknown users label an image with the same tag, then that tag might be a good label (von Ahn & Dabbish, 2004) (see Figure 2).

- **Peekaboom:** It is an improvement of the ESP game, because the input items are all taken from the images database of the previous game. The two players are randomly paired and they cover the roles of *Peek* and *Boom*. Boom sees an image in his/her screen and a related word, Peek tries to guess it by discovering little parts of the image that are slowly revealed by Boom. To help his/her partner, Boom can say if Peek attempts are "hot" or "cold" and, if Peek guesses the word associated to the image, both users gather points. Unlike the ESP game, this GWAP is useful to determine where is each specific object in the image (von Ahn, 2006).

- **Verbosity:** Inspired by the popular game Taboo™, Verbosity is a game designed to

Figure 1. Screenshot with the description of the GWAP approach's aim (© 2011, www.gwap.com/gwap/about/, used with permission)

collect common-sense facts about the given input word. According to the GWAP paradigm, people play because the game is appealing and fun, and not because they want to be helpful to computer science. In fact, the collection of useful knowledge for the artificial intelligence is only the side effect of the success of the online game. The two-paired players are chosen randomly and the roles of *Guesser* and *Narrator* (also called *Describer*) are alternatively assigned to each one. The guesser try to deduce the "input word" given to the narrator, thanks to the help of the descriptions provided by his/her partner according to a defined template of possible statements. The strength of Verbosity is related to the cooperation between the players. If the describer is able to give correct information to the guesser, they have more possibility to gather points by guessing the correct input word. Hence,

thanks to the fact that the describer has all the advantages to type accurate and helpful common-sense facts about the input, the system can infer that they are also useful for computer science necessities (von Ahn, et al., 2006) (see Figures 3 and 4).

- **TagATune**: The aim of this GWAP is to collect tags (users' personal annotations to describe an item) about music and sound clips. As input, an audio clip is given to the two randomly paired users. This audio clip could be the same or not. Based on the tags given by the other player, each of them have to say if they are listening the same music or not. If players correctly guess the nature of the music, the system can use these tags as *good describers*, because these tags revealed to be helpful for users to recognize the equality or inequality of the input (Law, et al., 2007).

Figure 2. Screenshots of the ESP game; an example of a matching output (© 2011, http://www.gwap.com/gwap/gamesPreview/espgame/, used with permission)

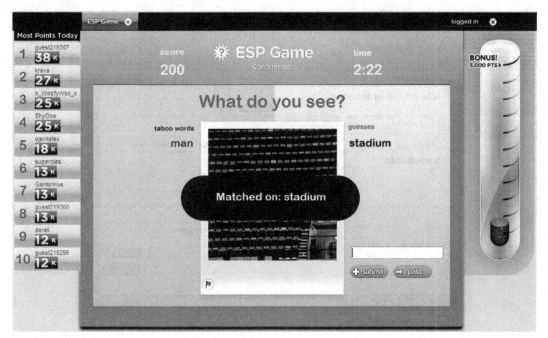

Design of GWAPs

According to von Ahn and Dabbish (2008), there are three main templates that are useful to classify the basic design of existent GWAPs: (1) *output-agreement games*, (2) *inversion-problem games,* and (3) *input-agreement games*. An additional design is described in Weng et al. (2011) and named *Chain Model*. All these templates are discussed in the following subsections.

- **Output-Agreement Games:** Two players are randomly chosen from all potential users that want to take part of the game, they do not know each other, and they cannot see what the other player is doing nor communicate each other. The aim of the game is to produce the same output as the other player after seeing a certain input displayed in the screen; for example, the same label for an image as in the *ESP game* (see previous paragraph for a detailed descrip-

tion of this game). In practice, what players must do is to try to *think like each other*: due to the fact that players cannot communicate which each other, the only strategy they have to make points is to provide an output that really describes the common input. Hence, thanks to the anonymity and the impossibility to cheat, it is possible to assume that outputs chosen by both stranger players are good ones for the given input (see Figure 5).

- **Inversion-Problem Games:** As before, two players are randomly chosen from all potential users that want to take part of the game, they do not know each other and they cannot see what the other player is doing nor communicate each other. One of them is called the *describer* and the other one is the *guesser*: the describer produces outputs for the guesser based on the given input; for example, an original input word as in *Verbosity*. The aim of the game is to

Figure 3. Screenshot of Verbosity game. The player is the guesser of the given input word (© 2011, http://www.gwap.com/gwap/gamesPreview/verbosity/, used with permission)

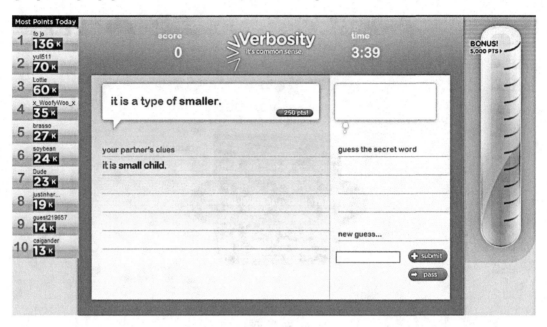

Figure 4. Screenshot of Verbosity game. The player is the describer of the given input word (© 2011, http://www.gwap.com/gwap/gamesPreview/verbosity/, used with permission)

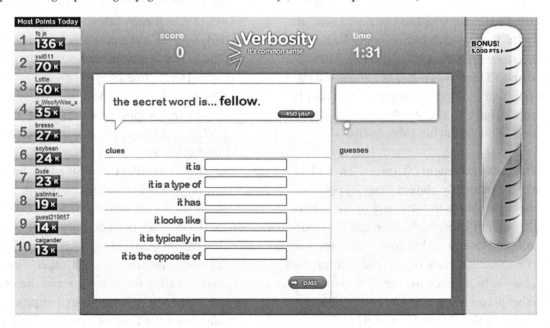

Figure 5. Output-agreement game. Players are given the same input and must agree on an appropriate output (adapted from von Ahn & Dabbish, 2008)

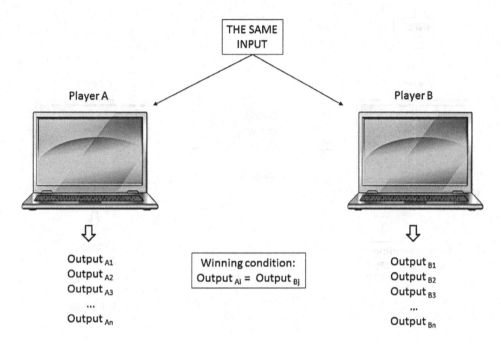

guess the original input of the other player. The only way they have to win is that the describer provides correct outputs describing the input. Otherwise, the guesser cannot guess it anymore. Hence, if the player guesses the correct input as received by the describer, it is possible to infer that chosen outputs by the describer are good ones for the given input. To improve enjoyment, it is possible to add some elements such as transparency and alternation. In the first case, the describer can see partner's guesses and eventually give a feedback, if they are "hot" or "cold" (e.g. in *Verbosity* and *Peekaboom*). The importance of this additional feature is related to the capability to increase the social connection between users, thanks to the awareness of being part of a team, as a result of the transparency. In the second case, the two players can alternate their roles switching from describer to guesser and vice versa, in order to balance the asymmetry of the interaction between them (see Figure 6).

- **Input-Agreement Games:** Here again, we have two randomly selected players with no communication rules as before. Both players receive an input produced by the system. Each player has to describe his/her input to the other player and the aim of the game is that both of them try to guess if they received the same input or a different one. An example of this template is *TagATune* (see previous paragraph for a detailed description of this game), where inputs are songs and outputs are labels to describe them. The only way they have to win is that both players provide correct and accurate outputs describing the personal input, otherwise the other one cannot guess whether the input was the same or not (see Figure 7).

- **Chain Model:** This model was described by Weng et al. (2011). It was introduced as

Figure 6. Inversion-problem game. Given an input, player A produces an output and player B guesses the input (adapted from von Ahn & Dabbish, 2008)

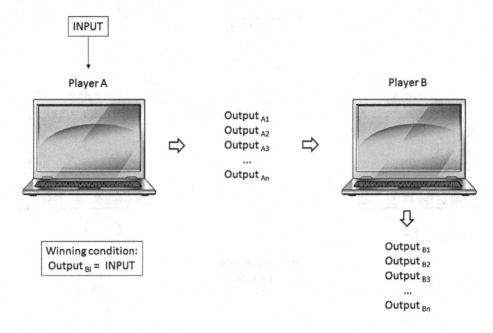

Figure 7. Input-agreement game. Players must determine whether they have been given the same input (adapted from von Ahn & Dabbish, 2008)

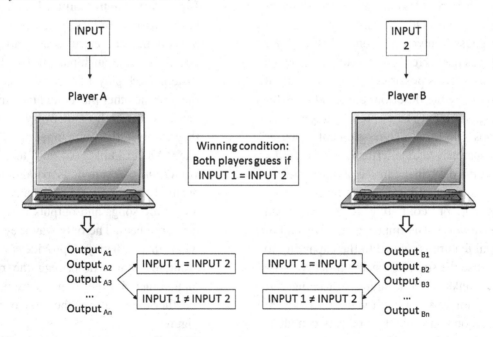

the basic template of two social tagging games, in order to create high-quality semantic associations of data. The knowledge base used for both games derives from the "GiveALink Project"[19]. To overcome the problem of irrelevant tags adding, both games show a control measure based on social comparison of tags annotated by other users. The first one is a browser game called *GiveALink Slider* (Weng & Menczer, 2010): the objects to be connected in the chain are Web pages, that the user is asked to tag in order to extend the chain; the system provides a set of pages related to user's tags among which he/she can choose the next object that must be annotated. The validity of each tag is monitored by the different category in which the chosen annotation is inserted by the system. If the tag is already used by someone for this page, it can be trusted as a relevant one. Otherwise, it needs further evidence to be categorized as a good one. The second one, instead, is an iPhone game called *Great Minds Think Alike* (Weng, et al., 2011). In this case, the *chain* is built by the player through semantic association of words. Given an origin, the user can add a related word or resource taken from the Web (such as photo, video, music, and so forth). Every tag is geo-located (obviously according to the user choice to turn on the location-aware option on the iPhone) and, thanks to the connection with his/her Facebook profile, the player can easily find like-minded friends in her/his proximity.

THE PROPOSED FRAMEWORK

Issues, Controversies, Problems

Considering the actual background scenario, we must collect and analyze data in order to understand human mobility. The goal is to discover if (and

eventually in which ways) the contextual environmental can influence how people move in the space. We are not interested in the mobility activity of the single user, but in the collective behavior of a community of people in its complexity. As discussed in González et al. (2008), understanding the laws that govern human mobility is a difficult task due to the common lack of information about the accurate location of a large amount of individuals. In the mobile domain, we have to take into account a new range of opportunities coming from *geo localization* of devices. In fact, previous cited geo-referenced applications constitute a source of interconnected information, such as locations, social links, user's interests, and tagging activity over multimedia resources that converge spontaneously into complex aggregated data.

As researchers, we look for tools that can help the governance of such complexity; relevant data must be collected, shared with the community, and analyzed to find dynamical patterns that correlate spatial-temporal information with user's preferences and objectives. This enables critical the understanding of dynamics of collective human behaviors, in terms of how the people socialize each other by way of new applications (Aiello, et al., 2012, 2010), and how human mobility can influence the efficiency of communication services (such as routing of information in a decentralized way (Panisson, et al., 2012)). When a researcher wants to empirically evaluate his/her approach, one of the most relevant issues is the access to significant data. In this context, we need to gather dynamic information about user locations, interests, friendships and so on. In collecting data, fundamental issues under consideration are the privacy of the users and the invasive gathering method. In the following section, we show how our proposed approach overcomes these limitations.

Solutions and Recommendations

Our proposed solution focuses on techno-social systems. It is a basic tool to design new services to users, and it provides a platform to gather behavioral data of collective behavior that can be analyzed to improve our understanding of social dynamics and human mobility patterns. Since it is important to study and analyze fresh data, we decided to adopt the GWAP paradigm to implement appealing applications that are widely spread throughout the user social network. Hence, the chosen framework consists of the implementation of a family of geo-social applications for mobile devices. These applications are based on the GWAP paradigm. The integration of a recommender system that provides suggestions to the users based also on their geographic position allowed us to design a platform able to gather a large amount of data, analyze these data and receive hints about future services for new applications that exploit the phenomena that emerge from the analysis (see Figure 8).

The approach is supposed to cycle back to the implementation of new games with a purpose of embedding original services in order to evaluate how the intuition emerged from observed dynamics is verified over real scenarios. A game-based paradigm addresses the mitigation of problems such as the violation of users' privacy and the intrusiveness of the process of gathering massive data. Indeed, if a user intentionally decides to take part of a game based on the geo-social localization, then the designer can assume his/her awareness of what he/she is sharing, that can be sensitive information such as geographic position, specific venue, people whom he/she is with and so on. In practice, users are willing to give useful data, if a form of entertainment is returned. Users voluntarily share geo-social data that the analyst can use them, in an anonymous form, for studying human mobility. The knowledge about visited places, chosen paths/routes, favorite locations and so forth, could be a really important source of data useful to discover emerging correlations between the environmental context and the mobility models that govern people dynamical patterns.

Architecture of the Framework

In Figure 9, we outline the architecture of the proposed platform with the main building blocks of the integrated framework.

The platform is based on a *client-server* architecture. In order to ensure a simple and practical communication between different tiers of the system, a server side API REST is assumed. This API provides functions that can be called by the client to access and modify the database in the back end of the framework.

Client-Side

Mobile clients (e.g. smartphone, tablet, laptop) are bound to a particular operative system, such as *iOS, Android, Blackberry, Symbian OS,* etc. For each one of these OS, a specific device environment is needed, in order to be able to implement a set of common applications above it. Thanks to this approach, the portability of the developed mobile apps can be guaranteed. The *device environment* is composed of different modules, each one delegated to manage a specific task of the system. In the following subsections, we describe all of these modules in order to give an overview of the whole functioning of our platform.

- **Social Module:** This block has the task to manage the social component of the platform; i.e., it deals with the friendship network of the users of the given application. Each user is seen as a node and takes part of his/her personal social network in which he/she is connected to his/her friends through a link relation. Combining all friendship networks, the module has the knowledge of the entire social interaction: who is his/her friend, how many friends a specific user has, who has been tagged in that location and by whom, and so forth.

- **Geo-Localization Module:** This module is able to manage the geographic position

Figure 8. The GWAP-based model to collect and analyze geo-social data

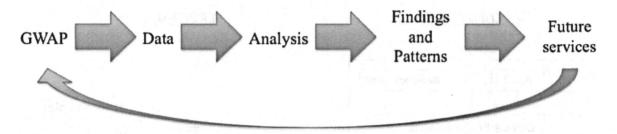

of the user and to track his/her current actions. Due to the position detection techniques implemented in systems such as GPS, it is able to know where the user is every time (s)he wants to share his(her) location. Hence, it is the primary module useful to track check-in events and users mobility behaviors.

- **Recommender Filter Module:** This module includes the recommender engine of the platform. Through its filtering task, the system can provide to the user only information inferred as the most interesting and useful (among the whole collection of gathered ones), personalized taking into account his/her preferences and the behavior of similar users around him.
- **Communication Module:** This module can be seen as the central module of the system, because it has the task to concretely integrate the communication among the different elements of the system. This block is supposed to implement a transport service, so that it works as a gateway for the information exchanged with the server through calls to API function. Such information is received or delivered from/to the other modules of the client-side application.

Server-Side

The server responds to clients' requests in order to provide requested information or to update the geo-social information of the given user in a timely manner.

- **Database:** Information can be stored (and retrieved on request) as tuples in the back end-database. Such tuples have the form $<u_i, l_i, r_i, i_i, ts_i>$, where u_i identifies the user, l_i the location, r_i the resource (that can be every multimedia or physical object chosen by the user, e.g. a photo or an environmental element), i_i the additional and contextual information (e.g., a feedback, a rating or a tag assigned to a resource or a location). Such tuple is sent to the server at time ts_i.
- **API of external services:** The server through third parties APIs can retrieve other requested information. Third parties APIs provide a set of functions that can be called to query the database of a specific service. For example, the Facebook API is useful, when the client wants to know the friendship network of the user inside this social network. The Twitter API is also needed, when the required information is the tweets or the followers of a specific user. Hence, APIs enable the server querying social applications.

Figure 9. The architecture of our integrated framework

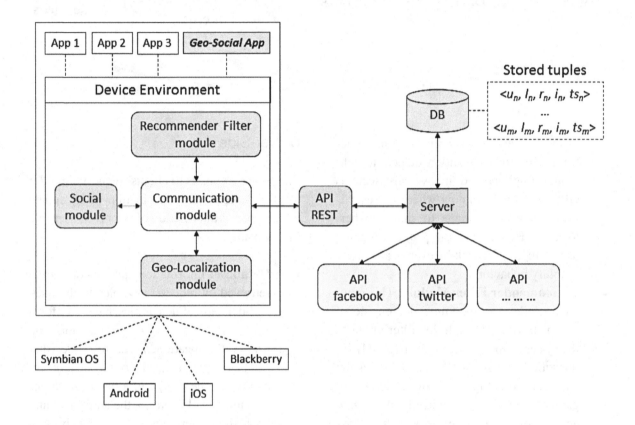

- **API REST:** This block has the task to connect the two sides of the framework, having the intermediary role between the clients and the server. REST (*Representational State Transfer*) is an architectural style for distributed hypermedia systems (Fielding, 2000).

The following use case is a minimal example of the described functioning.

Use case: *Finding Alice friends.*

1. The social module asks the communication module.

2. The communication module sends a request to the server through a function of the API REST.

3. The server receives the request and executes a query to the DB.

4. The DB returns a representation of Alice's friendship network.

5. The information is returned to the client's communication module that forwards it to the social module.

6. The social module processes the information and sends the result to the application level (that, for example, can show a graphical representation of the friendship network to the user).

After describing the global architecture of our framework, we explain the functioning of the *push notifications* of the system. We assume that *context-aware* notifications can be sent on the communication channel between client and server. The client periodically connects to the server to know if there is information relevant to the user current position or activity. If the server finds something to communicate, a notification is pushed to the client.

Figure 10 shows the blocks involved in this process, followed by a detailed description of the interactions among them.

Push notifications can be constrained to three elements of the $<u_i, l_i, r_i, i_i, ts_i>$ tuple: the *user* (the message takes the form of $<u_i, n_i>$), the *location* ($<l_i, n_i>$) and the *resource* ($<r_i, n_i>$). The aim of a push notification is to inform the user about what is happening (or what happened previously) around him/her. Some use cases of push notifications (listed from the most common to the most innovative ones) could be:

- Alice tagged her friend Charles in a photo: the geo-social application informs him.
- Jacob has made a check-in action in a restaurant: the geo-social application tells his friends where he is.
- Bob is in Turin on holiday: when he is near the Egyptian museum, the geo-social application tells him that his friend Mary has been there two years ago.
- David wants to dedicate a song to Emily in their favorite location: the geo-social application tells Emily to go to that specific location to listen the song from David.

To some extent, we can imagine that sophisticated form of geo-caching can be implemented in practice. Indeed, we do not necessarily need that a physical object has been placed to a specific location, but that an electronic "treasure" (a tagged video, a song, or any kind of electronic gift) can be discovered only if given geo-social constraints are met. Starting from the tuples (stored in the DB), the server pushes several data to the communication module of client devices. The recommender/filter module, upon request of the communication unit, analyzes all this information. Relying on recommendation criteria of the chosen approach, it sends back only relevant (context-aware and personalized for the single user) information. The communication module forwards the filtered information as push notifications to the user application.

There are several application fields of this service: from traffic information and transport advices to advanced mobile RS for tourists and voyagers, passing through augmented reality scenarios. For example, a simple functioning could be related to a tourist in Turin who receives a push notification with the recommendation to see the Egyptian museum: considering his/her location and similar users in the network, the system provides this specific recommendation as the most likely appreciated because he/she is in the neighborhood of the museum and lots of similar users have rated it with positive evaluations.

A more complex and interesting implementation of our model could be a GWAP that integrates public transport information (e.g., buses timetable, availability of sharing cars and bikes, taxies parking) with touristic information (e.g., points of interest – POIs, sightseeing, museums' opening hours and visits' duration), all related to the geographic position of the user. The purpose of the game for the user could be the creation of a personal touristic guide based on his/her favorite typology of transport, available time to spend for the visits, preferred POIs (such as churches, plazas, gardens) and so forth. The system could create a sequence of location-based activities based on user's interests, and the user may reward points for each scheduled location that he/she visits through the check-in paradigm, share multimedia data related to a specific venue with friends and so on.

Figure 10. Workflow of the notification system

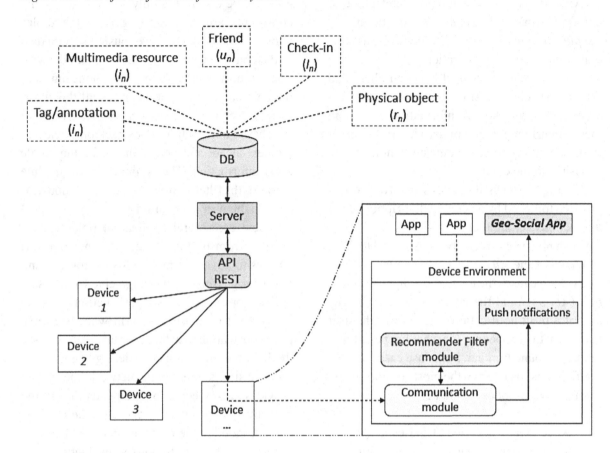

FUTURE RESEARCH DIRECTIONS

As shown from the previous examples, there is a huge amount of dimensions on which it is possible building location-based and user's interests-based recommendations. Therefore, by taking into account the fundamental users' entertainment aim of a GWAP approach, a researcher could create an enormous number of applications useful to gather mobility data as a side effect of the delightfulness of the game. A deep knowledge of users' geographic movements is the necessary starting point for the understanding of the complex laws governing human mobility. Finding patterns that describe mobility behaviors of the whole population enables researchers to use these models for very different purpose. The simplest aim could

be the creation of new geo-social applications for users, developed as a result of the cycle functioning of our approach. In addition to these direct services, the discovery of *correlations* between human mobility and environmental factors could bring enormous contributions to the design of a new concept of the city. In particular, the current diffusion of the *ecological urbanization* idea has highlighted the necessity of the realization of new urban concepts, such as smart and eco cities. Understanding human mobility patterns of the specific city will be a fundamental resource for the future development of urban infrastructures with the purpose to manage ecological problems (for example new transport concept to minimize urban traffic and improve users mobility).

CONCLUSION

In this chapter, we provided an overview of the state of the art of the three main components of our framework: (1) the *Geo-Social Networks*; (2) the *Recommender Systems;* and (3) the crowdsourcing approach with a particular interest on *Games With A Purpose (GWAP)*. Starting from these three paradigms, we proposed a framework to integrate all of them, in order to create a complete platform able to gather and manage a big amount of data related to user information (such as preferences, ratings on resources, and so forth), his/her social network and his/her geographic position. By using this framework to develop geo-located applications, researchers can collect extremely useful information about users, their locations, and their geographic actions. The GWAP approach enables the possibility to overcome privacy problems related to data sharing, because the user is a voluntary player that autonomously decides to communicate his/her information, in return to his/her satisfaction to be entertained. The availability of this amount of geo-located data is the crucial starting point for future research on human mobility that currently is just difficult because of the lack of accurate information about users' geographic movements.

REFERENCES

Aiello, L. M., Barrat, A., Cattuto, C., Ruffo, G., & Schifanella, R. (2010). Link creation and profile alignment in the aNobii social network. In *Proceedings of the 2nd IEEE International Conference on Social Computing (SocialCom 2010)*, (pp. 249-256). Minneapolis, MN: IEEE Press.

Aiello, L. M., Barrat, A., Schifanella, R., Cattuto, C., Markines, B., & Menczer, F. (2012). Friendship prediction and homophily in social media. *ACM Transactions on the Web, 6*(2).

Billsus, D., & Pazzani, M. (1998). Learning collaborative information filters. In *Proceedings of the 15th International Conference on Machine Learning (ICML 1998)*, (pp. 46-54). San Francisco, CA: Morgan Kaufmann Publishers Inc.

Braunhofer, M., Kaminskas, M., & Ricci, F. (2011). Recommending music for places of interest in a mobile travel guide. In *Proceedings of the Fifth ACM Conference on Recommender Systems (RecSys 2011)*, (pp. 253-256). New York, NY: ACM Press.

Burke, R. (2002). Hybrid recommender systems: Survey and experiments. *User Modeling and User-Adapted Interaction, 12*(4), 331–370. doi:10.1023/A:1021240730564

Calandrino, J. A., Kilzer, A., Narayanan, A., Felten, E. W., & Shmatikov, V. (2011). You might also like: Privacy risks of collaborative filtering. In *Proceedings of the Security and Privacy (SP) IEEE Symposium*, (pp. 231-246). IEEE Press.

Chang, J., & Sun, E. (2011). Location3: How users share and respond to location-based data on social networking sites. In *Proceedings of the Fifth International Conference on Weblogs and Social Media*, (pp. 74-80). IEEE.

Chee, S. H., Han, J., & Wang, K. (2001). RecTree: An efficient collaborative filtering method. In *Proceedings of the 3rd International Conference on DataWarehousing and Knowledge Discovery*, (pp. 141-151). IEEE.

Chen, L. J., Wang, B. C., & Chen, K. (2010). The design of puzzle selection strategies for GWAP systems. *Journal of Concurrency and Computation: Practice & Experience, 22*(7), 890–900.

Claypool, M., Gokhale, A., & Miranda, T. (1999). Combining content-based and collaborative filters in an online newspaper. In *Proceedings of the SIGIR-99 Workshop on Recommender Systems: Algorithms and Evaluation*. ACM Press.

Cooper, S., Khatib, F., Treuille, A., Barbero, J., Lee, J., & Beenen, M. (2010). Predicting protein structures with a multiplayer online game. *Nature, 466*, 756–760. doi:10.1038/nature09304

Cotter, P., & Smyth, B. (2000). PTV: Intelligent personalized TV guides. In *Proceedings of the Twelfth Conference on Innovative Applications of Artificial Intelligence*, (pp. 957-964). Austin, TX: IEEE.

Deerwester, S., Dumais, S. T., Furnas, G. W., Landauer, T. K., & Harshman, R. (1990). Indexing by latent semantic analysis. *Journal of the American Society for Information Science American Society for Information Science, 41*(6), 391–407. doi:10.1002/(SICI)1097-4571(199009)41:6<391::AID-ASI1>3.0.CO;2-9

Derryberry, A. (2012). *Serious games: Online games for learning*. Retrieved from http://www.adobe.com/resources/elearning/pdfs/serious_games_wp.pdf

Fielding, R. (2000). *Architectural styles and the design of network-based software architectures*. (PhD Thesis). University of California. Irvine, CA.

Goldberg, D., Nichols, D., Oki, B., & Terry, D. (1992). Using collaborative filtering to weave an information tapestry. *Communications of the ACM, 35*(12), 61–70. doi:10.1145/138859.138867

González, M. C., Hidalgo, C. A., & Barabási, A. L. (2008). Understanding individual human mobility patterns. *Nature, 453*, 779–782. doi:10.1038/nature06958

Guy, I. (2011). *Social recommender systems*. Paper presented at WWW 2011. Hyderabad, India.

Hooper, C. J., & Rettberg, J. W. (2011). Experiences with geographical collaborative systems: Playfulness in geosocial networks and geocaching. In *Proceedings of MobileHCI 2011*. Stockholm, Sweden: MobileHCI.

Howe, J. (2006, June). The rise of crowdsourcing. *Wired.*

Howe, J. (2012). *Crowdsourcing: A definition*. Retrieved January 31, 2012 from http://crowdsourcing.typepad.com

Kim, B. M., & Li, Q. (2004). Probabilistic model estimation for collaborative filtering based on items attributes. In *Proceedings of the IEEE/WIC/ACM International Conference on Web Intelligence (WI 2004)*, (pp. 185–191). Beijing, China: IEEE/ACM Press.

Kincaid, J. (2008). *A peek at brightkite for the iPhone*. Retrieved January 15, 2012 from http://techcrunch.com/2008/10/16/a-peek-at-brightkite-for-the-iphone/

Law, E., Ahn, L. V., Dannenberg, R., & Crawford, M. (2007). Tagatune: A game for music and sound annotation. In *Proceedings of the 8th International Conference on Music Information Retrieval ISMIR 2007*, (pp. 361–364). Vienna, Austria: ISMIR.

Linden, G., Smith, B., & York, J. (2003). Amazon.com recommendations: Item-to-item collaborative filtering. *IEEE Internet Computing, 7*(1), 76–80. doi:10.1109/MIC.2003.1167344

Melville, P., Mooney, R. J., & Nagarajan, R. (2002). Content-boosted collaborative filtering for improved recommendations. In *Proceedings of the Eighteenth National Conference on Artificial Intelligence (AAAI 2002)*, (pp. 187-192). Edmonton, Alberta: AAAI.

Melville, P., & Sindhwani, V. (2010). Recommender systems. In Sammut, C., & Webb, G. (Eds.), *Encyclopedia of Machine Learning*. Berlin, Germany: Springer.

Murphy, S. (2011). *The culture of recommendation: Has it gone too far?* Retrieved Jan 31, 2012 from http://www.suzemuse.com/2011/02/the-culture-of-recommendation-has-it-gone-too-far/

NDH. (2012, January 8). *Netflix's blog*. Retrieved Jan 15, 2012 from http://blog.netflix.com/

O'Connor, M., & Herlocker, J. (1999). Clustering items for collaborative filtering. In *Proceedings of the ACM SIGIR Workshop on Recommender Systems (SIGIR 1999)*. ACM Press.

Panisson, A., Barrat, A., Cattuto, C., Ruffo, G., Schifanella, R., & Van den Broeck, W. (2011). On the dynamics of human proximity for data diffusion in ad-hoc networks. *Ad Hoc Networks, 10*(8), 1532–1543. doi:10.1016/j.adhoc.2011.06.003

Papadimitriou, A., Symeonidis, P., & Manolopoulos, Y. (2011). Geo-social recommendations. In *Proceedings of the RecSys Workshop Personalization on Mobile Applications (PeMA 2011)*. Chicago, IL: PeMA.

Pazzani, M., & Billsus, D. (1997). Learning and revising user profiles: The identification of interesting web sites. *Machine Learning, 27*(3), 313–331. doi:10.1023/A:1007369909943

Pazzani, M. J. (1999). A framework for collaborative, content-based and demographic filtering. *Artificial Intelligence Review, 13*(5-6), 393–408. doi:10.1023/A:1006544522159

Quercia, D., Lathia, N., Calabrese, F., Di Lorenzo, G., & Crowcroft, J. (2010). Recommending social events from mobile phone location data. In *Proceedings of the 2010 IEEE International Conference on Data Mining (ICDM 2010)*, (pp. 971-976). Washington, DC: IEEE Press.

Quinn, A. J., & Bederson, B. (2011). Human computation: A survey and taxonomy of a growing field. In *Proceedings of the 2011 Annual Conference on Human Factors in Computing Systems (CHI 2011)*. Vancouver, Canada: ACM Press.

Resnick, P., & Varian, H. (1997). Recommender systems. *Communications of the ACM, 40*(3), 56–58. doi:10.1145/245108.245121

Ricci, F. (2011). Mobile recommender systems. *International Journal of Information Technology and Tourism, 12*(3), 205–231. doi:10.3727/109830511X12978702284390

Sarwar, B. M., Karypis, G., Konstan, J. A., & Riedl, J. (2000). Analysis of recommendation algorithms for E-commerce. In *Proceedings of the ACM E-Commerce*, (pp.158–167). Minneapolis, MN: ACM Press.

Sarwar, B. M., Karypis, G., Konstan, J. A., & Riedl, J. (2001). Item-based collaborative filtering recommendation algorithms. In *Proceedings of the 10th International Conference on World Wide Web (WWW 2001)*, (pp. 285–295). ACM Press.

Sarwar, B. M., Karypis, G., Konstan, J. A., & Riedl, J. (2002). Recommender systems for large-scale e-commerce: Scalable neighborhood formation using clustering. In *Proceedings of the 5th International Conference on Computer and Information Technology (ICCIT 2002)*. ICCIT.

Scellato, S. (2011, Autumn). Beyond the social web: The geo-social revolution. *SIGWEB Newsletter*.

Scellato, S., Mascolo, C., Musolesi, M., & Latora, V. (2010). Distance matters: Geo-social metrics for online social networks. In *Proceedings of the 3rd Conference on Online Social Networks (WOSN 2010)*. Berkeley, CA: USENIX Association.

Schein, A. I., Popescul, A., Ungar, L. H., & Pennock, D. M. (2002). Methods and metrics for cold-start recommendations. In *Proceedings of the 25th Annual International ACM SIGIR Conference on Research and Development in Information Retrieval (SIGIR 2002)*. ACM Press.

Schifanella, R., Panisson, A., Gena, C., & Ruffo, G. (2008). Mobhinter: Epidemic collaborative filtering and self-organization in mobile ad-hoc networks. In *Proceedings of the 2008 ACM Conference on Recommender Systems (RecSys 2008)*, (pp. 27–34). New York, NY: ACM Press.

Shiller, J. (2003). *Mobile communications*. Reading, MA: Addison-Wesley.

Su, X., & Khoshgoftaar, T. M. (2009). A survey of collaborative filtering techniques. *Advances in Artificial Intelligence*. Retrieved from http://www.hindawi.com/journals/aai/2009/421425/

Von Ahn, L. (2005). *Human computation*. (Doctoral Thesis). CMU.

Von Ahn, L. (2006, June). Games with a purpose. *IEEE Computer Magazine*, 96–98.

Von Ahn, L., & Dabbish, L. (2004). Labeling images with a computer game. *ACM SIGCHI Conference on Human Factors in Computing Systems*, (pp. 319-326). New York, NY: ACM Press.

Von Ahn, L., & Dabbish, L. (2008). Designing games with a purpose. *Communications of the ACM,51*(8),58–67.doi:10.1145/1378704.1378719

Von Ahn, L., Kedia, M., & Blum, M. (2006). Verbosity: A game for collecting common-sense facts. *ACM SIGCHI Conference on Human Factors in Computing Systems*, (pp. 75-78). New York, NY: ACM Press.

Weng, L., & Menczer, F. (2010). GiveALink tagging game: An incentive for social annotation. In *Proceedings of the ACM SIGKDD Workshop on Human Computation (HCOMP 2010)*, (pp. 26-29). New York, NY: ACM Press.

Weng, L., Schifanella, R., & Menczer, F. (2011). Design of social games for collecting reliable semantic annotations. In *Proceedings of the 16th International Conference on Computer Games (CGAMES)*, (pp. 185-192). Louisville, KY: CGAMES.

Weng, L., Schifanella, R., & Menczer, F. (2011). The chain model for social tagging game design. In *Proceedings of the 6th International Conference on Foundations of Digital Games (FDG 2011)*, (pp. 295-297). New York, NY: ACM Press.

Xue, G., Lin, C., & Yang, Q. (2005). Scalable collaborative filtering using cluster-based smoothing. In *Proceedings of the ACM SIGIR Conference*, (pp. 114–121). Salvador, Brazil: ACM Press.

Yuen, M.-C., King, I., & Leung, K.-S. (2011). A survey of crowdsourcing systems. In *Proceedings of Privacy, Security, Risk and Trust (PASSAT), 2011 IEEE Third International Conference on and 2011 IEEE Third International Confernece on Social Computing (SocialCom)*, (pp. 766-773). Boston, MA: IEEE Press.

Zyda, M. (2005). From visual simulation to virtual reality to games. *Computer*, *38*(9), 25–32. doi:10.1109/MC.2005.297

ADDITIONAL READING

Kapitsaki, G., Prezerakos, G., Tselikas, N., & Venieris, I. (2009). Context-aware service engineering: A survey. *Journal of Systems and Software*, *82*(8), 1285–1297. doi:10.1016/j.jss.2009.02.026

Rasch, K., Li, F., Sehic, S., Ayani, R., & Dustdar, S. (2011). Context-driven personalized service discovery in pervasive environments. *World Wide Web (Bussum)*, *14*(4), 295–319. doi:10.1007/s11280-011-0112-x

Sheng, Q. Z., Li, X., & Zeadally, S. (2008). Enabling next-generation RFID applications: Solutions and challenges. *IEEE Computer*, *41*(9), 21–28. doi:10.1109/MC.2008.386

Sheng, Q. Z., Yu, J., & Dustdar, S. (Eds.). (2010). *Enabling context-awareweb services: Methods, architectures, and technologies*. Boca Raton, FL: CRC Press. doi:10.1201/EBK1439809853

Welbourne, E., Battle, L., Cole, G., Gould, K., Rector, K., & Raymer, S. (2009). Building the internet of things using RFID: The RFID ecosystem experience. *IEEE Internet Computing*, *13*(3), 48–55. doi:10.1109/MIC.2009.52

KEY TERMS AND DEFINITIONS

Check-In: The user's action to share with friends his/her position in a specific location.

Collaborative Filtering: A category of information filtering that analyzes the behavior of the other users, in order to find similarities with the given user.

Crowdsourcing: The paradigm used to take advantage from the power of a group of people to solve computational problems that a computer cannot yet resolve.

Geo-Social Network: A friendship network created by a group of users in which their geographic position is a primary feature.

GWAP (Game With A Purpose): The paradigm thanks to which is possible to obtain useful data as a side effect of the game.

Recommender System: A tool able to provide suggestions to the given user, based on a variety of related features (e.g., preferences, rates, geographic position).

Tag: A personal annotation added to an item by the user to describe it.

ENDNOTES

[1]   http://www.facebook.com/about/location
[2]   http://www.foursquare.com
[3]   http://www.gowalla.com
[4]   http://www.geocaching.com
[5]   http://www.twitter.com
[6]   http://www.livejournal.com
[7]   http://www.brightkite.com
[8]   http://fireeagle.yahoo.net/
[9]   https://www.netflix.com
[10]   http://www.amazon.com/
[11]   http://en.wikipedia.org
[12]   http://answers.yahoo.com
[13]   http://delicious.com/
[14]   http://fold.it/portal/info/science
[15]   http://www.gwap.com/gwap/gamesPreview/espgame/
[16]   http://www.gwap.com/gwap/gamesPreview/tagatune/
[17]   http://www.gwap.com/gwap/gamesPreview/verbosity/
[18]   http://fold.it/portal/info/science
[19]   http://www.givealink.org

Section 6
Intelligent Image Processing and Image Retrieval

Chapter 17
Estimating the Level of Noise in Digital Images

Jakub Peksinski
West Pomerania University of Technology Szczecin, Poland

Michal Stefanowski
West Pomerania University of Technology Szczecin, Poland

Grzegorz Mikolajczak
West Pomerania University of Technology Szczecin, Poland

ABSTRACT

One of the significant problems in digital signal processing is the filtering and reduction of undesired interference. Due to the abundance of methods and algorithms for processing signals characterized by complexity and effectiveness of removing noise from a signal, depending on the character and level of noise, it is difficult to choose the most effective method. So long as there is specific knowledge or grounds for certain assumptions as to the nature and form of the noise, it is possible to select the appropriate filtering method so as to ensure optimum quality. This chapter describes several methods for estimating the level of noise and presents a new method based on the properties of the smoothing filter.

INTRODUCTION

The dynamic development of computer techniques that has been observed over the past twenty years and the development of digital algorithms for signal processing accompanying it allows for significant improvement of the quality of obtained images and purposeful interference in the image structure for bringing out certain qualities.

Improvement of image quality makes it possible to obtain a significantly greater amount of useful information and also to create a better aesthetic impression.

Images documented as primary—model—are created, in principle, as a representation of the reality surrounding us in a form intelligible by the senses or are created as a creative manmade act. In the first case, representation is effected with the use of an available form of electromagnetic energy (radar, X-Ray apparatus, television camera, photographic camera), mechanical energy—ultra-

DOI: 10.4018/978-1-4666-2833-5.ch017

sound (ultrasound, echo sounding), or other forms of energy such as heat (thermal vision), or the energy of an electron beam (electron microscope).

It is clear that images obtained by means of technical devices as well as those that are purely manmade—especially those created using technical tools—are burdened with distortions. Distortions pertain to geometric changes of the obtained image relative to the model and also changes resulting from the superimposition of unnecessary information, constituting noise, onto the model image.

The causes of distortions are also to be sought in the interference of measuring energy in the observed environment and in imperfections of the apparatus (nonlinearity of processing, the influence of sampling, quantization, the limited band of signal transfer, limited by the capabilities of fixing the image on monitor screens or on paper). Distortions also appear as a result of the influence of external electromagnetic fields and other fields during processing and transmission of images.

A significant practical matter is the search for methods of improvement of image quality and removal of distortions being the effect of noise. The main tasks in this scope are tasks of searching for methods and algorithms of image analysis for:

- Removal of undesired noise from the image,
- Removal of specific image defects (e.g. geometrical),
- Improvement of images with low technical quality,
- Reconstruction of images that have been partially destroyed.

The computer technology used for achieving one of the above listed goals is called "digital image processing" technology. This technology is incomparable to other technologies (e.g. retouching in conventional photography) in terms of the achieved effects. In this technology, an image is defined as an NxM matrix with values in the range [0-255] corresponding to 2^8 distinct levels. A basic operation on a so-defined set of data, having the purpose of removing undesired qualities of the image or to influence its new properties, is the filtering of this data.

Effectiveness of filtering, expressed for example by the noise reduction coefficient, is a function of many factors including: (1) the selected filtering algorithm; (2) certain information with noise qualities; and (3) also certain information about the model image. Of special significance is information on the qualities of noise—random or determined, the distribution of power spectral density, variance, etc. In most cases, it is not possible to obtain full data on the noise and attempts at estimation are undertaken—assessment of the level of noise expressed by variance through analysis of image data. Using the information on the level of noise in the image allows for obtainment of an optimal filtering quality, especially for realization of problems of image reconstruction, edge detection, and others. It is also among the information necessary for the creation and operation of adaptive filtering algorithms.

The applications of noise level estimation in images are very wide and include, among others:

- Removal of noise from astronomical photographs (Murtagh & Starck, 1998, 2000; Starck & Murtagh, 1998; Murtagh, et al., 1995; Starck & Murtagh, 2001; Starck, et al., 1995).
- Image reconstruction (Ranham & Katsaggelos, 1997; Gamier & Bilbro, 1995).
- Edge detection (Canny, 1986; Rosin, 1998, 1997).
- Image segmenting (Rosin, 1998; Spann & Wilson, 1985; Zhu & Yuille, 1996).
- Motion estimation (Mitiche, 1994).

- Image smoothing (Amer & Schroder, 1996; de Haan, et al., 1996; Jostschulte & Amer, 1998; Meer, et al., 1994).
- Removal of noise in TV receivers (Grafe & Scheffler, 1988; de Haan, et al., 1996, 1998).
- Reduction of noise in teledetection photographs (remote sensing imagery) (Comer, et al., 2003; Narayanan, et al., 2001, 2003).
- Reduction of noise in photographs made using magnetic resonance technology (MRI) (Dixon, 1988; Henkelman, 1985; Kaufman, et al., 1989; Murphy, et al., 1993; Sijbers, et al., 1998, 1999).
- Reduction of noise in video signals (Amer, 2001; Amer, et al., 2002).
- Segmenting of sonar images (Schmitt, et al., 1996).

The problem of estimation of noise level is most often based on determination of its variance σ^2_n (or standard deviation σ_n) from the digital image under the assumption that noise is an additive and stationary process not correlated with the useful signal, with an average value equal to zero.

Methods of estimation of noise level can be fundamentally divided into smoothing methods and block methods (Olsen, 1993). Smoothing methods are based on the filtering of the image with noise and assume that the image obtained after filtering is the original image. The difference between the signal representing the image with noise and the signal of the image after filtering leads to the determination of e.g. the standard deviation σ_n of the noise. In block methods, the value of noise standard deviation is estimated on the basis of analysis of the values of so-called local variances, determined from so-called "flat areas." It is assumed that only noise is decisive in regard to the variability of pixel values in these areas (the values of useful signal pixels is constant in these images—"flat"). A significant difficulty in these methods is the search for the above-defined "flat" areas in the image, and more specifically, the dif-

ficulty is based on the selection of the criterion according to which areas are designated as "flat."

In the group of smoothing methods, averaging and median filters with a *3x3* size window are often used. However, using an averaging filter leads to blurring of edges and a greater change in the image structure, which, in consequence, gives unsatisfactory results. That is why only certain filtered pixels are selected for estimation (Olsen, 1993). The selection can be made based on the value of the gradient of individual pixels, which is less than the set threshold value. In practice, it is difficult to determine the threshold value, which is why only a certain percentage of pixels with the lowest values are selected—10% in the case of this study. In the case where a median filter is used, this is not necessary, because it interferes with the image structure in a lesser degree.

In the works found under (Lee, 1980, 1981; Mastin, 1985; Olsen, 1993), the so-called block method of estimation of noise level is presented. This is a method based on determination of local variances for small area and then averaging a certain amount of the smallest values of these variances. Lee (1981) proposes a division of the entire image into *7x7* blocks. However, Mastin (1985) proposed an initial "manual" selection of the area acknowledged by the user as a "flat" area, that is, characterized by a small amount of details, and it is this area that is then divided into blocks. Then, the local variance is determined for every pixel in the area of a given block, that is, the variance of the pixels directly neighboring it. The estimate of the variance of a given block can be assumed to be the lowest value from the local variances. However, in relation to the small number of samples (pixels) participating in the determination of this value, Lee (1981) suggests averaging the five lowest local variances and to only consider this value as the estimate of noise variation in a given block. The last step is to determine the noise variation for the entire image. In Mastin (1985) and Olsen (1993), a selection of 10% of the lowest values among the variances determined for *7x7* blocks and their averaging is proposed. In this way, an estimate for noise variance of the entire image is obtained.

The method presented by Meer et al. (1990) is also based on determining the noise level through calculation of variances in individual blocks with dimensions of $2^l x 2^l$ for *l=1,2,...,n*, where $2^n x 2^n$ -image size. Next, the four lowest variance estimates for a given level (up to *l≤n-1* is possible) are selected, with the correct variance value being selected on the basis of statistical inference based on Dixon's test.

Mencattini et al. (2003) and Salmeri et al. (2001) present an estimation algorithm of the correct noise level for a situation where the type of noise probability distribution is known. The main idea of the method is to find the image blocks for which the probability distribution of samples (pixels) is close to the probability distribution of the noise, and then to determine the noise level estimate on their basis. The authors limited themselves to the case where the noise has a normal distribution. The parameter determining the measure of distribution similarity is the χ^2 parameter (Frieden, 1983; Taylor, 1982). In order to determine this parameter, the range of pixel values in the image must first be divided into *N* subranges. Then, if the probability distribution of the noise is defined with the same range of valued, it can be determined what number of image samples is assigned to a given subrange. The authors selected a division into 8 subranges, the bounds of which are marked with the points:

$$\Pi_{1,2,3} = \pm 0.3186 \cdot \sigma, \pm 0.6745 \cdot \sigma, \pm 1.1503 \cdot \sigma \quad (1)$$

By designating the expected number of samples in the subrange k as E_k and the actual number of samples in this subrange as *Ok.*, parameter χ^2 is defined as:

$$\chi^2 = \sum_{k=1}^{N} \frac{\left(O_k - E_k\right)^2}{E_k} \quad (2)$$

The closer the value of χ^2 is to zero, the more the distribution of pixels in a given block will be closer to the noise distribution and the better it will be suited for estimating the noise level. The number of ranges, N, cannot be too low, because then the differences between distributions would be imperceptible; this number also cannot be too large - it should be much lower than the number of samples (pixels) in a given block. Besides the χ^2 parameter, two other parameters are determined for each block:

- *m:* This is the average value of pixels in a given block and should be as close as possible to the median luminance value, which is 128 for 8-bit images;
- p_{oi}: This is the number of pixels in saturation for a given block, which, for 8-bit images, is a number of pixels equal to 0 or 255—the lower this value is, the better.

On the basis of a Lena test image, it was determined how noise level estimation error is distributed depending on the values of these three parameters. A similar estimation error distribution was accepted for all other images. Therefore, on the basis of these parameters, the accuracy of estimation of noise variance σ^2_i can be determined for the i-th block. For this purpose, a *fuzzy logic* system with Sugeno type inference is used. Therefore, the value $h_i \epsilon [0,1]$ is obtained for each block. Rules of inference as seen in Box 1 were used.

Where L, M, and H represent, respectively, membership function, a *fuzzy logic* system, the Low, Medium, and High values with respect to the input variables.

The value of σ^2_i and h_i is determined for N_w non-overlapping blocks with a size of 8x8 pixels covering the entire image. Finally the noise level estimate σ^2 is determined according to the following formula:

Box 1.

$$if \left(\chi^2 \ is \ L \right) \ and \ \left(p_{01} \ is \ L \right) \ and \ \left(\overline{x} \ is \ M \right) \ then \ \left(h_i \ is \ H \right)$$

$$if \left(\chi^2 \ is \ L \right) \ and \ \left(p_{01} \ is \ L \right) \ and \ \left(\overline{x} \ is \ H \right) \ then \ \left(h_i \ is \ H \right)$$

$$if \left(\chi^2 \ is \ L \right) \ and \ \left(p_{01} \ is \ L \right) \ and \ \left(\overline{x} \ is \ L \right) \ then \ \left(h_i \ is \ H \right)$$

$$if \left(\chi^2 \ is \ L \right) \ and \ \left(p_{01} \ is \ M \right) \ and \ \left(\overline{x} \ is \ L \right) \ then \ \left(h_i \ is \ H \right) \quad (3)$$

$$if \left(\chi^2 \ is \ L \right) \ and \ \left(p_{01} \ is \ M \right) \ and \ \left(\overline{x} \ is \ M \right) \ then \ \left(h_i \ is \ L \right)$$

$$if \left(\chi^2 \ is \ M \right) \ and \ \left(p_{01} \ is \ L \right) \ and \ \left(\overline{x} \ is \ H \right) \ then \ \left(h_i \ is \ L \right)$$

$$if \left(\chi^2 \ is \ M \right) \ and \ \left(p_{01} \ is \ M \right) \ and \ \left(\overline{x} \ is \ M \right) \ then \ \left(h_i \ is \ L \right)$$

$$\sigma^2 = \frac{\sum_{i=1}^{N_W} h_i \cdot \sigma_i^2}{\sum_{i=1}^{N_W} h_i} \quad (4)$$

The shape of the input membership functions is triangular, and the degree of overlapping is equal to 2. Furthermore, the membership function for χ^2 accounts for the dependency of this parameter on the number of subranges N. The rules and parameters of the membership function were adjusted on the basis of the average error distribution determined using a set of test images disrupted with noise of various levels of strength.

The authors also propose increasing the efficiency of the method—so as not to process each block, which, in the case of large images causes a significant increase in calculation time, initial selection of blocks having a "flat" pixel value distribution is to be executed and further calculations are to be done using this small number of blocks. For this purpose, the SUD parameter, defined as the maximum of the average pixel values calculated for each frame entering into a given block, is determined for each block. Figure 1 presents the division of a block into frames.

First, the average from the four middle pixels of a given block, designated as q_m is determined,

and then, for each square frame around these four pixels, the q_j value is determined according to the following formula:

$$q_j = \frac{1}{P_j} \sum_{k=1}^{P_j} \left| p_{kj} - q_m \right| \quad (5)$$

The *SUD* (Spatial Uniformity Degree) value is determined as:

$$SUD = \max_{\forall j} \left(q_j \right) \quad (6)$$

If the SUD value determined as above is greater than a certain set SUT (Spatial Uniformity Threshold) threshold value, the given block is omitted in further calculations. As simulations show, the selection of an ideal SUT threshold value is dependent on the level of image noise, which in turn, is to be estimated using the SUT parameter. The authors have determined the dependency between these values as:

$$SUT = f\left(\sigma^2 \right) = 86 \cdot \sigma^2 + 0.729 \quad (7)$$

The solution of this problem is the iterative noise level estimation algorithm proposed by the authors. In the first step, a small value of $SUT_0 = 2.2$

Figure 1. Division of the block into frames in the estimation method using fuzzy logic

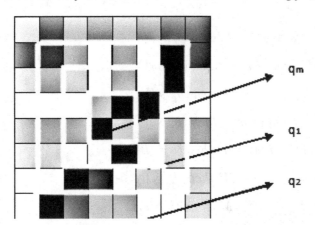

is selected, which ensures the selection of a small amount of blocks for further calculations in this step. Next, the noise variance σ_k^2 is estimated, and on its basis, the next value of parameter SUT_{k+1} is determined. Calculations are repeated cyclically and end after a certain set number of steps (the authors suggest $k=10$) or when the difference between successive determined values of the SUT parameter are sufficiently small (the authors suggest 0.005).

The method presented in Amer (2001) and Amer et al. (2002) estimates noise variance on the basis of the variance of a set of regions (blocks) classified as the most homogenous (uniform) in terms of pixel intensity, that is, those that exhibit the smallest oscillations in structure.

The image is divided into WxW size blocks. Next, homogeneity ζB_h is determined for each of them using a local analyzer using high frequency operators calculating homogeneity in eight directions. The coefficients of these operators are $\{-1,-1...(W-1)...-1,-1\}$. For each pixel, the operator operates in all of the directions presented in Figure 2.

If the distribution of intensiveness is "flat" in one of the directions, the operator gives a result that is close to zero. The summed results from all eight directions determine the value of ζB_h. Next, m blocks with the lowest ζB_h values are selected,

that is, those blocks with the best homogeneity, and local variance σ_{Bh}^2 is calculated for each of them. Finally, the obtained results are averaged. The result is an estimated variance of image noise:

$$\sigma^2 = \frac{\sum_{h=1}^{m} \sigma_{B_h}^2}{m} \tag{8}$$

In relation to the fact that the most homogenous areas may have significantly differing values of local variances, only the areas with local variances near the reference values are to be selected for averaging. Similarity to the reference value occurs if the following is true:

$$\left| \sigma_{Br}^2 - \sigma_{B_h}^2 \right| < t_\sigma \tag{9}$$

where t_σ is the acceptable difference between the actual and the estimated value of noise variance. This parameter is not dependent on the input image and is rather simple to determine experimentally. For example, in television receiver noise reducers, this parameter assumes values between 3 and 5 (Blume, et al., 1997; de Haan, et al., 1996). The author accepted the value of $t_\sigma=3$. However, reference variance σ_{Br}^2 is determined as the median of the three most homogenous areas.

Figure 2. Directions of operation of the homogeneity analyzer for W=3

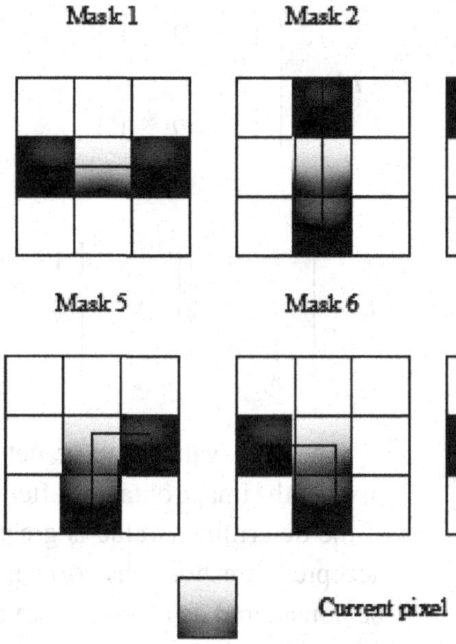

The method presented above is strongly dependent on the size *W* of the window used. The larger the window, the better the results for large noise level values, but results are worse for small values. For example, by using a window with *W=3* the best results are achieved for small noise values, but by using a size of *W=5*, results are better for larger noise level values *(PSNR>40dB)*. The author tested various window sizes *W=3,5,7,9,11*. A *5x5* size window was finally selected on the basis of obtained results, as a compromise between the effectiveness and efficiency of the method.

Immerkaer (1996) presents a fast and simple algorithm for noise level estimation in images. It is based on the assumption that image structures, such as edges, have strong differential summands of the second order, which suggests the use of a filter mask removing these types of structures in the estimation process. The author uses mask *N*, being the difference between two Laplace masks *L1* and *L2* (see Box 2.)

The average value of *N* mask coefficients is zero, and the noise amplification coefficient *q=36*, therefore, if the input signal will be noise with

variance of σ^2, then after filtration with the *N* mask, noise variance σ_n^2 will be equal to:

$$\sigma_n^2 = 36 \cdot \sigma^2 \qquad (11)$$

By designating the filtering operation of image *I* using mask *N* at point *(x, y)* as $(x,y)_*N$, the noise variance σ^2 of image *I* can be determined as:

$$\sigma^2 = \frac{1}{36 \cdot (W-2) \cdot (H-2)} \sum_I (I(x,y) * N)^2 \quad (12)$$

where W and H designate the width and height of the image in pixels, respectively.

In order to speed up the estimation process, the author suggests that the operation of raising to the second power be omitted by using another formula for the estimation of noise standard deviation σ:

$$\sigma^2 = \sqrt{\frac{\pi}{2}} \cdot \frac{1}{6 \cdot (W-2) \cdot (H-2)} \sum_I |I(x,y) * N| \quad (13)$$

Conducted simulations did not show significant differences of results for the application of both

Box 2.

$$L1 = \begin{bmatrix} 0 & 1 & 0 \\ 1 & -4 & 1 \\ 0 & 1 & 0 \end{bmatrix} \quad L2 = \frac{1}{2}\begin{bmatrix} 1 & 0 & 1 \\ 0 & -4 & 0 \\ 1 & 0 & 1 \end{bmatrix} \quad N = 2(L2 - L1) = \begin{bmatrix} 1 & -2 & 1 \\ -2 & 4 & -2 \\ 1 & -2 & 1 \end{bmatrix} \quad (10)$$

of the above dependencies. Despite the fact that the method works for a wide range of noise level values, it is not suitable for textures and for images containing many details due to the fact that the results are burdened with very large errors.

The noise level estimation technique used in Immerkaer (1996) was also used in Comer et al. (2003). Using the Laplace filter, the authors suggest separating the structure of the original image from the noise, the level of which can then be determined. They propose that this method is to be used in the process of de-noising remote detection images. These types of images are characterized by a high degree of correlation between a given pixel and the pixels adjacent to it. The filter mask proposed to be the difference between two Laplace masks (Immerkaer, 1996) performs excellently for such images by removing most of the original image. However the authors have observed that part of the image structure, especially edges, remain in the image even after filtering with a Laplace filter. Therefore, in order to obtain a precise result, elements of the original image that remain after Laplace filtering are to be additionally detected and removed. The authors suggest the used of the so-called Sobel edge detector (Jensen, 1996) as the tool for edge detection, which can be presented as:

$$|\nabla f| = \sqrt{f_x^2 + f_y^2} \quad (14)$$

where masks f_x and f_y:

$$f_x = \begin{bmatrix} -1 & 0 & 1 \\ -2 & 0 & 2 \\ -1 & 0 & 1 \end{bmatrix} \quad f_y = \begin{bmatrix} 1 & 2 & 1 \\ 0 & 0 & 0 \\ -1 & -2 & -1 \end{bmatrix} \quad (15)$$

The $|\nabla f|$ value is to be determined for each pixel of the image obtained after Laplace filtering. If the determined value is greater than a certain accepted threshold, the corresponding pixel is to be considered as an edge. The authors acknowledged that the optimal threshold value for the edge detector is the value of *3.5*. In this way, a binary edge map can be created for the processed image, and using it, pixels recognized as edges can be removed from the image, and the effect of edge variance on the noise level estimate can be decreased.

The final step of estimation is the determination of local standard deviations for every *9x9* pixel block. Pixels recognized as edges are omitted in the process of determining standard deviations; therefore, part of them will be calculated on the basis of a number of pixels that is less than the maximum number. The authors suggest that the median of the local deviation values is to be accepted as the final estimated noise standard deviation value.

The above method gives good results for small and medium noise level values, but for large values, worsening of results was observed. These errors can be mitigated by using blocks of sizes greater than *9x9* pixels in the process of calculation of local noise levels. Then, a greater number of pixels participates in the calculations, which contributes to better representation of standard deviation and more accurate results. Unfortunately, this causes a worsening of results for small noise levels. A solution to this problem could be dynamic selection of optimal block size on the basis of an initially estimated noise level value.

The algorithm for estimation of noise level σ_n^2 presented by Rank et al. (1999) is based on an iterative analysis of a histogram presenting local noise level estimates. Be designating the original image with a width of M and a height of N with x, and additive noise with n, image with noise, y, is defined as seen in Box 3.

The algorithm is divided into three steps. In the first step, the disrupted image is subjected to filtering for the purpose of removing the original image structure. This is to ensure separation of noise n and the original image, x, and at the same time, a more exact noise level estimate. The filter that is used is a cascade combination of two one-dimensional differentiating operators y_1 and y_2 (see Box 4).

Due to the fact that the variance of the sum of two independent disruptions is equal to the sum of the variances of these disruptions, it results that operations y_1 and y_2 do not have an effect on the variance of noise n, because it is assumed that it is white noise that is not correlated with the original signal, x. Therefore, the noise variance of signals y and y_2 is the same. It should be noted,

that as a result of operations y_1 and y_2, a small part of the pixels of the original image will be omitted because the width and height of signal y_2 is M-1 and N-1, respectively.

An advantage of differentiating operations y_1 and y_2 is that they transform areas exhibiting a linear increase in grayness into areas with a constant grayness, thanks to which local variances of the noise level $\sigma^2(m,n)$ can be estimated more accurately.

The second step is based on determining a histogram for local noise standard deviations $\sigma^2(m,n)$. First, for each pixel (m,n), its local variance is determined from the area adjacent to it. If the size of this area is equal to LxL, where $L=2K+1$, then the local variance (see Box 5).

Next, histogram $h(k)$ is determined for the value of $\sigma^2(m,n)$ according to Equation 19 (see Box 6.

The final step is the determination of noise variance σ^2 on the basis of histogram h(k). By calculating the average histogram values as:

$$s^2 = \frac{\sum_{k=0}^{k_{max}} k^2 \cdot h(k)}{\sum_{k=0}^{k_{max}} h(k)} \tag{20}$$

Box 3.

$$y(m,n) = x(m,n) + n(m,n) \quad \forall(m,n) \in D_0 = \{(m,n) | m \in [1,M] \cap n \in [1,N]\} \tag{16}$$

Box 4.

$$y_1(m,n) = \frac{1}{\sqrt{2}}[y(m+1,n) - y(m,n)] \quad \forall(m,n) \in D_1 = \{(m,n) | m \in [1,M-1] \cap n \in [1,N]\}$$

$$y_2(m,n) = \frac{1}{\sqrt{2}}[y_1(m,n+1) - y_1(m,n)] \quad \forall(m,n) \in D_2 = \{(m,n) | m \in [1,M-1] \cap n \in [1,N-1]\} \tag{17}$$

Box 5.

$$\sigma^2(m,n) = \frac{1}{L^2-1} \cdot \sum_{i=-K}^{K} \sum_{j=-K}^{K} \left[y_2(m+i, n+j) - \mu(m,n) \right]^2$$

$$\mu(m,n) = \frac{1}{L^2} \cdot \sum_{i=-K}^{K} \sum_{j=-K}^{K} y_2(m+i, n+j)$$

(18)

Box 6.

$$h(k) = \begin{cases} \left\| \left\{ (m,n) \middle| k - \frac{1}{2} \leq \alpha \cdot \sigma(m,n) < k + \frac{1}{2} \right\} \right\| & if \quad k = 1, \dots, k_{\max} \\ 2 \cdot \left\| \left\{ (m,n) \middle| 0 \leq \alpha \cdot \sigma(m,n) < \frac{1}{2} \right\} \right\| & if \quad k = 0 \end{cases}$$

(19)

where the symbol \\{·}\\ designates the number of elements in set {·}. The values of $k_{\max}$ and α must be selected so that an appropriate number of samples is assigned to every range of the histogram.

The value of the noise variance estimate can be determined as:

$$\sigma_n^2 = \frac{s^2}{\alpha^2}$$

(21)

This value is usually higher relative to the actual value of noise level in the image. This is the result of a slight shift of the histogram to the right, caused due to the omission of part of the pixels of the original image during differentiating operations y_1 and y_2. In order to minimize estimation errors, the authors propose iterative determination of successive values by using the function in Box 7.

$$s_{l+1}^2 = \frac{\sum_{k=0}^{k_{\max}} k^2 \cdot g_l(k) \cdot h(k)}{\sum_{k=0}^{k_{\max}} g_l(k) \cdot h(k)}$$

(23)

Parameter β should be selected experimentally, and the authors used a value equal to 2. They

also defined the maximum number of iterations as $l_{max}=4$. Finally, the value of noise variance is equal to:

$$\sigma_n^2 = \frac{s_{l_{\max}}^2}{\alpha^2}$$

(24)

A disadvantage of this method is its relatively high complexity of calculations and the necessity of selecting several parameters: α, β, k_{max}, and l_{max}.

An interesting approach to the problem of estimation has been presented in Brancho and Sanderson (1985). Assuming that the noise has a Gaussian distribution, the authors observed that the amplitude of the gradient of intensiveness has a Rayleigh distribution. Because the probability density function of this distribution has a maximum for values equal to standard deviation σ_n, therefore, this value can be determined from the gradient amplitude histogram. In the case where the image contains structures such as edges, the gradient values of these structures are much higher than the noise gradient values. Therefore,

Box 7.

$$
g_l(k) = \begin{cases} 1 & if \quad k \le s_l \\ \dfrac{1}{2} \cdot \left(1 - \cos\left(\dfrac{\beta - \dfrac{k}{s_l}}{1 - \beta} \cdot \pi \right) \right) & if \quad s_l < k < \beta \cdot s_l \\ 0 & if \quad k \ge \beta \cdot s_l \end{cases}
\tag{22}
$$

as long as areas with a constant level of intensity dominate in the image, the influence of the edge gradient on the gradient amplitude histogram will be slight—these values are situated at the end of the graph, well after the maximum determining the noise standard deviation. It is problematic when a larger part of the image is dominated by textures or edges. Then, the influence of the gradients of these structures causes equalization of the histogram and difficulties in locating the

Table 4. Relative error values calculated on the basis of averaged results of tests carried out using various methods

	Av1	Med.	Av2	Blok	Grad	Pyr	Cov		
σ_n							9x9	19x19	29x29
1	6.135	4.737	0.466	1.521	2.574	1.128	0.891	1.157	1.108
2	2.665	1.988	0.134	0.501	1.110	0.350	0.214	0.362	0.371
3	1.541	1.118	0.060	0.234	0.579	0.217	0.024	0.108	0.173
4	1.007	0.713	0.033	0.107	0.430	0.238	0.021	0.056	0.102
5	0.702	0.490	0.019	0.029	0.363	0.163	0.031	0.035	0.076
6	0.512	0.356	0.012	0.008	0.296	0.093	0.182	0.003	0.059
7	0.383	0.265	0.007	0.032	0.261	0.065	0.080	0.003	0.064
8	0.295	0.200	0.004	0.051	0.269	0.053	0.110	0.040	0.009
9	0.231	0.160	0.003	0.054	0.205	0.035	0.110	0.021	0.008
10	0.183	0.124	0.001	0.077	0.194	0.032	0.072	0.037	0.030
11	0.144	0.102	0.001	0.090	0.132	0.015	0.139	0.054	0.014
12	0.115	0.080	0.001	0.089	0.208	0.009	0.152	0.066	0.004
13	0.087	0.069	0.002	0.097	0.187	0.000	0.098	0.079	0.022
14	0.072	0.058	0.003	0.103	0.096	0.005	0.081	0.036	0.007
15	0.056	0.046	0.003	0.110	0.109	0.021	0.136	0.015	0.004
16	0.042	0.037	0.003	0.114	0.149	0.089	0.137	0.051	0.028
17	0.032	0.030	0.004	0.113	0.172	0.036	0.204	0.064	0.008
18	0.020	0.026	0.004	0.117	0.051	0.087	0.110	0.062	0.028
19	0.015	0.018	0.004	0.116	0.123	0.136	0.089	0.075	0.007
20	0.009	0.017	0.004	0.123	0.166	0.131	0.072	0.077	0.006

maximum. Vorhees and Poggio (1987) propose the use of the first derivative of the histogram in such a case, and thanks to the extrapolation of the descending part of the graph, finding of the point in which it crosses zero, which corresponds to the maximum of the histogram.

In the works found under Sijbers et al. (1998, 1999) methods of noise level estimation of images made using the nuclear technique of magnetic resonance have been described. This technique finds applications mainly in medical diagnostics as the MRI imaging technique (Magnetic Resonance Imaging). The advantage of this technique over X-ray images is two-fold. Firstly, neither a magnetic field nor radio waves (these are usually used in resonance imaging) are as harmful as X rays, and secondly, MRI gives the capability of imaging any cross-section, not just a projection onto plates as in the case of a conventional x-ray. By using magnetic resonance, any fragment of an observed object can enlarged with practically no limitations. Knowledge of the noise level in these types of images is used in the process of their reconstruction (Ranham & Katsaggelos, 1997; Gamier & Bilbro, 1995) or filtering (Gerig, et al., 1992; Yang, et al., 1995).

The technique of magnetic resonance enables the registration of signals in the form of complex numbers. This signal is disrupted by white noise with a Gaussian distribution (Wang & Lei, 1994; Wang, et al., 1996). However, the module of the registered signal is most often used for visualization. The module determined for two independent variables with a Gaussian distribution (the real and imaginary part of the registered signal) has a different distribution, called the *Rice distribution* (Papoulis, 1984). Therefore, if the real and imaginary part of the signal, the average value of which is equal to A_R and A_I, respectively, is disrupted by noise with Gaussian distribution with an average value of zero and standard deviation of σ, and the module is expressed as:

$$M = \sqrt{\sum_{k=1}^{2} x_k^{\,2}} \tag{25}$$

then the probability density function of module M has the Rice distribution:

$$p(M|A) = \frac{M}{\sigma^2} \cdot e^{-\frac{M^2 + A^2}{2 \cdot \sigma^2}} \cdot I_0\left(\frac{A \cdot M}{\sigma^2}\right) \tag{26}$$

where I_0 is a zero order Bessel function, and

$$A = \sqrt{A_R^{\,2} + A_I^{\,2}} \tag{27}$$

The value of noise variance $\sigma_M^{\,2}$ (or standard deviation σ_M) can be estimated on the basis of the moment of the Rice distribution function:

$$E\left[M^2\right] = 2 \cdot \sigma^2 + A^2 \tag{28}$$

If an area is selected for calculation for which it is known that it does not contain a useful signal (so-called image background), then it can be assumed that $A=0$, and then:

$$\sigma_M^{\,2} = \frac{1}{2 \cdot N} \sum_{i=1}^{N} M_i^{\,2} \tag{29}$$

where N is the number of averaged modules of pixels in the area of the selected area.

Many works pertaining to MRI image processing (Henkelman, 1985; Kaufman, et al., 1989; Sijbers, et al., 1998) use so-called background areas in the process of estimating noise level, that is, those regions of the image where there is no useful signal. In these types of areas, the Rice distribution of the module is transformed into a Rayleigh distribution. The moment of this distribution is equal to:

$$E[M] = \sigma \cdot \sqrt{\frac{\pi}{2}} \tag{30}$$

therefore the standard deviation can be presented (Gudbjartsson & Patz, 1995), as:

$$\sigma_C = \sqrt{\frac{2}{\pi}} \cdot \frac{1}{N} \cdot \sum_{i=1}^{N} M_i \tag{31}$$

Sijbers et al. (1998) compared both estimators σ_M and σ_C. They proved that for large values of N>50, the mean square error of estimator σ_C is greater than the mean square error of estimator σ_M by 9 percent.

Both of the above methods for estimating noise level require the user's interaction, who must "manually" select the area acknowledged by him to be the image "background." Sijbers et al. (1998) propose, therefore, an alternative scheme of estimation using two images made under identical conditions, that is using so-called double image acquisition. In this case, two equations are obtained, from which one pertains to the module of a single image, Ms, and the second pertains to the module of two averaged images, M_a:

$$E\left[\left\langle M_s^2 \right\rangle\right] = \frac{1}{N} \cdot \sum_{n=1}^{N} A_n^2 + 2 \cdot \sigma^2$$

$$E\left[\left\langle M_a^2 \right\rangle\right] = \frac{1}{N} \cdot \sum_{n=1}^{N} A_n^2 + 2 \cdot \left(\frac{\sigma}{\sqrt{2}}\right)^2 \tag{32}$$

where the symbol $\langle \cdot \rangle$ signifies spatial averaging over the entire image. The following estimate of nose variance σ^2 is obtained from the above equations:

$$\sigma^2 = \left\langle M_s^2 \right\rangle - \left\langle M_a^2 \right\rangle \tag{33}$$

The advantage of this approach is a lack of interference from the side of the user, because there is no need for selection of the "background"

area and because all image data participates in the estimation process, which significantly improves the precision of the estimator. Other methods of noise level estimation in MRI images using double image acquisition are described in Dixon (1988) and Murphy et al. (1993).

In Murtagh and Starck (2000), Starck and Murtagh (1998), and Starck et al. (1995), the authors present a method of automatic noise estimation in astronomical photographs. This method uses the so-called Continuous Wavelet pyramid, which is obtained by executing the Continuous Wavelet transform several times. The *"a trous"* algorithm was used for calculating the Continuous Wavelet transform (Holdschneider, et al., 1989). The original image, c_0, can be presented as the sum of coefficients w_j on all levels and c_p values:

$$c_0 = c_p + \sum_{j=1}^{p} w_j \tag{34}$$

Therefore the pixel at *(x,y)* can be defined as:

$$c_0(x,y) = c_p(x,y) + \sum_{j=1}^{p} w_j(x,y) \tag{35}$$

The Continuous Wavelet pyramid of image M can be considered as a three-dimensional table, the third dimension of which is the index defining the weight of the bit. Level j corresponds to the index in Box 8.

Therefore, the Continuous Wavelet pyramid logically defines whether a given image contains significant information at level j and position *(x,y)*. In the case where image noise has a Gaussian distribution, the comparison of its value with parameter $k\sigma_j$ is decisive of whether or not coefficient w_j is significant or not, where σ_j defines the standard deviation of noise for a given j level, and the value of parameter k is selected and usually equal to 3. If coefficient $w_j(x,y)$ is small, then its value is not significant because it is most

Box 8.

$$M(j,x,y) = \begin{cases} 1 & when & w_j(x,y) & & is & significant \\ 1 & when & w_j(x,y) & is & not & significant \end{cases} \qquad (36)$$

likely a result of the noise. If, however, $w_j(x,y)$ is large, then it contains significant information about the image (see Box 9).

Therefore, for determining the Continuous Wavelet pyramid, it is necessary to have knowledge of the noise standard deviation for every j level. As a result of the properties of the Continuous Wavelet transform, the noise standard deviation σ_j is equal to the image noise standard deviation σ_I multiplied by the noise standard deviation for level j of Continuous Wavelet transform σ_j^e:

$$\sigma_j = \sigma_I \cdot \sigma_j^e \qquad (38)$$

σ_j^e values are determined experimentally. Therefore, the value of σ_I must be found. The authors suggest the use of the Continuous Wavelet pyramid for this purpose. Estimate σ_I is obtained by taking into account the set of image pixels, S, with noise being the only deciding factor of their values and determining their standard deviation. Set S can be easily obtained from the Continuous Wavelet pyramid—these are all of the pixels (x,y) fulfilling the condition:

$$M(j,x,y) = 0 \; for \; all \;\; j \qquad (39)$$

Therefore, for these pixels, there is no significant coefficient w_j on any j level. The Authors propose and iterative algorithm for estimating σ_I:

1. Estimation of the initial value of σ_I^0 with any method.
2. Setting $n=0$.
3. Determining Continuous Wavelet pyramid M_n using $\sigma_I^{(n)}$.
4. Determining pixels belonging to set S on the basis of wavelet pyramid M_n.
5. For all pixels from set S, determination of the values of $I(x,y)-c_p(x,y)$ (c_p designates the so-called "background" of the image that is to be omitted during determination of the noise level), and subsequent determination of the value of $\sigma_I^{(n+1)}$.
6. $n=n+1$ incrementation.
7. When $|\sigma_I^{(n)}.- \sigma_I^{(n-1)}|>\varepsilon$, go to step 3.

During the last step of estimation, the obtained $\sigma_I^{(n)}$ value is scaled by an experimentally determined parameter equal to 0.974. This scaling results from the fact that a certain small part of coefficients w_j, the value of which solely results from noise, will be greater than parameter $k\sigma_I$ and will be recognized as significant, which will cause the omission of these coefficients in the process of determining standard deviation $\sigma_I^{(n)}$ and thus, a certain error that is corrected through this scaling.

Box 9.

$$\begin{aligned} when & \quad |w_j| \geq k \cdot \sigma_j & then & \quad w_j & is & significant \\ when & \quad |w_j| < k \cdot \sigma_j & then & \quad w_j & is \; not \; significant \end{aligned} \qquad (37)$$

The methods described here do not exhaust the entire scope of the matter. A comparison of various algorithms for estimating noise level can be found in Amer et al. (2002), Mencattini et al. (2003), Olsen (1993), Rank et al. (1999), and Starck and Murtagh (1998).

NOISE LEVEL ESTIMATION USING SMOOTHING FILTER

One of the significant problems in digital signal processing is the filtering and reduction of undesired interference. Due to the abundance of methods and algorithms for processing signals characterized by complexity and effectiveness of removing noise from a signal, depending on the character and level of noise, it is difficult to choose the most effective method. So long as there is specific knowledge or grounds for certain assumptions as to the nature and form of the noise, it is possible to select the appropriate filtering method so as to ensure optimum quality. For example, the moving average filter has greater noise reduction coefficient than the medium filter with the same mask size, so it will be more suited to removing "large" noise. However, the median is better for maintaining edges and interferes in the signal structure in a lesser degree, which, with a smaller noise reduction coefficient, is more suitable for filtering signals with "low" noise.

The problem is significantly more complex when the character of a given signal cannot be determined. Without additional information, it is often difficult to assess the level and "type of noise," and sometimes, it is not possible to state whether a random or deterministic course is being dealt with. Due to this, at this point, a test analysis of the properties of smoothing filters and their influence on noise level in the output signal will be conducted.

Noise Level Estimation Using Exponential Smoothing Filter

The values of variances of noise signals for individual smoothing methods are derived below. Exponential smoothing of the signal is given by the formula:

$$y_m = (1-a) \cdot y_{m-1} + a \cdot x_m \quad where \quad y_0 = (1-a) \cdot x_0 \tag{40}$$

With the acceptance of the following assumptions:

$$x = s + n \qquad V(N) = \sigma_n^2 \tag{41}$$

where: x – disrupted signal; s – useful signal; n – noise.

Taking into account that the noise and useful signal are not correlated, the variance of the input signal is equal to:

$$V(X) = \sigma_s^2 + \sigma_z^2 \quad where: \quad V(S) = V_s = \sigma_s^2 \tag{42}$$

Assuming that smoothing reduces only the noise variance, a dependency of the output signal variance can be written as:

Table 1. Results of estimation of noise variance using formula (2.9), where: σ^2 – noise variance, a – smoothing parameter, N – number of samples per period

σ^2	a=0.99		a=0.9	
	N=100	N=200	N=100	N=200
0.01	0.015	0.012	0.014	0.011
0.04	0.041	0.034	0.036	0.036
0.09	0.068	0.094	0.123	0.092
0.16	0.167	0.172	0.136	0.158
0.25	0.298	0.247	0.243	0.254

$$V(Y) = \sigma_s^2 + \frac{a}{2-a}\sigma_n^2 \qquad (43)$$

By designating: $V(Y)=V_a$ – variance after exponential smoothing, V_0 – signal variance without smoothing, the following is obtained:

$$V_a = \sigma_s^2 + \frac{a}{2-a}\sigma_n^2 \qquad (44)$$

$$V_0 = \sigma_s^2 + \sigma_n^2 \Rightarrow \sigma_s^2 = V_0 - \sigma_n^2 \qquad (45)$$

By substituting (45) to (44), after transformations, the dependency for noise variance (46) is obtained, determined on the basis of knowledge of the variance of the disrupted signal and the variance of the signal after exponential smoothing.

$$\sigma_n^2 = \frac{2-a}{2-2a}(V_0 - V_a) \qquad (46)$$

The method of determining noise variance presented above (46) is characterized by a need for knowledge of only the variance of the input and output signal, the value of which is known. This makes it possible to easily determine the noise level in the case of both smoothing methods.

The results of the proposed noise level estimation method for exponential smoothing have been presented in Table 1. For analysis of the proposed method, a model signal of the cosine function was used, with N samples per period. This function was disrupted by a random signal with a normal distribution (Gaussian).

During the process of noise estimation using exponential smoothing, the following values of the smoothing parameter were applied: a=0.99 and a=0.9. The acceptance of such parameters ensured a minimal influence of filtering on the tested signal, due to the fact that the assumption pertaining only to noise attenuation was not fully valid. Next, the variance value according to the given formula (46) was determined. As the results in Table 1 show, satisfactory values of noise estimation were obtained, especially for a greater noise level. These results seem to confirm the correctness of the accepted assumption as to the method of estimation of noise in the tested signal.

In general, the idea of the method is based on the knowledge of the noise variance reduction coefficient, the value of input and output signal variance, and on the assumption that, during filtering, only the noise in the output signal is attenuated.

The above establishments find their reflection in dependency (47), which is the basis for estimation of the noise level in the analyzed signal.

$$\sigma_n^2 = \frac{V(Y) - V_q(Y)}{1-q} \qquad (47)$$

where: $V(Y)$ – signal variance without filtering; $V_q(Y)$ – signal variance after filtering; q – noise reduction coefficient.

Table 2. Results of noise variance estimation in the Lena test image for individual areas where a-coefficient filter mask (48)

Area	I		II		III		IV	
σ^2	a=0.01	a=0.02	a=0.01	a=0.02	a=0.01	a=0.02	a=0.01	a=0.02
25	113.4	111.9	313.1	309.3	30.6	30.0	30.4	30.3
100	180.4	179.5	383.5	376.9	99.7	99.3	98.4	97.8
225	282.7	280.4	487.5	486.4	213.1	212.8	214.7	214.8

Estimation Noise Variance Using Averaging Filter

For a two-dimensional weighted average filter with a 3x3 mask:

$$\begin{bmatrix} a & a & a \\ a & 1 & a \\ a & a & a \end{bmatrix} \quad where \quad 0 < a \leq 1 \qquad (48)$$

The noise reduction coefficient is expressed by formula:

$$q = \frac{1 + 8a^2}{(1 + 8a)^2} \qquad (49)$$

Due to the fact that the assumption pertaining to attenuation of only noise is not completely fulfilled, that is, the filter also interferes in the signal structure, small weight values are to be selected (a~0.01), and the area for estimation should be characterized by low variability of the useful signal.

Table 2 presents the results of noise level estimation for areas marked in Figure 3, presenting a well-known image with 256 shades of gray and a size of 256x256 pixels, often used for analysis of the effectiveness of image processing algorithms—Lena.

The fragments from which noise variance was estimated are marked in Figure 3. The entire image was subjected to additive noise with normal distribution $N(0,\sigma)$. During the selection of areas, the leading consideration was that the assumption pertaining to the removal of only noise from the filtered signal was inaccurate. This is shown by the results of noise estimation in areas I and II. These areas contain a large amount of small details and a large range of gray levels, which makes estimation in these areas ineffective. However, in areas III and IV, the estimated variance value shows good conformance with the variance of the

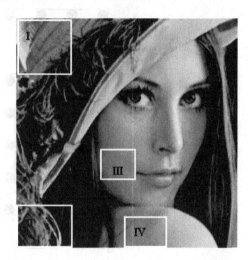

Figure 3. Test image Lena I-IV – areas from which noise variance was estimated

noise to which the test image was subjected. This is an effect of the fact that the selected areas exhibit high uniformity of gray levels and a lack of small details. Test results have been presented in Table 2.

The results shown confirm the correctness of the accepted assumption, on the basis of which the formula for noise level estimation was derived (47). The only source of doubts is the fact that selection of the area is done subjectively, i.e., the area that seems the most appropriate for evaluating noise level is selected. A conclusion can be made, that a method for finding an area from which noise level estimation would give satisfactory results should be elaborated.

Finding the Area for Noise Level Estimation

It is proposed for selection of the area to be made based on image analysis due to the correlation coefficient determined based on auto-covariance, because the correlation between individual pixels of distinguished "good" areas is small.

The auto-covariance function defines the dependency between values of the same process

Figure 4. The placement of points relative to which auto-covariance is calculated (52)

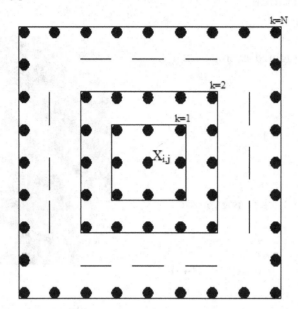

(signal) at instant t and t+τ. For a stationary random process, X(t) is defined by the equation:

$$C_x(t) = \int\limits_{-\infty}^{\infty} \int\limits_{-\infty}^{\infty} (x_1 - m)(x_2 - m)p(x_1, x_2, \tau)dx_1 dx_2 \quad (50)$$

where: x1 and x2 are values of random function X(t) in time instants t and t+τ, p(x1, x2, τ) designates the probability density function for random function X(t), and m, the average value.

However, the estimator of the auto-covariance function determined for set N of the samples representing the tested signal x is equal to:

$$C(r, \Delta t) = \frac{1}{N - r} \sum_{k=1}^{N-r} (x_k - m)(x_{k+r} - m) \quad (51)$$

where: the kth sample of signal x, k=1,2, ..., N-1, and Δt is the time interval between neighboring samples.

In the case of searching for an area with the lowest values of correlation coefficients, the following procedure was used:

- The average value – m of disrupted pixels in a given area is determined by moving an NxN window over the analyzed image.
- Next, for each pixel from the given area, the auto-covariance coefficient is calculated relative to pixels found in the kth layer (52) (see Box 10). The placement of individual layers is shown in Figure 4.
- The variance of pixels of the k-th layer is defined as:

$$(V_{i,j})_k = \frac{1}{(2k+1)^2} \sum_{r=-k}^{k} \sum_{s=-k}^{k} (x_{i+r,j+s} - m)^2 \quad (53)$$

- On the basis of the obtained auto-covariance (52) and variance (53) values in the

Table 3. Results of noise variance estimation in the case of finding of an NxN area using an autocorrelation coefficient

σ^2	29x29	39x39
0	8.57	7.48
25	29.30	33.46
100	97.56	104.50
225	197.25	203.88

Figure 5. Areas found, from which noise level was estimated, with sizes: I-29x29, II-39x39

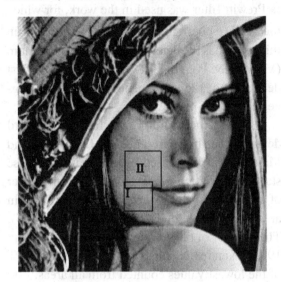

analyzed area, the correlation coefficients ρ can be determined for each layer:

$$(\rho_{i,j})_k = \frac{(C_{i,j})_k}{(V_{i,j})_k} \qquad (54)$$

- Next, the obtained values of correlation coefficients are averaged for individual points relative to successive layers. In this way, values of correlation coefficients R_k are obtained in the given area for a given layer:

$$R_k = \frac{1}{N^2} \sum_{r=1}^{N} \sum_{s=1}^{N} (\rho_{r,s})_k \qquad (55)$$

- The selection of the area for estimation of the noise level is done based on the lowest value of the slope of the straight line determined from linear approximation (linear regression) carried out for correlation coefficients R_k.

The above method of searching for the appropriate area for estimation of the level of noise variance was subjected to verification during the experiment. The level of noise with a normal distribution (Gaussian) was estimated on the Lena test image. Tests were conducted for varying noise levels. An area of various sizes was searched for. The size of the area was accepted to be relatively large (29x29 and above). The results of estimation have been presented in Table 3; however, the areas found by means of the procedure described above have been marked in Figure 5.

The results seem to be promising. Only for a lack of noise can an incorrect result be stated, because the noise estimator shows a certain noise value. However, in the case of such a value, it can be said that the noise level is insignificant, which is sufficient for interpreting the noise properties of the image. Practically, such a noise level corresponds to an SNR coefficient value of SNR>50dB (SNR – Signal to Noise Ratio), which signifies that it is best not to use any filtering because the filtering process itself will cause greater deformation of the structure of the processed image.

The presented methods of noise estimation are very simple to realize and give satisfactory results, as shown by the results of simulations.

Box 10.

$$(C_{i,j})_k = \frac{1}{8k}(x_{i,j} - m)\left[\sum_{r=-k}^{k}(x_{i-k,j+r} - m) + \sum_{r=-k}^{k}(x_{i+k,j+r} - m) + \ldots \right.$$

$$\left. \ldots + \sum_{r=-k+1}^{k-1}(x_{i+r,j-k} - m) + \sum_{r=-k+1}^{k-1}(x_{i+r,j+k} - m) \right] \qquad (52)$$

EXPERIMENT AND DISCUSSION OF RESULTS

Noise level estimation methods can be divided into two groups. The first is based on filtering of the disrupted image with assumption that the filtered signal is the original. This leads to the determination of noise standard deviation σ_n on the basis of the difference between the disrupted signal and the filtered signal. The second method of noise estimation is based on determining σ_n from the variance of a „flat" area found in the disrupted image. These methods are based on the assumption, that noise is the deciding factor for pixel values in the determined area (Olsen, 1993).

Several noise level estimation methods including the above mentioned groups have been presented in publications (Brancho & Sanderson, 1985; Lee, 1981; Mastin, 1985; Meer, et al., 1994; Olsen, 1993; Vorhees & Poggio, 1987).

Averaging and median filters with a 3x3 window are often used in the first group. However, using an averaging filter leads to blurring of edges and a greater change in the image structure, which, in consequence, gives unsatisfactory results. That is why only certain filtered pixels are selected for estimation (Olsen, 1993). The selection can be made based on the value of the gradient of individual pixels, which is less than the set threshold value. In practice, it is difficult to determine the threshold value, which is why only a certain percentage of pixels with the lowest values are selected—10% in the case of this study. In the case where a median filter is used, this is not necessary, because it interferes with the image structure in a lesser degree.

A different approach to the problem of estimation was presented by Bracho and Sanderson (1985). Assuming that the noise has a Gaussian distribution, it can be observed that the amplitude of the gradient has a Rayleigh distribution. The maximum of the histogram corresponds to the value of the estimated noise standard distribution σ_n, scaled by a constant dependent on the applied method of gradient calculation, $\|\nabla\| = \sqrt{f_x^2 + f_y^2}$ (a Prewitt filter was used in the work, for which the scaling coefficient is $-\sqrt{6}$). A problem in this method is the determination of the maximum (Vorhees & Poggio, 1987), the use of the first derivative of the smoothened histogram is suggested for this purpose.

Another method for determining standard deviation has been presented by Lee (1981) and Mastin (1985). It is based on determining the noise standard deviation by averaging a certain number of the lowest values of variances calculated in areas with a size of 7x7 pixels (Mastin, 1985). The number of blocks suggested in work (Olsen, 1993) taken for calculation of variances was 10% of the lowest values counted from all areas.

The method presented by Meer et al. (1994) is also based on determining the noise level through calculation of variances in individual blocks with dimensions of $2^l x 2^l$ for $l=1,2,...,n$, where $2^n x 2^n$ - image size. Next, the four lowest variance estimates for a given level (up to $l \leq n-1$ is possible) are selected, with the correct variance value being selected on the basis of statistical inference based on Dixon's test. More details on this method can be found in Meer et al. (1995).

The proposed method for estimating the level of noise (47) has been subjected to verification during the experiment. A level of noise with a normal distribution (Gaussian) was estimated, in the range of change $\sigma_n = 1 \div 20$, on test images lena, clown, birdge_sf. An area for estimation of various sizes was searched for.

The results of estimation, in the form of error calculated from average values of standard deviations obtained during individual tests were presented in Table 4.

The described method was compared to the methods shown in Olsen (1993). Individual methods have been marked as:

- **Av1:** Estimation using an averaging filter, where σ_n was calculated from all pixels.
- **Med.:** As above, only using a median filter.

- **Av2:** Similar to Av1 only 10% of pixels with the lowest values were used for estimation.
- **Block:** σn calculated from averaging 10% of the variances calculated in areas of 7x7 pixels (Lee, 1981; Mastin, 1985).
- **Gradient:** On the basis of the gradient histogram (Dixon, 1988; Vorhees & Poggio, 1987).
- **Pyramid:** Through calculation of variances in individual blocks, with sizes $2^l x 2^l$, for $l=1,2,...,n$, where $2^n x 2^n$: Image size (Meer, et al., 1994).
- **Cov:** On the basis of the method described by formula (47).

The obtained test results shown in Table 4 show, that the most accurate results—16 (the lowest errors) were obtained using the method designated as Av2, especially for $\sigma_n<8$. However, the method shown by (50-53) placed second, mainly for a *29x29* area. For a *9x9* area, no advantage of this method over the others was visible. The conclusion that comes to mind is such, that an even bigger area should be selected for estimation. However, there exists a risk that in the case of a lack of a flat area of such a large size, there will be many small details in it, which may worsen the results.

REFERENCES

Amer, A. (2001). *Object and event extraction for video processing and representation in online video applications*. (Doctoral Dissertation). Universite du Quebec. Quebec, Canada. Retrieved from http://users.encs.concordia.ca/~amer/paper/phd/amer_diss.pdf

Amer, A., Mitiche, A., & Dubois, E. (2002). Reliable and fast structure-oriented video noise estimation. In *Proceedings IEEE International Conference on Image Processing, ICIP 2002*, (vol. 1, pp. 840-843). Rochester, NY: IEEE Press.

Amer, A., & Schroder, H. (1996). A new video noise reduction algorithm using spatial sub-bands. In *Proceedings of IEEE International Conference on Electronics, Circuits and Systems*, (vol. 1, pp. 45-48). Rodos, Greece: IEEE Press.

Blume, H., Amer, A., & Schroder, H. (1997). Vector-based postprocessing of MPEG-2 signals for digital TV-receivers. In *Proceedings of Conference Visual Communications and Image Processing*, (vol. 3024, pp. 1176-1187). San Jose, CA: IEEE.

Brancho, R., & Sanderson, A. C. (1985). Segmentation of images based on intensity gradient information. In *Proceedings of Conference on Computer Vision and Pattern Recognition, CVPR 1985*, (pp. 341-347). San Francisco, CA: CVPR.

Canny, J. (1986). A computational approach to edge detection. *IEEE Transactions on Pattern Analysis and Machine Intelligence, 9*, 679–698. doi:10.1109/TPAMI.1986.4767851

Comer, B. R., Narayanan, R. M., & Reichenbach, S. E. (2003). Noise estimation in remote sensing imagery using data masking. *International Journal of Remote Sensing, 24*(4), 689–702. doi:10.1080/014311602210164271

de Haan, G., Kwaaitaal-Spassova, T. G., & Ojo, O. A. (1996). Automatic 2-D and 3-D noise filtering for high-quality television receivers. In *Proceedings of International Workshop on Signal Processing and HDTV*, (vol. 6, pp. 221-230) Turin, Italy: IEEE.

de Hann, G., Kwaaitaal-Spassova, T. G., Larragy, M. M., Ojo, O. A., & Schutten, R. J. (1998). Television noise reduction IC. *IEEE Transactions on Consumer Electronics, 44*(1), 143–154. doi:10.1109/30.663741

Dixon, R. L. (Ed.). (1988). *MRI: Acceptance testing and quality control - The role of the clinical medical physicist*. Madison, WI: Medical Physics Publishing Corporation.

Frieden, B. R. (1983). *Probability, statistical optics and data testing.* Berlin, Germany: Springer-Verlag. doi:10.1007/978-3-642-96732-0

Gamier, S. J., & Bilbro, G. L. (1995). Magnetic resonance image restoration. *Journal of Mathematical Imaging and Vision, 5,* 7–19. doi:10.1007/BF01250250

Gerig, G., Kubler, O., Kikinis, R., & Jolesz, F. A. (1992). Nonlinear anisotropic filtering of MRI data. *IEEE Transactions on Medical Imaging, 11*(2), 221–232. doi:10.1109/42.141646

Grafe, T., & Scheffler, G. (1988). Inter field noise and cross color reduction IC for flicker free TV receivers. *IEEE Transactions on Consumer Electronics, 34*(3), 402–408. doi:10.1109/30.20134

Gudbjartsson, H., & Patz, S. (1995). The Rician distribution of noisy MRI data. *Magnetic Resonance in Medicine, 34,* 910–914. doi:10.1002/mrm.1910340618

Henkelman, R. M. (1985). Measurement of signal intensities in the presence of noise in MR images. *Medical Physics, 2*(2), 232–233. doi:10.1118/1.595711

Holdschneider, M., Kronland-Martinet, R., Morlet, J., & Tchamitchian, P. (1989). A real-time algorithm for signal analysis with the help of the wavelet transform. In Combes, J.-M., Grossmann, A., & Tchamitchian, P. (Eds.), *Time-Frequency Methods and Phase Space* (pp. 286–297). Berlin, Germany: Springer-Verlag.

Immerkaer, J. (1996). Fast noise variance estimation. *Computer Vision and Image Understanding, 64*(2), 300–302. doi:10.1006/cviu.1996.0060

Jensen, J. R. (1996). *Introductory digital image processing: A remote sensing perspective* (2nd ed.). Englewood Cliffs, NJ: Prentice Hall. doi:10.1080/10106048709354084

Jostschulte, K., & Amer, A. (1998). A new cascaded spatio-temporal noise reduction scheme for interlaced video. *IEEE Proceedings International Conference on Image Processing, ICIP 1998,* (vol. 2, pp. 493-497). Chicago, IL: IEEE Press.

Kaufman, L., Kramer, D. M., Crooks, L. E., & Ortendahl, D. A. (1989). Measuring signal-to-noise ratios in MR imaging. *Radiology, 173,* 265–267.

Lee, J. S. (1980). Digital image enhancement and noise filtering by use of local statistics. *IEEE Transactions on Pattern Analysis and Machine Intelligence, 2*(2), 165–168. doi:10.1109/TPAMI.1980.4766994

Lee, J. S. (1981). Refined filtering of image noise using local statistics. *Computer Vision Graphics and Image Processing, 15,* 380–389. doi:10.1016/S0146-664X(81)80018-4

Mastin, G. A. (1985). Adaptive filters for digital noise smoothing: An evaluation. *Computer Vision Graphics and Image Processing, 31,* 103–121. doi:10.1016/S0734-189X(85)80078-5

Meer, P., Jolion, J., & Rosenfeld, A. (1990). A fast parallel algorithm for blind estimation of noise variance. *IEEE Transactions on Pattern Analysis and Machine Intelligence, 12*(2), 216–223. doi:10.1109/34.44408

Meer, P., Park, R., & Cho, K. (1994). Multiresolution adaptive image smoothing. *Graphical Models and Image Processing, 44,* 140–148. doi:10.1006/cgip.1994.1013

Mencattini, A., Salmeri, M., Bertazzoni, S., & Salsano, A. (2003). Noise variance estimation in digital images using iterative fuzzy procedure. *WSEAS Transactions on Systems, 4*(2), 1048–1056.

Mitiche, A. (1994). *Computational analysis of visual motion.* New York, NY: Plenum Press.

Murphy, B. W., Carson, P. L., Ellis, J. H., Zhang, Y. T., Hyde, R. J., & Chenevert, T. L. (1993). Signal-to-noise measures for magnetic resonance imagers. *Magnetic Resonance Imaging, 11*, 425–428. doi:10.1016/0730-725X(93)90076-P

Murtagh, F., & Starck, J. L. (1998). Noise detection and filtering using multiresolution transform methods. *Astronomical Data Analysis Software and Systems, 145*, 449–456.

Murtagh, F., & Starck, J. L. (2000). Image processing through multiscale analysis and measurement noise modeling. *Statistics and Computing Journal, 10*, 95–103. doi:10.1023/A:1008938224840

Murtagh, F., Starck, J. L., & Bijaoui, A. (1995). Multiresolution in astronomical image processing: A general framework. *International Journal of Imaging Systems and Technology, 6*, 332–338. doi:10.1002/ima.1850060406

Narayanan, R., Ponnappan, S., & Reichenbach, S. (2001). Effects of uncorrelated and correlated noise on image information content. In *Proceedings of the IEEE International Geoscience and Remote Sensing Symposium*, (pp. 1898-1900). IEEE Press.

Narayanan, R., Ponnappan, S., & Reichenbach, S. (2003). Effects of noise on information content of remote sensing images. *Geocarto International, 18*(2), 15–26. doi:10.1080/10106040308542269

Olsen, S. I. (1993). Estimation of noise in images: An evaluation. *Graphical Models and Image Processing, 55*(4), 319–323. doi:10.1006/cgip.1993.1022

Papoulis, A. (1984). *Probability, random variables and stochastic processes* (2nd ed.). Tokyo, Japan: McGraw-Hill.

Ranham, M. R., & Katsaggelos, A. K. (1997). Digital image restoration. *IEEE Signal Processing Magazine, 3*, 24–41.

Rank, M., Lendl, M., & Unbehauen, R. (1999). Estimation of image noise variance. *Proceedings of the IEEE Visual Signal Processing, 146*, 80–84. doi:10.1049/ip-vis:19990238

Rosin, P. (1997). Edges: Saliency measures and automatic thresholding. *Machine Vision and Applications, 9*, 139–159. doi:10.1007/s001380050036

Rosin, P. (1998). Thresholding for change detection. In *Proceedings of International Conference on Computer Vision*, (pp. 274-279). Bombay, India: IEEE Press.

Salmeri, M., Mencattini, A., Ricci, E., & Salsano, A. (2001). Noise estimation in digital images using fuzzy processing. In *Proceedings of the IEEE International Conference on Image Processing, ICIP 2001*. Thessaloniki, Greece: IEEE Press.

Schmitt, F., Mignotte, M., Collet, C., & Thourel, P. (1996). Estimation of noise parameters on sonar images. In *Proceedings of the Conference Signal and Image Processing, SPIE 1996*, (vol. 2823, pp. 1-12). Denver, CO: SPIE.

Sijbers, J., den Dekker, A. J., Raman, E., & Van Dyck, D. (1999). Parameter estimation for magnitude MR images. *International Journal of Imaging Systems and Technology, 10*, 109–114. doi:10.1002/(SICI)1098-1098(1999)10:2<109::AID-IMA2>3.0.CO;2-R

Sijbers, J., den Dekker, A. J., Van Dyck, D., & Raman, E. (1998). Estimation of signal and noise from rician distributed data. In *Proceedings of the International Conference on Signal Processing and Communications*, (pp. 140-142). Gran Canada, Spain: IEEE.

Spann, M., & Wilson, R. (1985). A quad-tree approach to image segmentation which combines statistical and spatial information. *Pattern Recognition, 18*(3/4), 257–269. doi:10.1016/0031-3203(85)90051-2

Starck, J. L., & Murtagh, F. (1998). Automatic noise estimation from the multiresolution support. *Publications of the Astronomical Society of the Pacific, 110*, 193–199. doi:10.1086/316124

Starck, J. L., & Murtagh, F. (2001). Astronomical image and signal processing. *IEEE Signal Processing Magazine, 18*(2), 30–40. doi:10.1109/79.916319

Starck, J. L., Murtagh, F., & Bijaoui, A. (1995). Multiresolution support applied to image filtering and restoration. *Graphical Models and Image Processing, 57*, 420–431. doi:10.1006/gmip.1995.1036

Taylor, J. R. (1982). *An introduction to error analysis: The study of uncertainties in physical measurements.* New York, NY: University Science Books. doi:10.1063/1.882103

Vorhees, H., & Poggio, T. (1987). Detecting textons and texture boundaries in natural images. In *Proceedings of the First International Conference on Computer Vision*, (pp. 250-258). London, UK: IEEE.

Wang, Y., & Lei, T. (1994). Statistical analysis of MR imaging and its applications in image modeling. In *Proceedings of the IEEE International Conference on Image Processing and Neural Networks,* (vol. 1, pp. 866-870). IEEE Press.

Wang, Y., Lei, T., Sewchand, W., & Mun, S. K. (1996). MR imaging statistic and 1st application in image modeling. In *Proceedings SPIE Conference on Medical Imaging*, (vol. 2708, pp. 706-717). Newport Beach, CA: SPIE.

Yang, G. Z., Burger, P., Firrnin, D. N., & Undrwood, S. R. (1995). Structure adaptive anisotropic filtering for magnetic resonance image enhancement. [CAIP.]. *Proceedings of Computer Analysis of Images and Patterns, CAIP, 1995*, 384–391. doi:10.1007/3-540-60268-2_320

Zhu, S., & Yuille, A. (1996). Region competition: Unifying snakes, region growing, and Bayes MDL for multiband image segmentation. *IEEE Transactions on Pattern Analysis and Machine Intelligence, 18*, 884–900. doi:10.1109/34.537343

ADDITIONAL READING

Aja-Fernández, S., Vegas-Sánchez-Ferrero, G., Martín-Fernández, M., & Alberola-López, C. (2009). Automatic noise estimation in images using local statistics: Additive and multiplicative cases. *Image and Vision Computing, 27*(6), 756–770. doi:10.1016/j.imavis.2008.08.002

Hemami, S., & Reibman, A. (2010). No-reference image and video quality estimation: Applications and human-motivated design. *Signal Processing Image Communication, 25*(7), 469–481. doi:10.1016/j.image.2010.05.009

Juang, L.-H., & Wu, M.-N. (2010). Image noise reduction using Wiener filtering with pseudoinverse. *Measurement, 43*(10), 1649–1655. doi:10.1016/j.measurement.2010.09.021

Lee Hurt, S., & Rosenfeld, A. (1984). Noise reduction in three-dimensional digital images. *Pattern Recognition, 17*(4), 407–421. doi:10.1016/0031-3203(84)90069-4

Wong, A., & Wang, X. Y. (2012). Monte Carlo cluster refinement for noise robust image segmentation. *Journal of Visual Communication and Image Representation, 23*(7), 984–994. doi:10.1016/j.jvcir.2012.06.006

Wu, W.-Y., Wang, M.-J. J., & Liu, C.-M. (1992). Performance evaluation of some noise reduction methods. *CVGIP: Graphical Models and Image Processing, 54*(2), 134–146. doi:10.1016/1049-9652(92)90061-2

Yang, S.-M., & Tai, S.-C. (2012). A design framework for hybrid approaches of image noise estimation and its application to noise reduction. *Journal of Visual Communication and Image Representation*, *23*(5), 812–826. doi:10.1016/j.jvcir.2012.04.007

KEY TERMS AND DEFINITIONS

Filter: It is a device or process that removes from a signal some unwanted component or feature.

Image Noise: It is random (not present in the object imaged) variation of brightness or color information in images. The standard model of amplifier noise is additive, Gaussian, independent at each pixel, and independent of the signal intensity.

Image Processing: It is any form of signal processing for which the input is an image. The output of image processing may be either an image or a set of characteristics or parameters related to the image. Most image-processing techniques involve treating the image as a two-dimensional signal and applying standard signal-processing techniques to it.

Noise: It is the colloquialism for recognized amounts of unexplained variation in a sample.

Noise Estimation: It is the process of finding an estimate, or approximation, noise parameters such as noise variance, standard deviation, distribution type, etc.

Noise Reduction Coefficient: It is a scalar quantity characterizing noise attenuation in the process of filtration.

Signal Processing: It is an area of systems engineering, electrical engineering and applied mathematics that deals with operations on or analysis of signals, in either discrete or continuous time.

Smoothing: In statistics and image processing, to smooth a data set is to create an approximating function that attempts to capture important patterns in the data, while leaving out noise or other fine-scale structures/rapid phenomena.

Chapter 18

Context–Aware Medical Image Retrieval for Improved Dementia Diagnosis

Melih Soydemir
Bahçeşehir University, Turkey

Devrim Unay
Bahçeşehir University, Turkey

ABSTRACT

Progress in medical imaging technology together with the increasing demand for confirming a diagnostic decision with objective, repeatable, and reliable measures for improved healthcare have multiplied the number of digital medical images that need to be processed, stored, managed, and searched. Comparison of multiple patients, their pathologies, and progresses by using image search systems may largely contribute to improved diagnosis and education of medical students and residents. Supporting image content information with contextual knowledge will lead to increased reliability, robustness, and accuracy in search results. To this end, the authors present an image search system that permits search by a multitude of image features (content), and demographics, patient's medical history, clinical data, and ontologies (context). Moreover, they validate the system's added value in dementia diagnosis via evaluations on publicly available image databases.

INTRODUCTION

The human brain, although protected by the skull, suspended in cerebrospinal fluid, and isolated from the bloodstream by the blood-brain barrier, has a delicate nature susceptible to damages and

diseases. Among the various disorders affecting the brain, dementias usually caused by a neuro-degenerative disease, such as Alzheimer's, are a devastating subgroup because they are increasingly prevalent and incurable (Williams, 2002).

For the clinical diagnosis of dementia, the practice guidelines recommend the usage of medical imaging together with cognitive and behavioral assessment, laboratory investigations and genetic

DOI: 10.4018/978-1-4666-2833-5.ch018

testing. The guidelines further advice that Magnetic Resonance (MR) imaging, when available, should be preferred over computed tomography or X-ray, because it provides greater contrast between different soft tissues of the body and uses no harmful ionizing radiation (Waldemar, et al., 2000).

As there exists high variation in the characteristics of different dementia subtypes, and their causes and progress are still not yet fully known, a promising way to improve their diagnosis is to compare multiple patients, their pathologies, and progresses by Content-Based Image Retrieval (CBIR) systems among large repositories of brain MR images.

CBIR in the multimedia domain is known to have a challenging nature (Lew, et al., 2006), while its application in the medical domain demands also consideration for domain or modality-specific differences and challenges, such as ever-increasing quantities of digital images used for diagnosis and therapy that need to be stored, managed, addressed, and searched, subtle focal differences that need to be detected for some pathologies, and image-related variations such as bias field and spatial misalignment of images that need to be handled properly.

Recent research in medical image retrieval has shown improved search capabilities when content-based search is combined with context information. Consequently, to assist the medical experts in the diagnosis of dementia, we are working towards a context-aware medical image retrieval system that allows improved case retrieval via augmenting image-based features with contextual information captured by ontologies. The retrieval system permits the expert to select a query from previously diagnosed cases, to define the search space, to specify the features to be extracted, and to view images of the search result along with the diagnosis-relevant metadata.

In the proposed search system, search space can be defined by restricting the search in two ways: 1) specifying a Volume-Of-Interest (VOI) via an atlas registered on the query image, through a structured list of anatomical entities (anatomical ontology), or by a manual delineation, and 2) confining the search to a disease group selected from a structured list (disease ontology). Moreover, a combination of intensity, texture and shape features from the VOI can be extracted and exploited for retrieval.

In this work, we will introduce our search system from both functional and architectural aspects and validate its added value in dementia diagnosis via evaluations on publicly available image databases.

The chapter is organized as follows. First, we give a general background on image-based diagnosis via search and retrieval, followed by a concise literature review and our contributions. Then we present a clinically-relevant and robust brain MR feature—specifically lateral ventricle shape change—used in this work and explain how it is employed in a search scheme for population analysis. In the following section, the architecture of the system is detailed with emphasis on the system sub-components such as query formation level and the relevance feedback scheme. The proposed system is evaluated in a neuroimaging search and retrieval scenario, where the task is to differentiate sub-groups of neurodegenerative diseases, specifically dementia and Alzheimer's. Accordingly, subjects and imaging data used in this study are introduced, and the corresponding experimental results are presented. The chapter is concluded with future research directions in medical image search and retrieval, and conclusion.

BACKGROUND

In the medical domain, a diagnosis by a specialist often requires a visit to a radiology department to obtain various images that highlight the suspected pathology. Despite the high resolution of the acquired images, image based diagnosis often utilizes a considerable amount of qualitative measures. To improve the diagnosis and efficiency, the research in medical image analysis has focused on the computation of quantitative measures by automating some of the error-prone and excruciatingly time-consuming tasks, such as segmentation of a structure. A less emphasized approach for improving diagnosis has been comparison of multiple patients, their pathologies, and progresses by search and retrieval systems. This should especially improve the diagnosis of diseases whose causes and progress have not yet been completely unraveled, and diseases with high prevalence. The area of neurology can greatly benefit from such a methodology because diagnosis of neurodegenerative diseases from one patient data has limitations.

Content-based search and retrieval of medical images is a difficult task, due to the presence of domain or even modality-specific differences and challenges. The captured images are usually grayscale lacking the valuable color information. In MR, contrary to some medical modalities, intensities vary due to factors like imperfect magnetic field, scanning parameters, etc. Furthermore, the characteristics of relevant and irrelevant segments of the image data are very close to each other for the databases that are specific to a modality and an organ, such as brain MR databases. Researchers working on content-based search and retrieval of MR images mainly focused on the shape information. For example, Hsu et al. (1996) described brain ventricles by geometric features, while Huang et al. (2005) combined geometric features and Fourier descriptors to represent brain images of pediatric patients. In order to circumvent MR-related problems, Unay et al. (2010) proposed a spatially enhanced texture-based technique for automated retrieval of the relevant slice from a brain MR volume in response to a given sample slice. Furthermore, researchers also worked on content-based retrieval of images from multiple modalities (Felipe, et al., 2003).

Indexing of digital images in the healthcare centers is conventionally done using textual information or keywords. Accordingly, (con)text-based search and retrieval of medical images has long been a popular research area (Müller, et al., 2004). In such systems, images in the database are matched with representative keywords and search is done by matching at this keyword-space. More complicated systems try to also use semantic relations between representative keywords by exploiting context-free grammars (Ogiela & Tadeusiewicz, 2001), standardized dictionaries (Jaulent, et al., 2000), and specialized image representation languages (Traina, et al., 1997). For example, Hersh et al. (2001) converted the text from radiology reports to the concepts defined in Unified Medical Language System meta-thesaurus dictionary (UMLS) and performed image search with these concepts. More recent studies focus on using ontologies—representations of concepts and the between-concept relationships in a domain using formal logic-based languages, such as hierarchically structured standard dictionaries—for medical image indexing and retrieval (Kulikowski, 2006).

Even though context-based image search and retrieval supports strong semantic description of image content, such methods will probably be insufficient in describing visual features of images. Furthermore, in medical information retrieval there are new trends emerging, such as combining structured reporting (via ontologies) with image annotation methods, and exploiting this semantic image information to improve interpretation performance of radiologists (Rubin & Napel, 2010). Accordingly, recent research attempts focus on combining content and context information to provide a stronger solution to the image search and retrieval problem. For example, in the work of Lacoste et al. (2007) support vector machines are trained using concepts from UMLS meta-thesaurus dictionary with colour, texture and shape features to perform image search from databases of different modalities. Another study (Hu, et al., 2003) presents a medical image annotation system for breast cancer diagnosis where texture and shape features from mammography images are combined with ontology-based representations.

As presented, conventional solutions for medical image search and retrieval problem focus on image content or context information separately. Because integrating these two pieces of information for image description will lead to more powerful search systems, recent research efforts focus on this task. However due to not only modality-specific difficulties in medical image analysis and large diversity in application areas, but also difficulty in customizing wide context knowledge to the application area, such efforts have been limited and specific to some modalities and organs (e.g. ontology-based search and retrieval of chest X-ray images (Wei & Bamaghi, 2007)). To the best of our knowledge, there is no study in the literature that performs content and context based indexing and retrieval of brain MR images. The study in this proposal aims at integrating content and context information for brain MR image search and retrieval, which simultaneously processes different MR contrast images, permits disease-specific search, and is specialized to neurology-radiology area. With these features, the proposal targets in fulfilling the related gap in the literature.

SEARCH AND RETRIEVAL OF MEDICAL IMAGES

One of the crucial steps of an efficient CBIR system is the extraction of the most descriptive/discriminative image features. In diagnosing neurodegenerative diseases, medical experts commonly use visual or qualitative interpretation of images, which may suffer from large inter-rater variability. To overcome this subjectivity problem, the research focus has shifted to quantitative approaches such as voxel-based morphometry (Thompson, et al., 2007) and more recently machine learning methods (Lao, et al., 2004; Kloppel, et al., 2008). Besides, in order to provide

diagnostic support to the clinician by search and retrieval of medical images, researchers mainly exploited shape information (Hsu, et al., 1996; Huang, et al., 2005). However, MR and disease specific descriptors have been less investigated. Accordingly, in this work we have designed and implemented a clinically relevant and robust MR feature describing the shape change (deformation) of lateral ventricles.

Nevertheless, the proposed search system is specifically designed to accommodate any image feature describing intensity, local structure, or edge information, and furthermore these features can be linked to any spatial region, such as the whole organ and structures in the organ-of-interest.

Content-Based Image Description

Clinically Relevant Features: Splines-Based Ventricular Shape Representation

The presented search tool allows the user the flexibility to define the search space as an atlas-defined region, an arbitrary shaped volume-of-interest, or the whole image. As an example, in this study manual delineations of lateral ventricles are used (see Figure 1). An expert marks some boundary points from the structure under investigation (lateral ventricles), and the tool provides a delineation of the structure via spline-based curve fitting on the marked points.

In addition to shape features as explained above, the presented search system also permits searching images using intensity or texture features, which are robust to variations in MR. For example, one can employ Local Binary Patterns (LBP) (Ojala, et al., 2002), which is a grayscale invariant local texture descriptor with low computational complexity and robustness to intensity variations of MR, such as intensity inhomogeneity or bias field (Unay, et al., 2007). Note that, only the results with the presented shape feature are included in the current study.

Figure 1. Lateral ventricle of a subject manually delineated on an axial image by defining control points (left) and automatically connecting those points using splines (middle), and two such contours spatially registered and displayed together

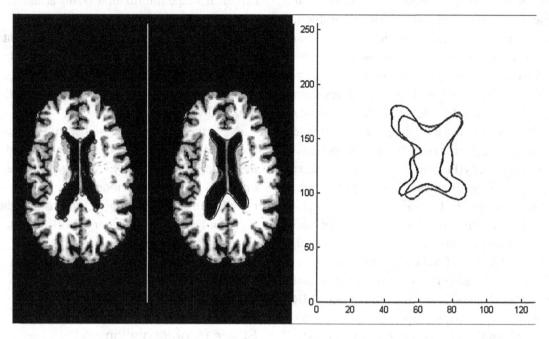

Quantification of Intrasubject Ventricular Shape Change

In order to overcome variations between the scans of the same subject (intra-subject variation) as well as the scans of different subjects (inter-subject variation), global alignment via affine registration with 12 degrees-of-freedom is employed. Moreover, for quantifying intra-subject ventricular shape change, lateral ventricle segmentations are further aligned by (1) matching their geometric centers (translation invariance), and (2) computing their major/minor axes using singular value decomposition and rotating them such that their major axes are oriented in the vertical direction.

Figure 1 displays a lateral ventricle of a subject manually delineated on an axial slice using splines, and two such delineations spatially registered and superimposed (to be used in computing similarity of the two shapes).

Similarity between any lateral ventricle boundary pair is computed in the polar space (r,θ) with respect to the geometric center. The polar space is divided into N equal angles, and for each angle range the Euclidean distance between the closest two points (one from each boundary) falling in that range is computed. Similarity is then defined as the average distance over all possible angles. In this work N is taken as 180, because higher values (over-quantization) resulted in increased computation but no improvement in accuracy while lower values (under-quantization) degraded the accuracy.

Population Analysis Based on Ventricular Shape Change

Abnormal changes in an anatomical structure, such as aberrant volume loss or shape deformation by time, may be related to the onset or progression of diseases. Hence, the search tool is designed to permit detecting such changes. Based on the fact that change in lateral ventricles is observed to be greater in dementia cases than in healthy controls, this work focuses on shape deformation of lateral ventricles to discriminate different disease states. To this end, lateral ventricle similarities of subjects from different disease states (quantified in the previous step) are normalized by time (between the two scanning sessions) and compared.

Exploiting Contextual Information

Combining contextual knowledge like demographics or clinical data with content information, such as neuroimaging findings, will lead to more accurate or robust search of relevant cases and will eventually improve diagnosis or therapy planning. For this purpose, the proposed search system permits exploiting contextual information by allowing search with respect to demographics (e.g. search by age, gender, socio-economic status, etc.), patient's medical history (e.g. search by family history of the disease at hand), clinical data (e.g. search by blood test results or neuropsychological test results), ontologies (e.g. search via input from a disease or anatomical ontology), or a combination of these. Two open-source ontologies are incorporated in the current system: the Human Disease ontology and the Foundational Model of Anatomy ontology. The above-mentioned contextual information is encoded in the proposed search system using Extensible Markup Language (XML) format so that it is platform independent, portable, and extendible.

Efficient Relevant Cases Search

An effective relevant cases search requires taking full advantage of both content and context information extracted from the data. To this purpose, one can fuse information of the two either at the feature-level or at the similarity-level. In the former fusion scheme a new feature is constructed from individual features (extracted from both content and context information) and used for search, while in the latter search by each individual feature is realized and the resulting similarities are fused (e.g. by weighted averaging) to come up with a single retrieval result. In the proposed system, similarity-level fusion scheme is adopted due to its implementation simplicity.

System Architecture

The presented search system employs the *query-by-example* paradigm, where the user defines the query in the same representation as the database items. Figure 2 shows the architecture of our search system that consists of query formation, feedback, search engine, and data access layers. First, via the graphical user interface the user defines the query (e.g. image of a subject under investigation) and the search parameters (such as the search space, and features to be extracted) in the query formation layer. This query information together with the related relevance feedback input (if any) is employed by the system to construct a search action. Then the search engine accesses the databases, fetches related content and context information of each case in the database, computes pair wise similarities between each case and the query, sorts the cases in the database by their similarity to the query, and presents the retrieval results to the user. Furthermore, following the retrieval result the user can input his/her feedback to initiate another round of search and obtain a new, improved retrieval result.

The entire search system is implemented as an in-house built, stand-alone software application written in Matlab and C. In the current implementation content- and context-based representations of the database items (feature extraction) are carried out offline, whereas search action and display of results are realized on-the-fly (within a few seconds with the current un-optimized software). The aforementioned representations are stored (indexed) in a relational database using the open-source MySQL language, which provide multi-user access to multiple databases.

For an effective search system, in the current work ontologies are pruned to include only the

Figure 2. Architecture of the proposed search system. Meta data in the data access layer refers to contextual information such as demographics, medical history, and clinical tests

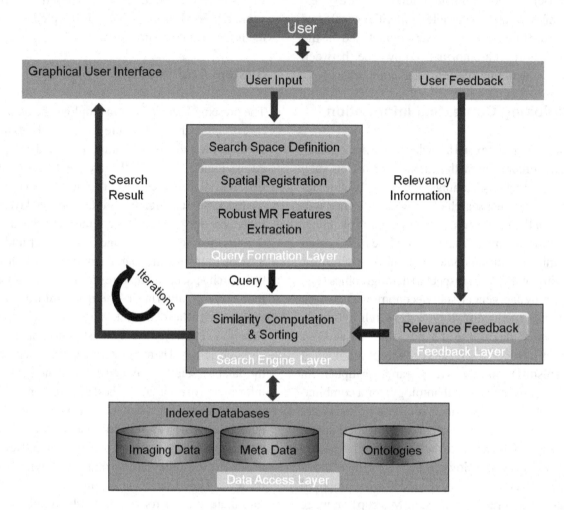

data of the anatomical structures (those of the brain) and the (brain) diseases under investigation, and the cases in the database are indexed offline for each feature (lateral ventricle shape change). Note that, the system is built such that it can easily be extended to include anatomical structures, diseases, and features other than stated above.

User Interaction

The interaction between the user and a search and retrieval system—e.g. via relevance feedback mechanism—is one of the main appeals of such systems. In the proposed system, the medical

expert's domain knowledge is integrated to the system through explicit relevance feedback. The expert (user) marks the relevant and the irrelevant cases among the retrieval returns. By using this relevance information the system updates the query, and then retrieves a new, improved set of cases using the updated query (Q_u) based on the classic Rocchio algorithm (Manning, et al., 2008):

$$Q_u = \alpha Q_o + \beta \frac{1}{|D_r|} \sum_{\vec{d}_i \in D_r} \vec{d}_i - \gamma \frac{1}{|D_{ir}|} \sum_{\vec{d}_j \in D_{ir}} \vec{d}_j$$

where Q_o is the original query vector, D_r and D_{ir} are the set of relevant and irrelevant markings respectively, and α, β, and γ are the corresponding weights.

Experimental Results

Subjects and Imaging Data

Two longitudinal MR datasets are used to evaluate the predictive power of the proposed search system. First one is a subset of the Open Access Series of Imaging Studies (OASIS, http://www. oasis-brains.org/) longitudinal repository, consisting of the neuroimaging data from thirty age- and gender-matched individuals: ten healthy controls, ten converters, and ten demented cases. In OASIS, each subject is scanned on two or more visits, separated by at least one year. For each scanning session, 3-4 individual T1-weighted MR images are obtained at 1.5 Tesla signal strength with 1mm x 1mm x 1.25mm voxel size, 256 axial slices, and 128x256 matrix dimensions. Note that, for this dataset the multiple within-session acquisitions are motion-corrected and averaged using the FreeSurfer v4.5.0 software package (Dale, et al., 1999) to create a single image with high contrast-to-noise ratio for each session (Unay, 2010).

The second dataset consists of the neuroimaging data of thirty age- and gender-matched individuals (ten healthy controls, ten mild cognitive impairment cases, and ten Alzheimer cases) from the Alzheimer Disease Neuroimaging Initiative (ADNI, http://adni.loni.ucla.edu/) cohort. Regarding the ADNI subset used in this study, each subject data contains a baseline scan and three follow-up scans at 6, 12, and 24 months after the baseline. T1-weighted MR data obtained at 1.5 Tesla signal strength with 1.2mm x 1.25mm x 1.25mm voxel size, 192 axial slices, and 160 x 192 matrix dimensions are used.

Table 1 displays the demographics of the subjects.

Results

First, the predictive power of ventricular shape change in identifying dementia is explored on the OASIS dataset. Figure 3 displays the results achieved with and without time-correction as box

Table 1. Demographic information on the database. Neuropsychological test scores reported correspond to the evaluations at the first visit.

OASIS subset			
	Control	**Converted**	**Demented**
Sex (F/M)	5/5	5/5	5/5
Age: mean (range)	77.1 (60-97)	79.8 (65-92)	76.3 (61-98)
MMSE score: mean (range)	29.2 (26-30)	28.7 (24-30)	24.2 (0-30)
CDR score: mean (range)	0.01 (0-0.5)	0.26 (0-0.5)	0.67 (0.5-2.0)
ADNI subset			
	Control	MCI	AD
Sex (F/M)	5/5	5/5	5/5
Age: mean (range)	79.9 (77-82)	79.5 (77-82)	79.6 (78-80)
MMSE score: mean (range)	28.8 (26-30)	27.5 (24-30)	23.9 (21-27)
CDR score: mean (range)	0 (0)	0.5 (0.5)	0.8 (0.5-1.0)

F=Female; M=Male; MMSE=Folstein Mini Mental State Examination; CDR=Clinical Dementia Rating; MCI=Mild Cognitive Impairment; AD=Alzheimer's Disease

Figure 3. Graphical representations of lateral ventricle shape similarity among each patient group in the OASIS subset. The graphic below corresponds to the time-corrected results

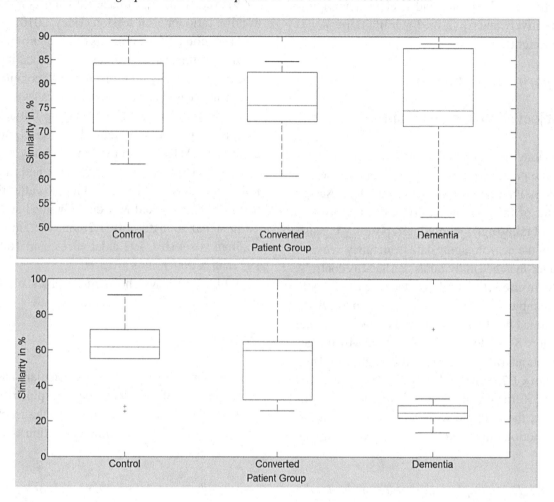

plots. As observed, time-correction (normalizing shape change by elapsed time between two scanning sessions) unveils that the distribution of demented cases is distinct from those of converters and healthy controls. Whereas the distributions are largely overlapping when time-correction is not employed. Furthermore, lateral ventricle similarities decrease with dementia severity (from healthy controls to demented cases), which is consistent with the fact that as severity rises affected parts of the brain are increasingly altered (relative to healthy aging).

The same experiment was repeated on a subset of the ADNI database. In contrast to the OASIS dataset, the time differences between scanning sessions in ADNI are controlled. The ADNI subset employed in this study includes baseline and follow-up (three sessions at 6, 12, and 24 months after the baseline) MR scans. As the time differences in this dataset are well structured, time-correction was not necessary. Figure 4 shows the results achieved for each time difference data. Regardless of the time difference, lateral ventricle shape change alone does not qualify as a predictive measure for differentiating AD, MCI, and control groups, which can be attributed to the complexity of the diagnosis task at hand.

Figure 4. Graphical representations of lateral ventricle shape similarity among each patient group in the ADNI subset for time lapses of 6 months (top), 12 months (center), and 24 months (bottom)

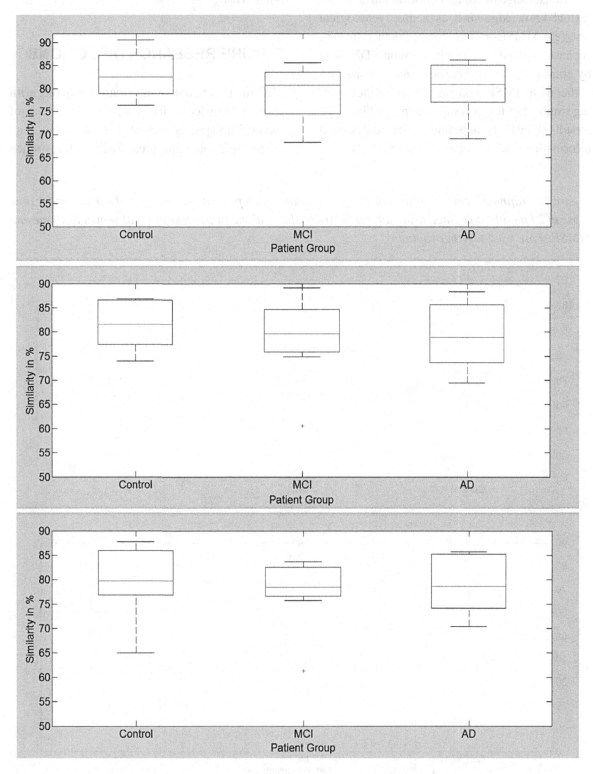

For improved diagnosis, one can go for a combined usage of content information and contextual knowledge such as neuropsychological tests (e.g. MMSE or CDR scores). For this purpose, we have realized a preliminary test on ADNI data by employing lateral ventricle shape change together with MMSE scores and observed the resulting similarities for each sub-group (see Figure 5). Results show that combining content and context information leads to better (but not perfect) separation of the sub-groups than its content-only counterpart.

FUTURE RESEARCH DIRECTIONS

Making use of content information together with context knowledge for search and retrieval of medical images, as presented in this study, is an emerging application area. Such systems when

Figure 5. Graphical representations of similarity among each patient group in the ADNI subset for time lapse of 24 months computed using lateral ventricle shape alone (top) versus lateral ventricle shape and MMSE score used together (bottom)

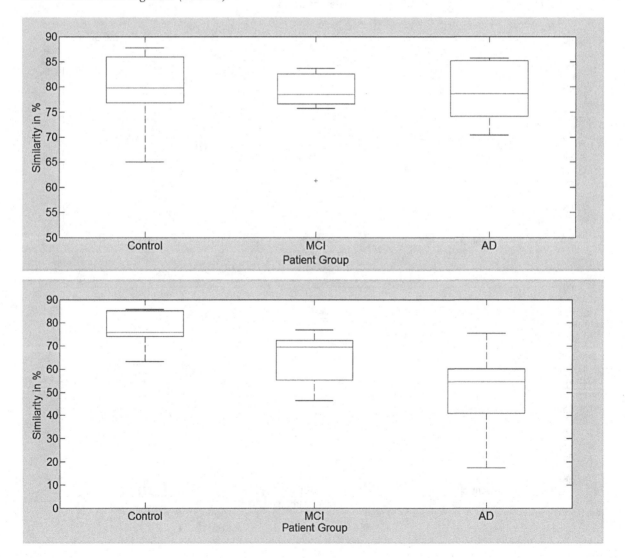

tailored to specific anatomies or diseases will find easier acceptance in the clinical as well as educational settings. In the framework of the presented study, some possible research opportunities are 1) extraction of clinically relevant, disease, and imaging modality specific features, 2) intelligent fusion of information for improved search, 3) research of efficient algorithms for simultaneous, rapid mining of both content and context data, and 4) efficacious, privacy-preserving indexing of the data at hand.

Search systems—like the one presented in this study—have utmost importance especially in the case of information scattered between multiple locations (e.g. searching thru the repositories of different clinical sites), where network-related issues such as bandwidth, privacy, cost, efficiency have to be addressed. It is the authors sincere hope that the research efforts in information retrieval (involving progress from the computer vision area, i.e. content-based, as well as the context-based information retrieval domain) and networking (including those presented in this book) will further improve the efficacy and practical usage of such systems, and eventually lead to improved patient care.

CONCLUSION

In this chapter, we presented a novel image search system for large medical image repositories that strengthens content information with contextual knowledge from demographic and clinical data, and ontologies for improved retrieval. As an exemplary usage scenario, search and retrieval of neuroimaging data from dementia and Alzheimer cases is presented. Predictive power of the proposed system is evaluated by only content information first, and it is shown that while lateral ventricle shape change can discriminate demented cases from healthy controls and converter, it is not discriminative enough for the Alzheimer cases. When content information is supported with con-

textual knowledge, such as neuropsychological tests, the discriminative capability of the search system is improved.

The proposed system can be useful in applications such as diagnosis and education, where experts or medical students and residents from the same clinical site as well as from different distant locations can benefit from the system. In the more common case of users located at different clinical sites, effective usage (and integration to the diagnostic or educational process) of such a search system will require efficient and low-cost networking solutions that need to handle technical issues like limited bandwidth and protection of data privacy.

ACKNOWLEDGMENT

This work is supported by the European Commission under the Grant FP7-PEOPLE-2009-RG-249253.

REFERENCES

Dale, A. M., Fischl, B., & Sereno, M. I. (1999). Cortical surface-based analysis: I: Segmentation and surface reconstruction. *NeuroImage*, *9*(2), 179–194. doi:10.1006/nimg.1998.0395

Felipe, J. C., Traina, A. J. M., & Traina, C., Jr. (2003). Retrieval by content of medical images using texture for tissue identification. In *Proceedings of 16th IEEE Symposium on Computer-Based Medical Systems*, (pp. 175-180). IEEE Press.

Hersh, W., Mailhot, M., Arnott-Smith, C., & Lowe, H. (2001). Selective automated indexing of findings and diagnoses in radiology reports. *Journal of Biomedical Informatics*, *34*, 262–273. doi:10.1006/jbin.2001.1025

Hsu, C. C., Chu, W. W., & Taira, R. K. (1996). A knowledge-based approach for retrieving images by content. *IEEE Transactions on Knowledge and Data Engineering*, *8*(4), 522–532. doi:10.1109/69.536245

Hu, B., Dasmahapatra, S., Lewis, P., & Shadbolt, N. (2003). Ontology-based medical image annotation with description logics. In *Proceedings 15th IEEE International Conference on Tools with Artificial Intelligence*, (pp. 77- 82). Sacramento, CA: IEEE Press.

Huang, H. K., Nielsen, J. F., Nelson, M. D., & Liu, L. (2005). Image-matching as a medical diagnostic support tool (DST) for brain diseases in children. *Computerized Medical Imaging and Graphics*, *29*(2-3), 195–202. doi:10.1016/j.compmedimag.2004.09.008

Jaulent, M.-C., Le Bozec, C., Cao, Y., Zapletal, E., & Degoulet, P. (2000). A property concept frame representation for flexible image content image content retrieval in histopathology databases. In *Proceedings of Symposium of the American Society for Medical Informatics*, (pp. 379-383). Los Angeles, CA: ASMI.

Kloppel, S. (2008). Accuracy of dementia diagnosis-a direct comparison between radiologists and a computerized method. *Brain*, *131*(11), 2969–2974. doi:10.1093/brain/awn239

Kulikowski, J. L. (2006). Interpretation of medical images based on ontological models. *Lecture Notes in Artificial Intelligence*, *4029*, 919–924.

Lacoste, C., Lim, J.-H., Chevallet, J.-P., & Le, D. T. H. (2007). Medical-image retrieval based on knowledge-assisted text and image indexing. *IEEE Transactions on Circuits and Systems for Video Technology*, *17*(7), 889–900. doi:10.1109/TCSVT.2007.897114

Lao, Z. (2004). Morphological classification of brains via high-dimensional shape transformations and machine learning methods. *NeuroImage*, *21*, 46–57. doi:10.1016/j.neuroimage.2003.09.027

Lew, M. S. (2006). Content-based multimedia information retrieval: State of the art and challenges. *ACM Transactions on Multimedia Computing, Communications, and Applications*, *2*(1), 1–19. doi:10.1145/1126004.1126005

Manning, C. D., Raghavan, P., & Schutze, H. (2008). *Introduction to information retrieval*. Cambridge, UK: Cambridge University Press. doi:10.1017/CBO9780511809071

Müller, H. (2004). A review of content-based image retrieval systems in medical applications - Clinical benefits and future directions. *International Journal of Medical Informatics*, *73*(1), 1–23. doi:10.1016/j.ijmedinf.2003.11.024

Ogiela, M. R., & Tadeusiewicz, R. (2001). Semantic-oriented syntactic algorithms for content recognition and understanding of images in medical databases. In *Proceedings of IEEE Computer Society International Conference on Multimedia and Exposition*, (pp. 621-624). Tokyo, Japan: IEEE Press.

Ojala, T., Piettikäinen, M., & Mäenpää, T. (2002). Multiresolution gray-scale and rotation invariant texture classification with local binary patterns. *IEEE Transactions on Pattern Analysis and Machine Intelligence*, *24*(7), 971–987. doi:10.1109/TPAMI.2002.1017623

Rubin, D. L., & Napel, S. (2010). Imaging informatics: Toward capturing and processing semantic information in radiology images. *Yearbook of Medical Informatics*, 34–42.

Thompson, P. M. (2007). Tracking Alzheimer's disease. *Annals of the New York Academy of Sciences*, *1097*, 183–214. doi:10.1196/annals.1379.017

Traina, C. Jr, Traina, J. M., dos Santos, R. R., & Senzako, E. Y. (1997). A support system for content-based medical image retrieval in object oriented databases. *Journal of Medical Systems*, *21*(6), 339–352. doi:10.1023/A:1022868128573

Unay, D. (2010). Augmenting clinical observations with visual features from longitudinal MRI data for improved dementia diagnosis. In *Proceedings of the International Conference on Multimedia Information Retrieval*, (pp. 193-200). New York, NY: ACM Press.

Unay, D., et al. (2007). Robustness of local binary patterns in brain MR image analysis. In *Proceedings of the IEEE-EMBS Conference*, (pp. 2098-2101). IEEE Press.

Unay, D., Ekin, A., & Jasinschi, R. (2010). Local structure-based region-of-interest retrieval in brain MR images. *IEEE Transactions on Information Technology in Biomedicine*, *14*(4), 897–903. doi:10.1109/TITB.2009.2038152

Waldemar, G. (2000). Diagnosis and management of Alzheimer's disease and other disorders associated with dementia: The role of neurologists in Europe. *European Journal of Neurology*, *7*, 133–144. doi:10.1046/j.1468-1331.2000.00030.x

Wei, W., & Barnaghi, P. M. (2007). Semantic support for medical image search and retrieval. In: *Proceedings of the Fifth IASTED International Conference: Biomedical Engineering*, (pp. 315-319). Anaheim, CA: ACTA Press.

Williams, A. (2002). Defining neurodegenerative diseases. *British Medical Journal*, *324*(7352), 1465–1466. doi:10.1136/bmj.324.7352.1465

ADDITIONAL READING

Deserno, T. M., Antani, S., & Long, R. (2009). Ontology of gaps in content-based image retrieval. *Journal of Digital Imaging*, *22*(2), 202–215. doi:10.1007/s10278-007-9092-x

Halle, M. W., & Kikinis, R. (2004). Flexible frameworks for medical multimedia. In *Proceedings of the 12th Annual ACM International Conference on Multimedia*, (pp. 768-775). New York, NY: ACM Press.

Jin, H., Sun, A., Zheng, R., He, R., & Zhang, Q. (2007). Ontology-based semantic integration scheme for medical image grid. In *Proceedings of the Seventh IEEE International Symposium on Cluster Computing and the Grid (CCGRID 2007)*, (pp. 127-134). Washington, DC: IEEE Computer Society.

Lehmann, T., Güld, M., Deselaers, T., Keysers, D., Schubert, H., & Spitzer, K. (2005). Automatic categorization of medical images for content-based retrieval and data mining. *Computerized Medical Imaging and Graphics*, *29*(2-3), 143–155. doi:10.1016/j.compmedimag.2004.09.010

Lehmann, T. M., Güld, M. O., Thies, C., Fischer, B., Spitzer, K., & Keysers, D. (2004). Content-based image retrieval in medical applications. *Methods of Information in Medicine*, *43*(4), 354–361.

Liu, Y., Zhang, D., Lu, G., & Ma, W.-Y. (2007). A survey of content-based image retrieval with high-level semantics. *Pattern Recognition*, *40*(1), 262–282. doi:10.1016/j.patcog.2006.04.045

Möller, M., & Sintek, M. (2007). A generic framework for semantic medical image retrieval. In T. Bürger, S. Dasiopoulou, C. Eckes, S. J. Perantonis, J. Pereira, & V. Tzouvaras (Eds.), *KAMC*. Retrieved from http://ceur-ws.org

Wang, H., Liu, S., & Chia, L.-T. (2006). Does ontology help in image retrieval? A comparison between keyword, text ontology and multimodality ontology approaches. In *Proceedings of the 14th Annual ACM International Conference on Multimedia (MULTIMEDIA 2006)*, (pp. 109-112). New York, NY: ACM Press.

Zhou, X. S., & Huang, T. S. (2003). Relevance feedback in image retrieval: A comprehensive review. *Multimedia Systems*, *8*(6), 536–544. doi:10.1007/s00530-002-0070-3

KEY TERMS AND DEFINITIONS

Alzheimer's Disease: Most common form of dementia.

Content-Based Image Retrieval: (1) Searching for digital images in large databases by using computer vision techniques; (2) Searching for digital images in large databases by making use of approaches other than computer vision techniques, such as mining related textual data and/or using ontologies.

Dementia: Loss of cognitive abilities, such as memory, attention, language, and problem solving, beyond what might be expected from normal aging. It is mostly observed in the elderly.

Lateral Ventricles: They are two curved-shaped cavities located in the cerebrum (part of the human brain) that provide a pathway for the circulation of the cerebrospinal fluid.

Longitudinal Data: Medical imaging data acquired over time.

Magnetic Resonance Imaging: A medical imaging technique, which makes use of nuclear magnetic resonance to image nuclei of atoms, to visualize detailed internal structures inside the body.

Medical Diagnosis: Identification of a possible disease or disorder by analyzing multitude of data, such as demographics, patient's medical history, clinical tests, imaging results, etc.

Neuroimaging: Imaging the structure or function of the brain.

Ontology: Formal representation of knowledge as a set of concepts within a domain and the relationships between those concepts.

Relevance Feedback: A feature of information retrieval systems, where initial returns from a given query are identified as relevant or not and this relevancy information is used to perform a new (typically improved) search.

Compilation of References

3 GPP TR 23.829. (2010). *Local IP access and selected IP traffic offload, release 10, V1.3.0.* 3GPP Technical Report.

3 GPP TS 23.246. (2011a). Multimedia broadcast/multicast service (MBMS) architecture and functional description, release 10, V10.1.0. 3GPP Technical Specification.

3 GPP TS 23.402. (2011b). Architecture enhancements for non-3GPP accesses, release 10, V10.4.0. 3GPP Technical Specification.

3 GPP. (2007). *Internet protocol (IP) multimedia call control protocol based on session initiation protocol (SIP) and session description protocol (SDP): Stage 3.* Technical Report. Retrieved from http://www.3gpp.org

Abdullah-Al-Wadud, M., & Oksam, C. (2007). Region-of-interest selection for skin detection based applications. In *Proceedings of the International Conference on Convergence Information Technology, 2007*, (pp. 1999-2004). IEEE.

Abhari, A., & Soraya, M. (2010). Workload generation for YouTube. *Multimedia Tools and Applications*, *46*, 91–118. doi:10.1007/s11042-009-0309-5

Abousleman, G. P. (2009). Target-tracking-based ultra-low-bit-rate video coding. In *Proceedings of Military Communications Conference, 2009*, (pp. 1-6). IEEE Press.

ACM SIGCHI. (2002). *Curricula for human-computer interaction.* New York, NY: The Association for Computing Machinery.

Adeyeye, M. (2009a). *The TransferHTTP controller.* Retrieved October 13, 2011, from http://transferhttp.berlios.de

Adeyeye, M. (2009b). *The TransferHTTP extension.* Retrieved October 13, 2011, from http://transferhttp.mozdev.org

Adeyeye, M., Ventura, N., & Humphrey, D. (2009). Control services for the HTTP session mobility service. In *Proceedings of the 3rd IEEE Conference on New Technologies, Mobility and Security (NTMS).* Cairo, Egypt: IEEE Press.

Adeyeye, M., & Ventura, N. (2010). A SIP-based web client for HTTP session mobility and multimedia services. *Computer Communications*, *33*(8), 954–964. doi:10.1016/j.comcom.2010.01.015

Adeyeye, M., Ventura, N., & Foschini, L. (2012). Converged multimedia services in emerging web 2.0 session mobility scenarios. *Wireless Networks*, *18*(2), 185–197. doi:10.1007/s11276-011-0394-z

Adibelli, Y., Parlak, M., & Hamzaoglu, I. (2011). Energy reduction techniques for H.264 deblocking filter hardware. *IEEE Transactions on Consumer Electronics*, *57*(3), 1399–1407. doi:10.1109/TCE.2011.6018900

Admob. (2010). *Mobile metrics report.* Retrieved from http://metrics.admob.com/wp-content/uploads/2010/06/May-2010-AdMob-Mobile-Metrics-Highlights.pdf

Adobe Flash Platform. (2012). *Website.* Retrieved March 2012 from www.adobe.com/flashplatform

Adobe. (2010). *HTTP dynamic streaming on the Adobe flash platform.* Retrieved from http://www.adobe.com/products/httpdynamicstreaming/pdfs/httpdynamicstreaming_wp_ue.pdf

Adzic, V., Kalva, H., & Furht, D. (2012). Content aware video encoding for adaptive HTTP streaming. In *Proceedings of the IEEE International Conference on Consumer Electronics*, (pp. 94-95). IEEE Press.

Adzic, V., Kalva, H., & Furht, B. (2011). A survey of multimedia content adaptation for mobile devices. *Multimedia Tools and Applications*, *51*(1), 379–396. doi:10.1007/s11042-010-0669-x

Aggarwal, A., Biswas, S., Singh, S., Sural, S., & Majumdar, A. K. (2006). Object tracking using background subtraction and motion estimation in MPEG videos. *Lecture Notes in Computer Science*, *3852*, 121–130. doi:10.1007/11612704_13

Agi, I., & Gong, L. (1996). An empirical study of MPEG video transmissions. In *Proceedings of the Internet Society Symposium on Network and Distributed System Security*, (pp. 137-144). IEEE.

Agilent-Technologies. (2008). *Validating IPTV service quality under multiplay network conditions*. White Paper. Agilent-Technologies.

Ahlswede, R., Cai, N., Li, S.-Y., & Yeung, R. W. (2000). Network information flow. *IEEE Transactions on Information Theory*, *46*(4), 1204–1216. doi:10.1109/18.850663

Ahmadi Aliabad, H., Moiron, S., Fleury, M., & Ghanbari, M. (2010). No-reference H.264/AVC statistical multiplexing for DVB-RCS. In *Proceedings of the 2nd International ICST Conference on Personal Satellite Services*, (pp. 163-178). ICST.

Ahmadi, H., & Kong, J. (2008). Efficient web browsing on small screens. In *Proceedings of the Working Conference on Advanced Visual Interfaces (AVI 2008)*, (pp. 23-30). ACM Press.

Ahmad, I., Wei, X., Sun, Y., & Zhang, Y. W. (2005). Video transcoding: An overview of various techniques and research issues. *IEEE Transactions on Multimedia*, *7*(5), 793–804. doi:10.1109/TMM.2005.854472

Ahson, S. A., & Ilyas, M. (2010). *Fixed-mobile convergence handbook*. Boca Raton, FL: CRC Press. doi:10.1201/EBK1420091700

Aiello, L. M., Barrat, A., Cattuto, C., Ruffo, G., & Schifanella, R. (2010). Link creation and profile alignment in the aNobii social network. In *Proceedings of the 2nd IEEE International Conference on Social Computing (SocialCom 2010)*, (pp. 249-256). Minneapolis, MN: IEEE Press.

Aiello, L. M., Barrat, A., Schifanella, R., Cattuto, C., Markines, B., & Menczer, F. (2012). Friendship prediction and homophily in social media. *ACM Transactions on the Web, 6*(2).

Akogrimo Project. (2008). *Access to knowledge through the grid in the mobile world*. Retrieved October 13, 2011, from http://www.mobilegrids.org

Akyol, E., Mukherjee, D., & Liu, Y. (2007). Complexity control for real-time video coding. In *Proceedings of the 2007 IEEE International Conference on Image Processing, ICIP 2007*, (vol. 1, pp. 77-80). IEEE Press.

Alcock, S., & Nelson, R. (2011). Application flow control in YouTube video streams. *ACM SIGCOMM Computer Communication Review*, *41*(2), 25–30. doi:10.1145/1971162.1971166

Alexandraki, C., & Akoumianakis, D. (2010). Exploring new perspectives in network music performance: The DIAMOUSES framework. *Computer Music Journal*, *34*(2), 66–83. doi:10.1162/comj.2010.34.2.66

Alfonso, D. (2010). *Proposals for video coding complexity assessment*. Paper presented at the JCTVC-A026, Joint Collaborative Team on Video Coding (JCT-VC) of ITU-T SG16 WP3 and ISO/IEC JTC1/SC29/WG11 1st Meeting. Dresden, Germany.

Ali, I., Fleury, M., & Ghanbari, M. (2012). Content-aware intra-refresh for video streaming over lossy links. In *Proceedings of the IEEE International Conference on Consumer Electronics*. IEEE Press.

Al-Jobouri, L., Fleury, M., & Ghanbari, M. (2012a). Versatile IPTV for broadband wireless with adaptive channel coding. In *Proceedings of the IEEE International Conference on Consumer Electronics*, (pp. 344-345). IEEE Press.

Al-Jobouri, L., Fleury, M., & Ghanbari, M. (2012). Comprehensive protection of data-partitioned video for broadband wireless IPTV streaming. *Mobile Information Systems*, *8*(2), 1–23.

Allison, D., & Gwaltney, T. (1991). How people use electronic interactives: Information age - People, information & technology. [Pittsburgh, PA: ICHIM.]. *Proceedings of ICHIM, 1991*, 62–73.

Amer, A. (2001). *Object and event extraction for video processing and representation in on-line video applications*. (Doctoral Dissertation). Universite du Quebec. Quebec, Canada. Retrieved from http://users.encs.concordia.ca/~amer/paper/phd/amer_diss.pdf

Amer, A., & Schroder, H. (1996). A new video noise reduction algorithm using spatial sub- bands. In *Proceedings of IEEE International Conference on Electronics, Circuits and Systems*, (vol. 1, pp. 45-48). Rodos, Greece: IEEE Press.

Amer, A., Mitiche, A., & Dubois, E. (2002). Reliable and fast structure-oriented video noise estimation. In *Proceedings IEEE International Conference on Image Processing, ICIP 2002*, (vol. 1, pp. 840-843). Rochester, NY: IEEE Press.

Anderson, C. (2006). *The long tail*. New York, NY: Hyperion Press.

Andrews, J. G., Ghosh, A., & Muhamed, R. (2007). *Fundamentals of WiMAX: Understanding broadband wireless networking*. Upper Saddle River, NJ: Prentice Hall.

Android Developers. (2010). *What is Android?* Retrieved from http://developer.android.com/guide/basics/what-is-android.html

Android Developers. (2012). *Download the Android SDK*. Retrieved from http://developer.android.com/sdk/index.html

Angehrn, A., Nabeth, T., & Roda, C. (2001). *Towards personalised, socially aware and active e-learning systems*. CALT White Paper. Washington, DC: CALT.

Anselmo, T., & Alfonso, D. (2010). Constant quality variable bit-rate control for SVC. In *Proceedings of the 2010 11th International Workshop on Image Analysis for Multimedia Interactive Services (WIAMIS)*, (pp. 1-4). WIAMIS.

Apteker, R. T., Fisher, J. A., Kisimov, V. S., & Neishlos, H. (1995). Video acceptability and frame rate. *IEEE MultiMedia*, 2(3), 32–40. doi:10.1109/93.410510

Archer, W., Garrison, R. D., Anderson, T., & Rourke, L. (2001). *A framework for analysing critical thinking in computer conferences*. Paper presented at the European Conference on Computer-Supported Collaborative Learning. Maastricht, The Netherlands.

Ardito, C., Costabile, M. F., Lanzilotti, R., & Simeone, A. L. (2010). Combining multimedia resources for an engaging experience of cultural heritage. In *Proceedings of SAPMIA 2010*. Firenze, Italy: SAPMIA.

Arkko, J., Vogt, C., & Haddad, W. (2007, May). *Enhanced route optimization for mobile IPv6*. IETF RFC 4866. Retrieved from http://www.ietf.org/rfc/rfc4866.txt

Arsenio, A. (2004). *Cognitive-developmental learning for a humanoid robot: A caregiver's gift*. (PhD Thesis). MIT. Cambridge, MA.

Arsenio, A. (2011). *Method and apparatus for adaptation of a multimedia content*. International Patent Number: EP2341680 A1. Nokia Siemens Networks GMBH & CO.

Aryananda, L. (2002). Recognizing and remembering individuals: Online and unsupervised face recognition for humanoid robot. In *Proceedings of the International IEEE/RSJ Conference on Intelligent Robots and Systems*. IEEE.

Asadi, K. M., & Dufourd, J.-C. (2005). Context-aware semantic adaptation of multimedia presentations. In *Proceedings of the IEEE International Conference on Multimedia and Expo (ICME 2005)*, (pp. 362 – 365). Amsterdam, The Netherlands: IEEE Computer Society.

Assunção, P. A. A., & Ghanbari, M. (1997). Transcoding of single-layer MPEG video into lower rates. *Proceedings on the Institute of Electronics Engineers: Vision, Image, and Signal Processing*, 144(6), 377–383. doi:10.1049/ip-vis:19971558

Assunção, P. A. A., & Ghanbari, M. (1998). A frequency domain video transcoder for dynamic bit-rate reduction of MPEG-2 bit streams. *IEEE Transactions on Circuits and Systems for Video Technology*, 8(8), 953–967. doi:10.1109/76.736724

Aucouturier, J. J., & Pachet, F. (2002). Music similarity measures: What's the use. In *Proceedings of the 3rd International Conference on Music Information Retrieval (ISMIR 2002)*, (pp. 157-163). Paris, France: ISMIR.

Aviv, R. (2000). Education performance of ALN via content analysis. *Journal of Asynchronous Learning Networks*, 14(2).

Babbie, E. (2004). *The practice of social research* (10th ed.). Belmont, CA: Thomson/Wadsworth Learning.

Babu, R. V., Ramakrishnan, K. R., & Srinivasan, S. H. (2004). Video object segmentation: A compressed domain approach. *IEEE Transactions on Circuits and Systems for Video Technology*, *14*(4), 462–474. doi:10.1109/TCSVT.2004.825536

Bales, R. F., & Strodbeck, F. L. (1951). Phases in group problem-solving. *Journal of Abnormal and Social Psychology*, *46*, 485–495. doi:10.1037/h0059886

Balkrishnan, H., Padmanabhan, V., Seshan, S., & Katz, R. (2007). A comparison of mechanisms for improving TCP performance over wireless links. *IEEE/ACM Transactions on Networking*, *5*(6), 756–769. doi:10.1109/90.650137

Bansal, D., & Balakrishnan, H. (2001). Binomial congestion control algorithms. In *Proceedings of IEEE INFOCOM* (pp. 631–640). IEEE Press.

Barak, M., Hersoviz, O., Kaberman, Z., & Dori, Y. (2009). MOSAICA: A web-2.0 based system for the preservation and presentation of cultural heritage. *Computers & Education*, *53*(3), 841–852. doi:10.1016/j.compedu.2009.05.004

Barmada, B., Ghandi, M., Jones, E., & Ghanbari, M. (2005). Prioritized transmission of data partitioned H. 264 video with hierarchical QAM. *IEEE Signal Processing Letters*, *12*(8), 577–580. doi:10.1109/LSP.2005.851261

Barrington, L., O'Malley, D., Turnbull, D., & Lanckriet, G. (2009). User-centered design of a social game to tag music. In *Proceedings of the ACM SIGKDD Workshop on Human Computation*. New York, NY: ACM.

Bates, A. W. (1995). *Technology, open learning and distance education*. London, UK: Routledge.

Bauer, M., Bosch, P., Khrais, N., Samuel, L. G., & Schefczik, P. (2007). The UMTS base station router. *Bell Labs Technical Journal, I. Wireless Network Technology*, *11*(4), 93–111.

Behnke, W. (2003). *Online/eLearning - An overview*. Vancouver, Canada: Vancouver Community College.

Belenky, M. F., Clinchy, B. M., Goldberger, N. R., & Tarule, J. M. (1986). *Women's ways of knowing: The development of self, voice, and mind* (2nd ed.). New York, NY: Basic Books.

Belhumeur, P., Hespanha, J., & Kriegman, D. (1997). Eigenfaces vs. fisherfaces: Recognition using class specific linear projection. *IEEE Transactions on Pattern Analysis and Machine Intelligence*, *19*(7), 711–720. doi:10.1109/34.598228

Benzougar, A., Bouthemy, P., & Fablet, R. (2001). MRF-based moving object detection from MPEG coded video. In *Proceedings of the IEEE International Conference on Image Processing*, (vol. 3, pp. 402-405). IEEE Press.

Bernardos, C. (2012). *Proxy mobile IPv6 extensions to support flow mobility*. Retrieved from http://draft-ietf-netext-pmipv6-flowmob-03

Bernardos, C., Oliva, A. D., Giust, F., Melia, T., & Costa, R. (2012). *A PMIPv6-based solution for distributed mobility management*. Retrieved from http://draft-bernardos-dmm-pmip-01

Bertone, F., Menkovski, V., & Liotta, A. (2012). Adaptive P2P streaming. In M. Fleury & N. Qadri (Eds.), *Streaming Media with Peer-to-Peer Networks: Wireless Perspectives*. Hershey, PA: IGI Global.

Bhanu, B., Dudgeon, D. E., Zelnio, E. G., Rosenfeld, A., Casasent, D., & Reed, I. S. (1997). Guest editorial introduction to the special issue on automatic target detection and recognition. *IEEE Transactions on Image Processing*, *6*(1), 1–6. doi:10.1109/TIP.1997.552076

Bhat, R. R., & Tait, D. (2001). JAVA APIs for integrated networks. In Jespen, T. (Ed.), *Java in Telecommunications: Solutions for Next Generation Networks* (p. 193). New York, NY: Wiley & Sons.

Biersdorfer, J. D., & Pogue, D. (2010). *IPod: The missing manual*. Sebastopol, CA: O'Reilly Media.

Billsus, D., & Pazzani, M. (1998). Learning collaborative information filters. In *Proceedings of the 15th International Conference on Machine Learning (ICML 1998)*, (pp. 46-54). San Francisco, CA: Morgan Kaufmann Publishers Inc.

Bing, B. (2010). *3D and HD broadband video networking*. Norwood, MA: Artech House.

Blume, H., Amer, A., & Schroder, H. (1997). Vector-based postprocessing of MPEG-2 signals for digital TV-receivers. In *Proceedings of Conference Visual Communications and Image Processing*, (vol. 3024, pp. 1176-1187). San Jose, CA: IEEE.

Bobick, A., & Pinhanez, C. (1995). Using approximate models as source of contextual information for vision processing. In *Proceedings of the ICCV 1995 Workshop on Context-Based Vision*, (pp. 13-21). Cambridge, MA: ICCV.

Bohn, A., Feinerer, I., Hornik, K., & Mair, P. (2011). Content-based social network analysis of mailing lists. *Remediation Journal*, *3*(1), 11–18.

Bokor, L., Faigl, Z., & Imre, S. (2011). Flat architectures: Towards scalable future internet mobility. *Lecture Notes in Computer Science*, *6656*, 35–50. doi:10.1007/978-3-642-20898-0_3

Bolchini, C., Curino, C. A., Quintarelli, E., Schreiber, F. A., & Tanca, L. (2007). A data-oriented survey of context models. *SIGMOD Record*, *36*(4), 19–26. doi:10.1145/1361348.1361353

Bolchini, C., Curino, C., Quintarelli, E., Schreiber, F., & Tanca, L. (2006). *Context-ADDICT. Technical Report*. Milan, Italy: Politecnico di Milano.

Bolchini, C., Orsi, G., Quintarelli, E., Schreiber, A., & Tanca, L. (2011). Context modeling and context awareness: Steps forward in the Context-ADDICT project. *A Quarterly Bulletin of the Computer Society of the IEEE Technical Committee on Data Engineering*, *34*(2), 47–54.

Boletsis, C., Gratsani, A., Chasanidou, D., Karydis, I., & Kermanidis, K. (2011). Comparative analysis of content-based and context-based similarity on musical data. In *Proceedings of the EANN/AIAI (2) 2011*, (pp.179-189). Corfu, Greece: EANN/AIAI.

Bondy, J., & Murty, U. (2008). *Graph theory*. Berlin, Germany: Springer. doi:10.1007/978-1-84628-970-5

Böröczy, L., Ngai, A. Y., & Westermann, E. F. (1999). Statistical multiplexing using MPEG-2 video encoders. *IBM Journal of Research and Development*, *43*(4), 511–520. doi:10.1147/rd.434.0511

Boswell, D., King, B., Oeschger, I., Collins, P., & Murphy, E. (2002). *Creating applications with Mozilla*. New York, NY: O'Reilly Press.

Boulton, C., Melanchuk, T., & McGlashan, S. (2011). *Media control channel framework*. RFC 6230. Retrieved from http://www.rfc-editor.org/rfc/rfc6230.txt

BoxGrinder. (2011). *The BoxGrinder*. Retrieved October 13, 2011, from http://boxgrinder.org/

Braden, R., Zhang, L., Berson, S., Herzog, S., & Jamin, S. (1997). *Resource reservation protocol (RSVP)*. IETF RFC 2205. Retrieved from http://www.ietf.org/rfc/rfc2205.txt

Brancho, R., & Sanderson, A. C. (1985). Segmentation of images based on intensity gradient information. In *Proceedings of Conference on Computer Vision and Pattern Recognition, CVPR 1985*, (pp. 341-347). San Francisco, CA: CVPR.

Braunhofer, M., Kaminskas, M., & Ricci, F. (2011). Recommending music for places of interest in a mobile travel guide. In *Proceedings of the Fifth ACM Conference on Recommender Systems (RecSys 2011)*, (pp. 253-256). New York, NY: ACM Press.

Breazeal, C., & Aryananda, L. (2002). Recognition of affective communicative intent in robot-directed speech. *Autonomous Robots*, *12*(1), 83–104. doi:10.1023/A:1013215010749

Broadband Forum. (2007). *ADSL2/ADSL2plus performance test plan*. Technical Report, TR-100. Broadband Forum.

Browser, I. D. (2012). *The BrowserID project*. Retrieved October 13, 2011, from https://browserid.org

Buchholz, S., Hamann, T., & Hübsch, G. (2004). Comprehensive structured context profiles (CSCP): Design and experiences. In I. C. Society (Ed.), *Proceedings of the Second IEEE Annual Conference on Pervasive Computing and Communications Workshops, PERCOMW 2004*, (p. 43). Orlando, FL: IEEE Press.

Bulterman, D., Jack, J., Pablo, C., Sjoerd, M., Eric, H., Marisa, D., et al. W3C. (2008). *Synchronized multimedia integration language (SMIL 3.0), W3C recommendation*. Retrieved from http://www.w3.org/TR/SMIL/

Buono, A., Castaldi, T., Miniero, L., & Romano, S. P. (2007). Design and implementation of an open source IMS enabled conferencing architecture. In *Proceedings of the 7th International Conference on Next Generation Teletraffic and Wired/Wireless Advanced Networking (NEW2AN 2007)*. St. Petersburg, Russia: NEW2AN.

Burke, R. (2002). Hybrid recommender systems: Survey and experiments. *User Modeling and User-Adapted Interaction, 12*(4), 331–370. doi:10.1023/A:1021240730564

Bystrom, M., Richardson, I., & Zhao, Y. (2008). Efficient mode selection for H.264 complexity reduction in a Bayesian framework. *Signal Processing Image Communication, 23*(2), 71–86. doi:10.1016/j.image.2007.11.001

Cai, L., Shen, X., Pan, J., & Mark, J. W. (2005). Performance analysis of TCP-friendly AIMD algorithms for multimedia applications. *IEEE Transactions on Multimedia, 7*(2), 339–355. doi:10.1109/TMM.2005.843360

Calandrino, J. A., Kilzer, A., Narayanan, A., Felten, E. W., & Shmatikov, V. (2011). You might also like: Privacy risks of collaborative filtering. In *Proceedings of the Security and Privacy (SP) IEEE Symposium*, (pp. 231-246). IEEE Press.

Calderón, M., Bernardos, C. J., Bagnulo, M., Soto, I., & Oliva, A. D. (2006). Design and experimental evaluation of a route optimization solution for NEMO. *IEEE Journal on Selected Areas in Communications, 24*(9), 1702–1716. doi:10.1109/JSAC.2006.875109

Cali, F., Conti, M., & Gregori, E. (1998). IEEE 802.11 wireless LAN: Capacity analysis and protocol enhancement. In *Proceedings of the 1998 IEEE Seventeenth Annual Joint Conference of the IEEE Computer and Communications Societies (INFOCOM 1998)*, (vol. 1, pp. 142-149). IEEE Press.

Camarillo, G. (2006). *Session description protocol (SDP) format for binary floor control protocol (BFCP) streams*. RFC 4583. Retrieved from http://tools.ietf.org/html/rfc4583

Camarillo, G., Ott, J., & Drage, K. (2006). *The binary floor control protocol (BFCP)*. RFC 4582. Retrieved from http://tools.ietf.org/html/rfc4582

Camarillo, G., & García-Martín, M. A. (2011). *The 3G IP multimedia subsystem (IMS): Merging the internet and the cellular worlds* (3rd ed.). New York, NY: Wiley.

Candy, P. (1991). *Self-direction for lifelong learning*. San Francisco, CA: Jossey-Bass.

Canfora, G., Di Santo, G., Venturi, G., Zimeo, E., & Zito, M. V. (2005). Proxy-based handoff of web sessions for user mobility. In *Proceedings of the Second Annual International Conference on Mobile and Ubiquitous Systems: Networking and Services (MobiQuitous 2005)*. IEEE.

Canny, J. (1986). A computational approach to edge detection. *IEEE Transactions on Pattern Analysis and Machine Intelligence, 9*, 679–698. doi:10.1109/TPA-MI.1986.4767851

Canovas, A., Bri, D., Sendra, S., & Lloret, J. (2011). Vertical WLAN handover algorithm and protocol to improve the IPTV QoS of the end user. In *Proceedings of the IEEE International Conference on Communications (ICC 2011)*, (pp. 10-15). Ottawa, Canada: IEEE Press.

CAP. (2004). *E-guide: Using computer mediated communication in learning and teaching*. Retrieved November 8, 2004 from http://www2.warwick.ac.uk/services/cap/resources/eguides/cmc/cmclearning/

Casey, M. A., Veltkamp, R., Goto, M., Leman, M., Rhodes, C., & Slaney, M. (2008). Content-based music information retrieval: Current directions and future challenges. *Proceedings of the IEEE, 96*(4), 668–696. doi:10.1109/JPROC.2008.916370

Cavendish, D., Murakami, K., Yun, S.-H., Matsuda, O., & Nishihara, M. (2002). New transport services for next-generation SONET/SDH systems. *IEEE Communications Magazine, 40*(5), 80–87. doi:10.1109/35.1000217

Celma, O. (2006). Foafing the music: Bridging the semantic gap in music recommendation. *International Journal of Web Semantics: Science. Services and Agents on the World Wide Web, 6*(4), 250–256. doi:10.1016/j.websem.2008.09.004

CERN. (2003). *10 years public domain*. Retrieved February 7, 2012, from http://tenyears-www.web.cern.ch/tenyears-www/Welcome.html

Chakrabarti, S., Van den Berg, M., & Dom, B. (1999). Focused crawling: A new approach to topic-specific web resource discovery. *Computer Networks, 31*(11-16), 1623-1640.

Chang, C.-C., & Lin, C.-J. (2003). *LIBSVM: A library for support vector machines.* Retrieved from http://www. csie.ntu.edu.tw/~cjlin/libsvmtools

Chang, J., & Sun, E. (2011). Location3: How users share and respond to location-based data on social networking sites. In *Proceedings of the Fifth International Conference on Weblogs and Social Media,* (pp. 74-80). IEEE.

Chang, S.-F., & Eleftheriadis, A. (1994). Error accumulation of repetitive image coding. In *Proceedings of the IEEE Symposium on Circuits and Systems,* (Vol. 3, pp. 201-204). IEEE Press.

Chang, J.-W., & Kim, Y.-J. (2006). XML document retrieval system based on document structure and image content for digital museum. *Lecture Notes in Computer Science, 3842,* 107–111. doi:10.1007/11610496_12

Chang, S. F., & Messerschmidt, D. G. (1995). Manipulation and compositing of MC-DCT compressed video. *IEEE Journal on Selected Areas in Communications, 13*(1), 1–11. doi:10.1109/49.363151

Chee, S. H., Han, J., & Wang, K. (2001). RecTree: An efficient collaborative filtering method. In *Proceedings of the 3rd International Conference on Data Warehousing and Knowledge Discovery,* (pp. 141-151). IEEE.

Chellappa, R., Wilson, C., & Sirohey, S. (1995). Human and machine recognition of faces: A survey. *Proceedings of the IEEE, 83,* 705–741. doi:10.1109/5.381842

Chen, H., Han, Z., Hu, R., & Ruan, R. (2008). Adaptive FMO selection strategy for error resilient H.264 coding. In *Proceedings of the International Conference on Audio, Lang. and Image Processing, ICALIP 2008,* (pp. 868-872). Shanghai, China: ICALIP.

Chen, J., Qu, Y., & He, Y. (2004). A fast mode decision algorithm in H.264. In *Proceedings of the Picture Coding Symposium,* (pp. 46-49). San Francisco, CA: IEEE.

Chen, M.-J., Chi, M.-C., Hsu, C.-T., & Chen, J.-W. (2003). ROI video coding based on H.263+ with robust skin-color detection technique. In *Proceedings of the 2003 IEEE International Conference on Consumer Electronics, ICCE 2003,* (pp. 44-45). IEEE Press.

Chen, Q.-H., Xie, X.-F., Guo, T.-J., Shi, L., & Wang, X.-F. (2010). The study of ROI detection based on visual attention mechanism. In *Proceedings of the 2010 6th International Conference on Wireless Communications Networking and Mobile Computing (WiCOM),* (pp. 1-4). IEEE.

Chen, T.-C., Lian, C.-J., & Chen, L.-G. (2006). Hardware architecture design of an H.264/AVC video codec. In *Proceedings of the 2006 Asia and South Pacific Design Automation Conference (ASP-DAC 2006),* (pp. 750-757). Piscataway, NJ: IEEE Press.

Cheng, W.-H., Chu, W.-T., Kuo, J.-H., & Wu, J.-L. (2005). Automatic video region-of-interest determination based on user attention model. In *Proceedings of IEEE International Symposium on Circuits and Systems (ISCAS 2005),* (pp. 3219-3222). IEEE Press.

Chen, L. J., Wang, B. C., & Chen, K. (2010). The design of puzzle selection strategies for GWAP systems. *Journal of Concurrency and Computation: Practice & Experience, 22*(7), 890–900.

Chen, M., & Zakhor, A. (2005). Rate control for streaming video over wireless. *IEEE Wireless Communications, 12*(4), 32–41. doi:10.1109/MWC.2005.1497856

Cheong, H.-Y., & Ortega, A. (2007). Distance quantization method for fast nearest neighbor search computations with applications to motion estimation, signals, systems and computers, 2007. In *Proceedings of the Conference Record of the Forty-First Asilomar Conference,* (pp. 909-913). ACSSC.

Chevalier, M., Soulé-Dupuy, C., & Tchienehom, P. (2006). Semantics-based profiles modeling and matching for resources access. *Journal des Sciences pour l'Ingénieur, 1*(7), 54–63.

Chiang, C.-K., & Lai, S.-H. (2009). Fast multi-reference motion estimation via statistical learning for H.264/AVC. In *Proceedings of the IEEE International Conference on Multimedia and Expo, 2009, ICME 2009,* (pp. 61-64). IEEE Press.

Chiang, C.-K., Pan, W.-H., Hwang, C., Zhuang, S.-S., & Lai, S.-H. (2011). Fast H.264 encoding based on statistical learning. *IEEE Transactions on Circuits and Systems for Video Technology, 21*(9), 1304–1315. doi:10.1109/TCSVT.2011.2147250

Chi, M., Bassok, M., Lewis, M., Reimann, P., & Glaser, R. (1989). Self-explanations: How students study and use examples in learning to solve problems. *Cognitive Science, 13*, 145–182. doi:10.1207/s15516709cog1302_1

Chou, P. A., Wu, Y., & Jain, K. (2003). Practical network coding. In *Proceedings of 51st Allerton Conference on Communication, Control and Computing.* Allerton Conference on Communication, Control, and Computing.

Chow, C.-O., & Ishii, H. (2007). Enhancing real-time video streaming over mobile ad hoc networks using multipoint-to-point communication. *Computer Communications, 30*(8), 1754–1764. doi:10.1016/j.comcom.2007.02.004

Chuang, C.-H., Chen, B.-N., Chang, C.-C., & Cheng, S.-C. (2011). Computation-aware fast motion estimation for H.264/AVC using image indexing. *Journal of Visual Communication and Image Representation, 22*(6), 451–464. doi:10.1016/j.jvcir.2011.05.003

Chung, H., Romacho, D., & Ortega, A. (2003). Fast long-term motion estimation for H.264 using multiresolution search. In *Proceedings of the International Conference on Image Processing,* (pp. 905-908). Barcelona, Spain: IEEE.

Cicco, L. D., Mascolo, S., Bari, P., & Orabona, V. (2010). An experimental investigation of the Akamai adaptive video streaming. In *Proceedings of the 6th International Conference on HCI in Work and Learning, Life and Leisure: Workgroup Human-Computer Interaction and Usability Engineering,* (pp. 447-464). HCI.

Cisco. (2011). *Cisco visual networking index: Forecast and methodology, 2010-2011.* White Paper. San Jose, CA: Cisco.

Cisco, V. N. I. (2011). *Global mobile data traffic forecast update, 2010-2015.* New York, NY: Cisco.

Claypool, M., Gokhale, A., & Miranda, T. (1999). Combining content-based and collaborative filters in an online newspaper. In *Proceedings of the SIGIR-99 Workshop on Recommender Systems: Algorithms and Evaluation.* ACM Press.

Clinchy, B. M. (1989). The development of thoughtfulness in college women: Integrating reason and care. *The American Behavioral Scientist, 32*, 647–657. doi:10.1177/0002764289032006005

Clinchy, B. M. (1996). Connected and separate knowing: Toward a marriage of two minds. In Goldberger, N., Tarule, J., Clinchy, B., & Belenky, M. (Eds.), *Knowledge, Difference, and Power: Essays Inspired by Women's Ways of Knowing.* New York, NY: Basic Books.

Cohen, W. W., & Fan, W. (2000). Web-collaborative filtering: Recommending music by crawling the web. *WWW9/Computer Networks, 33*(1–6), 685–698.

Comer, B. R., Narayanan, R. M., & Reichenbach, S. E. (2003). Noise estimation in remote sensing imagery using data masking. *International Journal of Remote Sensing, 24*(4), 689–702. doi:10.1080/01431160210164271

Conklin, G., Greenbaum, G., Lillevold, K., Lippman, A., & Reznik, Y. (2001). Video coding for streaming media delivery over the Internet. *IEEE Transactions on Circuits and Systems for Video Technology, 11*(3), 269–281. doi:10.1109/76.911155

Cooper, S., Khatib, F., Treuille, A., Barbero, J., Lee, J., & Beenen, M. (2010). Predicting protein structures with a multiplayer online game. *Nature, 466*, 756–760. doi:10.1038/nature09304

Correia, A., Silva, J., Souto, N., Silva, L., Boal, A., & Soares, A. (2007). Multi-resolution broadcast/multicast systems for MBMS. *IEEE Transactions on Broadcasting, 53*(1), 224–234. doi:10.1109/TBC.2007.891705

Costa, A., Quadros, C., Melo, A., Cerqueira, E., Abelém, A., Neto, A., et al. (2011). QoE-based packet dropper controllers for multimedia streaming in WiMAX networks. In *Proceedings of the 6th Latin America Networking Conference, LANC 2011,* (pp. 12-19). Quito, Ecuador: ACM Press.

Côté, G., & Kossentini, F. (1999). Optimal intra coding of blocks for robust video communication over the Internet. *EUROSIP Journal of Image Communication.* Retrieved from http://wftp3.itu.int/av-arch/video-site/9811_Seo/q15f38attach.pdf

Cotter, P., & Smyth, B. (2000). PTV: Intelligent personalized TV guides. In *Proceedings of the Twelfth Conference on Innovative Applications of Artificial Intelligence*, (pp. 957-964). Austin, TX: IEEE.

Cover, T. M. (1991). *Information theory*. New York, NY: Wiley-Interscience.

Cruz, R., Nunes, M., Patrikakis, C., & Papaoulakis, N. (2010). SARACEN: A platform for adaptive, socially aware multimedia distribution over P2P networks. In *Proceedings of the 4th IEEE Workshop on Enabling the Future Service-Oriented Internet: Towards Socially-Aware Networks, GLOBECOM 2010*. IEEE Press.

Culture.gr. (2012). *Museum of the history of the Olympic games of antiquity*. Retrieved February 01, 2012 from http://odysseus. culture. gr/h/1/eh155. jsp?obj_id=3488

Cunningham, D., Lane, B., & Lane, W. (1999). *Gigabit ethernet networking*. Indianapolis, IN: Macmillan Publishing Co., Inc.

Curwen, R., & Blake, A. (1992). Dynamic contours: Real-time active splines. In *Active Vision*. Cambridge, MA: MIT Press.

Dai, Q., Zhu, D., & Ding, R. (2004). Fast mode decision for inter prediction in H.264. In *Proceedings of the IEEE International Conference on Image Processing*, (pp. 119-122). Singapore, Singapore: IEEE Press.

Dale, A. M., Fischl, B., & Sereno, M. I. (1999). Cortical surface-based analysis: I: Segmentation and surface reconstruction. *NeuroImage*, *9*(2), 179–194. doi:10.1006/nimg.1998.0395

Damasio, A. (1994). *Descartes error: Emotion, reason, and the brain*. London, UK: Bard.

De Angeli, A., & Sue, K. (2005). Learning conversations: A case study into e-learning communities. In *Proceedings of the Interact 2005 eLearning and Human-Computer Interaction Workshop*. IEEE.

de Haan, G., Kwaaitaal-Spassova, T. G., & Ojo, O. A. (1996). Automatic 2-D and 3-D noise filtering for high-quality television receivers. In *Proceedings of International Workshop on Signal Processing and HDTV*, (vol. 6, pp. 221-230) Turin, Italy: IEEE.

de Hann, G., Kwaaitaal-Spassova, T. G., Larragy, M. M., Ojo, O. A., & Schutten, R. J. (1998). Television noise reduction IC. *IEEE Transactions on Consumer Electronics*, *44*(1), 143–154. doi:10.1109/30.663741

de Prycker, M. (1993). *Asynchronous transfer mode: Solutions for broadband ISDN*. Upper Saddle River, NJ: Prentice Hall.

De Rango, F., Tropea, M., Fazio, P., & Marano, S. (2008). Call admission control for aggregate MPEG-2 traffic over multimedia geo-satellite networks. *IEEE Transactions on Broadcasting*, *54*(3), 612–622. doi:10.1109/TBC.2008.2002716

Debaecker, J., & El Hadi, W. M. (2011). Music indexing and retrieval: Evaluating the social production of music metadata and its use. In *International Society for Knowledge Organisation*. Dublin, Ireland: Emerald Group Publishing Ltd.

December, J. (1997). Notes on defining of computer-mediated communication. *Computer-Mediated Communication Magazine*, *3*(1).

December, J. (2004). *What is computer-mediated communication?* Retrieved October 19, 2004, from http://www.december.com/john/study/cmc/what.html

Deering, S., & Hinden, R. (1998). *Internet protocol, version 6 (IPv6) specification*. IETF RFC 2460. Retrieved from http://www.ietf.org/rfc/rfc2460.txt

Deering, S., Fenner, W., & Haberman, B. (1999). *Multicast listener discovery (MLD) for IPv6*. IETF RFC 2710. Retrieved from http://www.ietf.org/rfc/rfc2710.txt

Deerwester, S., Dumais, S. T., Furnas, G. W., Landauer, T. K., & Harshman, R. (1990). Indexing by latent semantic analysis. *Journal of the American Society for Information Science American Society for Information Science*, *41*(6), 391–407. doi:10.1002/(SICI)1097-4571(199009)41:6<391::AID-ASI1>3.0.CO;2-9

Derryberry, A. (2012). *Serious games: Online games for learning*. Retrieved from http://www.adobe.com/resources/elearning/pdfs/serious_games_wp.pdf

Deruelle, J. (2008). JSLEE and SIP-servlets interoperability with mobicents communication platform. In *Proceedings of the 2nd International Conference on Next Generation Mobile Applications, Services and Technologies,* (pp. 634-639). Cardiff, UK: IEEE.

Devarapalli, V., Wakikawa, R., Petrescu, A., & Thubert, P. (2005). *Network mobility (NEMO) basic support protocol.* IETF RFC 3963. Retrieved from http://tools.ietf.org/html/rfc3963

Díaz, A., Merino, P., & Rivas, F. J. (2010). QoS analysis of video streaming service in live cellular networks. *Computer Communications, 33*(3), 322–335. doi:10.1016/j.comcom.2009.09.007

Dimakis, A., Godfrey, P., Wu, Y., Wainwright, M., & Ramchandran, K. (2010). Network coding for distributed storage systems. *IEEE Transactions on Information Theory, 56*(9), 4539–4551. doi:10.1109/TIT.2010.2054295

DiNucci, D. (1999). Fragmented future. *Design & New Media, 53*(4).

Dixon, R. L. (Ed.). (1988). *MRI: Acceptance testing and quality control - The role of the clinical medical physicist.* Madison, WI: Medical Physics Publishing Corporation.

DLBOIS. (2002). *Distance learning benefits organizations, individuals and society.* Retrieved November 4, 2002, from http://www.ciscoworldmagazine.com/monthly/2001/04/distance.shtml

Dogan, S. (2002). Personalised multimedia services for real-time video over 3G mobile networks. In *Proceedings of the Third International Conference on 3G Mobile Communication Technologies.* IEEE.

Dohler, M., Watteyne, T., & Alonso-Zárate, J. (2010). Machine-to-machine: An emerging communication paradigm. In *Proceedings of GlobeCom2010.* IEEE.

Dougiamas, M., & Taylor, P. (2003). Moodle: Using learning communities to create an open source course management system. In *Proceedings of EDMEDIA 2003.* EDMEDIA.

Dovrolis, C., Ramanathan, P., & Moore, D. (2004). Packet dispersion techniques and a capacity-estimation methodology. *IEEE/ACM Transactions on Networking, 12*(6), 963–977. doi:10.1109/TNET.2004.838606

Downie, J. S. (2008). The music information retrieval evaluation exchange (2005–2007): A window into music information retrieval research. *Acoustical Science and Technology, 29*(4), 247–255. doi:10.1250/ast.29.247

Droms, R. (2003). *Dynamic host configuration protocol for IPv6 (DHCPv6).* IETF RFC 3315. Retrieved from http://www.ietf.org/rfc/rfc3315.txt

Du, Q., Shang, S., Lu, H., & Tang, X. (2008). A fast MPEG-2 to H.264 downscaling transcoder. In *Proceedings of the 8th Conference on Signal Processing, Computational Geometry, and Artificial Vision,* (pp. 230-233). IEEE.

Duchowski, A. T. (2000). Acuity-matching resolution degradation through wavelet coefficient scaling. *IEEE Transactions on Image Processing, 9*(8), 1437–1440. doi:10.1109/83.855439

Eck, D., Lamere, P., Bertin-Mahieux, T., & Green, S. (2007). Automatic generation of social tags for music recommendation. In *Proceedings of the 21st Annual Conference on Neural Information Processing Systems.* Vancouver, Canada: IEEE.

Eleftheriadis, A., & Anastassiou, D. (1995). Constrained and general dynamic rate shaping of compressed digital video. In *Proceedings of the IEEE International Conference on Image Processing,* (vol. 3, pp. 396–399). IEEE Press.

Eleftheriadis, A., Civanlar, R., & Shapiro, O. (2006). Multipoint videoconferencing with scalable video coding. *Journal of Zhejiang University – Science A, 7*(5), 696-705.

Engelke, U., Zepernick, H.-J., & Maeder, A. (2009). Visual attention modeling: Region-of-interest versus fixation patterns. In *Proceedings of the Picture Coding Symposium, 2009, PCS 2009,* (pp. 1-4). PCS.

Fahy, P. J. (2003). Indicators of support in online interaction. *International Review of Research in Open and Distance Learning, 4*(1).

Fang, Z., Guoliang, Z., & Zhanjiang, S. (2001). Comparison of different implementations of MFCC. *Journal of Computer Science and Technology, 16*(6), 582–589. doi:10.1007/BF02943243

Färber, N., Steinbach, E., & Girod, B. (1996). Robust H.263 compatible transmission over wireless channels. In *Proceedings of the International Coding Picture Symposium*, (pp. 575-578). IEEE.

Färber, N., Döhla, S., & Issing, J. (2006). Adaptive progressive download based on the MPEG-4 file format. *Journal of Zhejiang University Science A, 7*(1), 106–111. doi:10.1631/jzus.2006.AS0106

Faugeras, O. (1993). *Three - dimensional computer vision: A geometric viewpoint.* Cambridge, MA: MIT Press.

Felder, R., & Brent, R. (2001). Effective strategies for cooperative learning. *Journal of Cooperation & Collaboration in College Teaching, 10*(2), 69–75.

Felipe, J. C., Traina, A. J. M., & Traina, C., Jr. (2003). Retrieval by content of medical images using texture for tissue identification. In *Proceedings of 16th IEEE Symposium on Computer-Based Medical Systems*, (pp. 175-180). IEEE Press.

FemtoForum. (2010). *Femtocells – Natural solution for offload – a Femto forum brief.* FemtoForum.

Fenner, B., Handley, M., Holbrook, H., & Kouvelas, I. (2006). *Protocol independent multicast - sparse mode (PIM-SM): Protocol specification (revised).* IETF RFC 4601. Retrieved from http://tools.ietf.org/html/rfc4601

Ferman, A., Errico, J., van Beek, P., & Sezan, M. (2002). Content-based filtering and personalization using structured metadata. In *Proceedings of the 2nd ACM/IEEE-CS Joint Conference on Digital Libraries (JCDL 2002)*, (pp. 393-393). New York, NY: ACM Press.

Ferrell, C. (1998). Emotional robots and learning during social exchanges. In *Proceedings of Agents in Interaction- Acquiring Competence through Imitation: Papers from a Workshop at the Second International Conference on Autonomous Agents (Autonomous Agents 1998).* Autonomous Agents.

Ferscha, A., Hechinger, M., Riener, A., Schmitzberger, H., Franz, M., Rocha, M., & Zeidler, A. (2006). Context-aware profiles. In *Proceedings of the Autonomic and Autonomous Systems, ICAS 2006.* ICAS.

FFmpeg. (2011). *Website.* Retrieved January 20, 2012, from http://ffmpeg.org/

Fielding, R. (2000). *Architectural styles and the design of network-based software architectures.* (PhD Thesis). University of California. Irvine, CA.

Fischer, W. (2008). *Digital video and audio broadcasting technology* (2nd ed.). Berlin, Germany: Springer-Verlag.

Florido, G., Liberal, F., & Fajardo, J. O. (2009). QoS-oriented admission control in HSDPA networks. *Network Protocols and Algorithms, 1*(1), 52–61. doi:10.5296/npa.v1i1.178

Foo, B., Andreopoulos, Y., & Van der Schaar, M. (2008). Analytical rate-distortion-complexity modeling of wavelet-based video coders. *IEEE Transactions on Signal Processing, 56*, 797–815. doi:10.1109/TSP.2007.906685

Foote, J., & Uchihashi, S. (2001). The beat spectrum: A new approach to rhythm analysis. In *Proceedings of the IEEE International Conference on Multimedia and Expo.* Tokyo, Japan: IEEE Press.

Ford, L. R., & Fulkerson, D. R. (1956). Maximal flow through a network. *Canadian Journal of Mathematics.* Retrieved from http://www.cs.yale.edu/homes/lans/readings/routing/ford-max_flow-1956.pdf

Forough, S., & Reza, J. (2012). A comparative study of context modeling approches and applying in an infrustructure. *Canadian Journal on Data Information and Knowledge Engineering, 3*(1).

Forum, U. M. T. S. (2010). *Recognising the promise of mobile broadband.* White Paper. Washington, DC: UMTS.

Forum, W. (2001). *WAG UAProf, wireless application protocol.* Retrieved from http://www.wapforum.org

Freeman, W. T., & Adelson, E. H. (1991). The design and use of steerable filters. *IEEE Transactions on Pattern Analysis and Machine Intelligence, 13*(9), 891–906. doi:10.1109/34.93808

Frieden, B. R. (1983). *Probability, statistical optics and data testing.* Berlin, Germany: Springer-Verlag. doi:10.1007/978-3-642-96732-0

Frossard, P., & Vercheure, O. (2001). Joint source/FEC rate selection for quality optimal MPEG-2 video delivery. *IEEE Transactions on Image Processing, 10*(12), 1815–1825. doi:10.1109/83.974566

Fu, D., Hammond, K., & Swain, M. (1994). Vision and navigation in man-made environments: Looking for syrup in all the right places. In *Proceedings of CVPR Workshop on Visual Behaviors,* (pp. 20-26). Seattle, WA: IEEE Press.

Fuente, Y., et al. (2011). iDASH: Improved dynamic adaptive streaming over HTTP using scalable video coding. In *Proceedings of the Second Annual ACM Conference on Multimedia Systems (MMSys 2011)*, (pp. 257-264). New York, NY: ACM Press.

Fukunaga, K. (1990). Introduction to statistical pattern recognition. In *Computer Science and Scientific Computing*. New York, NY: Academic Press.

Fu, Z., Lu, G., Ting, K., & Zhang, D. (2011). A survey of audio-based music classification and annotation. *IEEE Transactions on Multimedia, 2*(13), 303–319. doi:10.1109/TMM.2010.2098858

Gabidulin, E. M. (1985). Theory of codes with maximum rank distance. *Problemy Peredachi Informatsii, 21*(1), 3–16.

Galotti, K. M., Clinchy, B. M., Ainsworth, K., Lavin, B., & Mansfield, A. F. (1999). A new way of assessing ways of knowing: The attitudes towards thinking and learning survey (ATTLS). *Sex Roles, 40*(9/10), 745–766. doi:10.1023/A:1018860702422

Gamier, S. J., & Bilbro, G. L. (1995). Magnetic resonance image restoration. *Journal of Mathematical Imaging and Vision, 5*, 7–19. doi:10.1007/BF01250250

Gao, W., Reader, C., Wu, F., He, Y., Yu, L., & Lu, H. … Pan, X. (2004). AVS—The Chinese next-generation video coding standard. In *Proceedings of the National Association of Broadcasters (NAB) Conference*. Las Vegas, NV: NAB.

Garton, L., Haythorthwaite, C., & Wellman, B. (1997). Studying on-line social networks. In Jones, S. (Ed.), *Doing Internet Research*. Thousand Oaks, CA: Sage.

Geleijnse, G., Schedl, M., & Knees, P. (2007). The quest for ground truth in musical artist tagging in the social web era. In *Proceedings of the Eighth International Conference on Music Information Retrieval (ISMIR 2007)*, (pp. 525-530). Vienna, Austria: ISMIR.

Gerig, G., Kubler, O., Kikinis, R., & Jolesz, F. A. (1992). Nonlinear anisotropic filtering of MRI data. *IEEE Transactions on Medical Imaging, 11*(2), 221–232. doi:10.1109/42.141646

Gershenfeld, N. (1999). *The nature of mathematical modeling*. Cambridge, UK: Cambridge University Press.

Ghanbari, M., Fleury, M., Khan, E., et al. (2006). *Future performance of video codecs*. London, UK: Office of Communications (Ofcom).

Ghanbari, M. (2011). *Standard codecs: Image compression to advanced video coding* (3rd ed.). Stevenage, UK: IET Press. doi:10.1049/PBTE054E

Ghinea, G., & Thomas, J. P. (1998). QoS impact on user perception and understanding of multimedia video clips. In *Proceedings of the 6th ACM International Conference on Multimedia*, (pp. 49-54). Bristol, UK: ACM Press.

Gill, P., Arlitt, M., Li, Z., & Mahant, A. (2007). YouTube traffic characterization: A view from the edge. In *Proceedings of the 7th ACM SIGCOMM Conference on Internet Measurement,* (pp. 15-28). ACM Press.

Girma, B., Brunie, L., & Pierson, J.-M. (2006). Planning-based multimedia adaptation services composition for pervasive computing. In *Proceedings of the 2nd International Conference on Signal-Image Technology and Internet-Based Systems (SITIS 2006)*. Hammamet, Tunisia: ACM/IEEE.

Girma, B., Lionel, B., & Jean-Marc, P. (2005). Content adaptation in distributed. *Journal of Digital Information Management, 3*(2), 95–100.

Girod, B., & Färber, N. (1999). Feedback-based error control for mobile video transmission. *Proceedings of the IEEE, 87*(10), 1707–1723. doi:10.1109/5.790632

Gkantsidis, C., & Rodriguez, P. (2005). Network coding for large scale content distribution. In *Proceedings of 24th Annual Joint Conference of the IEEE Computer and Communications Societies*, (Vol. 4, pp. 2235-2245). IEEE Press.

Goldberg, D., Nichols, D., Oki, B., & Terry, D. (1992). Using collaborative filtering to weave an information tapestry. *Communications of the ACM, 35*(12), 61–70. doi:10.1145/138859.138867

González, M. C., Hidalgo, C. A., & Barabási, A. L. (2008). Understanding individual human mobility patterns. *Nature, 453*, 779–782. doi:10.1038/nature06958

Google Corporation. (2008). *Google browser sync*. Retrieved October 13, 2011, from http://www.google.com/tools/firefox/browsersync/

Google Corporation. (2009a). *The wave protocol*. Retrieved October 13, 2011, from http://www.waveprotocol.org

Google Corporation. (2009b). *The Google wave project*. Retrieved October 13, 2011, from http://wave.google.org

Google Corporation. (2009c). *The Google wave click-to-dial*. Retrieved October 13, 2011, from http://googlewavedev.blogspot.com/2009/06/twiliobot-bringing-phone-conversations.html

Gopalan, R. (2009). *Exploiting region-of-interest for improved video coding*. (PhD Thesis). The Ohio State University. Columbus, OH.

Goto, M. (2012). Grand challenges in music information research. In Mueller, M., Goto, M., & Schedl, M. (Eds.), *Multimodal Music Processing* (pp. 217–226). Schloss Dagstuhl-Leibniz-Zentrum fuer Informatik.

Goyal, V. K. (2001). Multiple description coding: Compression meets the network. *IEEE Signal Processing Magazine, 18*(5), 74–93. doi:10.1109/79.952806

GPAC. (2011). *MP4Box | GPAC*. Retrieved January 10, 2012, from http://gpac.wp.mines-telecom.fr/mp4box/

Grafe, T., & Scheffler, G. (1988). Inter field noise and cross color reduction IC for flicker free TV receivers. *IEEE Transactions on Consumer Electronics, 34*(3), 402–408. doi:10.1109/30.20134

Greco, C., & Cagnazzo, M. (2011). A cross-layer protocol for cooperative content delivery over mobile ad-hoc networks. *International Journal of Communication Networks and Distributed Systems, 7*(1-2), 49–63. doi:10.1504/IJCNDS.2011.040977

Green, S. J., Lamere, P., Alexander, J., Maillet, F., Kirk, S., & Holt, J. … Mak, X. W. (2009). Generating transparent, steerable recommendations from textual descriptions of items. In *Proceedings of the Third ACM Conference on Recommender Systems, (RecSys 2009)*, (pp. 281–284). New York, NY: ACM Press.

Greengrass, J., Evans, J., & Begen, A. (2009). Not all packets are equal, part 2: The impact of network packet loss on video quality. *IEEE Internet Computing, 13*(2), 74–82. doi:10.1109/MIC.2009.40

Grois, D., & Hadar, O. (2011a). Efficient adaptive bit-rate control for scalable video coding by using computational complexity-rate-distortion analysis. In *Proceedings of the 2011 IEEE International Symposium on Broadband Multimedia Systems and Broadcasting (BMSB)*, (pp. 1-6). Nuremberg, Germany: IEEE Press.

Grois, D., & Hadar, O. (2011b). Recent advances in region-of-interest coding. In J. del ser Lorente (Ed.), *Recent Advances on Video Coding*, (pp. 49-76). Intech.

Grois, D., & Hadar, O. (2011c). Complexity-aware adaptive bit-rate control with dynamic ROI pre-processing for scalable video coding. In *Proceedings of the 2011 IEEE International Conference on Multimedia and Expo (ICME)*, (pp. 1-4). Barcelona, Spain: IEEE Press.

Grois, D., & Hadar, O. (2011d). Complexity-aware adaptive spatial pre-processing for ROI scalable video coding with dynamic transition region. In *Proceedings of the International Conference on Image Processing (ICIP 2011)*. Brussels, Belgium: ICIP.

Grois, D., Kaminsky, E., & Hadar, O. (2009). Buffer control in H.264/AVC applications by implementing dynamic complexity-rate-distortion analysis. In *Proceedings of the 2009 IEEE International Symposium on Broadband Multimedia Systems and Broadcasting, BMSB 2009*, (pp. 1-7). IEEE Press.

Grois, D., Kaminsky, E., & Hadar, O. (2010a). ROI adaptive scalable video coding for limited bandwidth wireless networks. In *Proceedings of Wireless Days (WD), 2010 IFIP*, (pp. 1-5). IFIP.

Grois, D., Kaminsky, E., & Hadar, O. (2010b). Adaptive bit-rate control for region-of-interest scalable video coding. In *Proceedings of the 2010 IEEE 26th Convention of Electrical and Electronics Engineers in Israel (IEEEI)*, (pp. 761-765). IEEE Press.

Grois, D., Kaminsky, E., & Hadar, O. (2010c). Optimization methods for H.264/AVC video coding. In Angelides, M. C., & Agius, H. (Eds.), *The Handbook of MPEG Applications: Standards in Practice*. Chichester, UK: John Wiley & Sons, Ltd.doi:10.1002/9780470974582.ch7

Grois, D., Kaminsky, E., & Hadar, O. (2011a). Efficient real-time video-in-video insertion into a pre-encoded video stream. *ISRN Signal Processing, 2011*, 1–11. doi:10.5402/2011/975462

Grosskurth, A., & Godfrey, M. (2005). A reference architecture for web browsers. In *Proceedings of the 21st IEEE International Conference on Software Maintenance (ICSM 2005)*, (pp. 661-664). IEEE Press.

Gudbjartsson, H., & Patz, S. (1995). The Rician distribution of noisy MRI data. *Magnetic Resonance in Medicine, 34*, 910–914. doi:10.1002/mrm.1910340618

Gunawardena, C., Lowe, C., & Anderson, T. (1997). Analysis of a global online debate and the development of an interaction analysis model for examining social construction of knowledge in computer conferencing. *Journal of Educational Computing Research, 17*(4), 397–431. doi:10.2190/7MQV-X9UJ-C7Q3-NRAG

Gundavelli, S., Leung, K., Devarapalli, V., Chowdhury, K., & Patil, B. (2008). *Proxy mobile IPv6*. IETF RFC 5213. Retrieved from http://tools.ietf.org/html/rfc5213

Gurbani, V. K., & Sun, X.-H. (2004). Terminating telephony services on the Internet. *IEEE/ACM Transactions on Networking, 12*(4), 471–481. doi:10.1109/TNET.2004.833145

Gürler, C. G., Görkemli, B., Saygili, G., & Tekalp, M. (2011). Flexible transport of 3D video over networks. *Proceedings of the IEEE, 99*(4), 694–707. doi:10.1109/JPROC.2010.2100010

Guy, I. (2011). *Social recommender systems*. Paper presented at WWW 2011. Hyderabad, India.

Hamburg, I., Lindecke, C., & Thij, H. (2003). Social aspects of e-learning and blending learning methods. In *Proceedings of the 4th European Conference E-COMM-LINE*. Bucharest, Romania: E-COMM-LINE.

Han, S., & Vasconcelos, N. (2008). Object-based regions of interest for image compression. In *Proceedings of the Data Compression Conference, 2008*, (pp. 132-141). IEEE.

Handley, M., Floyd, S., Padhye, J., & Widmer, J. (2003). TCP friendly rate control (TFRC) protocol specification. RFC 3448. Retrieved from http://www.ietf.org/rfc/rfc3448.txt

Handley, M., Jacobson, V., & Perkins, C. (2006). *SDP: Session description protocol*. RFC 4566. Retrieved from http://tools.ietf.org/html/rfc4566

Handley, M., Schulzrinne, H., Schooler, E., & Rosenberg, J. (1999). *SIP: Session initiation protocol*. IETF RFC 2543. Retrieved from http://www.ietf.org/rfc/rfc2543.txt

Hanfeng, C., Yiqiang, Z., & Feihu, Q. (2001). Rapid object tracking on compressed video. In *Proceedings of the 2nd IEEE Pacific Rim Conference on Multimedia*, (pp. 1066-1071). IEEE Press.

Hannuksela, M. M., Wang, Y. K., & Gabbouj, M. (2004). Isolated regions in video coding. *IEEE Transactions on Multimedia, 6*(2), 259–267. doi:10.1109/TMM.2003.822784

Haritaoglu, I., Harwood, D., & Davis, L. S. (2000). W^4: Real-time surveillance of people and their activities. *IEEE Transactions on Pattern Analysis and Machine Intelligence, 22*(8), 809–830. doi:10.1109/34.868683

Hartung, F., & Girod, B. (1998). Watermarking of uncompressed and compressed video. *IEEE Transactions on Signal Processing, 66*(3), 283–301.

Haskell, P., & Messerschmitt, D. (1992). Resynchronization of motion compensated video affected by ATM cell loss. In *Proceedings of IEEE International Conference on Acoustics, Speech, and Signal Processing*, (pp. 545-548). IEEE Press.

Heeren, E. (1996). *Technology support for collaborative distance learning*. (Doctoral Dissertation). University of Twente. Enschede, The Netherlands.

Hellberg, C., Greene, D., & Boyes, T. (2007). *Broadband network architectures: Designing and deploying triple-play services.* Upper Saddle River, NJ: Prentice Hall.

Hellge, C., Gomez-Barquero, D., Schierl, T., & Wiegand, T. (2011). Layer-aware forward error correction for mobile broadcast of layered media. *IEEE Transactions on Multimedia, 13*(3), 551–562. doi:10.1109/TMM.2011.2129499

Henkelman, R. M. (1985). Measurement of signal intensities in the presence of noise in MR images. *Medical Physics, 2*(2), 232–233. doi:10.1118/1.595711

Henri, F. (1992). Computer conferencing and content analysis. In Kaye, A. R. (Ed.), *Collaborative Learning through Computer Conferencing: The Najaden Papers* (pp. 117–136). Berlin, Germany: Springer-Verlag. doi:10.1007/978-3-642-77684-7_8

Hersh, W., Mailhot, M., Arnott-Smith, C., & Lowe, H. (2001). Selective automated indexing of findings and diagnoses in radiology reports. *Journal of Biomedical Informatics, 34*, 262–273. doi:10.1006/jbin.2001.1025

He, Z., Cheng, W., & Chen, X. (2008). Energy minimization of portable video communication devices based on power-rate-distortion optimization. *IEEE Transactions on Circuits and Systems for Video Technology, 18*(5), 596–608. doi:10.1109/TCSVT.2008.918802

Hinden, R., & Deering, S. (2006). *IP version 6 addressing architecture.* IETF RFC 4291. Retrieved from http://tools.ictf.org/html/rfc4291

Ho, T., Koetter, R., Médard, M., Karger, D. R., & Effros, M. (2003). The benefits of coding over routing in a randomized setting. In *Proceedings of IEEE International Symposium on Information Theory*, (p. 442). IEEE Press.

Hoffman, D., Fernando, G., Goyal, V., & Civanlar, M. (1998) *RTP payload format for MPEG1/MPEG2 video.* RFC 2250. Retrieved from http://tools.ietf.org/rfc/rfc2250.txt

Holdschneider, M., Kronland-Martinet, R., Morlet, J., & Tchamitchian, P. (1989). A real-time algorithm for signal analysis with the help of the wavelet transform. In Combes, J.-M., Grossmann, A., & Tchamitchian, P. (Eds.), *Time-Frequency Methods and Phase Space* (pp. 286–297). Berlin, Germany: Springer-Verlag.

Holme, P., Edling, C. R., & Liljeros, F. (2004). Structure and time evolution of an internet dating community. *Social Networks, 26*, 155–174. doi:10.1016/j.socnet.2004.01.007

Hooper, C. J., & Rettberg, J. W. (2011). Experiences with geographical collaborative systems: Playfulness in geosocial networks and geocaching. In *Proceedings of MobileHCI 2011*. Stockholm, Sweden: MobileHCI.

Horn, B. K. (1986). *Robot vision.* Cambridge, MA: MIT Press.

Howe, J. (2006, June). The rise of crowdsourcing. *Wired.*

Howe, J. (2012). *Crowdsourcing: A definition.* Retrieved January 31, 2012 from http://crowdsourcing.typepad.com

Hsieh, M.-D., Wang, T.-P., Tsai, C.-S., & Tseng, C.-C. (2006). Stateful session handoff for mobile WWW. *Information Sciences, 176*(9), 1241–1265. doi:10.1016/j.ins.2005.02.009

Hsu, C. C., Chu, W. W., & Taira, R. K. (1996). A knowledge-based approach for retrieving images by content. *IEEE Transactions on Knowledge and Data Engineering, 8*(4), 522–532. doi:10.1109/69.536245

HTML5. (2009). *HTML 5 differences from HTM.* Retrieved October 13, 2011, from http://www.w3.org/TR/html5-diff/

Hu, B., Dasmahapatra, S., Lewis, P., & Shadbolt, N. (2003). Ontology-based medical image annotation with description logics. In *Proceedings 15th IEEE International Conference on Tools with Artificial Intelligence*, (pp. 77- 82). Sacramento, CA: IEEE Press.

Huang, H. K., Nielsen, J. F., Nelson, M. D., & Liu, L. (2005). Image-matching as a medical diagnostic support tool (DST) for brain diseases in children. *Computerized Medical Imaging and Graphics, 29*(2-3), 195–202. doi:10.1016/j.compmedimag.2004.09.008

Huang, Y.-W., Hsieh, B.-Y., Chien, S.-Y., Ma, S.-Y., & Chen, L.-G. (2006). Analysis and complexity reduction of multiple reference frames motion estimation in H.264/AVC. *IEEE Transactions on Circuits and Systems for Video Technology, 16*(4), 507–522. doi:10.1109/TCSVT.2006.872783

Hu, Y., Rajan, D., & Chia, L. (2008). Detection of visual attention regions in images using robust subspace analysis. *Journal of Visual Communication and Image Representation*, *19*(3), 199–216. doi:10.1016/j.jvcir.2007.11.001

Hwang, J.-N. (2009). *Multimedia networking: From theory to practice*. Cambridge, UK: Cambridge University Press. doi:10.1017/CBO9780511626654

IEEE. (2009). *IEEE standard for local and metropolitan area networks- Part 21: Media independent handover*. IEEE Std 802.21-2008. Retrieved from http://ieeexplore.ieee.org/xpl/articleDetails.jsp?tp=&arnumber=4769367&contentType=Standards&sortType%3Dasc_p_Sequence%26filter%3DAND%28p_Publication_Number%3A4769363%29

IFPI. (2011). Digital music report 2011. *International Federation of the Phonographic Industry*. Retrieved February 7, 2012, from http://www.ifpi.org/content/library/DMR2011.pdf

Immerkaer, J. (1996). Fast noise variance estimation. *Computer Vision and Image Understanding*, *64*(2), 300–302. doi:10.1006/cviu.1996.0060

inCode Telecom Group Inc. (2006). *The quad-play – The first wave of the converged services evolution*. White paper. Retrieved from http://www.techrepublic.com/whitepapers/the-quad-play-the-first-wave-of-the-converged-services-evolution/288116

Indulska, J., Robinson, R., Rakotonirainy, A., & Henricksen, K. (2003). Experiences in using CC/PP in context-aware systems. In *Proceedings of Mobile Data Management 4th International Conference (MDM 2003)*, (pp. 247-261). Melbourne, Australia: Springer.

Information Society SA. (2012). *Digital "wreath" for ancient Olympia*. Retrieved February 01, 2012 from http://www. infosoc. gr/infosoc/en-UK/culture/specials/default. htm

International Telecommunication Union. Telecommunication Standardization Sector. (2002). *Recommendation BT 500-11, methodology for the subjective assessment of the quality of television pictures*. Geneva, Switzerland: ITU-T.

IPTV Focus Group. (2012). *Website*. Retrieved from http://www.itu.int/ITUT/IPTV/

ISO/IEC. (2000). International standard 13818-2 information technology — Generic coding of moving pictures and associated audio information: Video (mpeg-2/h.262) video buffering verifier. In *Annex C* (2nd ed.). Washington, DC: ISO.

Issariyakul, T., & Hossain, E. (2009). *Introduction to the ns2 simulator*. Berlin, Germany: Springer Verlag. doi:10.1007/978-0-387-71760-9

ITU-T Rec J.281. (2005). *Website*. Retrieved from http://www.itu.int/rec/T-REC-J.281-200503-I

ITU-T. (1997). *Hypothetical reference decoder: Video coding for low bit rate communication*. Annex B. Retrieved from http://www.itut.int

Ivanov, Y. V., & Bleakley, C. J. (2010). Real-time H.264 video encoding in software with fast mode decision and dynamic complexity control. *ACM Transactions on Multimedia Computer Communication Applications*, *6*(1).

Iverson, P., & Krumhansi, C. (1993). Isolating the dynamic attributes of musical timbre. *The Journal of the Acoustical Society of America*, *94*(5), 2595–2603. doi:10.1121/1.407371

Iwatsuki, K., Kani, J., Suzuki, H., & Fujiwara, M. (2004). Access and metro networks based on WDM technologies. *Journal of Lightwave Technology*, *22*(11), 2623. doi:10.1109/JLT.2004.834492

Jack, K. (2007). *Video demystified* (5th ed.). Amsterdam, The Netherlands: Newnes.

Jaggi, S., Sanders, P., Chou, P., Effros, M., Egner, S., & Jain, K. (2005). Polynomial time algorithms for multicast network code construction. *IEEE Transactions on Information Theory*, *51*(6), 1973–1982. doi:10.1109/TIT.2005.847712

Jain, R., Bakker, J., & Anjum, F. (2005). *Programming converged networks: Call control in Java, XML, and Parlay/OSA*. New York, NY: Wiley Interscience.

Jammeh, E., Fleury, M., & Ghanbari, M. (2008). Fuzzy logic congestion control of transcoded video streaming without packet loss feedback. *IEEE Transactions on Circuits and Systems for Video Technology*, *18*(3), 387–393. doi:10.1109/TCSVT.2008.918459

Jamrozik, M. L., & Hayes, M. H. (2002). A compressed domain video object segmentation system. In *Proceedings of the IEEE International Conference on Image Processing*, (vol. 1, pp. 113-116). IEEE Press.

Jannach, D., & Leopold, K. (2007). Knowledge-based multimedia adaptation for ubiquitous multimedia consumption. *Journal of Network and Computer Applications*, 30(3), 958–982. doi:10.1016/j.jnca.2005.12.007

Jarnikov, D., & Özçelebi, T. (2011). Client intelligence for adaptive streaming solutions. *Signal Processing Image Communication*, 26(7), 378–389. doi:10.1016/j.image.2011.03.003

Jaulent, M.-C., Le Bozec, C., Cao, Y., Zapletal, E., & Degoulet, P. (2000). A property concept frame representation for flexible image content image content retrieval in histopathology databases. In *Proceedings of Symposium of the American Society for Medical Informatics*, (pp. 379-383). Los Angeles, CA: ASMI.

Jedidi, A., Amous, I., & Sèdes, F. (2005). A contribution to multimedia document modeling and querying. *Multimedia Tools and Applications*, 25(3), 391–404. doi:10.1007/s11042-005-6542-7

Jensen, J. (2009). *Feature extraction for music information retrieval*. Aalborg, Denmark: Aalborg University.

Jensen, J. R. (1996). *Introductory digital image processing: A remote sensing perspective* (2nd ed.). Englewood Cliffs, NJ: Prentice Hall. doi:10.1080/10106048709354084

Jeon, D.-S., & Park, H.-W. (2009). An adaptive reference frame selection method for multiple reference frame motion estimation in the H.264/AVC, In *Proceedings of the 2009 16th IEEE International Conference on Image Processing (ICIP)*, (pp. 629-632). IEEE Press.

Jeong, C. Y., Han, S. W., Choi, S. G., & Nam, T. Y. (2006). An objectionable image detection system based on region of interest. In *Proceedings of the 2006 IEEE International Conference on Image Processing*, (pp. 1477-1480). IEEE Press.

Jeong, C.-Y., & Hong, M.-C. (2008). Fast multiple reference frame selection method using inter-mode correlation. In *Proceedings of the 14th Asia-Pacific Conference on Communications, 2008, APCC 2008*, (pp. 1-4). APCC.

Jiang, J., Guo, B., & Mo, W. (2008). Efficient intra refresh using motion affected region tracking for surveillance video over error prone networks. In *Proceedings of the International Conference on Intelligent System Design and Applications*, (pp. 242 –246). IEEE.

Jiao, L., Zhou, J., & Chen, R. (2011). Efficient parallel intra-prediction mode selection scheme for 4x4 blocks In *Proceedings of the 2011 International Conference on Intelligent Computation Technology and Automation (ICICTA)*, (vol. 2, pp. 527-530). ICICTA.

Ji, S., & Park, H. W. (2000). Moving object segmentation in DCT-based compressed video. *Electronics Letters*, 36(21). doi:10.1049/el:20001279

Johnson, D. W., & Johnson, R. T. (1999). *Learning together and alone, cooperative, competitive and individualistic learning*. Needham Heights, MA: Allyn and Bacon.

Jones, S. (1995). Computer-mediated communication and community: Introduction. *Computer-Mediated Communication Magazine*, 2(3), 38.

Jostschulte, K., & Amer, A. (1998). A new cascaded spatio-temporal noise reduction scheme for interlaced video. *IEEE Proceedings International Conference on Image Processing, ICIP 1998*, (vol. 2, pp. 493-497). Chicago, IL: IEEE Press.

Jouanot, F., Cullot, N., & Yétongnon, K. (2003). Context comparison for object fusion. *Lecture Notes in Computer Science*, 2681, 1031. doi:10.1007/3-540-45017-3_36

JSVM. (2009). *JSVM software manual, ver. JSVM 9.19 (CVS tag: JSVM_9_19), Nov. 2009*. JSVM.

Jun, D.-S., & Park, H.-W. (2010). An efficient priority-based reference frame selection method for fast motion estimation in H.264/AVC. *IEEE Transactions on Circuits and Systems for Video Technology*, 20(8), 1156–1161. doi:10.1109/TCSVT.2010.2057016

Kafle, V. P., Kamioka, E., & Yamada, S. (2006). MoRaRo: Mobile router-assisted route optimization for network mobility (NEMO) support. *IEICE Transactions in Information & Systems*, E89-D(1).

Kagal, L., Korolev, V., Chen, H., Anupam, J., & Finin, T. (2001). Centaurus: A framework for intelligent services in a mobile environment. In *Proceedings of the 21st International Conference on Distributed Computing System Workshop*. Mesa.

Kagal, L., Korolev, V., Avancha, S., Joshi, A., Finin, T., & Yesha, Y. (1995). Centaurus: An infrastructure for service management in ubiquitous computing environments. *Wireless Networks*, 1.

Kalva, H., Adzic, V., & Furht, B. (2012). Comparing MPEG AVC and SVC for adaptive HTTP streaming. In *Proceedings of the IEEE International Conference on Consumer Electronics*, (pp. 160-161). IEEE Press.

Kaminsky, E., Grois, D., & Hadar, O. (2008). Dynamic computational complexity and bit allocation for optimizing H.264/AVC video compression. *Journal of Visual Communication and Image Representation*, 19(1), 56–74. doi:10.1016/j.jvcir.2007.05.002

Kanellopoulos, D. (2011). Quality of service in networks supporting cultural multimedia applications. *Program: Electronic Library and Information Systems*, 45(1), 50–66. doi:10.1108/00330331111107394

Kannangara, C. S., Richardson, I. E., Bystrom, M., & Zhao, Y. (2009). Complexity control of H.264/AVC based on mode-conditional cost probability distributions. *IEEE Transactions on Multimedia*, 11(3), 433–442. doi:10.1109/TMM.2009.2012937

Kannangara, C. S., Richardson, I. E., & Miller, A. J. (2008). Computational complexity management of a real-time H.264/AVC encoder. *IEEE Transactions on Circuits and Systems for Video Technology*, 18(9), 1191–1200. doi:10.1109/TCSVT.2008.928881

Karczewicz, M., & Kurceren, R. (2003). The SP-and SI-frames design for H.264/AVC. *IEEE Transactions on Circuits and Systems for Video Technology*, 13(7), 637–644. doi:10.1109/TCSVT.2003.814969

Karlsson, L. S., & Sjostrom, M. (2005). Improved ROI video coding using variable Gaussian pre-filters and variance in intensity. In *Proceedings of the IEEE International Conference on Image Processing, 2005, ICIP 2005*, (vol. 2, pp. 313-316). IEEE Press.

Kasai, H., Nilsson, M., Jebb, T., Whybray, M., & Tominaga, H. (2002). The development of a multimedia transcoding system for mobile access to video conferencing. *IEICE Transactions on Communications*, 10(2), 2171–2181.

Kas, C., & Nicolas, H. (2009). Compressed domain indexing of scalable H.264/SVC streams. *Signal Processing Image Communication*, 484–498. doi:10.1016/j.image.2009.02.007

Kasprowski, R. (2010). Perspectives on DRM: Between digital rights management and digital restrictions management. *Bulletin of the American Society for Information Science and Technology*, 36(3), 49–54. doi:10.1002/bult.2010.1720360313

Kass, M., Witkin, A., & Terzopoulos, D. (1988). SNAKES: Active contour models. *International Journal of Computer Vision*, 1(4), 321–331. doi:10.1007/BF00133570

Katti, S., Rahul, H., Hu, W., Katabi, D., Medard, M., & Crowcroft, J. (2008). XORs in the air: Practical wireless network coding. *IEEE/ACM Transactions on Networking*, 16(3), 497–510. doi:10.1109/TNET.2008.923722

Kaufman, L., Kramer, D. M., Crooks, L. E., & Ortendahl, D. A. (1989). Measuring signal-to-noise ratios in MR imaging. *Radiology*, 173, 265–267.

Keiser, G. (2006). *FTTX concepts and application*. New York, NY: John Wiley & Sons, Inc. doi:10.1002/047176910X

Kent, S., & Seo, K. (2005). *Security architecture for the internet protocol*. IETF RFC 4301. Retrieved from http://tools.ietf.org/html/rfc4301

Kim, B. M., & Li, Q. (2004). Probabilistic model estimation for collaborative filtering based on items attributes. In *Proceedings of the IEEE/WIC/ACM International Conference on Web Intelligence (WI 2004)*, (pp. 185–191). Beijing, China: IEEE/ACM Press.

Kim, D.-K., & Wang, Y.-F. (2009). Smoke detection in video. In *Proceedings of the 2009 WRI World Congress on Computer Science and Information Engineering*, (vol. 5, pp. 759-763). WRI.

Kim, H., & Altunbasak, Y. (2004). Low-complexity macroblock mode selection for H.264/AVC encoders. In *Proceedings of the International Conference on Image Processing*, (pp. 765-768). Singapore, Singapore: IEEE.

Kim, J., & Jeong, J. (2011). Fast intra mode decision algorithm using the sum of absolute transformed differences. In *Proceedings of the 2011 International Conference on Digital Image Computing Techniques and Applications (DICTA)*, (pp. 655-659). DICTA.

Kim, J., & Kang, S. (2011). An ontology-based personalized target advertisement system on interactive TV. In *Proceedings of the IEEE International Conference on Consumer Electronics*, (pp. 895-896). IEEE Press.

Kim, T., Hwang, U., & Jeong, J. (2011). Efficient block mode decision and prediction mode selection for intra prediction in H.264/AVC high profile. In *Proceedings of the 2011 International Conference on Digital Image Computing Techniques and Applications (DICTA)*, (pp. 645-649). DICTA.

Kim, Y., Kim, W., & Jeong, J. (2011). Fast intra mode decision algorithm using sub-sampled pixels. In *Proceedings of the 2011 IEEE 15th International Symposium on Consumer Electronics (ISCE)*, (pp. 290-293). IEEE Press.

Kim, J.-H., & Kim, B.-G. (2011). Efficient intra-mode decision algorithm for inter-frames in H.264/AVC video coding. *IET Image Processing*, 5(3), 286–295. doi:10.1049/iet-ipr.2009.0097

Kim, W., You, J., & Jeong, J. (2010). Complexity control strategy for real-time H.264/AVC encoder. *IEEE Transactions on Consumer Electronics*, 56(2), 1137–1143. doi:10.1109/TCE.2010.5506050

Kincaid, J. (2008). *A peek at brightkite for the iPhone.* Retrieved January 15, 2012 from http://techcrunch.com/2008/10/16/a-peek-at-brightkite-for-the-iphone/

Klaue, J. (2007). *EvalVid - A video quality evaluation tool-set.* Retrieved January 10, 2012, from http://www2.tkn.tu-berlin.de/research/evalvid/

Kloppel, S. (2008). Accuracy of dementia diagnosis-a direct comparison between radiologists and a computerized method. *Brain*, 131(11), 2969–2974. doi:10.1093/brain/awn239

Klyne, G., & Carroll, J. (2004). *Resource description framework (RDF): Concepts and abstract syntax.* Retrieved from http://www.w3.org/TR/2004/REC-rdf-concepts-20040210/

Klyne, G., Reynolds, F., Woodrow, C., Hidetaka, O., Hjelm, J., Butler, M. H., & Tran, L. (2004). *Composite capability/preference profiles (CC/PP): Structure and vocabularies 1.0.* Retrieved from http://www.w3.org/Mobile/CCPP/

Knees, P., Pohle, T., Schedl, M., & Widmer, G. (2007a). A music search engine built upon audio-based and web-based similarity measures. In *Proceedings of the 30th Annual International ACM SIGIR Conference on Research and Development in Information Retrieval*, (pp. 447-454). New York, NY: ACM Press.

Knees, P., Schedl, M., Pohle, T., & Widmer, G. (2007b). Exploring music collections in virtual landscapes. *IEEE MultiMedia*, 14(3), 46–54. doi:10.1109/MMUL.2007.48

Knoke, D., & Kuklinski, J. H. (1982). *Network analysis.* Beverly Hills, CA: Sage Publications.

Kodikara Arachchi, H., Fernando, W. A. C., Panchadcharam, S., & Weerakkody, W. A. R. J. (2006). Unequal error protection technique for ROI based H.264 video coding. In *Proceedings of the Canadian Conference on Electrical and Computer Engineering*, (pp. 2033-2036). Ottawa, Canada: IEEE.

Koetter, R., & Kschischang, F. R. (2008). Coding for errors and erasures in random network coding. *IEEE Transactions on Information Theory*, 54(8), 2579–3591. doi:10.1109/TIT.2008.926449

Koetter, R., & Médard, M. (2003). An algebraic approach to network coding. *IEEE/ACM Transactions on Networking*, 11(5), 782–795. doi:10.1109/TNET.2003.818197

Koodli, R. (2009). *Mobile IPv6 fast handovers.* IETF RFC 5568. Retrieved from http://tools.ietf.org/html/rfc5568

Kornfeld, M., & Reimers, U. (2005). DVB-H — The emerging standard for mobile data communication. *EBU Technical Review.* Received from http://tech.ebu.ch/docs/techreview/trev_301-dvb-h.pdf

Korzenny, F. (1978). A theory of electronic propinquity: Mediated communication in organizations. *Communication Research*, 5, 3–23. doi:10.1177/009365027800500101

Koshizuka, N., & Sakamura, K. (2000). Tokyo university digital museum. In *Proceedings of the 2000 Kyoto International Conference on Digital Libraries: Research and Practice*, (pp. 85-92). IEEE.

Kosmach, J., Neff, R., Sherwood, G., Tapia, J., & Veselinovic, D. (2008). Introduction to the opencore video components used in the Android platform. In *Proceedings of the 34th International Conference: New Trends in Audio for Mobile and Handheld Devices*. IEEE.

Kovács, J., Bokor, L., & Jeney, G. (2011). Performance evaluation of GNSS aided predictive multihomed NEMO configurations. In *Proceedings of the 2011 11th International Conference on ITS Telecommunications*, (pp. 293-298). St. Petersburg, Russia: ITST.

Kovács, Á., Gódor, I., Rácz, S., & Borsos, T. (2010). Cross-layer quality-based resource reservation for scalable multimedia. *Computer Communications, 33*(3), 283–292. doi:10.1016/j.comcom.2009.09.006

Krause, E., et al. (1991). *Method and apparatus for refreshing motion compensated sequential video images*. US 5,057,916. Washington, DC: United States Patent Office.

Krebs, V. (2004). *An introduction to social network analysis*. Retrieved November 9, 2004 from http://www.orgnet.com/sna.html

Krechmer, K. (2009). Open standards: A call for change. *IEEE Communications Magazine, 47*(5), 88–94. doi:10.1109/MCOM.2009.4939282

Kuhn, M., & Antkowiak, J. (2000). *Statistical multiplex what does it mean for DVB-T?*Berlin, Germany: FKT Fachzeitschrift für Fernsehen, Film und Elektronische Medien.

Kulikowski, J. L. (2006). Interpretation of medical images based on ontological models. *Lecture Notes in Artificial Intelligence, 4029*, 919–924.

Kumar, A. (2007). *Mobile TV: DVB-H, DMB, 3G systems and rich media applications*. Amsterdam, The Netherlands: Focal Press.

Kumar, S., Xu, L., Mandal, M., & Panchanathan, S. (2006). Error resiliency schemes in H. 264/AVC standard. *Elsevier Journal of Visual Communication and Image Representation, 17*, 425–450. doi:10.1016/j.jvcir.2005.04.006

Kuo, T.-Y., & Lu, H.-J. (2008). Efficient reference frame selector for H.264. *IEEE Transactions on Circuits and Systems for Video Technology, 18*(3), 400–405. doi:10.1109/TCSVT.2008.918111

Kurhila, J., Miettinen, M., Nokelainen, P., & Tirri, H. (2004). The role of the learning platform in student-centered e-learning. In *Proceedings of the 4th IEEE International Conference on Advanced Learning Technologies*. IEEE Press.

Kurutepe, E., Aksay, A., Bilen, C., Gürler, C. G., Sikora, T., Akar, G. B., & Tekalp, A. M. (2007). A standards-based, flexibile, end-to-end multi-view streaming architecture. In *Proceedings of the International Packet Video Workshop*, (pp. 302-307). IEEE.

Kurutepe, E., Civanlar, M. R., & Tekalp, A. M. (2007a). Client-driven selective streaming of multiview video for interactive 3DTV. *IEEE Transactions on Circuits and Systems for Video Technology, 17*(11), 1558–1565. doi:10.1109/TCSVT.2007.903664

Kwon, H., Han, H., Lee, S., Choi, W., & Kang, B. (2010). New video enhancement preprocessor using the region-of-interest for the videoconferencing. *IEEE Transactions on Consumer Electronics, 56*(4), 2644–2651. doi:10.1109/TCE.2010.5681152

Laborie, S., Euzenat, J., & Layaïda, N. (2011). Semantic adaptation of multimedia documents. *Multimedia Tools and Applications, 55*(3), 379–398. doi:10.1007/s11042-010-0552-9

Lacoste, C., Lim, J.-H., Chevallet, J.-P., & Le, D. T. H. (2007). Medical-image retrieval based on knowledge-assisted text and image indexing. *IEEE Transactions on Circuits and Systems for Video Technology, 17*(7), 889–900. doi:10.1109/TCSVT.2007.897114

Laghos, A. (2005). FESNeL: A methodological framework for assessing the evolutionary structure of social networks in e-learning. In *Proceedings of the Junior Researchers of the European Association for Research on Learning and Instruction 11th Biennial JURE/EARLI Conference*. Nicosia, Cyprus: JURE.

Laghos, A., & Zaphiris, P. (2006). *Sociology of student-centred e-learning communities: A network analysis*. Paper presented at the e-Society 2006 Conference. Dublin, Ireland.

Laghos, A., & Zaphiris, P. (2007). Investigating student roles in online student-centered learning. *International Council for Educational Media Conference (ICEM 2007)*. Nicosia, Cyprus: ICEM.

Lakshman, T. V., Ortega, A., & Reibman, A. R. (1998). VBR video: Trade-offs and potentials. *Proceedings of the IEEE, 86*(5), 952–973. doi:10.1109/5.664282

Lambert, P., Schrijver, D. D., Van Deursen, D., De Neve, W., Dhondt, Y., & Van de Walle, R. (2006). A real-time content adaptation framework for exploiting ROI scalability in H.264/AVC. In *Proceedings of Advanced Concepts for Intelligent Vision Systems* (pp. 442–453). IEEE. doi:10.1007/11864349_40

Lambrecht, C., & Verscheure, O. (1996). Perceptual quality measure using a spatiotemporal model of the human visual system. In *Proceedings- SPIE The International Society for Optical Engineering* (Vol. 2668, pp. 450–461). San Jose, CA: SPIE.

Lamere, P. (2008). Social tagging and music information retrieval. *Journal of New Music Research, 37*(2), 101–114. doi:10.1080/09298210802479284

Lao, Z. (2004). Morphological classification of brains via high-dimensional shape transformations and machine learning methods. *NeuroImage, 21*, 46–57. doi:10.1016/j.neuroimage.2003.09.027

Laplace, S., Dalmau, M., & Roose, P. (2008). Kalinahia: Considering quality of service to design and execute distributed multimedia applications. In *Proceedings of the IEEE/IFIP International Conference on Network Management and Management Symposium*. Salvador de Bahia, Brazil: IEEE Press.

Laurier, C., Grivolla, J., & Herrera, P. (2008). Multimodal music mood classification using audio and lyrics. In *Proceedings of 7th International Conference on Machine Learning and Applications (ICMLA 2008)*, (pp. 688–693). ICMLA.

Law, E. L. M., Von Ahn, L., Dannenberg, R. B., & Crawford, M. (2007). Tagatune: A game for music and sound annotation. In *Proceedings of the 8th International Conference on Music Information Retrieval (ISMIR 2007)*, (pp. 361–364). Vienna, Austria: ISMIR.

Lee, H., Jung, B., Jung, J., & Jeon, B. (2009). Computational complexity scalable scheme for power-aware H.264/AVC encoding. In *Proceedings of the IEEE International Workshop on Multimedia Signal Processing, 2009, MMSP 2009*, (pp. 1-6). IEEE Press.

Lee, J., & Ahn, S. (2006). I-FHMIPv6: *A novel FMIPv6 and HMIPv6 integration mechanism*. Retrieved from http://draft-jaehwoon-mipshop-ifhmipv6-01.txt

Lee, Y.-M., & Lin, Y.-Y. (2006). A fast intermode mode decision for H.264 video coding. In *Proceedings of the 2006 8th International Conference on Signal Processing*, (vol. 2). IEEE.

Lee, J. S. (1980). Digital image enhancement and noise filtering by use of local statistics. *IEEE Transactions on Pattern Analysis and Machine Intelligence, 2*(2), 165–168. doi:10.1109/TPAMI.1980.4766994

Lee, J. S. (1981). Refined filtering of image noise using local statistics. *Computer Vision Graphics and Image Processing, 15*, 380–389. doi:10.1016/S0146-664X(81)80018-4

Lee, J.-B., & Kalva, H. (2008). *The VC-1 and H.264 video compression standards for broadband video services*. New York, NY: Springer. doi:10.1007/978-0-387-71043-3

Lee, J., & McKendree, J. (1999). Learning vicariously in a distributed environment. *Journal of Active Learning, 10*, 4–10.

Lee, J., & Park, H. (2011). A fast mode decision method based on motion cost and intra prediction cost for H.264/AVC. *IEEE Transactions on Circuits and Systems for Video Technology, 22*(3), 393–402. doi:10.1109/TCSVT.2011.2163460

Lee, K., Jeon, G., & Jeong, J. (2009). Fast reference frame selection algorithm for H.264/AVC. *IEEE Transactions on Consumer Electronics, 55*(2), 773–779. doi:10.1109/TCE.2009.5174453

Lefol, D., Bull, D., & Canagarajah, N. (2006). Performance evaluation of transcoding algorithms for H.264. *IEEE Transactions on Consumer Electronics, 25*(1), 215–222.

Lei, Y. Q., Wang, Y.-G., & Liang, F. (2010). Overlapping interval differences-based fast intra mode decision for H.264/AVC. In *Proceedings of the 2010 Sixth International Conference on Intelligent Information Hiding and Multimedia Signal Processing (IIH-MSP),* (pp. 659-663). IIH-MSP.

Lemlouma, T., & Layaïda, N. (2002). Content adaptation and generation principles for heterogeneous clients. In *Proceedings of the W3C Workshop on Device Independent Authoring Techniques.* St. Leon-Rot, Germany: W3C.

Lemlouma, T., & Layaïda, N. (2001). *NAC: A basic core for the adaptation and negotiation of multimedia services. Opera Project.* INRIA.

Lew, M. S. (2006). Content-based multimedia information retrieval: State of the art and challenges. *ACM Transactions on Multimedia Computing, Communications, and Applications, 2*(1), 1–19. doi:10.1145/1126004.1126005

Li, B., Sullivan, G. J., & Xu, J. (2011). Comparison of compression performance of HEVC working draft 4 with AVC high profile. In *Proceedings of 7th Meeting of Joint Collaborative Team on Video Coding (JCT-VC).* JCT-VC.

Li, Z. G., Yao, W., Rahardja, S., & Xie, S. (2007). New framework for encoder optimization of scalable video coding. In *Proceedings of the 2007 IEEE Workshop on Signal Processing Systems,* (pp. 527-532). IEEE Press.

Li, Z., Zhang, X., Zou, F., & Hu, D. (2010). Study of target detection based on top-down visual attention. In *Proceedings of the 2010 3rd International Congress on Image and Signal Processing (CISP),* (vol. 1, pp. 377-380). CISP.

Lian, S.-J., Chien, S.-Y., Lin, C.-P., Tseng, P.-C., & Chen, L.-G. (2007). Power-aware multimedia: Concepts and design perspectives. *IEEE Circuits and Systems Magazine, 7*(2), 26–34. doi:10.1109/MCAS.2007.4299440

Liao, J., & Villasenor, J. (1996). Adaptive intra update for video coding over noisy channels. In *Proceedings of IEEE International Conference on Image Processing (ICIP),* (vol. 3, pp. 763-766). Lausanne, Switzerland: IEEE Press.

Liem, C. S., Müller, M., Eck, D., & Tzanetakis, G. (2011). 1st International ACM workshop on music information retrieval with user-centered and multimodal strategies. In *Proceedings of the 2011 ACM Multimedia Conference and Co-Located Workshops,* (pp. 603-604). Scottsdale, AZ: ACM Press.

Liew, C. (2006). Online cultural heritage exhibitions: A survey of strategic issues. *Program: Electronic Library and Information Systems, 40*(4), 372–388. doi:10.1108/00330330610707944

Lim, K.-P., Sullivan, G., & Wiegand, T. (2005b). Text description of joint model reference encoding methods and decoding concealment methods. *Study of ISO/IEC 14496-10 and ISO/IEC 14496-5/ AMD6 and Study of ITU-T Rec. H.264 and ITU-T Rec. H.2.64.2, in Joint Video Team (JVT) of ISO/IEC MPEG and ITU-T VCEG,* Busan, Korea, Doc. JVT-O079.

Linden, G., Smith, B., & York, J. (2003). Amazon.com recommendations: Item-to-item collaborative filtering. *IEEE Internet Computing, 7*(1), 76–80. doi:10.1109/MIC.2003.1167344

Linner, D., Stein, H., Staiger, U., & Steglich, S. (2010). Real-time communication enabler for web 2.0 applications. In *Proceedings of the Sixth International Conference on Networking and Services (ICNS 2010),* (pp. 42-48). Cancun, Mexico: ICNS.

Lin, W., Panusopone, K., Baylon, D. M., & Sun, M.-T. (2010). A computation control motion estimation method for complexity-scalable video coding. *IEEE Transactions on Circuits and Systems for Video Technology, 20*(11), 1533–1543. doi:10.1109/TCSVT.2010.2077773

Lin, Y., & Bhanu, B. (2005). Object detection via feature synthesis using MDL-based genetic programming. *IEEE Transactions on Systems, Man, and Cybernetics. Part B, Cybernetics, 35*(3), 538–547. doi:10.1109/TSMCB.2005.846656

Li, S.-Y. R., Yeung, R. W., & Cai, N. (2003). Linear network coding. *IEEE Transactions on Information Theory, 49*(2), 371–381. doi:10.1109/TIT.2002.807285

Liu, B., Sun, M., Liu, Q., Kassam, A., Li, C.-C., & Sclabassi, R. J. (2006). Automatic detection of region of interest based on object tracking in neurosurgical video. In *Proceedings of the 27th International Conference of the Engineering in Medicine and Biology Society, 2005,* (pp. 6273-6276). IEEE Press.

Liu, L., Zhang, S., Ye, X., & Zhang, Y. (2005). Error resilience schemes of H.264/AVC for 3G conversational video services. In *Proceedings of the Fifth International Conference on Computer and Information Technology,* (pp. 657- 661). Binghamton, NY: IEEE.

Liu, Z., Wu, Z., Liu, H., & Stein, A. (2007). A layered hybrid-ARQ scheme for scalable video multicast over wireless networks. In *Proceedings of the Asilomar Conference on Signals, Systems and Computers,* (pp. 914-919). Asilomar.

Liu, Y., Li, Z. G., & Soh, Y. C. (2008). Rate control of H.264/AVC scalable extension. *IEEE Transactions on Circuits and Systems for Video Technology, 18*(1), 116–121. doi:10.1109/TCSVT.2007.903325

Li, X., Wien, M., & Ohm, J.-R. (2011, Jul.). Rate-Complexity-Distortion optimization for hybrid video coding. *IEEE Transactions on Circuits and Systems for Video Technology, 21*(7), 957–970. doi:10.1109/TCSVT.2011.2133750

Li, Z., Pan, F., Lim, K. P., Feng, G., Lin, X., & Rahardja, S. (2003). Adaptive basic unit layer rate control for JVT. In *Joint Video Team (JVT) of ISO/IEC MPEG and ITU-T VCEG (ISO/IEC JTC1/SC29/WG11 and ITU-T SG16 Q.6), Doc. JVT-G012.* Pattaya, Thailand: JVT.

Lloret, J., Atenas, M., Canovas, A., & Garcia, M. (2011b). A network management algorithm based on 3D coding techniques for stereoscopic IPTV delivery. In *Proceedings of the IEEE Global Communications Conference (IEEE Globecomm 2011).* Houston, TX: IEEE Press.

Lloret, J., Canovas, A., Rodrigues, J. J. P. C., & Lin, K. (2011c). A network algorithm for 3D/2D IPTV distribution using WiMAX and WLAN technologies. *Journal of Multimedia Tools and Applications.* Retrieved from http://rd.springer.com/article/10.1007/s11042-011-0929-4

Lloret, J., Garcia, M., Atenas, M., & Canovas, A. (2011). A QoE management system to improve the IPTV network. *International Journal of Communication Systems, 24*(1), 118–138. doi:10.1002/dac.1145

Lloret, J., Garcia, M., Atenas, M., & Canovas, A. (2011a). A stereoscopic video transmission algorithm for an IPTV network based on empirical data. *International Journal of Communication Systems, 24*(10), 1298–1329. doi:10.1002/dac.1196

Lloret, J., García, M., & Boronat, F. (2009b). *IPTV: La televisión por Internet.* Malaga, Spain: Publicaciones Vértice.

Lloret, J., Garcia, M., Edo, M., & Lacuesta, R. (2009a). *IPTV distribution network access system using WiMAX and WLAN technologies.* New York, NY: ACM Press.

Lockley, E., Pritchard, C., & Foster, E. (2004). Professional evaluation: Students supporting students – Lessons learnt from an environmental health peer support scheme. *Journal of Environmental Health Research, 3*(2).

Logan, B., & Salomon, A. (2001). A music similarity function based on signal analysis. In *Proceedings of the IEEE International Conference on Multimedia and Expo.* Tokyo, Japan: IEEE Press.

Lohmar, T., Einarsson, T., Fröjdh, P., Gabin, F., & Kampmann, M. (2011). Dynamic adaptive HTTP streaming of live content. In *Proceedings of IEEE Symposium on the World of Wireless, Mobile and Multimedia Networks,* (pp. 1-8). IEEE Press.

Lonsdale, P., Baber, C., Sharples, M., Byrne, W., Arvanitis, T. N., Brundell, P., & Beale, R. (2004). Context awareness for MOBIlearn: Creating an engaging learning experience in an art museum. In *Proceedings of the 3rd MLEARN Conference,* (pp. 115-118). Rome, Italy: MLEARN.

Lu, Z., Lin, W., Li, Z., Pang Lim, K., Lin, X., Rahardja, S., Ping Ong, E., & Yao, S. (2005b). *Perceptual region-of-interest (ROI) based scalable video coding.* JVT-O056, Busan, KR.

Lu, Z., Peng, W.-H., Choi, H., Thang, T. C., & Shengmei, S. (2005a). *CE8: ROI-based scalable video coding.* JVT-O308, Busan, KR.

Luby, M., Gasiba, T., Stockhammer, T., & Watson, M. (2007). Reliable multimedia download delivery in cellular broadcast networks. *IEEE Transactions on Broadcasting, 53*(1), 235–246. doi:10.1109/TBC.2007.891703

Luby, M., Stockhammer, T., & Watson, M. (2008). Application layer FEC in IPTV services. *IEEE Communications Magazine*, *45*(5), 95–101.

Ma, S., Gao, W., Wu, F., & Lu, Y. (2003). Rate control for JVT video coding scheme with HRD considerations. In *Proceedings of the IEEE International Conference on Multimedia and Expo*, (pp. 793-796). IEEE Press.

Mahedero, J. P. G., Martinez, A., Cano, P., Koppenberger, M., & Gouyon, F. (2005). Natural language processing of lyrics. In *Proceedings of the 13th Annual ACM International Conference on Multimedia, MULTIMEDIA 2005*, (pp. 475–478). New York, NY: ACM Press.

Ma, K. J., Bartos, R., Bhatia, S., & Nair, R. (2011). Mobile video delivery with HTTP. *IEEE Communications Magazine*, *49*(4), 166–175. doi:10.1109/MCOM.2011.5741161

Makris, C., Panagis, Y., Sakkopoulos, E., & Tsakalidis, A. (2004). *Efficient search algorithms for large scale web service data*. Technical Report No. CTI-TR-2004/09/01. Patras, Greece: Research Academic Computer Technology Institute.

Mandel, M., & Ellis, D. (2005). Song-level features and support vector machines for music classification. In *Proceedings of the 6th International Conference on Music Information Retrieval (ISMIR 2005)*, (pp. 594-599). Barcelona, Spain: ISMIR.

Manerba, F., Benois-Pineau, J., Leonardi, R., & Mansencal, B. (2008). Multiple object extraction from compressed video. *EURASIP Journal on Advances in Signal Processing*. Retrieved from http://asp.eurasipjournals.com/content/2008/1/231930

Manning, C. D., Raghavan, P., & Schutze, H. (2008). *Introduction to information retrieval*. Cambridge, UK: Cambridge University Press. doi:10.1017/CBO9780511809071

Manzalini, A. (2008). Tomorrow's open internet for telco and web federation. In *Proceedings of the International Conference on Complex, Intelligent and Software Intensive Systems (CISIS 2008)*, (pp. 567-572). Barcelona, Spain: CISIS.

Mao, S., Lin, S., Panwar, S., & Wang, Y. (2001). Reliable transmission of video over ad-hoc networks using automatic repeat request and multi-path transport. In *Proceedings of IEEE Vehicular Technology Conference*, (pp. 615-619). IEEE Press.

Marquet, A., Monteiro, J., Martins, N., & Nunes, M. (2010). Quality of experience vs. QoS in video transmission. In *Quality of Service Architectures for Wireless Networks* (pp. 352–376). Hershey, PA: IGI Global. doi:10.4018/978-1-61520-680-3.ch016

Martin, J., & Trummer, C. (2005). SCALEX – A personalized multimedia information system for museums and exhibitions. In *Proceedings of the First International Conference on Automated Production of Cross Media Content for Multi-Channel Distribution (AXMEDIS 2005)*. AXMEDIS.

Mason, R. (1991). Analyzing computer conferencing interactions. *Computers in Adult Education and Training*, *2*(3), 161–173.

Mastin, G. A. (1985). Adaptive filters for digital noise smoothing: An evaluation. *Computer Vision Graphics and Image Processing*, *31*, 103–121. doi:10.1016/S0734-189X(85)80078-5

Mate, S., Chandra, U., & Curcio, I. (2006). Moveble-multimedia: Session mobility in ubiquitous computing ecosystem. In *Proceedings of the 5th International Conference on Mobile and Ubiquitous Multimedia (MUM 2006)*. Stanford, CA: MUM.

McCann, J., Deering, S., & Mogul, J. (1996). *Path MTU discovery for IP version 6*. IETF RFC 1981. Retrieved from http://www.ietf.org/rfc/rfc1981.txt

McCarthy, J. D., Sasse, M. A., & Miras, D. (2004). Sharp or smooth? Comparing the effects of quantization versus frame rate for streamed video. In *Proceedings of the SIGCHI Conference on Human Factors Computing Systems*, (pp. 535-542). Vienna, Austria: ACM Press.

McCreary, E. (1990). Three behavioural models for computer-mediated communication. In Harasim, L. (Ed.), *Online Education: Perspectives on a New Environment*. New York, NY: Praeger.

McGlashan, S., Melanchuk, T., & Boulton, C. (2011). *An interactive voice response (IVR) control package for the media control channel framework*. RFC 6231. Retrieved from http://tools.ietf.org/html/rfc6231

McGlashan, S., Melanchuk, T., & Boulton, C. (2012). *A mixer control package for the media control channel framework*. RFC 6505. Retrieved from http://www.rfc-editor.org/rfc/rfc6505.txt

McLoughlin, C. (2004). *A learning conversations: Dynamics, collaboration and learning in computer mediated communication*. Washington, DC: TAFE Media Network.

McMillan, S. J. (2002). A four-part model of cyber-interactivity: Some cyber-places are more interactive than others. *New Media & Society, 4*(2), 271–291.

Mead, S., Hilton, D., & Curtis, L. (2001). Peer support: A theoretical perspective. *Psychiatric Rehabilitation Journal, 25*, 134–141.

MEDIACTRL. (2012). *IETF media server control prototype*. Retrieved from http://mediactrl.sourceforge.net

MediaPlayer Android. (2012). *Website*. Retrieved from http://developer.android.com/reference/android/media/MediaPlayer.html

Meer, P., Jolion, J., & Rosenfeld, A. (1990). A fast parallel algorithm for blind estimation of noise variance. *IEEE Transactions on Pattern Analysis and Machine Intelligence, 12*(2), 216–223. doi:10.1109/34.44408

Meer, P., Park, R., & Cho, K. (1994). Multiresolution adaptive image smoothing. *Graphical Models and Image Processing, 44*, 140–148. doi:10.1006/cgip.1994.1013

Meesters, L. M. J., Ijsselsteijn, W. A., & Seuntiems, P. J. H. (2004). A survey of perceptual evaluations and requirements of three-dimensional TV. *IEEE Transactions on Circuits and Systems for Video Technology, 14*(3), 381–391. doi:10.1109/TCSVT.2004.823398

Mehmood, M. O. (2009). Study and implementation of color-based object tracking in monocular image sequences. In *Proceedings of the 2009 IEEE Student Conference on Research and Development (SCOReD)*, (pp. 109-111). IEEE Press.

Melville, P., Mooney, R. J., & Nagarajan, R. (2002). Content-boosted collaborative filtering for improved recommendations. In *Proceedings of the Eighteenth National Conference on Artificial Intelligence (AAAI 2002)*, (pp. 187-192). Edmonton, Alberta: AAAI.

Melville, P., & Sindhwani, V. (2010). Recommender systems. In Sammut, C., & Webb, G. (Eds.), *Encyclopedia of Machine Learning*. Berlin, Germany: Springer.

Mencattini, A., Salmeri, M., Bertazzoni, S., & Salsano, A. (2003). Noise variance estimation in digital images using iterative fuzzy procedure. *WSEAS Transactions on Systems, 4*(2), 1048–1056.

Meng, B., Li, M.-Z., & Ren, Y.-H. (2010). Fast mode selection for H.264/AVC based on MB motion characteristics. In *Proceedings of the 2010 International Conference on Intelligent Computing and Integrated Systems (ICISS)*, (pp. 205-208). ICISS.

Meng, J., & Chang, S. F. (1998). Embedding visible video watermarks in the compressed domain. In *Proceedings of the IEEE International Conference on Image Processing*, (Vol. 1, pp. 474-477). IEEE Press.

Menn, J. (2003). *All the rave: The rise and fall of Shawn Fanning's Napster*. New York, NY: Crown Business.

Merkle, P., Müller, K., Smolic, A., & Wiegand, T. (2007). Efficient compression of multiview depth data based on MVC. In *Proceedings of the 3DTV Conference*, (pp. 1-4). 3DTV.

Merkle, P., Morvan, Y., Smolic, A., Farin, D., Müller, K., de With, P. H. N., & Wiegand, T. (2009). The effects of multiview depth video compression on multiviewrendering. *Signal Processing Image Communication, 24*(1–2), 73–88. doi:10.1016/j.image.2008.10.010

Merkle, P., Smolic, A., Müller, K., & Wiegend, T. (2007a). Efficient prediction structures for multiview video coding. *IEEE Transactions on Circuits and Systems for Video Technology, 17*(11), 1461–1473. doi:10.1109/TCSVT.2007.903665

Metcalfe, B. (1992). *Internet fogies to reminisce and argue at interop conference*. New York, NY: InfoWorld.

Meyer, E., Grussenmeyer, P., Perrin, J.-P., Durand, A., & Drap, P. (2007). A web information system for the management and the dissemination of cultural heritage data. *Journal of Cultural Heritage*, 8(4), 396–411. doi:10.1016/j.culher.2007.07.003

Miao, C.-H., & Fan, C.-P. (2011). Efficient mode selection with extreme value detection based pre-processing algorithm for H.264/AVC fast intra mode decision. In *Proceedings of TENCON 2011 - 2011 IEEE Region 10 Conference*, (pp. 316-320). IEEE Press.

Micheloni, C., Salvador, E., Bigaran, F., & Foresti, G. L. (2005). An integrated surveillance system for outdoor security. In *Proceedings of the IEEE Conference on Advanced Video and Signal Based Surveillance, 2005, AVSS 2005*, (pp. 480-485). IEEE Press.

Microsystems, S. (2001a). *JAIN SIP release 1.1 specification, 2001*. Retrieved October 13, 2011, from http://www.jcp.org/en/jsr/detail?id=289

Microsystems, S. (2001b). *JAIN SIP release 1.0 specification, 2001*. Retrieved October 13, 2011, from http://www.jcp.org/aboutJava/communityprocess/final/jsr032

Mims, C. (2010). *How itunes genius really works*. Retrieved February 7, 2012, from http://www.technology-review.com/blog/mimssbits/25267/

Miotto, R., Barrington, L., & Lanckriet, G. (2010). Improving auto-tagging by modelling semantic co-occurrences. In *Proceedings of the International Society of Music Information Retrieval Conference (ISMIR 2010)*, (pp. 297-302) Utrecht, The Netherlands: ISMIR.

Mirra, M. (2009). *Web sharing and RichDraw*. Retrieved October 13, 2011, from http://hyperstruct.net/2007/2/24/xml-sync-islands-let-the-Web-sharing-begin

Mitiche, A. (1994). *Computational analysis of visual motion*. New York, NY: Plenum Press.

Mobicents. (2009). *The Mobicents open source SLEE and SIP server*. Retrieved October 13, 2011, from http://www.mobicents.org/index.html

Moiron, S., Ali, I., Ghanbari, M., & Fleury, M. (2010). Limitations of multiple reference frames with cyclic intra-refresh line for H.264/AVC. *Electronics Letters*, 47(2), 103–104. doi:10.1049/el.2010.3018

Moody, N. M. (2011). iTunes great for Apple but was it for music biz? *Associated Press*. Retrieved February 7, 2012, from http://technology.inquirer.net/5209/itunes-great-for-apple-but-was-it-for-music-biz/

Mozilla Corporation. (2008). *A prototype of Mozilla weave*. Retrieved October 13, 2011, from http://labs.mozilla.com/2007/12/introducing-weave/

Mu, M., Gostner, R., Mauthe, A., Tyson, G., & Garcia, F. (2009). Visibility of individual packet loss on H. 264 encoded video stream--A user study on the impact of packet loss on perceived video quality. In *Proceedings of the Sixteenth Annual Multimedia Computing and Networking (MMCN 2009)*. San Jose, CA: SPIE.

Mu, M., Mauthe, A., & Garcia, F. (2008). A utility-based QoS model for emerging multimedia applications. In *Proceedings of the Second International Conference on Next Generation Mobile Applications, Services and Technologies (NGMAST 2008)*, (pp. 521-528). Cardiff, UK: IEEE Press.

Müller, H. (2004). A review of content-based image retrieval systems in medical applications - Clinical benefits and future directions. *International Journal of Medical Informatics*, 73(1), 1–23. doi:10.1016/j.ijmedinf.2003.11.024

Mulligan, C. E. A. (2009). Open API standardization for the NGN platform. *IEEE Communications Magazine*, 47(5), 108–113. doi:10.1109/MCOM.2009.4939285

Munkongpitakkun, W., Kamolphiwong, S., & Sae-Wong, S. (2007). Enhanced web session mobility based on SIP. In *Proceedings of the 4th International Conference on Mobile Technology, Applications and Systems (Mobility2007)*, (pp. 346-350). Singapore, Singapore: Mobility.

Murase, H., & Nayar, S. (1995). Visual learning and recognition of 3d objects from appearance. *International Journal of Computer Vision*, 14(1), 5–24. doi:10.1007/BF01421486

Murphy, S. (2011). *The culture of recommendation: Has it gone too far?* Retrieved Jan 31, 2012 from http://www.suzemuse.com/2011/02/the-culture-of-recommendation-has-it-gone-too-far/

Murphy, B. W., Carson, P. L., Ellis, J. H., Zhang, Y. T., Hyde, R. J., & Chenevert, T. L. (1993). Signal-to-noise measures for magnetic resonance imagers. *Magnetic Resonance Imaging, 11*, 425–428. doi:10.1016/0730-725X(93)90076-P

Murtagh, F., & Starck, J. L. (1998). Noise detection and filtering using multiresolution transform methods. *Astronomical Data Analysis Software and Systems, 145*, 449–456.

Murtagh, F., & Starck, J. L. (2000). Image processing through multiscale analysis and measurement noise modeling. *Statistics and Computing Journal, 10*, 95–103. doi:10.1023/A:1008938224840

Murtagh, F., Starck, J. L., & Bijaoui, A. (1995). Multiresolution in astronomical image processing: A general framework. *International Journal of Imaging Systems and Technology, 6*, 332–338. doi:10.1002/ima.1850060406

MusicBrainz. (2012). *General FAQ*. Retrieved February 7, 2012, from http://musicbrainz.org/doc/General_FAQ#Why_does_MusicBrainz_not_support_genre_information.3F

Musmann, H. G. (2006). Genesis of the MP3 audio coding standard. *IEEE Transactions on Consumer Electronics, 52*(3), 1043–1049. doi:10.1109/TCE.2006.1706505

Mustafah, Y. M., Bigdeli, A., Azman, A. W., & Lovell, B. C. (2009). Face detection system design for real time high resolution smart camera. In *Proceedings of the Third ACM/IEEE International Conference on Distributed Smart Cameras, 2009, ICDSC 2009*, (pp. 1-6). ACM/IEEE.

Nakajima, Y., Hori, H., & Kanoh, T. (1995). Rate conversion of MPEG coded video by requantization process. In *Proceedings of the IEEE International Conference on Image Processing*, (pp. 408-411). IEEE Press.

Narayanan, R., Ponnappan, S., & Reichenbach, S. (2001). Effects of uncorrelated and correlated noise on image information content. In *Proceedings of the IEEE International Geoscience and Remote Sensing Symposium*, (pp. 1898-1900). IEEE Press.

Narayanan, R., Ponnappan, S., & Reichenbach, S. (2003). Effects of noise on information content of remote sensing images. *Geocarto International, 18*(2), 15–26. doi:10.1080/10106040308542269

Narkhede, N. S., & Kant, N. (2009). The emerging H.264/AVC advanced video coding standard and its applications. In *Proceedings of International Conference on Advances in Computing, Communication and Control*, (pp. 300-305). IEEE.

Narten, T., Nordmark, E., Simpson, W., & Soliman, H. (2007). *Neighbor discovery for IP version 6 (IPv6)*. IETF RFC 4861. Retrieved from http://tools.ietf.org/html/rfc4861

National Research Council. (2010). Highway capacity manual. In *Transportation Research Board*. Washington, DC: National Research Council.

NDH. (2012, January 8). *Netflix's blog*. Retrieved Jan 15, 2012 from http://blog.netflix.com/

Ndili, O., & Ogunfunmi, T. (2006). On the performance of a 3D flexible macroblock ordering for H.264/AVC. *Digest of Technical Papers International Conference on Consumer Electronics*, 37-38.

Neal, J., Green, R., & Landovskis, J. (2001). Interactive channel for multimedia satellite networks. *IEEE Communications Magazine, 39*(3), 192–198. doi:10.1109/35.910607

Nemoianu, I., Greco, C., Cagnazzo, M., & Pesquet-Popescu, B. (2012). A framework for joint multiple description coding and network coding over wireless ad-hoc networks. In *Proceedings of IEEE International Conference on Acoustics, Speech and Signal Processing*. IEEE Press.

NetLMM. I. (2012). *Network-based localized mobility management (NetLMM) WG homepage*. Retrieved from http://datatracker.ietf.org/wg/netlmm/charter/

Neumayer, R., Lidy, T., & Rauber, A. (2005). Content-based organization of digital audio collections. In *Proceedings of the 5th Open Workshop of MUSICNETWORK*. Vienna, Austria: MUSICNETWORK.

Nguyen, K., Nguyen, T., & Cheung, S.-C. (2007). Peer-to-peer streaming with hierarchical network coding. In *Proceedings of the IEEE International Conference on Multimedia and Expo*, (pp. 396-399). IEEE Press.

Niazi, M. A., & Hussain, A. (2011). Social network analysis of trends in the consumer electronics domain. In *Proceedings of the IEEE International Conference on Consumer Electronics (ICCE)*. Las Vegas, NV: IEEE Press.

Nielsen, J. (2006). *Participation inequality: Encouraging more users to contribute.* Retrieved Oct 16 2011 from http://www.useit.com/alertbox/participation_inequality.html

Nightingale, J. M., Wang, Q., & Grecos, C. (2012). Benchmarking real-time HEVC streaming. In *Proceeding of SPIE Conference on Real-Time Imaging.* SPIE.

Ni, P., Eichhorn, A., Griwodz, C., & Halvorsen, P. (2010). Frequent layer switching for perceived video quality improvements of coarse-grained scalable video. *Multimedia Systems, 16*(3), 171–182. doi:10.1007/s00530-010-0186-9

Nishikawa, K., Munadi, K., & Kiya, H. (2008). No-reference PSNR estimation for quality monitoring of motion JPEG2000 video over lossy packet networks. *IEEE Transactions on Multimedia, 10*(4), 637–645. doi:10.1109/TMM.2008.921849

NS-2. (2011). *Website.* Retrieved January 12, 2012, from http://www.isi.edu/nsnam/ns/

O'Connor, M., & Herlocker, J. (1999). Clustering items for collaborative filtering. In *Proceedings of the ACM SIGIR Workshop on Recommender Systems (SIGIR 1999).* ACM Press.

O'Murchu, D. (2005). New teacher and student roles in the technology-supported language classroom. *International Journal of Instructional Technology & Distance Learning, 2*(2).

O'Neill, D. (2006). *ID3v1 – ID3.org.* Retrieved February 7, 2012, from http://www.id3.org/ID3v1

O'Reilly, T. (2005). *What is web 2.0?* Retrieved February 7, 2012, from http://oreilly.com/web2/archive/what-is-web-20.html

Ogiela, M. R., & Tadeusiewicz, R. (2001). Semantic-oriented syntactic algorithms for content recognition and understanding of images in medical databases. In *Proceedings of IEEE Computer Society International Conference on Multimedia and Exposition*, (pp. 621-624). Tokyo, Japan: IEEE Press.

Ojala, T., Piettikäinen, M., & Mäenpää, T. (2002). Multiresolution gray-scale and rotation invariant texture classification with local binary patterns. *IEEE Transactions on Pattern Analysis and Machine Intelligence, 24*(7), 971–987. doi:10.1109/TPAMI.2002.1017623

Oliva, A., & Torralba, A. (2001). Modeling the shape of the scene: a holistic representation of the spatial envelope. *International Journal of Computer Vision, 42*(3), 145–175. doi:10.1023/A:1011139631724

Olsen, S. I. (1993). Estimation of noise in images: An evaluation. *Graphical Models and Image Processing, 55*(4), 319–323. doi:10.1006/cgip.1993.1022

Open Mobile Alliance. (2011). *MMS architecture. Technical specification, OMA-AD-MMS-V1_3-20110913-A.* Washington, DC: Open Mobil Alliance.

OpenStack. (2012). *The OpenStack project.* Retrieved October 13, 2011, from http://openstack.org/

OPTICOM. (2010a). *PEVQ perceptual evaluation of video quality.* Retrieved January 16, 2012, from http://www.pevq.org/

OPTICOM. (2010b). *V-factor quality of experience platform.* Retrieved January 16, 2012, from http://www.pevq.org/

Ortega, A., & Wang, H. (2007). Mechanisms for adapting compressed multimedia to varying bandwidth conditions. In van der Schaar, M., & Chou, P. A. (Eds.), *Multimedia over IP and Wireless Networks* (pp. 81–116). Amsterdam, The Netherlands: Academic Press. doi:10.1016/B978-012088480-3/50005-9

Ostermann, J., Bormans, J., List, P., Marpe, D., Narroschke, M., & Pereira, F. (2004). Video coding with H.264/AVC: Tools, performance and complexity. *IEEE Circuits and Systems Magazine, 4*(1), 7–28. doi:10.1109/MCAS.2004.1286980

Ozbek, N., & Tumnali, T. (2005). A survey on the H. 264/AVC standard. *Turk Journal of Electrical Engineering, 13*, 287–302.

PacketVideo Corporation. (2009). *OpenCORE multimedia framework capabilities.* Retrieved from http://opencore.net/files/opencore_framework_capabilities.pdf

Padhye, J., Firoiu, V., Towsley, D., & Kurose, J. (1998). Modeling TCP throughput: A simple model and its empirical validation. In *Proceedings of ACM SIGCOMM*, (pp. 303–314). ACM Press.

Pampalk, E. (2006). Audio-based music similarity and retrieval: Combining a spectral similarity model with information extracted from fluctuation patterns. In *Proceedings of the 3rd Annual Music Information Retrieval eXchange (MIREX 2006)*. Victoria, Canada: MIREX.

Pampalk, E., Rauber, A., & Merkl, D. (2002). Content-based organization and visualization of music archives. In *Proceedings of the 10th ACM International Conference on Multimedia*, (pp. 570–579). New York, NY: ACM Press.

Pan, Z., & Kwong, S. (2011). A fast inter-mode decision scheme based on luminance difference for H.264/AVC. In *Proceedings of the 2011 International Conference on System Science and Engineering (ICSSE)*, (pp. 260-263). ICSSE.

Pang, C., Au, O. C., Dai, J., & Zou, F. (2011). Frame complexity guided Lagrange multiplier selection for H.264 intra-frame coding. *IEEE Signal Processing Letters*, *18*(12), 733–736. doi:10.1109/LSP.2011.2172940

Panisson, A., Barrat, A., Cattuto, C., Ruffo, G., Schifanella, R., & Van den Broeck, W. (2011). On the dynamics of human proximity for data diffusion in ad-hoc networks. *Ad Hoc Networks*, *10*(8), 1532–1543. doi:10.1016/j.adhoc.2011.06.003

Pantos, R., & May, W. (2011). *HTTP live streaming*. IETF Draft. Retrieved from http://tools.ietf.org/html/draft-pantos-http-live-streaming-06

Pao, I. M., & Sun, M. T. (2001). Encoding stored video for streaming applications. *IEEE Transactions on Circuits and Systems for Video Technology*, *11*(2), 199–209. doi:10.1109/76.905985

Papadimitriou, A., Symeonidis, P., & Manolopoulos, Y. (2011). Geo-social recommendations. In *Proceedings of the RecSys Workshop Personalization on Mobile Applications (PeMA 2011)*. Chicago, IL: PeMA.

Papageorgiou, C., & Poggio, T. (2000). A trainable system for object detection. *International Journal of Computer Vision*, *38*(1), 15–33. doi:10.1023/A:1008162616689

Papoulis, A. (1984). *Probability, random variables and stochastic processes* (2nd ed.). Tokyo, Japan: McGraw-Hill.

Park, S., Jeong, S.-H., & Hwang, C. (2008). Mobile IPTV expanding the value of IPTV. In *Proceedings of the 2008 Seventh International Conference on Networking (ICN 2008)*, (pp. 296-301). Cancun, Mexico: ICN.

Park, J. S., & Song, H. J. (2006). Selective intra-prediction mode decision for H.264/AVC encoders. *Transactions on Engineering. Computing and Technology*, *13*, 51–55.

Park, S., & Jeong, S. (2009). Mobile IPTV: Approaches, challenges, standards, and QoS support. *IEEE Internet Computing*, *13*(3), 23–31. doi:10.1109/MIC.2009.65

Park, S., Yoon, H., & Kim, J. (2006). Network-adaptive HD MPEG-2 video streaming with cross-layered channel monitoring in WLAN. *Journal of Zhejiang University Science A*, *7*(5), 885–893. doi:10.1631/jzus.2006.A0885

Parlak, M., Adibelli, Y., & Hamzaoglu, I. (2008). A novel computational complexity and power reduction technique for H.264 intra prediction. *IEEE Transactions on Consumer Electronics*, *54*(4), 2006–2014. doi:10.1109/TCE.2008.4711266

Pasi, G., & Villa, R. (2005) Personalized news content programming (PENG): A system architecture. In *Proceedings of the Sixteenth International Workshop on Database and Expert Systems Applications*, (pp. 1008-1012). IEEE.

Passani, L. (2007). *Wurfl (wireless universal resource file)*. Retrieved from http://wurfl.sourceforge.net

Paul, M., Frater, M. R., & Arnold, J. F. (2009). An efficient mode selection prior to the actual encoding for H.264/AVC encoder. *IEEE Transactions on Multimedia*, *11*(4), 581–588. doi:10.1109/TMM.2009.2017610

Paul, M., Lin, W., Lau, C. T., & Lee, B.-S. (2011). Direct intermode selection for H.264 video coding using phase correlation. *IEEE Transactions on Image Processing*, *20*(2), 461–473. doi:10.1109/TIP.2010.2063436

Paul, S. (2011). *Digital video distribution in broadband, television, mobile and converged networks*. Chichester, UK: John Wiley & Sons Ltd. doi:10.1002/9780470972915

Pazzani, M. J. (1999). A framework for collaborative, content-based and demographic filtering. *Artificial Intelligence Review*, *13*(5-6), 393–408. doi:10.1023/A:1006544522159

Pazzani, M., & Billsus, D. (1997). Learning and revising user profiles: The identification of interesting web sites. *Machine Learning*, 27(3), 313–331. doi:10.1023/A:1007369909943

Peng, Q., & Jing, J. (2003). H.264 codec system-on-chip design and verification. In *Proceedings of the 5th International Conference*, (vol. 2, pp. 922-925). IEEE.

Pérez-Costa, X., Schmitz, R., Hartenstein, H., & Liebsch, M. (2002). A MIPv6, FMIPv6 and HMIPv6 handover latency study: Analytical approach. In *Proceedings of the IST Mobile & Wireless Telecommunications Summit (IST Summit)*. Thessaloniki, Greece: IST Summit.

Pérez-Costa, X., Torrent-Moreno, M., & Hartenstein, H. (2003). A performance comparison of mobile IPv6, hierarchical mobile IPv6, fast handovers for mobile IPv6 and their combination. *ACM SIGMOBILE Mobile Computing and Communications Review*, 7(4), 5–19. doi:10.1145/965732.965736

Perkins, C., Johnson, D., & Arkko, J. (2011). *Mobility support in IPv6*. IETF RFC 6275. Retrieved from http://tools.ietf.org/html/rfc6275

Perkins, C. (2003). *RTP: Audio and video for the internet*. Boston, MA: Addison Wesley.

Pew Internet. (2008). *55% of adult Americans have home broadband connections*. Retrieved from http://www.pewinternet.org/Press-Releases/2008/55-of-adult-Americans-have-home-broadband-connections.aspx

Pike, T., Russell, C., Krumm-Heller, A., & Sivaraman, V. (2007). IPv6 and multicast filtering for high-performance multimedia application. In *Proceedings of the Australasian Telecommunication Networks and Applications Conference (ATNAC 2007)*, (pp. 146-150). ATNAC.

Ping, G., & Desheng, F. (2010). The discussions on implementing QoS for IPv6. In *Proceedings of the International Conference on Multimedia Technology (ICMT)*, (pp. 1-4). ICMT.

Pingdom. (2012). *Internet 2011 in numbers*. Retrieved February 3, 2012, from http://royal.pingdom.com/2012/01/17/internet-2011-in-numbers/

Pinson, M., & Wolf, S. (2004). A new standardized method for objectively measuring video quality. *IEEE Transactions on Broadcasting*, 50(3), 312–322. doi:10.1109/TBC.2004.834028

Plass, S., Richter, G., & Han Vinck, A. J. (2008). Coding schemes for crisscross error patterns. *Wireless Personal Communications*, 47(1), 39–49. doi:10.1007/s11277-007-9389-6

Plissonneau, L., En-Najjary, T., & Urvoy-Keller, G. (2008). Revisiting web traffic from a DSL provider perspective: The case of YouTube. In *Proceedings of the 19th ITC Specialist Seminar on Network Usage and Traffic*. IEEE.

Pohle, T., Schnitzer, D., Schedl, M., Knees, P., & Widmer, G. (2009). On rhythm and general music similarity. In *Proceedings of the 10th International Conference on Music Information Retrieval (ISMIR 2009)*. Kobe, Japan: ISMIR.

Policy, X. A. C. M. L. (2012). *Implementation in mobicents*. Retrieved October 13, 2011, from http://www.jboss.org/picketlink/XACML

Porter, L. R. (2004). *Developing an online curriculum: Technologies and techniques*. Hershey, PA: IGI Global.

Postel, J. (1980). *User datagram protocol*. RFC 768. Retrieved from http://tools.ietf.org/html/rfc768

Powers, R. A. (1995). Batteries for low power electronics. *Proceedings of the IEEE*, 83(4), 687–693. doi:10.1109/5.371974

Preece, J. (2000). *Online communities: Designing usability, supporting sociability*. Chichester, UK: John Wiley and Sons.

Psannis, K., & Ishibashi, Y. (2008). Enhanced H.264/AVC stream switching over varying bandwidth networks. *IEICE Electronics Express*, 5(19), 827–832. doi:10.1587/elex.5.827

Qayyum, U., & Javed, M. Y. (2006). Real time notch based face detection, tracking and facial feature localization. In *Proceedings of the International Conference on Emerging Technologies, 2006, ICET 2006*, (pp. 70-75). ICET.

Qi, B., Zhang, D., Song, Y., Du, G., & Zheng, Y. (2011). Design and implementation of a new pipelined H.264 encoder. In *Proceedings of the 2011 International Conference on Computer Science and Network Technology (ICCSNT),* (vol. 1, pp. 130-133). ICCSNT.

Qiao, L., Nahrstedt, K., & Tam, M.-C. (1997). Is MPEG encryption by using random list instead of zigzag order secure? In *Proceedings of the IEEE International Symposium on Consumer Electronics,* (pp. 226-229). IEEE Press.

Quercia, D., Lathia, N., Calabrese, F., Di Lorenzo, G., & Crowcroft, J. (2010). Recommending social events from mobile phone location data. In *Proceedings of the 2010 IEEE International Conference on Data Mining (ICDM 2010),* (pp. 971-976). Washington, DC: IEEE Press.

Quinn, A. J., & Bederson, B. (2011). Human computation: A survey and taxonomy of a growing field. In *Proceedings of the 2011 Annual Conference on Human Factors in Computing Systems (CHI 2011).* Vancouver, Canada: ACM Press.

Quyang, K., Chen, C., & Chen, J. (2010). Simplified directional mode selection for h.264/avc intra mode decision. In *Proceedings of the 2010 International Conference on Computational Intelligence and Software Engineering (CiSE),* (pp. 1-4). CiSE.

Rahda, H. M., van der Schaar, M., & Chen, Y. (2001). The MPEG-4 fine-grained scalable video coding method for multimedia streaming over IP. *IEEE Transactions on Multimedia, 3*(1), 53–68. doi:10.1109/6046.909594

Rajahalme, J., Conta, A., Carpenter, B., & Deering, S. (2004). *IPv6 flow label specification.* IETF RFC 3697. Retrieved from http://www.ietf.org/rfc/rfc3697.txt

Ramanathan, S., & Venkat Rangan, P. (1994). Architectures for personalized multimedia. *IEEE MultiMedia, 1*(1), 37–46. doi:10.1109/93.295266

Ramasubramonian, A., & Woods, J. (2010). Multiple description coding and practical network coding for video multicast. *IEEE Signal Processing Letters, 17*(3), 265–268. doi:10.1109/LSP.2009.2038110

Ramsden, P. (1992). *Learning to teach in higher education.* London, UK: Routledge. doi:10.4324/9780203413937

Ranham, M. R., & Katsaggelos, A. K. (1997). Digital image restoration. *IEEE Signal Processing Magazine, 3,* 24–41.

Rank, M., Lendl, M., & Unbehauen, R. (1999). Estimation of image noise variance. *Proceedings of the IEEE Visual Signal Processing, 146,* 80–84. doi:10.1049/ip-vis:19990238

Rao, R., & Ballard, D. (1997). Dynamic model of visual recognition predicts neural response properties in the visual cortex. *Neural Computation, 9*(4), 721–763. doi:10.1162/neco.1997.9.4.721

Razavi, R., Fleury, M., & Ghanbari, M. (2009). Adaptive packet-level interleaved FEC for wireless priority-encoded video streaming. *Advances in Multimedia.* Retrieved from http://www.hindawi.com/journals/am/2009/982867/

Razavi, R., Fleury, M., & Ghanbari, M. (2008). Energy efficient video streaming over Bluetooth using rateless coding. *Electronics Letters, 44*(22), 1309–1310. doi:10.1049/el:20080851

Redmill, D. W., & Kingsbury, N. G. (1996). The EREC: An error resilient technique for coding variable-length blocks of data. *IEEE Transactions on Image Processing, 5*(4), 565–574. doi:10.1109/83.491333

Reimers, U. (1998). Digital video broadcasting. *IEEE Communications Magazine, 36*(6), 104–110. doi:10.1109/35.685371

Rejaie, R., Handley, M., & Estrin, D. (1999). RAP: An end-to-end rate-based congestion control mechanism for realtime streams in the internet. In *Proceedings of the IEEE INFOCOM,* (pp. 1337–1345). IEEE Press.

Ren, F., & Dong, J. (2010). Fast and efficient intra mode selection for H.264/AVC. In *Proceedings of the Second International Conference on Computer Modeling and Simulation, 2010, ICCMS 2010,* (vol. 2, pp. 202-205). ICCMS.

Resnick, P., & Varian, H. (1997). Recommender systems. *Communications of the ACM, 40*(3), 56–58. doi:10.1145/245108.245121

Revés, X., Nafisi, N., Ferrús, R., & Gelonch, A. (2006). User perceived quality evaluation in a B3G network testbed. In *Proceedings of the IST Mobile Summit,* (pp. 1-5). IST.

Reynolds, A. D. (2009). Gaussian mixture models. In Li, S. Z., & Jain, A. K. (Eds.), *Encyclopedia of Biometrics* (pp. 659–663). New York, NY: Springer.

Rhee, C. E., Jung, J.-S., & Lee, H.-J. (2010). A real-time H.264/AVC encoder with complexity-aware time allocation. *IEEE Transactions on Circuits and Systems for Video Technology, 20*(12), 1848–1862. doi:10.1109/TCSVT.2010.2087834

Rhee, I., Ozdemir, V., & Yi, T. (2000). *TEAR: TCP emulation at receivers. Technical Report*. Raleigh, NC: North Carolina State University.

Ricci, F. (2011). Mobile recommender systems. *International Journal of Information Technology and Tourism, 12*(3), 205–231. doi:10.3727/109830511X12978702284390

Rice, R. (1994). Network analysis and computer mediated communication systems. In Galaskiewkz, S. W. J. (Ed.), *Advances in Social Network Analysis*. Newbury Park, CA: Sage. doi:10.4135/9781452243528.n7

Rice, R. E., Grant, A. E., Schmitz, J., & Torobin, J. (1990). Individual and network influences on the adoption and perceived outcomes of electronic messaging. *Social Networks, 12*, 17–55. doi:10.1016/0378-8733(90)90021-Z

Richardson, I. E. (2003c). *H.264/MPEG-4 part 10 white paper: Inter prediction*. Retrieved from http://www.vcodex.com/h264.html

Richardson, I. (2003). *H.264 and MPEG-4 video compression: Video coding for next-generation multimedia*. New York, NY: John Wiley & Sons Inc. doi:10.1002/0470869615

Richardson, I. E. (2010). *The H.264 advanced video compression standard* (2nd ed.). Chichester, UK: John Wiley & Sons, Ltd.doi:10.1002/9780470989418

Roisin, C. (1998). Authoring structured multimedia documents. In *Proceedings of the 25th Conference on Current Trends in Theory and Practice of Informatics: Theory and Practice of Informatics,* (pp. 222-239). Springer-Verlag.

Romaniak, P., Mu, M., Leszczuk, M., & Mauthe, A. (2008). Framework for the integrated video quality assessment. In *Proceedings of the 18th ITC Specialist Seminar on Quality of Experience,* (pp. 81-89). Karlskrona, Sweden: Blekinge Institute of Technology.

Roodaki, H., Rabiee, H. R., & Ghanbari, M. (2010). Rate-distortion optimization of scalable video codecs. *Signal Processing Image Communication, 25*(4), 276–286. doi:10.1016/j.image.2010.01.004

Rosenberg, J. (2004). *A presence event package for the session initiation protocol* (SIP). IETF RFC 3856. Retrieved from http://www.ietf.org/rfc/rfc3856.txt

Rosenberg, J., Peterson, J., Schulzrinne, H., & Camarillo, G. (2004). *Best current practices for third party call control (3PCC) in the session initiation protocol (SIP)*. RFC 3725. Retrieved from http://tools.ietf.org/html/rfc3725

Rosenberg, J., Schulzrinne, H., Camarillo, G., et al. (2002). *SIP: Session initiation protocol*. RFC 3261. Retrieved from http://www.ietf.org/rfc/rfc3261.txt

Rosin, P. (1998). Thresholding for change detection. In *Proceedings of International Conference on Computer Vision,* (pp. 274-279). Bombay, India: IEEE Press.

Rosin, P. (1997). Edges: Saliency measures and automatic thresholding. *Machine Vision and Applications, 9*, 139–159. doi:10.1007/s001380050036

Rouse, D., & Hemami, S. (2008). Understanding and simplifying the structural similarity metric. In *Proceedings of the 2008 15th IEEE International Conference on Image Processing (ICIP 2008),* (pp. 1188-1191). San Diego, CA: IEEE Computer Society Press.

RTCWeb-SIP. I. (2011). *IETF RTCWeb-SIP WG*. Retrieved October 13, 2011, from http://tools.ietf.org/html/draft-kaplan-rtcweb-sip-interworking-requirements-01

Rubenstein, D., Kurose, J., & Towsley, D. (1998). *Real-time reliable multicast using proactive forward error correction*. Technical Report 98-19. Amherst, MA: University of Massachusetts.

Rubin, D. L., & Napel, S. (2010). Imaging informatics: Toward capturing and processing semantic information in radiology images. *Yearbook of Medical Informatics,* 34–42.

Rupp, M. (Ed.). (2009). *Video and multimedia transmissions over cellular networks: Analysis, modelling and optimization in live 3G mobile networks*. Chichester, UK: John Wiley & Sons.

Sadykhov, R. K., & Lamovsky, D. V. (2008). Algorithm for real time faces detection in 3D space. In *Proceedings of the International Multiconference on Computer Science and Information Technology, 2008, IMCSIT 2008,* (pp. 727-732). IMCSIT.

Saint-Andre, P. (2004). *Extensible messaging and presence protocol (XMPP): Core.* IETF RFC 3920. Retrieved from http://www.ietf.org/rfc/rfc3920.txt

Salmeri, M., Mencattini, A., Ricci, E., & Salsano, A. (2001). Noise estimation in digital images using fuzzy processing. In *Proceedings of the IEEE International Conference on Image Processing, ICIP 2001.* Thessaloniki, Greece: IEEE Press.

Sandvine. (2011). *Sandvine's global internet phenomena report.* White Paper. New York, NY: Sandvine.

Sang-Jo, Y., & Seak-Jae, S. (2006). Fast handover mechanism for seamless multicasting services in mobile IPv6 wireless networks. *Wireless Personal Communications, 42*(4), 509–526.

Saponara, S., Blanch, C., Denolf, K., & Bormans, J. (2003). The JVT advanced video coding standard: Complexity and performance analysis on a tool-by-tool basis. In *Proceedings of the Packet Video Workshop (PV 2003).* Nantes, France: PV.

Sarin, A. (2007). The future of convergence in the communications industry. *IEEE Communications Magazine, 45*(9), 12–14. doi:10.1109/MCOM.2007.4342843

Sarwar, B. M., Karypis, G., Konstan, J. A., & Reidl, J. (2001). Item-based collaborative filtering recommendation algorithms. In *Proceedings of the World Wide Web,* (pp. 285–295). ACM Press.

Sarwar, B. M., Karypis, G., Konstan, J. A., & Riedl, J. (2000). Analysis of recommendation algorithms for E-commerce. In *Proceedings of the ACM E-Commerce,* (pp.158–167). Minneapolis, MN: ACM Press.

Sarwar, B. M., Karypis, G., Konstan, J. A., & Riedl, J. (2002). Recommender systems for large-scale e-commerce: Scalable neighborhood formation using clustering. In *Proceedings of the 5th International Conference on Computer and Information Technology (ICCIT 2002).* ICCIT.

Savas, S. S., Tekalp, A. M., & Gürler, C. G. (2011). Adaptive multi-view video streaming over P2P networks considering quality of experience. In *Proceedings of the 2011 ACM Workshop on Social and Behavioural Networked Media Access,* (pp. 53-58). ACM Press.

Saxtoft, C. (2008). *Convergence: User expectations, communications enablers and business opportunities.* London, UK: John Wiley & Sons Ltd.

Scaringella, N., Zoia, G., & Mlynek, D. (2006). Automatic genre classification of music content- A survey. *IEEE Signal Processing Magazine, 23*(2), 133–141. doi:10.1109/MSP.2006.1598089

Scellato, S. (2011, Autumn). Beyond the social web: The geo-social revolution. *SIGWEB Newsletter.*

Scellato, S., Mascolo, C., Musolesi, M., & Latora, V. (2010). Distance matters: Geo-social metrics for online social networks. In *Proceedings of the 3rd Conference on Online Social Networks (WOSN 2010).* Berkeley, CA: USENIX Association.

Schedl, M., Pohle, T., Knees, P., & Widmer, G. (2006). Assigning and visualizing music genres by web-based co-occurrence analysis. In *Proceedings of the 7th International Conference on Music Information Retrieval. ISMIR 2006,* (pp. 260-265). Victoria, Canada: ISMIR.

Schedl, M. (2011). Web-based and community-based music information extraction. In Li, T., Ogihara, M., & Tzanetakis, G. (Eds.), *Music Data Mining* (pp. 219–249). Boca Raton, FL: CRC Press. doi:10.1201/b11041-11

Schedl, M., Widmer, G., Knees, P., & Pohle, T. (2011). A music information system automatically generated via web content mining techniques. *Information Processing & Management, 47*(3), 426–439. doi:10.1016/j.ipm.2010.09.002

Schein, A. I., Popescul, A., Ungar, L. H., & Pennock, D. M. (2002). Methods and metrics for cold-start recommendations. In *Proceedings of the 25th Annual International ACM SIGIR Conference on Research and Development in Information Retrieval (SIGIR 2002).* ACM Press.

Schiele, B., & Crowley, J. (2000). Recognition without correspondence using multidimensional receptive field histograms. *International Journal of Computer Vision, 36*(1), 31–50. doi:10.1023/A:1008120406972

Schierl, T., & Narasimhan, S. (2011). Transport and storage systems for 3-D video using MPEG-2 systems, RTP, and ISO file format. *Proceedings of the IEEE, 99*(4), 671–683. doi:10.1109/JPROC.2010.2091370

Schifanella, R., Panisson, A., Gena, C., & Ruffo, G. (2008). Mobhinter: Epidemic collaborative filtering and self-organization in mobile ad-hoc networks. In *Proceedings of the 2008 ACM Conference on Recommender Systems (RecSys 2008),* (pp. 27–34). New York, NY: ACM Press.

Schmitt, F., Mignotte, M., Collet, C., & Thourel, P. (1996). Estimation of noise parameters on sonar images. In *Proceedings of the Conference Signal and Image Processing, SPIE 1996,* (vol. 2823, pp. 1-12). Denver, CO: SPIE.

Schneiderman, B. (1998). *Designing the user interface: Strategies for effective human-computer interaction.* Reading, MA: Addison Wesley Longman. doi:10.1145/25065.950626

Schoepflin, T., Chalana, V., Haynor, D. R., & Kim, Y. (2001). Video object tracking with a sequential hierarchy of template deformations. *IEEE Transactions on Circuits and Systems for Video Technology, 11*(11), 1171–1182. doi:10.1109/76.964784

Schonwalder, J., Fouquet, M., Rodosek, G. D., & Hochstatter, I. (2009). Future internet = content + services + management. *IEEE Communications Magazine, 47*(7), 27–33. doi:10.1109/MCOM.2009.5183469

Schreiber, G., Amin, A., van Assem, M., de Boer, V., Hardman, L., & Hildebrand, M. (2008). Semantic annotation and search of cultural-heritage collections: The MultimediaN e-culture demonstrator. *Web Semantics: Science, Services, and Agents on the World Wide Web, 6*(4), 243–249. doi:10.1016/j.websem.2008.08.001

Schreier, R. M., & Rothermel, A. (2006). Motion adaptive intra refresh for low-delay video coding. In *Proceedings of the IEEE International Conference on Consumer Electronics,* (pp. 453-454). IEEE Press.

Schulzrinne, H. (1992). *The real time transport protocol (RTP).* In *Proceedings of the MCNC 2nd Packet Video Workshop.* MCNC.

Schulzrinne, H., Casner, S., Frederick, R., & Jacobson, V. (1996). *RTP: A transport protocol for real-time applications.* RFC 1889. Retrieved from http://www.ietf.org/rfc/rfc1889.txt

Schulzrinne, H., Rao, A., & Lanphier, R. (1998). *Real time streaming protocol (RTSP).* RFC 2326. Retrieved from http://www.ietf.org/rfc/rfc2326.txt

Schwarz, H., & Wiegand, T. (2007). R-D optimized multilayer encoder control for SVC. In *Proceedings of the IEEE International Conference on Image Processing,* (pp. 281-284). IEEE Press.

Schwarz, H., Marpe, D., & Wiegand, T. (2007). Overview of the scalable video coding extension of the H.264/AVC standard. *IEEE Transactions on Circuits and Systems for Video Technology, 17*(9), 1103–1120. doi:10.1109/TCSVT.2007.905532

Scopus Video Networking. (2006). *Advanced encoding mechanism and statistical multiplexing.* White Paper. Sunnyvale, CA: Scopus Video Networking.

Scotcit. (2003). Enabling large-scale institutional implementation of communications and information technology (ELICIT). *Using Computer Mediated Conferencing.* Retrieved November 2, 2004 from http://www.elicit.scotcit.ac.uk/modules/cmc1/welcome.htm

Scott, J. (2002). *Social network analysis: A handbook* (2nd ed.). London, UK: Sage.

Sebe, N., & Tian, Q. (2007). Personalized multimedia retrieval: the new trend? In *Proceedings of the International Workshop on Multimedia Information Retrieval (MIR 2007),* (pp. 299-306). New York, NY: ACM Press.

Seeling, P., & Reisslein, M. (2005). The rate variability-distortion (VD) curve of encoded video and its impact on statistical multiplexing. *IEEE Transactions on Broadcasting, 51*(4), 473–492. doi:10.1109/TBC.2005.851121

Seferoglu, H., & Markopoulou, A. (2009). Video-aware opportunistic network coding over wireless networks. *IEEE Journal on Selected Areas in Communications, 27*(5), 713–728. doi:10.1109/JSAC.2009.090612

Segall, C. A., & Sullivan, G. J. (2007). Spatial scalability within the H.264/AVC scalable video coding extension. *IEEE Transactions on Circuits and Systems for Video Technology*, *17*(9), 1121–1131. doi:10.1109/TCSVT.2007.906824

Serral-Gracià, R., Cerqueira, E., Curado, M., Yannuzzi, M., Monteiro, E., & Masip-Bruin, X. (2010). An overview of quality of experience measurement challenges for video applications in IP networks. In *Wired/Wireless Internet Communications* (pp. 252–263). Luleå, Sweden: Springer. doi:10.1007/978-3-642-13315-2_21

Seyerlehner, K., Widmer, G., & Pohle, T. (2010). Fusing block-level features for music similarity estimation. In *Proceedings of the 13th International Conference on Digital Audio Effects (DAFx 2010)*. Graz, Austria: DAFx.

Shacham, R., Schulzrinne, H., Thakolsri, S., & Kellerer, W. (2007). *Session initiation protocol (SIP) Session mobility: Internet-draft: Draft-shacham-sipping-session-mobility-05*. Retrieved October 13, 2011, from ftp://ftp.rfc-editor.org/in-notes/internet-drafts/draft-shacham-sipping-session-mobility-05.txt

Shade, J., Gortler, S., He, L. W., & Szeliski, R. (1998). Layered depth images. [ACM Press.]. *Proceedings of SIGGRAPH*, *1998*, 231–242.

Shanableh, T., & Ghanbari, M. (2000). Heterogeneous video transcoding to lower spatio-temporal resolutions and different encoding formats. *IEEE Transactions on Multimedia*, *2*(2), 101–110. doi:10.1109/6046.845014

Shavitt, Y., & Weinsberg, U. (2009). Song clustering using peer-to-peer co-occurrences. In *Proceedings of the 11th IEEE International Symposium on Multimedia*, (pp. 471–476). San Diego, CA: IEEE Press.

Shen, X., Yu, H., Buford, J., & Akon, M. (Eds.). (2009). *Handbook of peer-to-peer networking*. New York, NY: Springer.

Shi, J., & Tomasi, C. (1994). Good features to track. In *Proceedings of the IEEE Computer Society Conference on Computer Vision and Pattern Recognition*, (pp. 593-600). IEEE Press.

Shiller, J. (2003). *Mobile communications*. Reading, MA: Addison-Wesley.

Shoaib, M., & Anni, C. (2010). Efficient residual prediction with error concealment in extended spatial scalability. In *Proceedings of the Wireless Telecommunications Symposium (WTS)*, (pp. 1-6). WTS.

Shokorallahi, A. (2006). Raptor codes. *IEEE Transactions on Information Theory*, *52*(6), 2551–2567. doi:10.1109/TIT.2006.874390

Shokurov, A., Khropov, A., & Ivanov, D. (2003). Feature tracking in images and video. In *Proceedings of the International Conference on Computer Graphics between Europe and Asia (GraphiCon-2003)*, (pp. 177-179). GraphiCon.

Sijbers, J., den Dekker, A. J., Van Dyck, D., & Raman, E. (1998). Estimation of signal and noise from rician distributed data. In *Proceedings of the International Conference on Signal Processing and Communications*, (pp. 140-142). Gran Canada, Spain: IEEE.

Sijbers, J., den Dekker, A. J., Raman, E., & Van Dyck, D. (1999). Parameter estimation for magnitude MR images. *International Journal of Imaging Systems and Technology*, *10*, 109–114. doi:10.1002/(SICI)1098-1098(1999)10:2<109::AID-IMA2>3.0.CO;2-R

Silvana, G. P., & Schulzrinne, H. (2008). SIP and 802.21 for service mobility and pro-active authentication. In *Proceedings of the Communication Networks and Services Research Conference (CNSR 2008)*, (pp. 176-182). Halifax, Canada: CNSR.

Simpson, W. (2008). *Video over wireless*. Burlington, MA: Focal Press.

Sirovich, L., & Kirby, M. (1987). Low dimensional procedure for the characterization of human faces. *Journal of the Optical Society of America*, *4*(3), 519–524. doi:10.1364/JOSAA.4.000519

Sisalem, D., & Wolisz, A. (2000). LDA+ TCP-friendly adaptation: A measurement and comparison study. In *Proceedings of the 10th International Workshop on Network and Operating Systems Support for Digital Audio and Video*, (pp. 25-28). IEEE.

Smith, J. R., Mohan, R., & Li, C.-S. (1999). Scalable multimedia delivery for pervasive computing. In *Proceedings of the Seventh ACM International Conference on Multimedia (Part 1) (MULTIMEDIA 1999)*, (pp. 131-140). New York, NY: ACM Press.

Solak, S. B., & Labeau, F. (2010). Complexity scalable video encoding for power-aware applications. In *Proceedings of the 2010 International Green Computing Conference*, (pp. 443-449). IEEE.

Soliman, H. (2009). *Mobile IPv6 support for dual stack hosts and routers*. IETF RFC 5555. Retrieved from http://tools.ietf.org/html/rfc5555

Soliman, H., Castelluccia, C., Elmalki, K. E., & Bellier, L. (2008). *Hierarchical mobile IPv6 mobility management (HMIPv6)*. IETF RFC 5380. Retrieved from http://tools.ietf.org/html/rfc5380

Soluciones & Tecnologia. (2007). *Tendencia del mercado IPTV mundial*. Retrieved from http://www.tecnologiahechapalabra.com/datos/soluciones/enlaces/articulo.asp?i=827

Son, N., & Jeong, J. (2008). An effective error concealment for H.264/AVC. In *Proceedings of IEEE 8th International Conference on Computer and Information Technology Workshops*, (pp. 385-390). IEEE Press.

Song, H. (2002). Browser session preservation and migration. In *Proceedings of WWW 2002*, (p. 2). Hawaii, HI: IEEE.

Spann, M., & Wilson, R. (1985). A quad-tree approach to image segmentation which combines statistical and spatial information. *Pattern Recognition, 18*(3/4), 257–269. doi:10.1016/0031-3203(85)90051-2

Spears, R., & Lea, M. (1992). Social influence and the influence of the social in computer-mediated communication. In Lea, M. (Ed.), *Contexts of Computer Mediated Communication* (pp. 30–65). London, UK: Harvester Wheatsheaf.

Starck, J. L., & Murtagh, F. (1998). Automatic noise estimation from the multiresolution support. *Publications of the Astronomical Society of the Pacific, 110*, 193–199. doi:10.1086/316124

Starck, J. L., & Murtagh, F. (2001). Astronomical image and signal processing. *IEEE Signal Processing Magazine, 18*(2), 30–40. doi:10.1109/79.916319

Starck, J. L., Murtagh, F., & Bijaoui, A. (1995). Multiresolution support applied to image filtering and restoration. *Graphical Models and Image Processing, 57*, 420–431. doi:10.1006/gmip.1995.1036

Stemler, S. (2001). An overview of content analysis. *Practical Assessment, Research & Evaluation, 7*(17).

Stockhammer, T. (2011). Dynamic adaptive streaming over HTTP – Design principles and standards. In *Proceedings of the Second Annual ACM Conference on Multimedia Systems*, (pp. 133-144). ACM Press.

Stockhammer, T., Fröjdh, P., Sodagar, I., & Rhyu, S. (Eds.). (2011). *Information technology — MPEG systems technologies — Part 6: Dynamic adaptive streaming over HTTP (DASH)*. ISO/IEC MPEG Draft International Standard. Retrieved from http://developer.longtailvideo.com/trac/export/1509/.../adaptive/.../dash.pdf

Stockhammer, T., & Zia, W. (2007). Error resilient coding and decoding strategies for video communication. In van der Schaar, M., & Chou, P. A. (Eds.), *Multimedia over IP and Wireless Networks* (pp. 13–58). Burlington, MA: Academic Press. doi:10.1016/B978-012088480-3/50003-5

Stoica, I., Morris, R., Karger, D., Kaashoek, M. F., & Balakrishnan, H. (2001). Chord: A scalable peer-to-peer lookup service for internet applications. In *Proceedings of the 2001 Conference on Applications, Technologies, Architectures, and Protocols for Computer Communications*, (pp. 149-160). New York, NY: ACM Press.

Stottrup-Andersen, J., Forchhammer, S., & Aghito, S. M. (2004). Rate-distortion-complexity optimization of fast motion estimation in H.264/MPEG-4 AVC. In *Proceedings of the IEEE International Conference on Image Processing*, (pp. 119-122). Singapore, Singapore: IEEE Press.

Strang, G., & Nguyen, T. (1996). *Wavelets and filter banks*. Cambridge, MA: Wellesley-Cambridge Press.

Styliaras, G. (2007). A web-based presentation framework for museums. In *Proceedings of the 2007 Euro American Conference on Telematics and Information Systems (EATIS 2007)*. New York, NY: ACM Press.

Su, X., & Khoshgoftaar, T. M. (2009). A survey of collaborative filtering techniques. *Advances in Artificial Intelligence*. Retrieved from http://www.hindawi.com/journals/aai/2009/421425/

Su, L., Lu, Y., Wu, F., Li, S., & Gao, W. (2009). Complexity-constrained H.264 video encoding. *IEEE Transactions on Circuits and Systems for Video Technology, 19*(4), 477–490. doi:10.1109/TCSVT.2009.2014017

Sullivan, G. J., & Ohm, J.-R. (2010). Recent developments in standardization of high efficiency video coding (HEVC). In *Proceedings of SPIE Applications of Digital Image Processing 23*. SPIE.

Sullivan, G. J., & Wiegand, T. (1998). Rate-distortion optimization for video compression. *IEEE Signal Processing Magazine, 15*(6), 74–90. doi:10.1109/79.733497

Sullivan, G., & Wiegand, T. (2005). Video compression—From concepts to the H.264/AVC standard. *Proceedings of the IEEE, 93*(1), 18–31. doi:10.1109/JPROC.2004.839617

Sumner, J., & Dewar, K. (2002). Peer-to-peer elearning and the team effect on course completion. In *Proceedings of ICCE*, (pp. 369-370). ICCE.

Sun, Q., Lu, Y., & Sun, S. (2010). A visual attention based approach to text extraction. In *Proceedings of the 2010 20th International Conference on Pattern Recognition (ICPR)*, (pp. 3991-3995). ICPR.

Sun, Z., & Sun, J. (2008). Tracking of dynamic image sequence based on intensive restraint topology adaptive snake. In *Proceedings of the 2008 International Conference on Computer Science and Software Engineering*, (vol. 6, pp. 217-220). IEEE.

Sun, H., Chen, X., & Chiang, T. (2005). *Digital video transcoders for transmission and storage*. Boca Raton, FL: CRC Press.

Sun, H., Kwok, W., & Zdepski, J. (1996). Architectures for MPEG compressed bitstream scaling. *IEEE Transactions on Circuits and Systems for Video Technology, 6*(2), 191–199. doi:10.1109/76.488826

Sun, H., Vetro, A., Bao, J., & Poon, T. (1997). A new approach for memory-efficient ATV decoding. *IEEE Transactions on Consumer Electronics, 43*(3), 517–525. doi:10.1109/30.628667

Swain, M., & Ballard, D. (1991). Colour indexing. *International Journal of Computer Vision, 7*(1), 11–32. doi:10.1007/BF00130487

T2i. (2012). *Models & tools for pervasive applications focusing on territory discovery (MOANO)*. Retrieved from http://moano.liuppa.univ-pau.fr

Tadayoni, R., & Sigurdsson, H. M. (2006). IPTV market development and regulatory aspects. In *Proceedings for the International Telecommunications Society Conference (ITS 2006)*. Beijing, China: ITS.

Takhteyev, Y., Gruzd, A., & Wellman, B. (2012). Geography of Twitter networks. *Social Networks, 34*(1), 73–81. doi:10.1016/j.socnet.2011.05.006

Tan, K., & Pearmain, A. (2010). An FMO-based error resilience method in H.264/AVC and its UEP application in DVB-H link layer. In *Proceedings of the IEEE International Conference on Multimedia and Expo*, (pp. 214 –219). IEEE Press.

Tan, Y. H., Lee, W. S., Tham, J. Y., Rahardja, S., & Lye, K. M. (2009). Complexity control and computational resource allocation during H.264/SVC encoding. In *Proceedings of the Seventeenth ACM International Conference on Multimedia*, (pp. 897-900). Beijing, China: ACM Press.

Tan, Y.-P., Liang, Y.-Q., & Yu, J. (2002). Video transcoding for fast forward/ reverse video playback. In *Proceedings of the IEEE International Conference Image Processing*, (vol. 1, pp. 713–716). IEEE Press.

Tandberg Television. (2008). *Reflex and data reflex statistical multiplexing system*. White Paper. Slough, UK: Tandberg Television.

Tang, L. (1996). Methods for encrypting and decrypting MPEG video data efficiently. In *Proceedings of the 4th ACM International Multimedia Conference and Exhibition*, (pp. 219-229). ACM Press.

Tanimoto, M., Tehrani, M. P., Fujii, T., & Yendo, T. (2011). Free-viewpoint TV. *IEEE Signal Processing Magazine, 28*(1), 67–76. doi:10.1109/MSP.2010.939077

Tan, W., & Zakhor, A. (1999). Multicast transmission of scalable video using layered FEC and scalable compression. *IEEE Transactions on Circuits and Systems for Video Technology, 11*(3), 373–386. doi:10.1109/76.911162

Tan, Y. H., Lee, W. S., Tham, J. Y., Rahardja, S., & Lye, K. M. (2010). Complexity scalable H.264/AVC encoding. *IEEE Transactions on Circuits and Systems for Video Technology, 20*(9), 1271–1275. doi:10.1109/TCSVT.2010.2058480

Tarak, C., Dejene, E., Frédérique, L., & Vasile-Marian, S. (2007). A comprehensive approach to model and use context for adapting applications in pervasive environments. *Journal of Systems and Software, 80*(12), 1973–1992. doi:10.1016/j.jss.2007.03.010

Taubman, D. S. (2000). High performance scalable image compression with EBCOT. *IEEE Transactions on Image Processing, 9*(7), 1158–1170. doi:10.1109/83.847830

Taylor, P., & Maor, D. (2000). Assessing the efficacy of online teaching with the constructivist on-line learning environment survey. In A. Herrmann & M. M. Kulski (Eds.), *Flexible Futures in Tertiary Teaching: Proceedings of the 9th Annual Teaching Learning Forum*. Perth, Australia: Curtin University of Technology.

Taylor, J. R. (1982). *An introduction to error analysis: The study of uncertainties in physical measurements*. New York, NY: University Science Books. doi:10.1063/1.882103

Terzopoulos, D., & Szeliski, R. (1992). Tracking with Kalman snakes. In *Active Vision*. Cambridge, MA: MIT Press.

Thang, T. C., Bae, T. M., Jung, Y. J., Ro, Y. M., Kim, J.-G., Choi, H., & Hong, J.-W. (2005). *Spatial scalability of multiple ROIs in surveillance video*. JVT-O037, Busan, KR.

Thomos, N., Argyropoulos, S., Boulgouris, N., & Strintzis, M. (2005). Error-resilient transmission of H.264/AVC streams using flexible macroblock ordering. In *Proceedings of Second European Workshop on the Integration of Knowledge, Semantic, and Digital Media Techniques*, (pp. 183-189). IEEE.

Thomos, N., Chakareski, J., & Frossard, P. (2011). Prioritized distributed video delivery with randomized network coding. *IEEE Transactions on Multimedia, 13*(4), 776–787. doi:10.1109/TMM.2011.2111364

Thompson, P. M. (2007). Tracking Alzheimer's disease. *Annals of the New York Academy of Sciences, 1097*, 183–214. doi:10.1196/annals.1379.017

Thomson, S., Narten, T., & Jinmei, T. (2007). *IPv6 stateless address autoconfiguration*. IETF RFC 4862. Retrieved from http://www.ietf.org/rfc/rfc4862.txt

Thubert, P., Wakikawa, R., & Devarapalli, V. (2006). *Global HA to HA protocol*. Retrieved from http://draft-thubert-nemo-global-haha-02

To, L., Pickering, M., Frater, M., & Arnold, J. (2004). A motion confidence measure from phase information. In *Proceedings of the 2004 International Conference on Image Processing, 2004,* (vol. 4, pp. 2583-2586). IEEE.

Torralba, A., Murphy, K. P., Freeman, W. T., & Rubin, M. A. (2003). Context-based vision system for place and object recognition. In *Proceedings of the IEEE International Conference on Computer Vision (ICCV)*, (Vol. 1, pp. 273-280). Nice, France: IEEE Press.

Torralba, A. (2003). Contextual priming for object detection. *International Journal of Computer Vision, 53*(2), 169–191. doi:10.1023/A:1023052124951

Traina, C. Jr, Traina, J. M., dos Santos, R. R., & Senzako, E. Y. (1997). A support system for content-based medical image retrieval in object oriented databases. *Journal of Medical Systems, 21*(6), 339–352. doi:10.1023/A:1022868128573

Tsirtsis, G., Soliman, H., Montavont, N., Giaretta, G., & Kuladinithi, K. (2011). *Flow bindings in mobile IPv6 and network mobility (NEMO) basic support*. IETF RFC 6089. Retrieved from http://tools.ietf.org/html/rfc6089

Tsui, C.-C., Lee, Y.-M., & Lin, Y. (2010). Selective multiple reference frames motion estimation for H.264/AVC video coding. In *Proceedings of the 2010 International Symposium on Information Theory and its Applications (ISITA)*, (pp. 237-242). ISITA.

Turk, M., & Pentland, A. (1991). Eigenfaces for recognition. *Journal of Cognitive Neuroscience, 3*(1). doi:10.1162/jocn.1991.3.1.71

Tzanetakis, G., & Cook, P. (2002). Musical genre classification of audio signals. *IEEE Transactions on Speech and Audio Processing, 10*(5), 293–302. doi:10.1109/TSA.2002.800560

Ugur, K., Andersson, K., Fuldseth, A., Bjontegaard, G., Endresen, L. P., & Lainema, J. (2010). High performance, low complexity video coding and the emerging HEVC standard. *IEEE Transactions on Circuits and Systems for Video Technology*, *20*(12), 1688–1697. doi:10.1109/TCSVT.2010.2092613

Unay, D. (2010). Augmenting clinical observations with visual features from longitudinal MRI data for improved dementia diagnosis. In *Proceedings of the International Conference on Multimedia Information Retrieval*, (pp. 193-200). New York, NY: ACM Press.

Unay, D., et al. (2007). Robustness of local binary patterns in brain MR image analysis. In *Proceedings of the IEEE-EMBS Conference*, (pp. 2098-2101). IEEE Press.

Unay, D., Ekin, A., & Jasinschi, R. (2010). Local structure-based region-of-interest retrieval in brain MR images. *IEEE Transactions on Information Technology in Biomedicine*, *14*(4), 897–903. doi:10.1109/TITB.2009.2038152

University of Idaho. (2002). *Distance education: An overview*. Retrieved November 13, 2002, from http://www.uidaho.edu/evo/dist1.html

Uskela, S. (2003). Key concepts for evolution toward beyond 3G networks. *IEEE Wireless Communications*, *10*(1), 43–48. doi:10.1109/MWC.2003.1182110

van der Auwera, G., David, P. T., & Reisslein, M. (2008b). Traffic and quality characterization of single-layer video streams encoded with the H.264/MPEG-4 advanced video coding standard and scalable video coding extension. *IEEE Transactions on Broadcasting*, *54*(3), 698–718. doi:10.1109/TBC.2008.2000422

van der Auwera, G., David, P. T., Reisslein, M., & Karam, L. J. (2008a). Traffic and quality characterization of the H.264/AVC scalable video coding extension. *Advances in Multimedia*, *2*, 1–27. doi:10.1155/2008/164027

van der Schaar, M., & Chou, P. A. (Eds.). (2007). *Multimedia over IP and wireless networks*. Amsterdam, The Netherlands: Academic Press.

Van Der Wal, T. (2007). *Folksonomy coinage and definition*. Retrieved February 7, 2012, from http://www.vanderwal.net/folksonomy.html

Vanam, R., Chon, J., Riskin, E. A., Ladner, R. E., Ciaramello, F. M., & Hemami, S. S. (2010). Rate-distortion-complexity optimization of an H.264/AVC encoder for real-time videoconferencing on a mobile device. In *Proceedings of the Fifth International Workshop on Video Processing and Quality Metrics for Consumer Electronics (VPQM 2010)*. Scottsdale, AZ: VPQM.

Varsa, V., & Karzcewicz, M. (2001). Long window rate control for video streaming. In *Proceedings of the 11th International Packet Video Workshop*. IEEE.

Veaceslav, C. (2011). *Applying next generation web technologies in the configuration of customer designed products*. Stockholm, Sweden: Royal Institute of Technology.

Veiga, J. (2009). *Virtual channels - Automatic TV/audio channel switching upon reception of information from special control channels*. European Patent Application EP2034639. Nokia Siemens Networks GMBH & CO.

Velasquez, J. (1997). Modeling emotions and other motivations in synthetic agents. In *Proceedings of the 1997 National Conference on Artificial Intelligence*, (pp. 10-15). IEEE.

Vetro, A., Christopoulos, C., & Sun, H. (2003). Video transcoding architectures and techniques: An overview. *IEEE Signal Processing Magazine*, *20*(2), 18–29. doi:10.1109/MSP.2003.1184336

Vetro, A., Wiegand, T., & Sullivan, G. J. (2011). Overview of the stereo and multiview video coding extensions of the H.264/MPEG-4 AVC extension. *Proceedings of the IEEE*, *99*(4), 626–642. doi:10.1109/JPROC.2010.2098830

Vezhnevets, M. (2002). Face and facial feature tracking for natural human-computer interface. In *Proceedings of the International Conference on Computer Graphics between Europe and Asia (GraphiCon-2002)*, (pp. 86-90). GraphiCon.

Video Group of MSU Graphics and Media Lab. (2011). *Website*. Retrieved January 10, 2012, from http://compression.ru/video/quality_measure/index_en.html

Viola, P., & Jones, M. (2001). *Robust real-time object detection. Technical Report*. Cambridge, MA: COMPAQ Cambridge Research Laboratory.

Von Ahn, L. (2005). *Human computation*. (Doctoral Thesis). CMU.

Von Ahn, L. (2006, June). Games with a purpose. *IEEE Computer Magazine*, 96–98.

Von Ahn, L., & Dabbish, L. (2004). Labeling images with a computer game. *ACM SIGCHI Conference on Human Factors in Computing Systems*, (pp. 319-326). New York, NY: ACM Press.

Von Ahn, L., Kedia, M., & Blum, M. (2006). Verbosity: A game for collecting common-sense facts. *ACM SIGCHI Conference on Human Factors in Computing Systems*, (pp. 75-78). New York, NY: ACM Press.

Von Ahn, L., & Dabbish, L. (2008). Designing games with a purpose. *Communications of the ACM*, *51*(8), 58–67. doi:10.1145/1378704.1378719

Vorhees, H., & Poggio, T. (1987). Detecting textons and texture boundaries in natural images. In *Proceedings of the First International Conference on Computer Vision*, (pp. 250-258). London, UK: IEEE.

Vrasidas, C. (2001). Studying human-human interaction in computer-mediated online environments. In Y. Manolopoulos & S. Evripidou (Eds.), *Proceedings of the 8th Panhellenic Conference on Informatics*, (pp. 118-127). Nicosia, Cyprus: IEEE.

Vrasidas, C. (2002). A working typology of intentions driving face-to-face and online interaction in a graduate teacher education course. *Journal of Technology and Teacher Education*, *10*(2), 273–296.

Vrasidas, C., & McIsaac, M. (2000). Principles of pedagogy and evaluation of web-based learning. *Educational Media International*, *37*(2), 105–111. doi:10.1080/095239800410405

Vrasidas, C., & McIsaac, S. M. (1999). Factors influencing interaction in an online course. *American Journal of Distance Education*, *13*(3), 22–36. doi:10.1080/08923649909527033

Vrasidas, C., & Zembylas, M. (2003). The nature of technology-mediated interaction in globalized distance education. *International Journal of Training and Development*, *7*(4), 1–16. doi:10.1046/j.1360-3736.2003.00186.x

Vukobratović, D., & Stanković, V. (2010). Unequal error protection random linear coding for multimedia communications. In *Proceedings of the IEEE International Workshop on Multimedia Signal Processing*, (pp. 280-285). IEEE Press.

Vygotsky, L. S. (1978). *Mind and society: The development of higher mental processes*. Cambridge, MA: Harvard University Press.

W3C (2007). *W3C ubiquitous web applications working group*. Retrieved from http://www.w3.org/2007/uwa/

W3C. (2001). *W3C device independence working group*. Retrieved from http://www.w3.org/2001/di/

W3C. (2012a). *W3C ubiquitous web domain*. Retrieved from http://www.w3.org/UbiWeb/

W3C. (2012b). *W3C web accessibility initiative*. Retrieved from http://www.w3.org/WAI/

Wakikawa, R. E. (2009). *Multiple care-of addresses registration*. IETF RFC 5648. Retrieved from http://tools.ietf.org/html/rfc5648

Waldemar, G. (2000). Diagnosis and management of Alzheimer's disease and other disorders associated with dementia: The role of neurologists in Europe. *European Journal of Neurology*, *7*, 133–144. doi:10.1046/j.1468-1331.2000.00030.x

Wallace, P. (1999). *The psychology of the internet*. Cambridge, UK: Cambridge University Press. doi:10.1017/CBO9780511581670

Wan, F., Cai, L., & Gulliver, A. (2008). Can we multiplex IPTV and TCP? In *Proceedings of the 2008 IEEE International Conference on Communications (ICC 2008)*, (pp. 5804-5808). IEEE Press.

Wand, J.-T., & Chang, P.-C. (1999). Error-propagation prevention technique for real-time video transmission over ATM networks. *IEEE Transactions on Circuits and Systems for Video Technology*, *9*(3), 513–523. doi:10.1109/76.754780

Wang, B., Kurose, J. F., Shenoy, P. J., & Towsley, D. F. (2004). Multimedia streaming via TCP: An analytic performance study. In *Proceedings of ACM Multimedia Conference*, (pp. 908-915). ACM Press.

Wang, H., Leng, J., & Guo, Z. M. (2002). Adaptive dynamic contour for real-time object tracking. In *Proceedings of the Image and Vision Computing New Zealand (IVCNZ 2002)*. IVCNZ.

Wang, H.-J., Wang, L.-L., & Li, H. (2007). A fast multiple reference frame selection algorithm based on H.264/AVC. In *Proceedings of the Third International Conference on Intelligent Information Hiding and Multimedia Signal Processing, 2007,* (vol. 1, pp. 525-528). IEEE.

Wang, J., & Hua, G. (2008). Implementing high definition video codec on TI DM6467 SOC. In *Proceedings of the 2008 IEEE International SOC Conference,* (pp. 193-196). IEEE Press.

Wang, J.-M., Cherng, S., Fuh, C.-S., & Chen, S.-W. (2008). Foreground object detection using two successive images. In *Proceedings of the IEEE Fifth International Conference on Advanced Video and Signal Based Surveillance, 2008, AVSS 2008,* (pp. 301-306). IEEE Press.

Wang, L.-L., Jia, K.-B., & Lu, Z.-Y. (2011). A multi-stage fast intra mode decision algorithm in H.264. In *Proceedings of the 2011 Seventh International Conference on Intelligent Information Hiding and Multimedia Signal Processing (IIH-MSP)*, (pp. 236-239). IIH-MSP.

Wang, M., & Li, B. (2007). Lava: A reality check of network coding in peer-to-peer live streaming. In *Proceedings of the 26th IEEE International Conference on Computer Communications*, (pp. 1082-1090). IEEE Press.

Wang, Y., & Lei, T. (1994). Statistical analysis of MR imaging and its applications in image modeling. In *Proceedings of the IEEE International Conference on Image Processing and Neural Networks,* (vol. 1, pp. 866-870). IEEE Press.

Wang, Y., Casares, M., & Velipasalar, S. (2009). Cooperative object tracking and event detection with wireless smart cameras. In *Proceedings of the 2009 Sixth IEEE International Conference on Advanced Video and Signal Based Surveillance (AVSS 2009)*, (pp. 394-399). Washington, DC: IEEE Press.

Wang, Y., Lei, T., Sewchand, W., & Mun, S. K. (1996). MR imaging statistic and 1st application in image modeling. In *Proceedings SPIE Conference on Medical Imaging*, (vol. 2708, pp. 706-717). Newport Beach, CA: SPIE.

Wang, Y.-K., Hannuksela, M. M., & Gabbouj, M. (2003). Error-robust inter/intra macroblock mode selection using isolated regions. In *Proceedings of the 13ᵗʰ Packet Video Workshop*. IEEE.

Wang, R., Yang, C., Xu, J., Yang, C., & Meng, X. (2009). An XML-based interface customization model. *Lecture Notes in Computer Science, 5940,* 190–202. doi:10.1007/978-3-642-11245-4_17

Wang, Y., Reibman, A. R., & Lin, S. (2005). Multiple description coding for video delivery. *Proceedings of the IEEE, 93*(1), 57–70. doi:10.1109/JPROC.2004.839618

Wang, Z., Lu, L., & Bovik, A. (2004). Video quality assessment based on structural distortion measurement. *Signal Processing Image Communication, 19*(2), 121–132. doi:10.1016/S0923-5965(03)00076-6

Watson, R. (2009). *Fixed-mobile convergence and beyond*. Amsterdam, The Netherlands: Elsevier (Newnes).

Watson, A. B. (1994). Image compression using the discrete cosine transform. *Mathematica Journal, 4*(1), 81–88.

Weber, R. P. (1990). *Basic content analysis* (2nd ed.). Newbury Park, CA: Sage.

WebRTC. (2011a). *WebRTC*. Retrieved October 13, 2011, from http://www.webrtc.org

WebRTC. (2011b). *IETF WebRTC*. Retrieved October 13, 2011, from http://tools.ietf.org/wg/rtcweb

Wei, W., & Barnaghi, P. M. (2007). Semantic support for medical image search and retrieval. In: *Proceedings of the Fifth IASTED International Conference: Biomedical Engineering*, (pp. 315-319). Anaheim, CA: ACTA Press.

Wei, Y., Bhandarkar, S., & Li, K. (2009). Client-centered multimedia content adaptation. *ACM Transactions on Multimedia Computing, Communications, and Applications, 5*(3).

Wei, Z., & Zhou, Z. (2010). An adaptive statistical features modeling tracking algorithm based on locally statistical ROI. In *Proceedings of the 2010 International Conference on Educational and Information Technology (ICEIT)*, (vol. 1, pp. 433-437). ICEIT.

Wei, D. X., Jin, C., Low, S. H., & Hedge, S. (2006). FAST TCP: Motivation, architecture, algorithm and performance. *IEEE/ACM Transactions on Networking, 4*(6), 1246–1259. doi:10.1109/TNET.2006.886335

Weihs, C., Ligges, U., Morchen, F., & Mullensiefen, D. (2007). Classification in music research. *Advances in Data Analysis and Classification, 1*(3), 255–291. doi:10.1007/s11634-007-0016-x

Wein, M., Schwarz, H., & Oelbaum, T. (2007). Performance analysis of SVC. *IEEE Transactions on Circuits and Systems for Video Technology, 17*(9), 1194–1203. doi:10.1109/TCSVT.2007.905530

Welch, J., & Clark, J. (2006). *A proposed media delivery index (MDI).* RFC 4445. Retrieved from http://tools.ietf.org/html/rfc4445

Wellman, B. (1982). Studying personal communities. In Lin, P. M. N. (Ed.), *Social Structure and Network Analysis*. Beverly Hills, CA: Sage.

Wellman, B. (1992). Which types of ties and networks give what kinds of social support? *Advances in Group Processes, 9*, 207–235.

Wellman, B. (1997). An electronic group is virtually a social network. In Kiesler, S. (Ed.), *Culture of the Internet* (pp. 179–205). Hillside, NJ: Lawrence Erlbaum.

Wen, W.-C., Hsiao, H.-F., & Yu, J.-Y. (2007). Dynamic FEC-distortion optimization for H.264 scalable video streaming. In *Proceedings of IEEE Workshop on Multimedia Signal Processing*, (pp. 147-150). IEEE Press.

Weng, L., & Menczer, F. (2010). GiveALink tagging game: An incentive for social annotation. In *Proceedings of the ACM SIGKDD Workshop on Human Computation (HCOMP 2010),* (pp. 26-29). New York, NY: ACM Press.

Weng, L., Schifanella, R., & Menczer, F. (2011). Design of social games for collecting reliable semantic annotations. In *Proceedings of the 16th International Conference on Computer Games (CGAMES),* (pp. 185-192). Louisville, KY: CGAMES.

Weng, L., Schifanella, R., & Menczer, F. (2011). The chain model for social tagging game design. In *Proceedings of the 6th International Conference on Foundations of Digital Games (FDG 2011),* (pp. 295-297). New York, NY: ACM Press.

Wenger, S. (2003). H. 264/AVC over IP. *IEEE Transactions on Circuits and Systems for Video Technology, 13*(7), 645–656. doi:10.1109/TCSVT.2003.814966

Wenger, S. (2003). H.264/AVC over IP. *IEEE Transactions on Circuits and Systems for Video Technology, 13*(7), 645–656. doi:10.1109/TCSVT.2003.814966

Wen, X., Au, O. C., Xu, J., Fang, L., Cha, R., & Li, J. (2011). Novel RD-optimized VBSME with matching highly data re-usable hardware architecture. *IEEE Transactions on Circuits and Systems for Video Technology, 21*(2), 206–219. doi:10.1109/TCSVT.2011.2106274

Wey, T., Blumstein, D. T., Shen, W., & Jordán, F. (2008). Social network analysis of animal behaviour: A promising tool for the study of sociality. *Animal Behaviour, 75*, 333–344. doi:10.1016/j.anbehav.2007.06.020

Whitman, B., & Lawrence, S. (2002). Inferring descriptions and similarity for music from community metadata. In *Proceedings of the 2002 International Computer Music Conference, ICMC 2002,* (pp. 591–598). Goteborg, Sweden: ICMC.

Widmer, J., Fragouli, C., & Le Boudec, J.-Y. (2005). Low-complexity energy-efficient broadcasting in wireless ad-hoc networks using network coding. In *Proceedings of Workshop on Network Coding, Theory, and Applications, (NetCod 2005)*. Riva del Garda, Italy: NetCod.

Widmer, J., Denda, R., & Mauve, M. (2001). A survey on TCP-friendly congestion control. *IEEE Network, 15*(3), 28–37. doi:10.1109/65.923938

Wiegand, T., & Sullivan, G. (2003). *Final draft ITU-T recommendation and final draft international standard of joint video specification (ITU-T Rec. H.264 ISO/IEC 14 496-10 AVC).* In Joint Video Team (JVT) of ITU-T SG16/Q15 (VCEG) and ISO/IEC JTC1/SC29/WG1, Annex C. Pattaya, Thailand, Doc. JVT-G050.

Wiegand, T., Noblet, L., & Rovati, F. (2009). Scalable video coding for IPTV services. *IEEE Transactions on Broadcasting, 55*(2), 527–538. doi:10.1109/TBC.2009.2020954

Wiegand, T., Schwarz, H., Joch, A., Kossentini, F., & Sullivan, G. (2003a). Rate-constrained coder control and comparison of video coding standards. *IEEE Transactions on Circuits and Systems for Video Technology, 13*(7), 688–703. doi:10.1109/TCSVT.2003.815168

Wiegand, T., Sullivan, G. J., Bjøntegaard, G., & Luthra, A. (2003). Overview of the H.264/AVC video coding standard. *IEEE Transactions on Circuits and Systems for Video Technology, 13*(7), 560–576. doi:10.1109/TCSVT.2003.815165

Wien, M., Cazoulat, R., Graffunder, A., Hutter, A., & Amon, P. (2007). Real-time system for adaptive video streaming based on SVC. *IEEE Transactions on Circuits and Systems for Video Technology, 17*(9), 1227–1237. doi:10.1109/TCSVT.2007.905519

Williams, A. (2002). Defining neurodegenerative diseases. *British Medical Journal, 324*(7352), 1465–1466. doi:10.1136/bmj.324.7352.1465

Wittmann, R., & Zitterbart, M. (2001). *Multicast communication: Protocols and applications.* San Francisco, CA: Morgan Kaufmann.

Wolfson, H., & Rigoutsos, I. (1997). Geometric hashing: An overview. *IEEE Computational Science & Engineering, 4*(4), 10–21. doi:10.1109/99.641604

World Wide Web Project. (1991). *Website.* Retrieved February 7, 2012, from http://www.w3.org/History/19921103-hypertext/hypertext/WWW/TheProject.html

Wu, C.-D., & Lin, Y. (2009). Efficient inter/intra mode decision for H.264/AVC inter frame transcoding. In *Proceedings of the 2009 16th IEEE International Conference on Image Processing (ICIP),* (pp. 3697-3700). IEEE Press.

Wu, X., & Schulzrinne, H. (2001). *Use SIP MESSAGE method for shared web browsing.* Retrieved October 13, 2011, from http://www3.tools.ietf.org/id/draft-wu-sipping-Webshare-00.txt

Wu, Y. (2009). Existence and construction of capacity-achieving network codes for distributed storage. In *Proceedings of the IEEE International Symposium on Information Theory,* (pp. 1150-1154). IEEE Press.

Wu, D., Hu, Y. T., & Zhang, Y.-Q. (2000). Transporting real-time video over the Internet: Challenges and approaches. *Proceedings of the IEEE, 88*(12), 1855–1875. doi:10.1109/5.899055

Wu, D., Pan, F., Lim, K. P., Wu, S., Li, Z. G., & Lin, X. (2005). Fast intermode decision in H.264/AVC video coding. *IEEE Transactions on Circuits and Systems for Video Technology, 15*(7), 953–958. doi:10.1109/TCSVT.2005.848304

Xiang, G. (2009). Real-time follow-up tracking fast moving object with an active camera. In *Proceedings of the 2nd International Congress on Image and Signal Processing, 2009, CISP 2009,* (pp. 1-4). IEEE.

Xiao, M., & Cheng, Y. (2011). A fast multi-reference frame selection algorithm for H.264/AVC. In *Proceedings of the 2011 IEEE International Conference on Computer Science and Automation Engineering (CSAE),* (vol. 4, pp. 615-619). IEEE Press.

Xin, J., Sun, M. T., & Chun, K. (2002). Motion re-estimation for MPEG-2 to MPEG-4 simple profile transcoding. In *Proceedings of the International Workshop on Packet Video.* IEEE.

Xue, G., Lin, C., & Yang, Q. (2005). Scalable collaborative filtering using cluster-based smoothing. In *Proceedings of the ACM SIGIR Conference,* (pp. 114–121). Salvador, Brazil: ACM Press.

Xu, L., Ma, S., Zhao, D., & Gao, W. (2005). Rate control for scalable video model. *Proceedings of SPIE: Visual Communication and Image Processing, 5960,* 525.

Yang, C.-C., Tan, K.-J., Yang, Y.-C., & Guo, J.-I. (2010). Low complexity fractional motion estimation with adaptive mode selection for H.264/AVC. In *Proceedings of the 2010 IEEE International Conference on Multimedia and Expo (ICME),* (pp. 673-678). IEEE Press.

Yang, G. Z., Burger, P., Firrnin, D. N., & Undrwood, S. R. (1995). Structure adaptive anisotropic filtering for magnetic resonance image enhancement. [CAIP.]. *Proceedings of Computer Analysis of Images and Patterns, CAIP, 1995,* 384–391. doi:10.1007/3-540-60268-2_320

Yang, M.-H., Kriegman, D. J., & Ahuja, N. (2002). Detecting faces in images: A survey. *IEEE Transactions on Pattern Analysis and Machine Intelligence, 24*(1), 34–58. doi:10.1109/34.982883

Yetgin, Z., & Seckin, G. (2008). Progressive download for 3G wireless multicasting. *International Journal of Hybrid Information Technology, 2*(2), 67–82.

Yokota, H., Kim, D., Sarikaya, B., & Xia, F. (2011). *Home agent initiated flow binding for mobile IPv6*. Retrieved from http://draft-yokota-mext-ha-init-flow-binding-01

You, W. (2010). *Object detection and tracking in compresses domain*. Retrieved from http://knol.google.com/k/wonsang-you/object-detection-and-tracking-in/3e2si9juvje7y/7# Yuan, L., & Mu, Z.-C. (2007). Ear detection based on skin-color and contour information. In *Proceedings of the 2007 International Conference on Machine Learning and Cybernetics,* (vol. 4, pp. 2213-2217). IEEE.

You, W., Sabirin, M. S. H., & Kim, M. (2009). Real-time detection and tracking of multiple objects with partial decoding in H.264/AVC bitstream domain. In N. Kehtarnavaz & M. F. Carlsohn (Eds.), *Proceedings of SPIE*. San Jose, CA: SPIE.

You, W., Sabirin, M. S. H., & Kim, M. (2007). Moving object tracking in H.264/AVC bitstream. *Lecture Notes in Computer Science, 4577*, 483–492. doi:10.1007/978-3-540-73417-8_57

Yu, A. C., & Martin, G. R. (2004). Advanced block size selection algorithm for inter frame coding in H.264/MPEG-4 AVC. In *Proceedings of the International Conference on Image Processing,* (pp. 95-98). Singapore, Singapore: IEEE.

Yucel, Z., & Ali Salah, A. (2009). Head pose and gaze direction estimation for joint attention modeling in embodied agents. In *Proceedings of the Annual Meeting of the Cognitive Science Society, COGSCI 2009*. COGSCI.

Yuen, M.-C., King, I., & Leung, K.-S. (2011). A survey of crowdsourcing systems. In *Proceedings of Privacy, Security, Risk and Trust (PASSAT), 2011 IEEE Third International Conference on and 2011 IEEE Third International Confernece on Social Computing (SocialCom),* (pp. 766-773). Boston, MA: IEEE Press.

Zafirovic-Vukotic, M., & Niemegeers, I. G. (1992). Multimedia communication system: Upper layers in the OSI reference model. *IEEE Journal on Selected Areas in Communications, 10*(9), 1397–1402. doi:10.1109/49.184869

Zambelli, A. (2009). IIS smooth streaming technical overview. *Microsoft Corporation*. Retrieved from http://users.atw.hu/dvb-crew/applications/documents/IIS_Smooth_Streaming_Technical_Overview.pdf

Zapater, M., & Bressan, G. (2007). A proposed approach for quality of experience assurance of IPTV. In *Proceedings of the 2007 First International Conference on the Digital Society (ICDS 2007),* (p. 25). Guadeloupe, French Caribbean: IEEE Computer Society Press.

Zeng, H., Cai, C., & Ma, K.-K. (2009). Fast mode decision for H.264/AVC based on macroblock motion activity. *IEEE Transactions on Circuits and Systems for Video Technology, 19*(4), 491–499. doi:10.1109/TCSVT.2009.2014014

Zeng, H., Ma, K.-K., & Cai, C. (2010). Hierarchical intra mode decision for H.264/AVC. *IEEE Transactions on Circuits and Systems for Video Technology, 20*(6), 907–912. doi:10.1109/TCSVT.2010.2045802

Zeng, W., Du, J., Gao, W., & Huang, Q. (2005). Robust moving object segmentation on H.264/AVC compressed video using the block-based MRF model. *Real-Time Imaging, 11*(4), 290–299. doi:10.1016/j.rti.2005.04.008

Zhang, H., & Malik, J. (2003). Learning a discriminative classifier using shape context distance. In *Proceedings of the International Conference on Computer Vision and Pattern Recognition,* (Vol. 1, pp. 242-247). IEEE.

Zhang, T., Liu, C., Wang, M., & Goto, S. (2009). Region-of-interest based H.264 encoder for videophone with a hardware macroblock level face detector. In *Proceedings of the IEEE International Workshop on Multimedia Signal Processing, 2009,* (pp. 1-6). IEEE Press.

Zhang, Q., Cheng, L., & Boutaba, R. (2010). Cloud computing: State of the art and research challenges. *Journal of Internet Services and Applications, 1*(1), 7–18. doi:10.1007/s13174-010-0007-6

Zhang, R., Regunthan, S. L., & Rose, K. (2000). Video coding with optimal inter/intra-mode switching for packet loss resilience. *IEEE Journal on Selected Areas in Communications, 18*(6), 966–976. doi:10.1109/49.848250

Zhang, Y., Mao, S., Yang, L. T., & Chen, T. M. (Eds.). (2008). *Broadband mobile multimedia*. Boca Raton, FL: Auerbach Publications.

Zhang, Z.-L., Kurose, J., Salehi, J. D., & Towsley, D. (1997). Smoothing, statistical multiplexing and call admission control for stored video. *IEEE Journal on Selected Areas in Communications, 15*(6), 1148–1166. doi:10.1109/49.611165

Zhao, L., & Kuo, C.-C. J. (2003). Buffer-constrained R-D optimized rate-control for video coding. In *Proceedings of the IEEE International Conference on Multimedia and Expo*, (pp. 377-380). IEEE Press.

Zheng, W. (2006). An efficient dynamic multicast protocol for mobile IPv6 networks. In *Proceedings of the 31st IEEE Conference on Local Computer Networks*, (pp. 913 - 920). IEEE Press.

Zhenhua, W., Qiong, S., Xiaohong, H., & Yan, M. (2010). IPv6 end-to-end QoS provision for heterogeneous networks using flow label. In *Proceedings of the 3rd IEEE International Conference on Broadband Network and Multimedia Technology (IC-BNMT)*, (pp. 130-137). IEEE Press.

Zhou, M., Zhang, R., Zeng, D., & Qian, W. (2010). Services in the cloud computing era: A survey. In *Proceedings of the 2010 4th International Universal Communication Symposium (IUCS)*, (pp. 40-46). Beijing, China: IEEE Computer Society Press.

Zhu, W., Ma, Y., & Wang, R. (2010). An adaptive fast algorithm for coding block mode selection. In *Proceedings of the 2010 Third International Symposium on Intelligent Information Technology and Security Informatics (IITSI)*, (pp. 402-406). IITSI.

Zhu, X., Agrawal, P., Singh, J. P., Alpcan, T., & Girod, B. (2007). Rate allocation for multi-user streaming over heterogeneous access networks. In *Proceedings of ACM Multimedia*, (pp. 37-46). ACM Press.

Zhu, S., & Yuille, A. (1996). Region competition: Unifying snakes, region growing, and Bayes MDL for multiband image segmentation. *IEEE Transactions on Pattern Analysis and Machine Intelligence, 18*, 884–900. doi:10.1109/34.537343

Zhu, W., Luo, C., Wang, J., & Li, S. (2011). Multimedia cloud computing. *IEEE Signal Processing Magazine, 28*(3), 59–69. doi:10.1109/MSP.2011.940269

Zink, M., Künzel, O., Scmitt, J., & Steinmetz, R. (2003). Subjective impression of variations in layer encoded videos. In *Proceedings of the 11ᵗʰ International Workshop on Quality of Service*, (pp. 137-154). IEEE.

Zitnick, C. L., Kang, S. B., Uyttendaele, M., Winder, S., & Szeliski, R. (2004). High-quality video view interpolation using a layered representation. *ACM Transactions on Graphics, 23*(3), 600–608. doi:10.1145/1015706.1015766

Zyda, M. (2005). From visual simulation to virtual reality to games. *Computer, 38*(9), 25–32. doi:10.1109/MC.2005.297

About the Contributors

Dimitris Kanellopoulos is a member of the Educational Software Development Laboratory (ESD-Lab) in the Department of Mathematics at the University of Patras, Greece. He received a Diploma in Electrical Engineering and a PhD in Computer Science (Multimedia Communication) from the University of Patras. Since 1990, he has been a Research Assistant in the Department of Electrical and Computer Engineering at the University of Patras and involved in several EU R&D projects. His research interests include: multimedia communication, multimedia networks, intelligent information systems, knowledge representation, and e-learning technologies. He has many publications to his credit in international journals and conferences in these areas. He serves as an editorial board member and reviewer in some refereed journals.

* * *

Antônio J. G. Abelém Antônio is a Doctor in Computer Science by PUC/RJ (2003) and Adjunct Professor of the Federal University of Pará, where he is the Director of the Center of Technology of Information and Communication of this Institution. He participates in the Post-Graduation Program (Master and Doctorate) of the Department of Electrical Engineering (PPGEE) and of the Post-Graduation Program (Master) of the Department of Computer Science (PPGCC) of UFPA, as a Researcher Professor ministering subjects and instructing dissertations and theses. His interests in research are in the areas of architectures of computer networks, selective diffusion (multicast), wireless networks and mobile systems, optical networks, QoS, and QoE. He acted in the work group that researched solutions in Wireless Mesh Network (GT-Mesh, fase 2), passed in the publication of the RNP of 2007/2008, in conjunction with researchers of UFF and of UFPR.

Michael Adeyeye is a Research Fellow at the Cape Peninsula University of Technology, South Africa. He is also a Research Associate at the University of Cape Town, South Africa, where he earned his second Master and PhD degrees. His research interests include Web and multimedia service mobility technologies and context awareness, multimodal and multi-channel access, Web 2.0 and mobile applications, and Next Generation Network (NGN) applications and services. He is a member of the IEEE.

Chrisoula Alexandraki is a Lecturer at the Department of Music Technology and Acoustics of the Technological Educational Institute of Crete and a PhD candidate in Systematic Musicology at the University of Hamburg. Her research interests are in Music Technology and specifically in the fields of Networked Music Performance and Machine Listening. She has participated in a number of National and European research projects, and she was the coordinator of the DIAMOUSES project. In the past, she has worked as an R&D Engineer at the Institute of Computer Science of FORTH as well as for Sony Research in Brussels. She received an MSc degree in Music Technology from the University of York (UK) in 1998, and a Degree in Physics from the University of Athens (Greece) in 1996.

Laith Al-Jobouri received his BSc in 1995 from the University of Technology, Iraq, and subsequently worked on a variety of wireless and microwave projects. He gained his MSc from the University of Mosul, Iraq, in 2001, and then returned to work as a project manager for national and international companies in the areas of wireless communication and cellular networks. He is now completing his research for a PhD degree at the University of Essex, UK. Laith's research interests are in video over wireless, channel error coding, and wireless technologies.

Ismail Ali received his MSc degree from the University of Duhok, Iraq, in 2005, in the area of Automatic Verification. He received his BSc degree in Electrical Engineering in 1993 from the University of Mosul, Iraq. Currently, he is pursuing a PhD degree at the School of Computer Science and Electronic Engineering, University of Essex, UK. His current research interests include video error resilience, cross-layer optimization, and mobile video communications.

Alessandro Amirante received both his MSc Degree in Telecommunications Engineering from the University of Napoli "Federico II" in 2007 and his Ph.D. in Computer Engineering and Systems in the fall of 2010. He is currently a Senior Researcher at the Computer Science Department of University of Napoli "Federico II." His research interests primarily fall in the field of networking, with special regard to Next Generation Network architectures and multimedia services over the Internet.

Artur Arsenio is Assistant Professor in Computer Science at Instituto Superior Técnico, Universidade Técnica de Lisboa. He led Innovation activities at Nokia Siemens Networks Portugal SA up to April 2012. Before, he worked as Senior Solution Architect leading several international R&D teams for Siemens, and as Chief Engineer represented Siemens Networks on IPTV Standardization forums. He collaborated on the creation of the interdisciplinary Institute for Human Studies and Intelligent Sciences in Cascais. He received his PhD in Computer Science from Massachusetts Institute of Technology (MIT) in 2004. He received both his MSc and Engineering Degrees in Electrical and Computer Engineering from Instituto Superior Técnico. He is the inventor of 6 international patent families, and authored/co-authored more than 80 scientific publications. He is the recipient of several scientific and innovation awards. He is the President of the MIT alumni association in Portugal.

Irene Bellver was born in Xátiva, Valencia (Spain), October 10, 1985. She received her M.Sc. in Technical Engineering in Administrative Data Processing in 2009; she did two specializations: Systems and Network Administration and Web-Based Services and Technologies. During one year, she worked as an intern in a domotics company. Later, she worked on a research fellowship in the Polytechnic University of Valencia. She is currently Programmer Applications Analyst in the Higher Polytechnic School of Gandia. She took part in several investigation projects. She is researching in the communications and remote sensing research group at the Research Institute (IGIC). She has some papers published in international journals, and she is in the local organization of several international conferences.

Paulo Bezerra graduated as a Technologist in Data Processing from the University of Amazonia UNAMA (1997). He has a specialization in Computer Networks completed (1999) at the same university. He earned a Master in Electrical Engineering from Federal University of Pará UFPA (2010) and PhD student of the Graduate Program in Electrical Engineering from the same university. He forms part of the research group on computer networks and multimedia communication (GERCOM) acting on the following topics: management of Quality of Service (QoS) and Quality of Experience (QoE), Resource Reservation, Mobility, and Multimedia Streaming in systems of wireless networks. He is a Professor at the Federal Institute of Education, Science and Technology – IFPA and Professor and Coordinator of Teaching Computer Science, University of Amazonia – UNAMA, and Delegate to the Institutional Brazilian Computer Society – SBC.

László Bokor graduated in 2004 with M.Sc. degree in Computer Engineering from the Budapest University of Technology and Economics (BME) at the Department of Telecommunicatons. In 2006, he got an M.Sc.+ degree in Bank Informatics from the same university's Faculty of Economic and Social Sciences. He is a Ph.D. candidate at BME, member of the IEEE, member of Multimedia Networks and Services Laboratory (MEDIANETS), and Mobile Innovation Centre Hungary (MIK), where he participates in researches of wireless protocols and works on advanced mobility management related projects (as FP6-IST PHOENIX and ANEMONE, EUREKA-Celtic BOSS, FP7-ICT OPTIMIX, EURESCOM P1857, EUREKA-Celtic MEVICO, FP7-ICT CONCERTO). His research interests include IPv6 mobility, mobile computing, next generation networks, mobile broadband networking architectures, network performance analyzing, and heterogeneous networks.

Tobia Castaldi received his degree in Telecommunications Engineering from the University of Napoli "Federico II," Italy, in 2006. He is currently a Senior Researcher at the Computer Science Department of the University of Napoli Federico II. The main topic of his research concerns real-time applications for the next-generation Internet with special regard to the IP Multimedia Subsystem (IMS) architecture and services.

Eduardo Cerqueira received his PhD and post-doc in Informatics Engineering from the University of Coimbra (UC) in 2008 and 2009, respectively. He is an Associate Professor at Faculty of Computer Engineering of the University of Para in Brazil. His research interests include Quality of Service, Quality of Experience, Sensor Wireless and Networks, Mobility, and Multimedia. His publications include four books, four patents, and over than 50 papers in national and international refereed journals/conferences. He is founder of a series of workshops on Future Multimedia Networking (FMN). He is coordinating or participating in several national and international projects such as UbiquiMesh, MultiMesh, OTIMA, Universal, and COMAH.

Martina Deplano is a PhD student of the Department of Computer Science at the University of Turin, Italy. She started her PhD program in January 2011, after a Master degree in "Communication in the Information Society" at the University of Turin with a thesis in User Experience Design. Her research interest is related to complex systems and networks, with particular attention to techno-social networks analysis, data visualization, user experience design, interaction design, and human-machine interaction. She is one of the members of the ARCS (Applied Research on Computational Complex Systems) Group in Turin (http://arcs.di.unito.it).

Cédric Dromzée is currently Ph.D. Student at the LIUPPA, the computer science laboratory of University of Pau (France). His thesis titled "Generic Profile for Multimedia Documents Adaptation in Pervasive Computing" is co-supervised by Sébastien Laborie and Philippe Roose. His research interests focus on context adaptation, Semantic Web technologies, Web Apps, multimedia document, profile modeling, open data, and interoperability. Since 2010, he is CEO of AEXIUM SAS, a French start-up that develops a context adaptation framework.

Martin Fleury holds a Maths/Physics degree from the Open University, Milton Keynes, UK. He was awarded an MSc in Astrophysics from QMW College, University of London, UK, in 1990, and an MSc from the University of South-West England, Bristol, UK, in Parallel Computing Systems, in 1991. He gained a PhD in Parallel Image Processing Systems from the University of Essex, Colchester, UK. He is currently a Visiting Fellow at the University of Essex, UK, having previously been employed as a Senior Lecturer and Director of Graduates (Research). He is an external examiner for the Arab Open University and for the Open University, UK. Martin has authored or co-authored over two hundred and thirty-five articles and book chapters as well as a book on high-performance computing. His edited book "Streaming Media with Peer-to-Peer Networks" appeared in May 2012. His current research interests center upon video communication over wired and wireless networks.

Miguel Garcia was born in Benissa, Alicante (Spain), December 29, 1984. He received his M.Sc. in Telecommunications Engineering in 2007 at Polytechnic University of Valencia and a postgraduate "Master en Tecnologías, Sistemas y Redes de Comunicaciones" in 2008. He is currently a Ph.D. student in the Department of Communications of the Universidad Politécnica de Valencia. He is a Cisco Certified Network Associate Instructor. He is working as a researcher in IGIC in the Higher Polytechnic School of Gandia, Spain. Until 2012, he had more than 45 scientific papers published in national and international conferences, and several educational papers. He had more than 30 papers published in international journals (most of them with Journal Citation Report). Mr. Garcia has been technical committee member in several conferences and journals. He has been in the organization of several conferences. Miguel is Associate Editor of *International Journal Networks, Protocols & Algorithms*. He is an IEEE graduate student member.

Mohammed Ghanbari is best known for his pioneering work on two-layer video coding for ATM networks (which earned him an IEEE Fellowship in 2001, now known as SNR scalability in the standard video codecs). He has served as an Associate Editor to *IEEE Transactions on Multimedia*. He has registered for eleven international patents on various aspects of video networking. He is the author of *Video Coding: An Introduction to Standard Codecs*, a book also published by IET press in 1999, which received the year 2000 best book award by the IEE, and the author of *Standard Codecs: Image Compression to Advanced Video Coding*, also published by the IET press in 2003. Prof. Ghanbari has authored or co-authored about 500 journal and conference papers, many of which have had a fundamental influence in this field.

Dan Grois received the B.Sc. degree at the Electrical and Computer Engineering Department, Ben-Gurion University (BGU), Beer-Sheva, Israel, 2002. He received the M.Sc. degree at the Communication Systems Engineering field (at the Electro-Optics Unit), BGU, 2006, and the Ph.D. degree at the Communication Systems Engineering Department, BGU, 2011. Starting from 2011, Dan is a senior post-doctoral researcher at the Communication Systems Engineering Department, BGU. Dan has a significant number of academic publications presented at peer-reviewed international conferences and published by IEEE and Elsevier scientific journals, Wiley publisher, etc. In addition, Dan is a referee of top-tier conferences and international journals, such as the *IEEE Transactions in Image Processing*, *Journal of Visual Communication and Image Representation*, Elsevier, *IEEE Sensors*, SPIE Optical Engineering, etc. Dan is a member of the ACM and a Senior Member of the IEEE. Dan's research interests include image and video coding and processing, video coding standards, region-of-interest scalability, computational complexity and bit-rate control, network communication and protocols, computer vision, and future multimedia applications and systems.

Ofer Hadar received the B.Sc., the M.Sc. (cum laude), and the Ph.D. degrees from the Ben-Gurion University of the Negev, Israel, in 1990, 1992, and 1997, respectively, all in Electrical and Computer Engineering. From August 1996 to February 1997, he was with CREOL at Central Florida University, Orlando, FL, as a Research Visiting Scientist, working on the angular dependence of sampling MTF and over-sampling MTF. From October 1997 to March 1999, he was Post-Doctoral Fellow in the Department of Computer Science at the Technion-Israel Institute of Technology, Haifa. Currently, he is a faculty member at the Communication Systems Engineering Department at Ben-Gurion University of the Negev. His research interests include: image and video compression, routing in ATM networks, flow control in ATM networks, packet video, transmission of video over IP networks, and video rate smoothing and multiplexing. Hadar also works as a consultant for several hi-tech companies, such as EnQuad Technologies Ltd. in the area of MPEG-4 and Scopus in the area of video compression and transmission over satellite networks. Hadar is a member of the SPIE and a Senior Member of the IEEE.

Sándor Imre was born in Budapest in 1969. He received the M.Sc. degree in Electrical Engineering from the Budapest University of Technology (BME) in 1993. Next, he started his PhD studies at BME and obtained Dr. Univ. degree in Probability Theory and Mathematical Statistics in 1996, Ph.D. degree in Telecommunications in 1999, and DSc degree from the Hungarian Academy of Sciences in 2007. Currently, he is carrying on his activities as Professor and Head of Dept. of Telecommunications. He is Chairman of Telecommunication Scientific Committee of the Hungarian Academy of Sciences. He participates on the Editorial Board of *Infocommunications Journal and Hungarian Telecommunications*. He was invited to join the Mobile Innovation Centre as R&D director in 2005. His research interest includes mobile and wireless systems, quantum computing, and communications. He has contributions on different wireless access technologies, mobility protocols, security and privacy, reconfigurable systems, quantum computing-based algorithms and protocols.

Zoltán Kanizsai is a member of IEEE. He graduated with M.Sc. degree in Computer Engineering from the Budapest University of Technology and Economics (BME) at the Department of Telecommunicatons in 2006. Now he is a Ph.D. student at the same department and a member of Mobile Communications and Computing Laboratory (MC2L) and Mobile Innovation Centre of BME. He assists in IPv6, mobility and multimedia related works, e.g. in former FP6-IST ANEMONE and FP7-ICT OPTIMIX projects. His research topics are: anycast-based IPv6 and MIPv6 optimization, Session Initiation Protocol with IMS over native IPv6 3G UMTS networks and computer simulations of wired and wireless IPv6 networks.

József Kovács received his MSc degree in Information Technology at Budapest University of Technology and Economics (BUTE). He had been the System Developer at Ramsys for years and was responsible for system integration management and protocol development as part of the work in the FP6-CVIS project. He has experience in communication protocol stack implementation and open-source OS kernel engineering. His interests are in IP-based Internet technologies with special focus on IPv6 protocols, communications protocol design, protocol and system architectures in special focus on mobility. He participated in several mobility-related research projects (FP6-IST ANEMONE, EUREKA-Celtic BOSS) as part of his job at Mobile Innovation Center (MIK) at BME. Currently, he is lead developer at MTA SZTAKI, working on the FP7-ITSSv6 project.

Sébastien Laborie defended his Ph.D. in Computer Science in 2008 at the Université Joseph Fourier (Grenoble, France). His thesis titled "Semantic Adaptation of Multimedia Documents" was jointly co-supervised by Jérome Euzenat (EXMO team, INRIA Grenoble Rhône-Alpes) and Nabil Layaïda (WAM team, INRIA Grenoble Rhône-Alpes). His research topics were (and are still) at the crossroad of multi-media document specification and adaptation, spatio-temporal quantitative and qualitative representation and reasoning, and Semantic Web technologies. Since 2010, he has been an Associate Professor at the Laboratoire d'Informatique de l'Université de Pau et des Pays de l'Adour (LIUPPA) working in the T2I team. He published several scientific papers in international conferences (ACM, IEEE, and Springer) and has co-supervised 2 postgraduate students.

Andrew Laghos is a Lecturer in the Department of Multimedia and Graphic Arts of the Cyprus University of Technology in Limassol, Cyprus. He holds a PhD from the Centre for Human-Computer Interaction Design of City University (United Kingdom) with specialization in e-learning, multimedia, and social networks. He holds an MSc in Interactive Multimedia from Middlesex University (United Kingdom), a BSc in Computer Science from Webster University (United States of America), and a certificate in Website Design and Development from the same university. His research interests include: online communities and social networks; multimedia; music; human-computer interaction; and communication via the Web.

Jaime Lloret received his M.Sc. in Physics in 1997, his M.Sc. in Electronic Engineering in 2003, and his Ph.D. in Telecommunication Engineering (Dr. Ing.) in 2006. He is a Cisco Certified Network Professional Instructor. He worked as a network designer and administrator in several enterprises. He is currently Associate Professor in the Polytechnic University of Valencia. He is the head of the research group "Communications and Remote Sensing" of the Integrated Management Coastal Research Institute, and he is the head of the "Active and Collaborative Techniques and Use of Technologic Resources in the Education (EITACURTE)" Innovation Group. He is the director of the University Expert Certificate "Redes y Comunicaciones de Ordenadores" and the University Master "Digital Post Production." He is currently Vice-Chair of the Internet Technical Committee (IEEE Communications Society and Internet society) and the Vice-Chair for the Europe/Africa Region of Cognitive Networks Technical Committee (IEEE Communications Society). He has 1 research book, and more than 165 research papers published in national and international conferences, international journals (most of them with Impact Factor), and books. He has 11 educational books, and more than 55 papers published in international conferences, journals, and books of education. He has been the co-editor of 15 conference proceedings and guest editor of several international books and journals. He is editor-in-chief of the international journal *Networks Protocols and Algorithms*, IARIA Journals Board Chair (8 Journals), and he is associate editor of several international journals. He has been involved in more than 160 program committees of international conferences and in many organization and steering committees. He led many national and international projects. He has been the general chair of SENSORCOMM 2007, UBICOMM 2008, ICNS 2009, ICWMC 2010, and eKNOW 2012, and co-chairman of ICAS 2009, INTERNET 2010, MARSS 2011, IEEE MASS 2011, SCPA 2011, and ICDS 2012. He is the co-chairman of IEEE SCPA 2012 and GreeNets 2012, and chair of MIC-WCMC 2013. He is an IEEE Senior and IARIA Fellow.

Adalberto Melo graduated as a Technologist in Data Processing from the University of Amazonia UNAMA (1997). He has a specialization in Computer Networks, completed (1999), at the same university. He is a Master student of the Graduate Program in Computer Science from Federal University of Para in Brazil (UFPA). He has experience in Computer Science with emphasis in Analysis and Development of Systems and Computer Networking. He also forms part of the research group on computer networks and multimedia communication (GERCOM) acting on the following topics: management of Quality of Service (QoS) and Quality of Experience (QoE), resource reservation, mobility, and multimedia streaming in systems of wireless networks.

Grzegorz Mikołajczak received his MSc degree in 1992 and his PhD in 2002 from the Faculty of Electrical, West Pomeranian University of Technology. His research interests include digital signal processing and image processing. He has written several papers on digital signal processing.

Lorenzo Miniero received his degree in Computer Engineering from the University of Napoli "Federico II," Italy, in 2006. He is currently a Senior Researcher at the Computer Science Department of the same university. His research interests mostly focus on next generation networks, network real-time applications, and communication protocols, with special emphasis on the related standardization efforts.

Irina-Delia Nemoianu received her M.Eng in 2009 from the "Politehnica" University of Bucharest, Romania. She is currently pursuing her Ph.D. degree in the Department of Signal and Image Processing of TELECOM ParisTech, Paris, France. Her research interests include multiple description video coding, mobile ad-hoc networking, and network coding.

Augusto Neto is currently an Associate Professor in the Department of Computer Engineering (DETI), Federal University of Ceará (UFC), working mainly in the area of Computer Networks and Distributed Systems. Previously, he was an Associate Professor at the Institute of Informatics (INF) of the Federal University of Goiás (UFG). He graduated from the course Technologist in Data Processing from the University of Amazonia (1996), post-graduated in Computer Networking from Federal University of Pará in Brazil (1998) and obtained his Masters degree in Computer Science from Federal University of Santa Catarina in Brazil (2001). He earned his doctorate at the University of Coimbra (UC), Portugal, Department of Informatics Engineering (DEI) in October 2008. He made postdoctoral fellow at the Institute of Telecommunications, Polo University of Aveiro (Portugal). He is also a member of the Group of Computer Networks, Software Engineering and Systems (GREat) the UFC, and the Institute of Telecommunications Polo University of Aveiro.

Mersini Paschou has graduated from the Department of Computer Engineering and Informatics of the University of Patras in 2007 and obtained a MSc degree in "Computer Science and Engineering" in 2009. Currently, she is a PhD candidate of the same department. She has been a member of the Graphics, Multimedia, and GIS Laboratory since 2006 and has participated in numerous national and European projects as a Software Engineer. Her research interests focus on the areas of Web services, information systems, Web technologies, and mobile Web applications.

Jakub Pęksiński holds a PhD and has graduated from the Technical University of Szczecin, Electrical division, in 1997. He obtained his EngD in Electrical Engineering in 2004. From 1997, he was employed on West Pomeranian University of Technology (before Technical University of Szczecin), in the Processing Signals and Multimedia Engineering Department. The main course of his study is related to signal processing. He has written several publications on digital signal processing.

Béatrice Pesquet-Popescu received the Engineering degree in Telecommunications from the "Politehnica" Institute in Bucharest in 1995 (highest honours) and the Ph.D. thesis from the Ecole Normale Superieure de Cachan in 1998. In 1998, she was a Research and Teaching Assistant at Universitı Paris XI, and in 1999, she joined Philips Research France, where she worked during two years as a research scientist, then project leader, in scalable video coding. Since Oct. 2000, she is with TELECOM ParisTech (formerly, ENST), first as an Associate Professor, and since 2007 as a Full Professor, Head of the Multimedia Group. She is also the Scientific Director of the UBIMEDIA common research laboratory between Alcatel-Lucent Bell Labs and Institut TELECOM. She was an EURASIP BoG (Board of Governors) member between 2003-2010, and an IEEE Signal Processing Society IVMSP TC (2008-2013) and IDSP TC (2010-2015, Vice-Chair since Jan 1st 2012) member, and an MMSP TC associate member. She is currently a member of the IEEE SPS Awards Board. Beatrice Pesquet-Popescu serves or has served as an Associate Editor for *IEEE Transactions on Multimedia, IEEE Transactions on Circuits and Systems for Video Technology, IEEE Transactions on Image Processing*, Area Editor for Elsevier *Image Communication Journal*, and Associate Editor for the Hindawi *International Journal of Digital Multimedia Broadcasting Journal*, Elsevier *Signal Processing*. She was or will be a Technical Co-Chair for the PCS2004 conference, and General Co-Chair for IEEE SPS MMSP2010, EUSIPCO 2012, and IEEE SPS ICIP 2014 conferences. In 2006, she was the recipient, together with D. Turaga and M. van der Schaar, of the *IEEE Transactions on Circuits and Systems for Video Technology* "Best Paper Award." In April 2012, Usine Nouvelle cited her among the "100 who matter in the digital world" in France. She holds 23 patents in wavelet-based video coding and has authored more than 230 book chapters, journal, and conference papers in the field. She is a co-editor of the book to appear *Emerging Technologies for 3D Video: Creation, Coding, Transmission, and Rendering*, Wiley, 2013. Her current research interests are in source coding, network coding, scalable, and robust and distributed video compression, multi-view video, and sparse representations.

Nadia N. Qadri was awarded with her PhD degree in 2010 at the University of Essex, UK. She received her M.E(Communication Systems and Networks) and B.E(Computer Systems), from Mehran University of Engineering and Technology,Jamshoro, Pakistan in 2004 and 2002, respectively. She has ten years of teaching and research experience at renowned universities of Pakistan. Dr. Nadia is presently working as Associate Professor and In-charge Telecommunication Engineering Program at CIIT,

Wah Campus. She is also leading a research group of Wireless Networks. Dr. Nadia is author of seven book chapters in different international books, nine international conference papers and Ten reputed international journal papers. Her research interests include video streaming for mobile ad hoc networks and vehicular ad hoc networks, along with P2P streaming.

Simon Pietro Romano received the degree in Computer Engineering from the University of Napoli "Federico II," Italy, in 1998. He obtained a PhD degree in Computer Networks in 2001. He is currently an Assistant Professor at the Computer Science Department of the University of Napoli. His research interests primarily fall in the field of networking, with special regard to QoS-enabled multimedia applications, network security, and autonomic network management. He is currently involved in a number of research projects, whose main objective is the design and implementation of effective solutions for the provisioning of services with quality assurance over premium IP networks. Simon Pietro Romano is member of both the IEEE Computer Society and the ACM.

Philippe Roose is currently Associate Professor at the LIUPPA Lab, University of Pau. He was chair of several national/international conferences, wrote and co-edited various books. He founded the ALCOOL research team (Software Architecture, Components, and Protocols) and held a first national project called TCAP – Multimedia Data flows on Wireless Sensor Networks. He is now responsible for the French National project MOANO (Models and Tools for Pervasive Applications focusing on Territory Discovery) involving several CNRS Labs and one INRIA Team. His research interests focus on software architecture, autonomic computing, adaptation in the large (document, deployments, and applications). He advised more than 8 postgraduate students and more than 8 PhD students.

Giancarlo Ruffo, PhD, is Associate Professor of Computer Science at the University of Turin, Italy, and Adjunct Professor at Schools of Informatics and Computing, Indiana University. His research interests range over the wide and multidisciplinary research area of complex systems and networks, with particular attention to techno-social systems. He investigated research issues on social media, Web and data mining, recommendation systems, networking analysis, distributed applications, peer-to-peer systems, security, and micro-payment systems. He is the principal investigator of ARCS Group in Turin (http://arcs.di.unito.it), and he is leading several ongoing projects. He has published about 40 peer-reviewed papers in international journals and conferences.

Evangelos Sakkopoulos was born in Greece in 1977. He holds a PhD from the Computer Engineering and Informatics Department, University of Patras, Greece, 2005. He has received the M.Sc. degree with honors and the diploma of Computer Engineering and Informatics at the same institution. His research interests include software engineering, information systems, mobile Web, Web services, Web engineering, Web usage mining, Web-based education, and intranets. He has more than 40 publications in international journals and conferences in these areas.

Melih Soydemir is pursuing a MS at the Electrical and Electronics Engineering Department at Bahçeşehir University, Turkey. His research interests include embedded systems, medical image analysis, and search and retrieval systems. Currently, he is working in a Marie Curie Project under the supervision of Dr. Unay towards creating a medical image retrieval platform. He earned a BS degree in Electronics Engineering (2009) from Sabancı University, Turkey.

Michał Stefanowski started his Master course in 1995 and occupied a Research Assistant position at the Chair of Signal Processing and Multimedia Engineering in the Szczecin University of Science. He holds a PhD in Electrical Engineering, Szczecin University of Science, Poland, 2005. After obtaining his PhD in "Estimation of the Amount of Noise in Images using Finite Differences," he suspended research activities at the university and joined Tieto Company, where he had a chance to develop his competences in embedded software development, especially in the telecommunication field (broadband access networks, radio networks). From 2006, he is Line Manager at Tieto Poland with responsibility for human resources management, technical project support, and business development.

Jesús Tomás graduated in Computer Science in 1993 at Polytechnic University of Valencia, getting the best ratings. He finished his Doctoral thesis in 2003. He worked as Software Programmer in several enterprises and freelance. From 1993, he is an Associate Professor at Polytechnic University of Valencia. He is member of the Integrated Management Coastal Research Institute. His research focuses on statistical translation, artificial intelligence, pattern recognition, and sensors networks. He has published multiple articles in national and international conferences and has multiple articles in international journals (more than 14 of these are included in the Journal Citation Report). He has been involved in several research projects related to public and private pattern recognition and artificial intelligence applied to multiple subjects (4 of them as principal investigator). He has been technical committee member in several conferences and journals. He is co-president of the International Congress INTERNET 2010.

Athanasios Tsakalidis, Computer-Scientist, Professor of the University of Patras. Born 27.6.1950 in Katerini, Greece. Studies: Diploma of Mathematics, University of Thessaloniki in 1973; Diploma of Informatics in 1980; and Ph.D. in Informatics in 1983, University of Saarland, Germany. Career: 1983-1989, Researcher in the University of Saarland. He has been student and cooperator (12 years) of Prof. Kurt Mehlhorn (Director of Max-Planck Institute of Informatics in Germany). 1989-1993 Associate Professor, and since 1993 Professor in the Department of Computer Engineering and Informatics of the University of Patras. 1993-1997 and 2001-today, Chairman of the same department. 1993-today, member of the Board of Directors of the Research Academic Computer Technology Institute (RACTI); 1997-today, Coordinator of Research and Development of RACTI; 2004-today Vice-Director of RACTI. He is one of the contributors to the writing of the *Handbook of Theoretical Computer Science* (Elsevier and MIT-Press 1990). He has published many scientific articles, having a special contribution to the solution of elementary problems in the area of data structures. Scientific interests: data structures, computational geometry, information retrieval, computer graphics, data bases, and bio-informatics.

Giannis Tzimas is currently an Assistant Professor in the Department of Applied Informatics in Management and Finance of the Technological Educational Institute of Mesolonghi. Since 1995, he is also an Adjoint Researcher in the Graphics, Multimedia, and GIS Lab, Department of Computer Engineering and Informatics of the University of Patras. He graduated from the Computer Engineering and Informatics Department in 1995 and has participated in the management and development of many research and development projects funded by national and EU resources, as well as the private sector. He also works in the Internet Technologies and Multimedia Research Unit of the Research Academic Computer Technology Institute (RACTI), as a Technical Coordinator, from 1997 until now. His research activity lies in the areas of Web engineering, Web modelling, bioinformatics, Web-based education, and intranets/extranets. He has published a considerable number of articles in prestigious national and international conferences and journals.

Devrim Unay is an Assistant Professor in the Department of Biomedical Engineering at Bahçeşehir University, Turkey. His research interests include medical image analysis, content-based information retrieval, pattern recognition and computer vision, machine learning, and machine vision and quality inspection. Unay has a PhD in Applied Sciences from Faculté Polytechnique de Mons, Belgium, and an MS in Biomedical Engineering, and a BS in Electrical and Electronics Engineering from Boğaziçi University, Turkey. During his post-doctoral years, Unay worked at Philips Research Europe as a Marie Curie Fellow. He was the co-guest editor of the special issue on "Model Selection and Optimization in Machine Learning" in the *Machine Learning Journal* (2011). He has served on the technical program committee of MMEDIA'09, ICPR'10, SIU'12, and co-organized the workshop on "medical multimedia analysis and retrieval" in ACM'11. Unay has written more than 50 publications and has 2 patents.

Manolis (Emmanouil) Viennas was born in Athens, Greece, in 1985. He received the Diploma of Computer Engineering and Informatics in 2008 and the Master degree in Computer Science and Engineering in 2010, both from University of Patras, Greece. His research interests lie in the areas of Web engineering, software engineering, mobile applications, and information systems. He is the author of more than 10 technical papers published in international journals and conferences. He is a member of the Institute of Electrical and Electronics Engineers (IEEE), the Association for Computing Machinery (ACM), the Technical Chamber of Greece, and the Hellenic Association of Computer Engineers and Communications.

Panagiotis Zervas is a Lecturer at the Department of Music Technology and Acoustics of the Technological Educational Institute of Crete. He received his PhD from the Wire Communication Lab, Department of Electrical Engineering and Computer Technology, Patras University, Greece, in 2008, and his MSc degree in "Signal and Image Processing, Theory, Applications," from the Department of Computer Engineering, Patras University, Greece, in 2000. His research interests are in the fields of speech processing and music information retrieval. In the past, he has worked as an R&D Engineer at ATMEL HELLAS S.A and has participated in a number of national and European research projects.

Index